AF324860

Engineering
Stem Cells
for Tissue Regeneration

Engineering
Stem Cells
for Tissue Regeneration

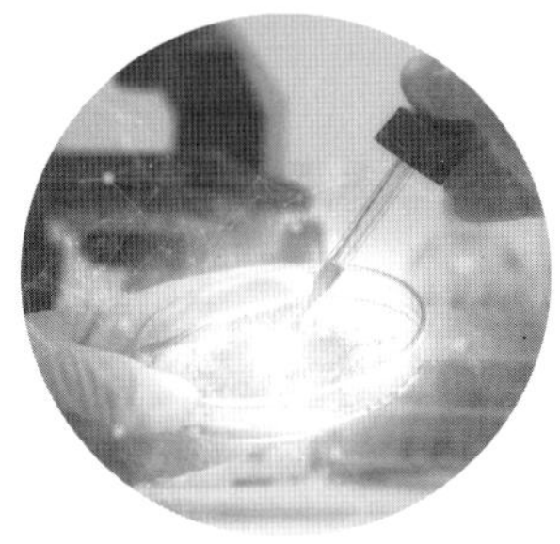

Editors

Ngan F. Huang
Stanford University, USA

Nicolas L'Heureux
University of Bordeaux, France

Song Li
University of California, Los Angeles, USA

World Scientific

NEW JERSEY · LONDON · SINGAPORE · BEIJING · SHANGHAI · HONG KONG · TAIPEI · CHENNAI · TOKYO

Published by

World Scientific Publishing Co. Pte. Ltd.

5 Toh Tuck Link, Singapore 596224

USA office: 27 Warren Street, Suite 401-402, Hackensack, NJ 07601

UK office: 57 Shelton Street, Covent Garden, London WC2H 9HE

Library of Congress Cataloging-in-Publication Data

Names: Li, Song, 1965–   editor. | L'Heureux, Nicolas, 1966–   editor. | Huang, Ngan, editor.

Title: Engineering stem cells for tissue regeneration / edited by Song Li,
    Nicolas L'Heureux, Ngan Huang.

Description: New Jersey : World Scientific, [2017] | Includes bibliographical references and index.

Identifiers: LCCN 2017030729 | ISBN 9789813147744 (hardcover : alk. paper)

Subjects: | MESH: Tissue Engineering--methods | Stem Cells | Regeneration

Classification: LCC R857.T55 | NLM QS 525 | DDC 610.28--dc23

LC record available at https://lccn.loc.gov/2017030729

British Library Cataloguing-in-Publication Data

A catalogue record for this book is available from the British Library.

For any available supplementary material, please visit
http://www.worldscientific.com/worldscibooks/10.1142/10228#t=suppl

Typeset by Stallion Press
Email: enquiries@stallionpress.com

Printed in Singapore

PREFACE

Broadly speaking, tissue engineering refers to the biological replacement, repair, or regeneration of a diseased or dysfunctional tissue. Tissue engineering includes *in vitro* engineering of functional tissue equivalents that can be transplanted *in vivo*, as well as *in situ* tissue engineering in which cells, biomaterials, and/or other factors are delivered *in vivo* to synergize with the intrinsic regeneration potential to promote tissue regeneration. Since the time tissue engineering was popularized by pioneers like Robert Langer and Joseph Vacanti in the 1990s, tissue engineering has had some initial successes in the generation of engineered skin, blood vessels, bladder and trachea. However, for the majority of tissues in the body that have complex function, geometry, and composition, the path towards successful tissue engineering has been less straightforward.

An important component of tissue engineering is the cell source. Stem cells are a promising candidate due to their ability to self-renew as well as give rise to more lineage-committed cells in a process known as differentiation. For this reason, stem cells have become a crucial cell source for tissue engineering research and thus are the focus of this textbook. As discussed in greater detail in this book, the stem cells include adult stem cells, such as those derived from the bone marrow or blood, that have limited ability to give rise to committed progeny. Alternatively, pluripotent stem cells, including embryonic stem cells and induced pluripotent stem cells, have greater flexibility to give rise to a more diverse pool of committed cells. However, these cells are prone to the generation of benign tumors known as teratomas and need to be carefully engineered for various differentiation lineages before being used for therapies. Embryonic stem cells are generated from the destruction of an embryo and are thus more ethically challenging. Although induced pluripotent stem cells do not involve the use of embryos, the process of artificially reprogramming somatic cells requires the activation of key genes, followed by a reprogramming process that is not well-understood. Owing to the successful generation of induced pluripotent stem cells along with improved differentiation protocols, researchers now have access to unprecedented numbers of cells for tissue engineering.

When choosing stem cells for tissue engineering, we need to consider the following important issues: (1) endogenous (native) or exogenous (implanted) cells; (2) autologous (from the patient) or allogenic (from another donor); (3) use of undifferentiated stem cells, partially committed progenitor cells, or fully committed differentiated cells; (4) delivery methods and biomaterials; (5) characterization and isolation of specific stem cells or progenitors; (6) reproducibility of stem cell differentiation *in vitro* and *in situ*; and (7) scalability for clinical application. These and other factors determine the product profile and may influence not only functional efficacy of the engineered tissue, but also the roadmap towards regulatory approval for clinical testing and commercialization.

To fully harness the potential of stem cells for tissue engineering applications, a critical gap in knowledge exists in our understanding of the microenvironmental factors that modulate stem cell differentiation, including biochemical and biophysical factors. Such factors influence the design of biomaterials to maximize differentiation efficiency and phenotype. As this area moves at a fast pace, there is an important need to summarize the recent progress. In this second edition, we have recruited many of the leaders in their respective fields to summarize the recent advances and provide insights into bone (Chapters 1 and 2), heart and blood vessels (Chapters 3–7), fat (Chapter 8), skeletal tissue engineering (Chapters 9–12), and nerve tissue engineering (Chapter 13). Following these chapters is a discussion of state-of-the-art technologies for stem cells and tissue engineering using genetic engineering (Chapter 14), high-throughput systems (Chapter 15), single-cell analysis (Chapters 16 and 17), microtechnologies (Chapter 18) and biomimetic materials (Chapter 19). In order to apply stem cell-based engineered tissues to clinical translation, Chapters 20 and 21 discuss quality control and regulatory issues. Although not all engineered tissues could be covered in this textbook, we hope that the concepts and approaches addressed will be generally applicable towards the engineering of most types of tissues.

We thank all the contributors for their valuable contribution. We also thank Joy Quek and other staff members of World Scientific Inc. for their tremendous effort in editing and organizing this book.

Ngan F. Huang, PhD
Stanford University

Nicolas L'Heureux, PhD
University of Bordeaux

Song Li, PhD
University of California Los Angeles

CONTENTS

CONTRIBUTORS

Aliza A. Allon
Department of Orthopedic Surgery
University of California San Francisco
San Francisco, CA 94143

Caroline Auclair-Daigle
Laboratoire d'Organogénèse Expérimentale (LOEX)
Centre de recherche du CHU de Québec-Université Laval,
    and Département de Chirurgie,
Université Laval, Quebec City, Canada

Jennifer Baker
Division of Plastic & Reconstructive Surgery
University of California San Diego
La Jolla, CA 92093

Hojae Bae
Harvard-MIT Division of Health Sciences and Technology
Center for Biomedical Engineering, Department of Medicine
Brigham and Women's Hospital, Harvard Medical School
Massachusetts Institute of Technology
Cambridge, MA 02139

Karthik Balakrishnan
Department of Mechanical Engineering
University of California Berkeley
Berkeley, CA 94720-1740

François Berthod
Laboratoire d'Organogénèse Expérimentale (LOEX)
Centre de recherche du CHU de Québec-Université Laval,
    and Département de Chirurgie
Université Laval, Quebec City, Canada

Sigurd Berven
Department of Orthopedic Surgery
University of California San Francisco
San Francisco, CA 94143

David A. Brafman
School of Biological and Health Systems Engineering
Arizona State University
Tempe AZ, 85287

Christopher K. Breuer
Pediatric Surgeon
Division of Pediatric Surgery — Department of Surgery
Nationwide Children's Hospital — Ohio State University
Columbus, OH 43215

Zorica Buser
Department of Orthopedic Surgery
University of California San Francisco
San Francisco, CA 94143

Lei Cai
Department of Materials Science and Engineering
Stanford University
Stanford, CA 94305

Bilal Cakir
Department of Genetics
Yale Stem Cell Center
Yale University School of Medicine
New Haven, CT 06520

Shu Chien
Departments of Bioengineering, Cellular and Molecular Medicine
Institute of Engineering in Medicine
University of California San Diego
San Diego, CA 92093

Karen L. Christman
Department of Bioengineering
University of California San Diego
La Jolla, CA 92093

Jeannine M. Coburn
Department of Biomedical Engineering
Worcester Polytechnic Institute
Worcester, MA 01609

Irina Conboy
Department of Bioengineering
University of California Berkeley
Berkeley, CA 94720-1762

Stephen Curran
Smith & Nephew Research Centre
York Science Park
York YO10 5DF, United Kingdom

Nathalie Dusserre
Cytograft Tissue Engineering, Suite 220
3 Hamilton Landing
Novato CA 94949

Jennifer H. Elisseeff
Department of Biomedical Engineering
John Hopkins University
Baltimore, MD 21218

Adam J. Engler
Department of Bioengineering
University of California San Diego
La Jolla, CA 92093

V. Thomas Eshraghi
Department of Periodontology
Oregon Health Sciences University
Portland, OR 97239

Michael S. Friedman
Novartis, Inc.
San Rafael, CA 94903

Sarah C. Heilshorn
Department of Materials Science and Engineering
Stanford University
Stanford, CA 94305

Luqia Hou
Cardiovascular Institute and Department of Cardiothoracic Surgery
Stanford University, Stanford, CA, 94305
Veterans Affairs Palo Alto Health Care System, Palo Alto, CA, 94304

Ngan F. Huang
Cardiovascular Institute and Department of Cardiothoracic Surgery
Stanford University, Stanford, CA, 94305
Veterans Affairs Palo Alto Health Care System, Palo Alto, CA, 94304

Ben P. Hung
Department of Biomedical Engineering
University of California, Davis
Davis, CA 95616

Corey Iyican
Cytograft Tissue Engineering
Novato, CA 94949

Eric Jabart
Department of Bioengineering
University of California Berkeley
Berkeley, CA 94720-1762

Nezamoddin N. Kachouie
Harvard-MIT Division of Health Sciences and Technology
Center for Biomedical Engineering, Department of Medicine
Brigham and Women's Hospital, Harvard Medical School
Massachusetts Institute of Technology
Cambridge, MA 02139

Ali Khademhosseini
Harvard-MIT Division of Health Sciences and Technology
Center for Biomedical Engineering, Department of Medicine
Brigham and Women's Hospital, Harvard Medical School
Massachusetts Institute of Technology
Cambridge, MA 02139

Kun-Yong Kim
Department of Genetics
Yale Stem Cell Center
Yale University School of Medicine
New Haven, CT 06520

Jennifer Kirkham
Biomineralisation Group
School of Dentistry, University of Leeds
Level 7, Wellcome Trust Brenner Building
St. James's University Hospital
Leeds, LS9 7TF, United Kingdom

J. Kent Leach
Departments of Biomedical Engineering and Orthopaedic Surgery
University of California, Davis,
Davis, CA 95616

Yong-Ung Lee
Research Scientist
Tissue Engineering Program
Nationwide Children's Hospital
Columbus, OH 43205

Nicolas L'Heureux
Laboratory for the Bioengineering of Tissues
BioTis U 1026 INSERM, University of Bordeaux,
146 Rue LEO SAIGNAT, Batiment 4A, 2ieme étage, Case 45

Ju Li
Department of Bioengineering
University of California Berkeley
Berkeley, CA 94720-1762

Song Li
Department of Bioengineering
University of California Los Angeles
Los Angeles, CA 90095

Jeffrey C. Lotz
Department of Orthopedic Surgery
University of California San Francisco
San Francisco, CA 94143

Brian Mailey
Division of Plastic & Reconstructive Surgery
Department of Surgery, School of Medicine
University of California San Diego
La Jolla, CA 92093

Mahdokht Masaeli
Harvard-MIT Division of Health Sciences and Technology
Center for Biomedical Engineering, Department of Medicine
Brigham and Women's Hospital, Harvard Medical School
Massachusetts Institute of Technology
Cambridge, MA 02139

Bradley McAllister
Department of Periodontology
Oregon Health Sciences University
Portland, OR 97239

Todd McAllister
Amnion Foundation
111 N. Chestnut Street
Suite 100, Winston-Salem, NC 27101
Email: todd@amnionfoundation.org

Edward McGillicuddy
Vascular Surgeon
Department of Vascular Surgery
Central Maine Heart and Vascular Institute
Lewiston, Maine 04240

Jason W. Nichol
Harvard-MIT Division of Health Sciences and Technology
Center for Biomedical Engineering, Department of Medicine
Brigham and Women's Hospital, Harvard Medical School
Massachusetts Institute of Technology
Cambridge, MA 02139

Ann Ouyang
Department of Orthopedic Surgery
University of California San Francisco
San Francisco, CA 94143

In-Hyun Park
Department of Genetics
Yale Stem Cell Center
Yale University School of Medicine
New Haven, CT 06520

Michael J. Paulsen
Department of Cardiothoracic Surgery
Stanford University
Stanford, CA 94305

Beth L. Pruitt
Department of Mechanical Engineering
Department of Molecular and Cellular Physiology
Department of Bioengineering
Stanford Cardiovascular Institute, Stanford University
Stanford, CA 94305

Sachin Rangarajan
Department of Bioengineering
University of California Berkeley
Berkeley, CA 94720-1762

Alexandre J. S. Ribeiro
Department of Mechanical Engineering
Stanford Cardiovascular Institute
Stanford University, Stanford, CA 94305,
Gladstone Institute of Cardiovascular Disease
San Francisco, CA 94158

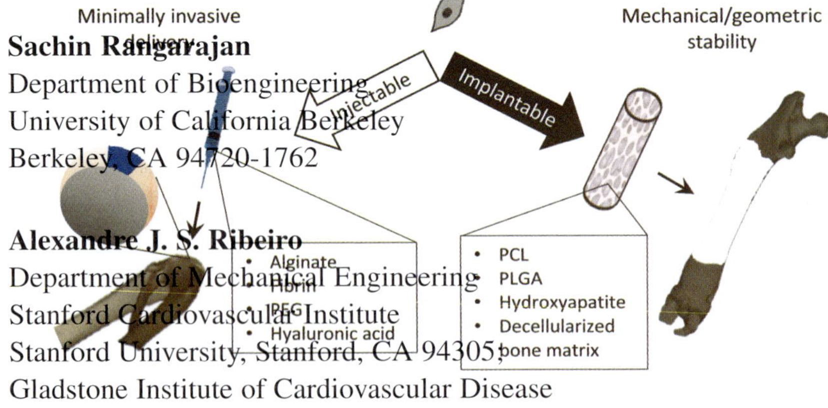

Sushmita Saha
Biomaterials and Tissue Engineering Group
School of Dentistry, University of Leeds
Level 7, Wellcome Trust Brenner Building
St. James's University Hospital
Leeds, LS9 7TF, United Kingdom

Rajendra Sawh-Martinez
Craniofacial Surgery Fellow | Clinical Instructor
Section of Plastic and Reconstructive Surgery
Yale University School of Medicine
New Haven, CT 06520

Toshiharu Shin'oka
Pediatric Cardiothoracic Surgeon
Division of Cardiothoracic Surgery — Department of Surgery
Nationwide Children's Hospital — Ohio State University
Columbus, OH 43205

Lydia L. Sohn
Department of Mechanical Engineering
University of California Berkeley
Berkeley, CA 94720-1740

Gustavo Villalona
Pediatric Surgeon
Department of Surgery
St. Louis University
St. Louis, MO 63104

Anne M. Wallace
Division of Plastic & Reconstructive Surgery
Department of Surgery, School of Medicine
University of California San Diego
La Jolla, CA 92093

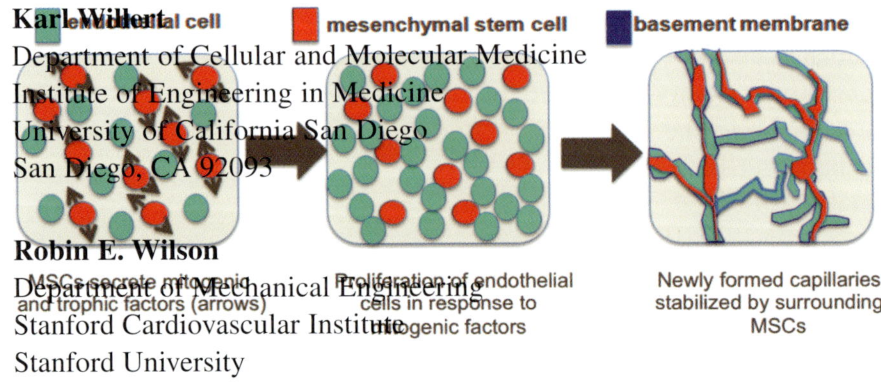

Karl Willert
Department of Cellular and Molecular Medicine
Institute of Engineering in Medicine
University of California San Diego
San Diego, CA 92093

Robin E. Wilson
Department of Mechanical Engineering
Stanford Cardiovascular Institute
Stanford University
Stanford, CA 94305

Y. Joseph Woo
Department of Cardiothoracic Surgery
Stanford University
Stanford, CA 94305

David Wood
Biomaterials and Tissue Engineering Group
School of Dentistry, University of Leeds
Level 7, Wellcome Trust Brenner Building
St. James's University Hospital
Leeds, LS9 7TF, United Kingdom

Guang Yang
Cardiovascular Institute and Department of Cardiothoracic Surgery,
Stanford University, Stanford, CA, 94305
Veterans Affairs Palo Alto Health Care System,
Palo Alto, CA, 94304

Xuebin B. Yang
Biomaterials and Tissue Engineering Group
School of Dentistry, University of Leeds
Level 7, Wellcome Trust Brenner Building
St. James's University Hospital
Leeds, LS9 7TF, United Kingdom

D. Adam Young
Department of Bioengineering
Sanford Consortium for Regenerative Medicine
University of California San Diego
La Jolla, CA 92037

Jennifer L. Young
Department of Cellular Biophysics
Max Planck Institute for Medical Research
69120 Heidelberg, Germany

Behnam Zamanian
Harvard-MIT Division of Health Sciences and Technology
Center for Biomedical Engineering, Department of Medicine
Brigham and Women's Hospital, Harvard Medical School
Massachusetts Institute of Technology
Cambridge, MA 02139

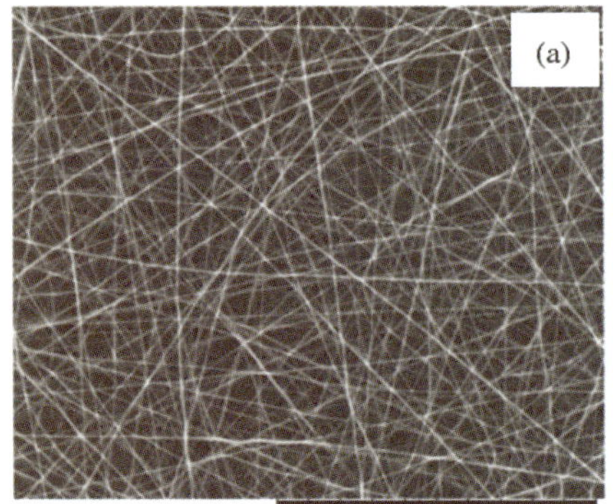
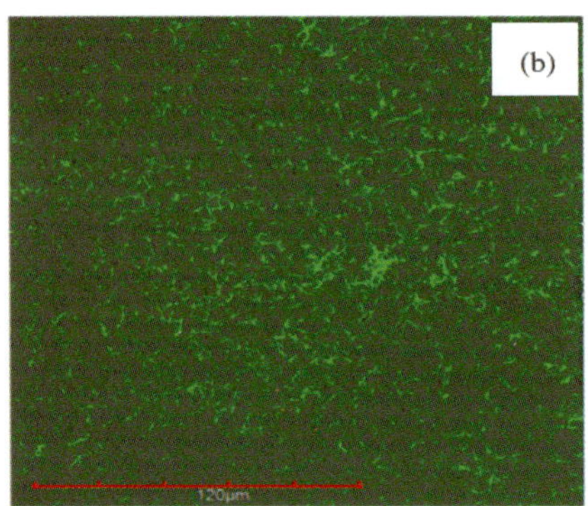
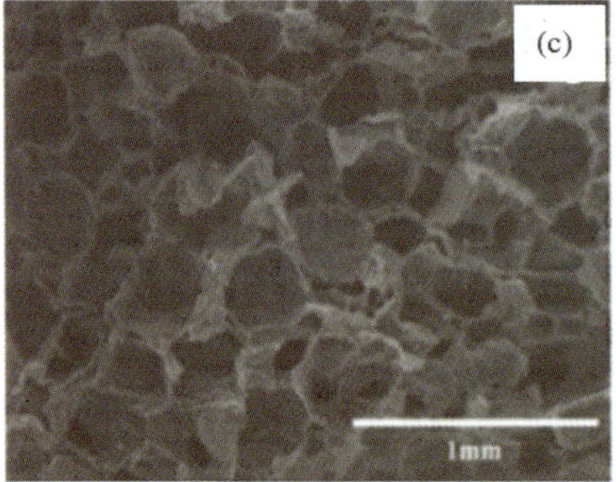

1

MESENCHYMAL STEM CELLS FOR TISSUE REGENERATION

Guang Yang, Song Li† and Ngan F. Huang‡*

1. Introduction

In the past two decades, significant progress has been made in the field of stem cell research. An important finding is that adult stem cells harbor greater regeneration potential and plasticity than what was previously thought. This discovery has led to tremendous research interest in developing methods to direct stem cell differentiation into lineages for the therapeutic delivery into diseased or dysfunctional tissues. Among the adult stem cells, mesenchymal stem cells (MSCs) are a promising therapeutic cell source due to the ease of isolation, high proliferative capacity, and multipotency.[1] MSCs can be found in numerous tissues of the adult mammal and can be harvested in large quantities by minimally invasive and reproducible approaches. Therapeutic MSCs can be delivered directly *in vivo* or incorporated into tissue engineered constructs *in vitro* before transplantation. Both of these two approaches have been explored for treating a wide range of diseases or traumatic events, including myocardial infarction, peripheral arterial disease, spinal cord injury, musculoskeletal system defects, and skin wounds. By developing robust methods of differentiating MSCs into therapeutic cells of interest and organizing the cells into functional three-dimensional tissues, it may be possible to fulfill the potential of MSCs for clinical use. This review aims to provide an overview of some therapeutic applications for MSCs in tissue engineering and regenerative medicine.

2. MSC Sources and Phenotype

MSCs can be generally defined as adherent and elongated cells that reside in mesenchymal tissues and can self-renew as well as produce progeny with more specialized functions. MSCs have been observed and purified from numerous origins, including bone marrow, adipose tissue, skeletal muscle, blood, liver spleen and dental pulp.[2–8] This review primarily focuses on MSCs derived from bone marrow and adipose tissues, as they are the most well characterized origins of MSCs among all known MSC sources.

2.1 *Bone marrow MSCs*

Although MSCs account for only 0.01% among total nucleated cells in the bone marrow, they have over a million-fold expansion capability and multilineage differentiation potential.[9] Bone marrow MSCs are easily harvested by aspiration from the iliac crest. Phenotypically, there is no unique single marker that specifically identifies bone marrow MSCs. Consequently, MSCs are characterized based on the positive expression of numerous cell surface antigens such as CD29 (integrin β1), CD44 (receptor for hyaluronic acid and matrix proteins), CD105 (endoglin), and CD166 (cell adhesion molecule).[10] On the other hand, they do not express markers typically associated with hematopoietic cells, such as CD14 (monocyte surface antigen), CD34 (hematopoietic stem cell surface antigen), and CD45 (leukocyte surface antigen).[1] Due to the differences in characterization methods, the International Society for Cellular Therapy suggested the recommended designation of these cells as multipotent mesenchymal stromal cells and proposed the following minimal criteria for MSCs: adherence to plastic dishes; phenotypic expression of CD105, CD73, and CD90; lack of surface molecule expression of CD45, CD34, CD14, or CD11b; CD79α or CD19 and class II major histocompatibility complex antigen (HLA-DR); and differentiation capacity into osteoblasts, adipocytes, and chondroblasts *in vitro*.[11]

Based on the cell surface antigens, bone marrow MSCs can be isolated using fluorescence-activated cell sorting (FACS) or magnetic-activated cell sorting (MACS). Other methods employed to purify MSCs include Percoll gradient centrifugation and selective adherence onto tissue-culture treated Petri dishes. To maintain their proliferative capacity, the purified bone marrow MSCs can be expanded in culture media containing defined serum-free components (i.e. StemPro® MSC Serum Free Medium, Invitrogen, Carlsbad, CA) or pre-screened fetal bovine serum. Under these growth conditions, bone marrow MSCs can be cultured for more than five passages with negligible changes in phenotype. However, the differentiation potential of bone marrow MSCs seems passage-dependent: loss of osteogenic and adipogenic potential was found along the passaging of cells.[12,13]

2.2 *Adipose-derived stem cells*

According to the definition established by the International Fat Applied Technology Society, adipose-derived stem cells (ASCs) are the plastic-adherent, multipotent cell population isolated from the stromal vascular fraction (SVF) of adipose tissue.[14] These cells were firstly found capable of differentiating into adipogenic, chondrogenic, and osteogenic lineages,[15] and were later confirmed to give rise to other cell types, including endothelial cells (ECs), smooth muscle cells (SMCs) and cardiomyocytes.[16–18] To isolate ASCs, adipose tissue derived from liposuction is digested with collagenase and then centrifuged to separate the SVF pellet from the adipocytes in the upper layer.[15] When cultured under standardized conditions, SVF is homogenized, giving rise to the pure population of ASCs.[19] Like bone marrow MSCs, ASCs can be purified based on the expression profile of surface marker antigens using FACS or MACS. ASCs can be maintained in defined serum-free medium (i.e. MesenPRO RS Media, Invitrogen, CA) as well as serum-containing media.

Although ASCs and bone marrow MSCs present more than 90% similarity in immunophenotype,[20] there are some reported differences in surface antigen expression. For example, CD14 and HLA-DR were reported to be absent in bone marrow MSCs, but have been identified in early-passage human ASCs at low frequency.[21] Furthermore, ASCs appear to have temporal changes in immunophenotype with subsequent passaging.[21] However, these differences in immunophenotype could also be attributed to differences in species or the methods of isolation, purification, or detection.

3. Differentiation of MSCs *in vitro*

The application of MSCs to tissue engineering and regenerative medicine often involves the pre-differentiation of cells *in vitro* into lineages of interest before delivering them *in vivo* for therapeutic treatment. Bone marrow MSCs and ASCs have been shown to differentiate into a variety of lineages, including myogenic, osteogenic, chondrogenic, and adipogenic lineages.[22] A number of strategies to direct their differentiation have been used, including the use of soluble factors, mechanical stimulation, extracellular matrix (ECM) factors, and genetic engineering approaches, which are briefly discussed below.

One of the most commonly used strategies to induce differentiation is to using soluble factors such as growth factors and small molecules. Using soluble factors, bone marrow MSCs and ASCs have a high propensity to differentiate into cells of mesenchymal lineage, including bone, adipose tissue, and cartilage. Osteogenic differentiation can be induced by culturing the cells in the presence of dexamethasone, ascorbic acid, and β-glycerophosphate.[1] Adipogenesis is usually accomplished by treatment with DMEM supplemented with FBS, dexamethasone, 3-isobutyl-1-methylxanthine, and 1x insulin-transferrin-selenium (ITS).[23] Chondro-genesis of high-density MSC pellet cultures is induced using serum-free DMEM supplemented with dexamethasone, L-proline, ascorbate, and transforming growth factor β1 or 3 (TGF-β1/3). In addition, bone morphogenetic protein 6 (BMP-6) is added to the chondrogenic medium to improve the chondrogenesis of ASC by rescuing the expression of TGF-β1.[24] Besides osteogenic, adipogenic, and chondrogenic lineages, MSCs have been shown to differentiate toward other lineages at lower yields. For example, platelet-derived growth factor (PDGF) and TGF-β3 stimulate smooth muscle phenotype,[25] whereas 5-azacytidine treatment induces the formation of cardiac-like cells that expresses cardiac markers β-myosin heavy chain, desmin, and α-cardiac actin.[26]

Besides soluble factors, mechanical stimulation is another potent regulator of cell behavior and function. Physiologically, mechanical stimulation plays an integral role in cell phenotype. For example, blood vessels experience fluid shear stress and cyclic strain due to the pulsatile blood flow generated by the beating heart. Likewise, bone cells are subjected to compressive loading from gravitational forces. Due to the importance of mechanical factors for physiological maintenance, the role of mechanical factors on MSCs and ASCs differentiation has become an area of research interest. We and others have shown that uniaxial strain promotes smooth muscle differentiation of bone marrow MSCs when cell orientation was restricted by microgrooves.[27,28] In terms of chondrogenesis, the application of cyclic mechanical compression to human bone marrow MSCs encapsulated in alginate beads promotes significant upregulation of chondrogenic markers sox9, type II collagen and aggrecan.[29] Taken together, these findings highlight the significance of presenting native mechanical environment as a design principle for eliciting desired cellular responses for tissue regeneration.

Another method of inducing MSC differentiation is using ECMs, which is a network of various structural proteins, polysaccharides, and other components that provide structural integrity as well as molecular cues that regulate cell behavior and function.[30,31] The biophysical and biochemical cues transmitted by the ECM to cells include matrix rigidity, matrix patterning, and matrix composition. Physiologically, matrix rigidity varies throughout different tissues of the body, from soft tissues of the brain to hard tissues in the bone. For example, Engler *et al.*

showed that physiological rigidity dictated the fate of human bone marrow-derived MSC differentiation.[32] When cultured on polyacrylamide gels of varying rigidities, the MSCs differentiated to osteogenic lineage on rigid matrix (34 kPa), but soft (0.1–1 kPa) gels promoted neurogenesis. Besides matrix rigidity, the spatial pattern of the ECM also regulates MSC differentiation by restricting cellular shape. For example, small (1024 μm^2) micropatterns of fibronectin stimulated human bone marrow-derived MSCs to differentiate toward adipogenic lineage and retain a rounded morphology, whereas large (10,000 μm^2) islands supported osteogenic differentiation and adherent morphology.[33] Finally, the matrix composition may also influence MSC differentiation. There is evidence that proteoglycans heparan sulfate and chondroitin sulfate could enhance neuronal differentiation of MSCs.[34] Porcine bone marrow MSCs cultured in three-dimensional patches of polyethylene glycol (PEG)-modified fibrin showed EC phenotype.[35] These studies suggest that the ECM plays a dynamic role of modulating MSC behavior and phenotype.

Besides soluble, mechanical, and ECM factors in the microenvironment, intracellular factors also play a role in inducing differentiation. MicroRNAs (miRs) are short endogenous nucleotide RNAs that post-transcriptionally regulate gene expression. In addition to modulating biological processes such as cell proliferation or cell death, they have recently been shown to direct stem cell fate lineage specification by negatively regulating gene expression.[36,37] For instance, overexpression of miR-21 was found to elevate the expression level of the adipogenesis-associated genes PPARγ, whereas inhibiting miR-21 reduced the adipogenesis of MSCs.[38] Furthermore, another group demonstrated that miR-148b, miR-27a, and miR-489 modulate osteogenesis in human bone marrow MSCs, even in the absence of osteogenic-inducing media.[39]

Together, these results suggest that bone marrow MSC and ASC differentiation are controlled by multiple microenvironmental cues and intracellular signaling. Further studies to elucidate the mechanism of microenvironmental factors on differentiation will be critical for directing cell differentiation with high purity for tissue engineering and regeneration purposes.

4. Tissue Engineering and Regeneration Using Bone Marrow MSCs and ASCs

MSCs and ASCs are promising stem cell candidates for the repair or regeneration of diseased tissues, and they have been frequently utilized in *in situ* or *in vitro* tissue engineering approaches to repair various tissues,[4] including the repair and regeneration of blood vessels, bone, skeletal muscle, periodontal tissue, and spinal cord (Table 1). For the *in situ* approach, the cells and other factors are delivered *in vivo* to elicit the therapeutic potential of the cells via mutual interactions with the host environment. For the *in vitro* approach, cells and matrix/scaffolds are used to engineer functional tissue constructs prior to implantation. Here we illustrate these different approaches of tissue engineering for the regeneration of heart, blood vessels, skin, cartilage, and tendon.

4.1 *In situ tissue regeneration*

4.1.1 *Cardiac repair*

Cardiovascular disease remains as one of the leading causes of mortality in the United States. Over 70 million people in the United States are symptomatic from or at risk of cardiovascular disease.[40] Stem cell-based approaches to repair or regenerate the heart after myocardial infarction are promising. The goal of stem cell therapy is to regenerate cardiac muscle, enhance angiogenesis, and ultimately improve cardiac function. The methods of delivering the therapeutic cells to the heart include systemic delivery, regional coronary infusion, or local myocardial injection.[41] In one study, autologous swine bone marrow MSCs were directly injected into the infarcted heart, and MSCs differentiated toward myogenic lineages as early as two weeks post-injection. Four weeks after cell delivery, the extent of aneurismal thinning and contractile dysfunction were significantly reduced in the cell-treated group.[42] Besides direct injection, regional and systemic infusion of bone marrow MSCs has also been demonstrated. When ^{99m}Tc-labeled rat bone marrow MSCs were transfused to infarcted rat hearts either into the left ventricular cavity or by intravenous delivery, intravenous delivery resulted in cell localization predominantly

Table 1. Overview of tissue engineering and regeneration applications using bone marrow MSCs or ASCs.

Organ/Tissue	Treatment	Example References
Heart	Cell delivery	43, 87, 88
	Cell delivery in ECM	50, 89
Blood vessel	Cell delivery	52, 54, 90
	Cell delivery in ECM	25, 56, 59, 91
Skeletal muscle	Cell delivery	92, 93
	Cell delivery in ECM	94, 95
Bone	Cell delivery	96, 97
	Cell delivery in ECM	98, 99
Cartilage	Cell delivery	100, 101
	Cell delivery in ECM	72, 73
Skin	Cell delivery	102, 103
	Cell delivery in ECM	64, 104
Teeth	Cell delivery	105, 106
	Cell delivery in ECM	107, 108
Spinal cord	Cell delivery	109, 110
	Cell delivery in ECM	111, 112

to the lungs and to a lesser degree to the heart.[43] In contrast, regional delivery produced better retention of cells within the heart, especially in the infarct region, and a lower uptake in the lungs.

Besides delivery of naive cells, MSCs and ASCs have also been genetically modified to improve cell survival or therapeutic effect. For instance, allogeneic VEGF-overexpressing MSCs injected into the peri-infarct area of ovine heart were found to reduce infarct size and induce tissue salvage predominantly in the subendocardial myocardium.[44] In another study, rat bone marrow MSCs overexpressing anti-apoptotic gene bcl-2 were injected into the infarct region. This treatment increased cell survival, decreased infarct scars, and promoted neovascularization to assess their survival, engraftment, and functional improvement after myocardial infarction.[45] These results suggest that genetic modification of MSCs can improve the therapeutic outcome.

Another approach to cardiac repair is to co-inject therapeutic cells with ECMs. ECM components alone such as collagen, fibrin, alginate, and matrigel have been reported to enhance neovascularization in the infarcted heart or attenuate infarct scar thinning.[46–49] The rationale behind the co-injection of cells with ECMs is that the ECMs may provide therapeutic enhancement of cardiac function and angiogenesis, while also providing structural support and favorable biochemical cues to localize the transplanted cells to the site of delivery. For example, we showed that co-injection of human bone marrow MSCs with fibrin significantly enhanced neovasculature formation following chronic myocardial infarction.[50]

4.1.2 *Peripheral vascular repair*

Peripheral artery disease (PAD) is typically due to atherosclerotic occlusive disease of the peripheral arteries of the limbs, causing symptoms such as intermittent claudication, painful ischemic ulcerations, and limb-threatening gangrene. A feature of PAD is dysfunction or damage to the vascular endothelium, the diaphanous layer of ECs that exerts control over vascular reactivity, remodeling and angiogenesis. Cell-based approaches to restore or regenerate the endothelium so as to enhance the angiogenic response to ischemia hold promise for the treatment of PAD.[51]

Bone marrow MSCs and ASCs have been examined in the setting of vascular repair for PAD, with the goal of generating and incorporating new functional vessels into existing vessels to remodel them. In one study, mice that underwent hindlimb ischemia, an experimental model of PAD, were treated with adductor muscle injections of murine bone marrow MSCs, mature ECs, or culture medium.[52] The animals treated with MSCs demonstrated significant improvement in limb blood perfusion and attenuation of muscle fibrosis after 21 days, compared to the other treatment groups. The cells did not appear to be incorporated into the mature collateral vessels, but appeared to increase the levels of vascular endothelial growth factor (VEGF) and basic fibroblast growth factor (bFGF) in the adductor muscle, suggesting a paracrine mechanism of enhancement. Besides bone marrow MSCs, ASCs have also shown therapeutic benefit for the treatment of hindlimb ischemia. For instance, tissue perfusion rates of mice ischemic limbs were found significantly higher in the group treated with human ASCs compared with those of the control mice as early as postoperative day 3.[53] In a comparative study of the therapeutic efficacy between human bone marrow MSCs and ASCs, equal numbers of cells were injected intramuscularly into the ischemic limb of immunodeficient mice.[54] Two weeks after cell transplantation, laser Doppler blood perfusion imaging showed significantly higher improvement in the ASC-treated group, compared to the bone marrow MSC and control groups. This finding is consistent with the more recent discovery that ASCs are more effective than bone marrow MSCs in promoting neovascularization and angiogenesis on a histological basis in a rat model of hindlimb ischemia.[55] The difference in therapeutic effects between bone marrow and MSC and ASCs might be attributed to higher levels of matrix metalloproteinases (MMPs) in ASCs than that in bone marrow MSCs. The therapeutic effect of ASCs was mitigated when ASCs were transfected with MMP3 or MMP9 siRNA prior to transplantation into the ischemic limb, suggesting that MMP3 and MMP9 were involved in the proangiogenic effect of ASCs.[54] In summary, these studies demonstrate that MSCs and ASCs significantly improve limb blood perfusion and are therefore promising candidates for treatment of PAD.

4.2 *In vitro tissue engineering*

4.2.1 *Tissue engineering of blood vessels*

Bypass surgeries are often used to treat obstructed coronary and peripheral arteries. However, synthetic vascular grafts with small diameters (less than 6 mm) have a high failure rate because of frequent clogging. Tissue-engineered vascular graft is a promising solution to this problem. MSCs can differentiate into SMCs and secrete growth factors to recruit ECs, and thus is a potential cell source for constructing vascular grafts. Interestingly, vascular grafts seeded with bone marrow MSCs have improved patency, which is attributed to the anti-platelet adhesion property of heparan sulfate proteoglycans on the surface of bone marrow MSCs.[56]

By engineering biochemical and mechanical factors (i.e. matrix proteins, soluble factors, and cyclic strain) in a bioreactor, the proliferation and differentiation of bone marrow MSCs can be controlled in vascular grafts.[57,58] *In vivo* study has also shown that bone marrow MSCs can differentiate into SMCs and promote endothelialization.[59] However, the electrophysiological profile of bone marrow MSC-differentiated SMCs appeared to be different from that of mature SMCs, suggesting that only partial differentiation of MSCs into SMCs was achieved. Interestingly, vascular grafts seeded with bone marrow-derived mononuclear cells (BM-MNCs) also have high patency. These grafts can mimic the cellular feature of native blood vessels comprised of SMCs and ECs.[60] However, whether bone marrow-derived MSCs or stromal cells and ASCs can replace SMCs and ECs as a cell source for vascular graft construction needs further investigation.

4.2.2 *Tissue engineering of biological skin equivalents for wound repair*

Skin wounds result from various conditions, including burns, traumatic accidents, or disease. Standard treatments for wound repair include autologous or cadaveric skin transplantations, but these methods face the potential risks of donor site morbidity and transmission of diseases. An alternative approach is tissue-engineered skin equivalents for replacing the damaged skin.[61] Physiologically, skin is comprised of a stratified and keratinized epidermis that physically protects the body, and below the epidermis is the dermal layer that provides strength and support.[62] For the treatment of wound injuries or disease, tissue-engineered skin replacements provide an off-the-shelf alternative that can minimize the potential complications of disease transmission or tissue harvesting. The desirable skin equivalents should quickly adhere to the wound, mimic the physiological function and mechanical properties of normal skin, accelerate wound healing, and do not undergo immune rejection.[63]

Tissue engineering of skin equivalents using bone marrow MSCs and ASCs are promising. In a porcine wound healing model, engineered skin constructs composed of autologous bone marrow MSCs seeded in collagen-glycosaminoglycan polymer scaffolds were grafted onto partial thickness wounds.[64] After four weeks, the wound contraction was significantly higher in grafted groups than the no-treatment control group. Histologically, the MSCs persisted in the epidermis and dermis up to four weeks, suggesting that the cells migrated from the scaffold to the neo-epidermis and dermis. In comparison to the acellular scaffold group after four weeks, the MSC-treated group showed significant enhancement of vascular density. In addition to the active role of bone marrow MSCs in skin wound healing, ASCs also participate in wound healing partly by differentiating into vascular lineages. Engineered constructs consisting of rat ASCs cultured in fibrin matrix were grafted onto full thickness burn wounds of rats for up to 14 days. The results showed that ASCs delivered in the dermal matrix persisted locally to the site of delivery, improved wound healing, and enhanced local angiogenesis.[65] Together, these studies suggest that engineered skin grafts containing bone marrow MSCs or ASCs accelerate wound healing, improve neovascularization, and may differentiate into vascular and epidermal lineages.

4.2.3 *Cartilage tissue engineering*

Articular cartilage plays an important physiological function of lubricating between diarthrodial joints and distributing mechanical loads. Due to the avascular nature of articular cartilage, any damage or disease can be difficult to heal because of limited regeneration capacity. Current treatments to promote cartilage repair include subchondral abrasion, microfracture, transplantation of osteochondral plugs and autologous chondrocyte transplantation, but these approaches do not successfully restore long-lasting healthy cartilage.[66,67] As a result, tissue engineering is gaining increasing research interest in an attempt to address the drawbacks of current treatment methods. MSCs and ASCs are ideal candidate cell sources for engineered cartilage because of their high expansion ratio, immunomodulatory effect, and propensity to differentiate toward chondrogenic lineages.[68,69]

A number of studies have investigated the role of engineered cartilage tissue constructs using MSCs or ASCs. In one study, the chondrogenic potential of human ASCs was tested in three-dimensional ECMs consisting of alginate, agarose, or gelatin.[70] In the presence of chondrogenic media containing TGF-β, the content of cartilage matrix such as hydroxyproline and sulfated glycosaminoglycans significantly increased with time for all three ECMs. To test their *in vivo* therapeutic potential, human ASCs were injected into the intra-articular space in rabbit osteoarthritis model, and were found to promote articular cartilage repair. Cartilage formation was verified 16 weeks post-surgery by histological stains for increased cartilage ECM collagen II, proteoglycan, and reduced MMP 13 level.[71]

In addition to using chondrogenic growth factors or defined ECM components, decellularized cartilage ECM is a potential element in MSC-based cartilage tissue engineering, because it already contains the structural and functional elements of native cartilage that constitute a pro-chondrogenic microenvironment. An acellular cartilage-derived matrix scaffold can be easily generated by decellularizing bovine articular cartilage, grinding the matrix into fine powder, and then remolding the matrix into porous three-dimensional cylindrical constructs.[72] To substantiate the chondrogenic effect of cartilage ECM on the differentiating MSCs *in vivo*, rabbit bone marrow MSCs were seeded into the decellularized ECM and then transplanted into full-thickness osteochondral defects in rabbits for up to 12 weeks. Based on histological assessment, the cell-treated group had significantly better cartilage quality at 12 weeks, compared to the acellular treatment group. Furthermore, the cell-treated group showed significant improvement at 12 weeks, compared to six weeks, which suggested temporal changes in chondrogenic repair. In a related study using ECMs derived from human cadaveric joints and canine bone marrow MSCs, the cell-seeded constructs were delivered subcutaneously into nude mice for four weeks.[73] The results demonstrated that the engineered constructs formed ectopic cartilage-like tissue, with evidence of collagen II deposition and positive safranin O staining. In summary, these studies illustrate the potential of cartilage tissue engineering using bone marrow MSCs or ASCs.

4.2.4 *Tendon tissue engineering*

Tendons are fibrous tissues composed of densely packed collagen fiber bundles that connect muscle to bone and function in force transmission. In cases of severe tendon injury, surgical interventions are employed to repair or replace the damaged tendon with autografts, allografts, xenografts, or prosthetic devices. In the United States alone, around 45% of the 32 million musculoskeletal injuries each year involve tendon, ligament, and joint capsule, in which tendon injuries are especially common and require over 120,000 surgical repairs annually.[74] Tendon tissue engineering that uses autologous cells represents the cutting-edge technique to heal tendon injuries. However, despite the emerging functionality of tendon-derived stem/progenitor cells in tendon tissue engineering, the relatively hypocellular nature of tendon tissue impedes the use of tendon cells on a large scale. Autologous MSCs have been considered as an ideal cell source to address this limitation. Bone marrow MSCs have been extensively used to populate various graft materials and demonstrated promising outcomes including significantly increased tendon-related gene expression, higher DNA content, and enhanced mechanical strength.[75,76]

However, due to the lack of standard differentiation protocols to induce robust tendon lineage commitment of MSCs, the potential of MSCs as the cellular component for tendon tissue engineering is not fully exploited. To date, MSCs are used for tendon tissue engineering primarily as a means to improve vascularization, matrix deposition, and implant integration. The potential ability of MSCs to recapitulate tendon cell phenotype has just begun to emerge when provided with proper tendon microenvironment including growth factors, topographical cues, and mechanical stimulation. For instance, murine and canine ASCs demonstrated up to six-fold of upregulation in scleraxis (SCX), the primary marker of early tendon differentiation, when treated with 100 ng/ml of GDF-5 for one week.[77,78] Likewise, Goncalves *et al.* succeeded in acquiring tendon phenotype of abdominal human ASCs (hASCs) along with aligned collagen pattern that resembles the native tendon ECM through TGF-β1 treatment. Meanwhile, this research group also reported the efficacy of a group of other growth factors on partial induction of tendon phenotype of hASCs.[79]

Scaffolds are another crucial element for MSC-based tendon tissue engineering. They provide temporary biomechanical support to the healing tissue until deposition of the newly synthesized native matrix, and thereby prevent tendon re-rupture. Moreover, the topographical cues presented by scaffolds are capable of modulating the behavior of MSCs. As far as tendon tissue engineering is concerned, alignment is an important topographical characteristic to create in order to reproduce the aligned fibrous feature of native tendon tissue. Placing cells on a substrate with alignment pattern leads to increased expression of tendon markers, such as SCX and tenomodulin (TNMD).[80,81] In addition, cell extension and matrix deposition are also found to align along the orientation of fibers that constitute scaffold mesh.[82]

In the context of tendon injury and repair, it is recognized that controlled motion of healing tendons is required to improve clinical outcomes. MSCs exposed to static or cyclic uniaxial tension manifested increased cellular proliferation, ECM production, and tenogensis.[83,84] More recently, the cellular and molecular basis for loading-induced tenogenic differentiation of MSCs has been explored. Integrin-dependent signaling is pivotal in conveying external mechanical stimulation to downstream intrinsic targets. For instance, Xu *et al.* revealed that RhoA/ROCK and focal adhesion kinase coordinate to drive tenogenesis of MSCs in response to mechanical stretch through the regulation of cytoskeletal organization.[85] Other signaling pathways, such as TGF-β, are believed to be relevant to force induced tenogenesis independently of cytoskeletal tension.[86]

5. Future Directions

In summary, bone marrow MSCs and ASCs are promising candidate cell sources for tissue engineering and regenerative medicine because of their ease and reproducibility of isolation, high proliferation capacity, and ability to differentiate into therapeutic cell lineages. However, substantial barriers will need to be overcome before these cells become a standard treatment for clinical care. These challenges include the optimal conditions for cell expansion, enhancement in cell viability and function *in vivo*, elucidation of the mechanisms underlying their repair and regeneration ability, and immune acceptance for allogeneic transplantation. Nevertheless, some clinical trials are already underway (http://www.clinicaltrials.gov). As these limitations become resolved, it is anticipated that bone marrow MSC and ASC therapy will fulfill their promise of clinical efficacy.

Acknowledgements

This work was supported in part by grants to Ngan F. Huang (NFH) from the US National Institutes of Health (R00HL098688, R01HL127113, and R21EB020235), Merit Review Award (1I01BX002310) from the Department of Veterans Affairs Biomedical Laboratory Research and Development, the Stanford Women and Sex Differences in Medicine

Center, and the Stanford Child Health Research Institute. NFH was also supported by a McCormick Gabilan fellowship. In addition, this work was supported in part by a grant from the US National Institutes of Health (NCATS-CTSA, UL1 TR001085).

References

1. M. F. Pittenger, A. M. Mackay, S. C. Beck, R. K. Jaiswal, R. Douglas, J. D. Mosca, M. A. Moorman, D. W. Simonetti, S. Craig and D. R. Marshak, Multilineage potential of adult human mesenchymal stem cells, *Science* **284**: 143–147 (1999).

2. C. Campagnoli, I. A. Roberts, S. Kumar, P. R. Bennett, I. Bellantuono and N M. Fisk, Identification of mesenchymal stem/progenitor cells in human first-trimester fetal blood, liver, and bone marrow, *Blood* **98**: 2396–2402 (2001).

3. P. S. In 't Anker, W. A. Noort, S. A. Scherjon, C. Kleijburg-van der Keur, A. B. Kruisselbrink, R. L. van Bezooijen, W. Beekhuizen, R. Willemze, H. H. Kanhai and W. E. Fibbe, Mesenchymal stem cells in human second-trimester bone marrow, liver, lung, and spleen exhibit a similar immunophenotype but a heterogeneous multilineage differentiation potential, *Haematologica* **88**: 845–852 (2003).

4. N. F. Huang, R. J. Lee and S. Li, Chemical and physical regulation of stem cells and progenitor cells: potential for cardiovascular tissue engineering, *Tissue Eng* **13**: 1809–1823 (2007).

5. S. H. Bernacki, M. E. Wall and E. G. Loboa, Isolation of human mesenchymal stem cells from bone and adipose tissue, *Methods Cell Biol* **86**: 257–278 (2008).

6. J. T. Williams, S. S. Southerland, J. Souza, A. F. Calcutt and R. G. Cartledge, Cells isolated from adult human skeletal muscle capable of differentiating into multiple mesodermal phenotypes, *Am Surg* **65**: 22–26 (1999).

7. A. Erices, P. Conget and J. J. Minguell, Mesenchymal progenitor cells in human umbilical cord blood, *Br J Haematol* **109**: 235–242 (2000).

8. S. Gronthos, M. Mankani, J. Brahim, P. G. Robey and S. Shi, Postnatal human dental pulp stem cells (dpscs) *in vitro* and *in vivo*, *Proc Natl Acad Sci USA* **97**: 13625–13630 (2000).

9. G. Chamberlain, J. Fox, B. Ashton and J. Middleton, Concise review: mesenchymal stem cells: their phenotype, differentiation capacity, immunological features, and potential for homing, *Stem Cells* **25**: 2739–2749 (2007).

10. F. J. Lv, R. S. Tuan, K. M. Cheung and V. Y. Leung, Concise review: the surface markers and identity of human mesenchymal stem cells, *Stem Cells* **32**: 1408–1419 (2014).

11. M. Dominici, K. Le Blanc, I. Mueller, I. Slaper-Cortenbach, F. Marini, D. Krause, R. Deans, A. Keating, D. Prockop and E. Horwitz, Minimal criteria for defining multipotent mesenchymal stromal cells, *Cytotherapy* **8**: 315–317 (2006).

12. H. Kagami, H. Agata, M. Inoue, I. Asahina, A. Tojo, N. Yamashita and K. Imai, The use of bone marrow stromal cells (bone marrow-derived multipotent mesenchymal stromal cells) for alveolar bone tissue engineering: basic science to clinical translation, *Tissue Eng Part B Rev* **20**: 229–232 (2014).

13. J. Lo Surdo and S. R. Bauer, Quantitative approaches to detect donor and passage differences in adipogenic potential and clonogenicity in human bone marrow-derived mesenchymal stem cells, *Tissue Eng Part C Methods* **18**: 877–889 (2012).

14. J. M. Gimble, A. J. Katz and B. A. Bunnell, Adipose-derived stem cells for regenerative medicine, *Circ Res* **100**: 1249–1260 (2007).

15. F. De Francesco, G. Ricci, F. d' Andrea, G. F. Nicoletti and G. A. Ferraro, Human adipose stem cells: from bench to bedside, *Tissue Eng Part B Rev* **21**: 572–584 (2015).

16. E. S. Jeon, W. S. Park, M. J. Lee, Y. M. Kim, J. Han and J. H. Kim, A rho kinase/myocardin-related transcription factor-a-dependent mechanism underlies the sphingosylphosphorylcholine-induced differentiation of mesenchymal stem cells into contractile smooth muscle cells, *Circ Res* **103**: 635–642 (2008).

17. V. Planat-Benard, C. Menard, M. Andre, M. Puceat, A. Perez, J. M. Garcia-Verdugo, L. Penicaud and L. Casteilla, Spontaneous cardiomyocyte differentiation from adipose tissue stroma cells, *Circ Res* **94**: 223–229 (2004).

18. V. Planat-Benard, J. S. Silvestre, B. Cousin, M. Andre, M. Nibbelink, R. Tamarat, M. Clergue, C. Manneville, C. Saillan-Barreau, M. Duriez, A. Tedgui, B. Levy, L. Penicaud and L. Casteilla, Plasticity of human adipose lineage cells toward endothelial cells: physiological and therapeutic perspectives, *Circulation* **109**: 656–663 (2004).

19. L. Zimmerlin, V. S. Donnenberg, M. E. Pfeifer, E. M. Meyer, B. Peault, J. P. Rubin and A. D. Donnenberg, Stromal vascular progenitors in adult human adipose tissue, *Cytometry A* **77**: 22–30 (2010).

20. P. A. Zuk, M. Zhu, P. Ashjian, D. A. De Ugarte, J. I. Huang, H. Mizuno, Z. C. Alfonso, J. K. Fraser, P. Benhaim and M. H. Hedrick, Human

adipose tissue is a source of multipotent stem cells, *Mol Biol Cell* **13**: 4279–4295 (2002).

21. K. McIntosh, S. Zvonic, S. Garrett, J. B. Mitchell, Z. E. Floyd, L. Hammill, A. Kloster, Y. Di Halvorsen, J. P. Ting, R. W. Storms, B. Goh, G. Kilroy, X. Wu and J. M. Gimble, The immunogenicity of human adipose-derived cells: temporal changes *in vitro*, *Stem Cells* **24**: 1246–1253 (2006).

22. S. J. Huang, R. H. Fu, W. C. Shyu, S. P. Liu, G. P. Jong, Y. W. Chiu, H. S. Wu, Y. A. Tsou, C. W. Cheng and S. Z. Lin, Adipose-derived stem cells: isolation, characterization, and differentiation potential, *Cell Transplant* **22**: 701–709 (2013).

23. R. Tuli, S. Tuli, S. Nandi, M. L. Wang, P. G. Alexander, H. Haleem-Smith, W. J. Hozack, P. A. Manner, K. G. Danielson and R. S. Tuan, Characterization of multipotential mesenchymal progenitor cells derived from human trabecular bone, *Stem Cells* **21**: 681–693 (2003).

24. T. Hennig, H. Lorenz, A. Thiel, K. Goetzke, A. Dickhut, F. Geiger and W. Richter, Reduced chondrogenic potential of adipose tissue derived stromal cells correlates with an altered tgfbeta receptor and bmp profile and is overcome by bmp-6, *J Cell Physiol* **211**: 682–691 (2007).

25. Z. Gong and L. E. Niklason, Small-diameter human vessel wall engineered from bone marrow-derived mesenchymal stem cells (hmscs), *FASEB J* **22**: 1635–1648 (2008).

26. W. Xu, X. Zhang, H. Qian, W. Zhu, X. Sun, J. Hu, H. Zhou and Y. Chen, Mesenchymal stem cells from adult human bone marrow differentiate into a cardiomyocyte phenotype *in vitro*, *Exp Biol Med (Maywood)* **229**: 623–631 (2004).

27. K. Kurpinski, J. Chu, C. Hashi and S. Li, Anisotropic mechanosensing by mesenchymal stem cells, *Proc Natl Acad Sci USA* **103**: 16095–16100 (2006).

28. N. Haghighipour, S. Heidarian, M. A. Shokrgozar and N. Amirizadeh, Differential effects of cyclic uniaxial stretch on human mesenchymal stem cell into skeletal muscle cell, *Cell Biol Int* **36**: 669–675 (2012).

29. T. Guo, L. Yu, C. G. Lim, A. S. Goodley, X. Xiao, J. K. Placone, K. M. Ferlin, B. N. Nguyen, A. H. Hsieh and J. P. Fisher, Effect of dynamic culture and periodic compression on human mesenchymal stem cell proliferation and chondrogenesis, *Ann Biomed Eng* **44**: 2103–2113 (2016).

30. K. H. Nakayama, L. Hou and N. F. Huang, Role of extracellular matrix signaling cues in modulating cell fate commitment for cardiovascular tissue engineering, *Adv Healthc Mater* **3**: 628–641 (2014).

31. J. K. Kular, S. Basu and R. I. Sharma, The extracellular matrix: structure, composition, age-related differences, tools for analysis and applications for tissue engineering, *J Tissue Eng* **5**: 2041731414557112 (2014).

32. A. J. Engler, S. Sen, H. L. Sweeney and D. E. Discher, Matrix elasticity directs stem cell lineage specification, *Cell* **126**: 677–689 (2006).

33. R. McBeath, D. M. Pirone, C. M. Nelson, K. Bhadriraju and C. S. Chen, Cell shape, cytoskeletal tension, and rhoa regulate stem cell lineage commitment, *Dev Cell* **6**: 483–495 (2004).

34. R. K. Okolicsanyi, L. R. Griffiths and L. M. Haupt, Mesenchymal stem cells, neural lineage potential, heparan sulfate proteoglycans and the matrix, *Dev Biol* **388**: 1–10 (2014).

35. G. Zhang, X. Wang, Z. Wang, J. Zhang and L. Suggs, A pegylated fibrin patch for mesenchymal stem cell delivery, *Tissue Eng* **12**: 9–19 (2006).

36. R. Yi, M. N. Poy, M. Stoffel and E. Fuchs, A skin microRNA promotes differentiation by repressing "stemness", *Nature* **452**: 225–229 (2008).

37. N. Xu, T. Papagiannakopoulos, G. Pan, J. A. Thomson and K. S. Kosik, MicroRNA-145 regulates oct4, sox2, and klf4 and represses pluripotency in human embryonic stem cells, *Cell* **137**: 647–658 (2009).

38. Y. Mei, C. Bian, J. Li, Z. Du, H. Zhou, Z. Yang and R. C. Zhao, Mir-21 modulates the erk-mapk signaling pathway by regulating spry2 expression during human mesenchymal stem cell differentiation, *J Cell Biochem* **114**: 1374–1384 (2013).

39. A. Schoolmeesters, T. Eklund, D. Leake, A. Vermeulen, Q. Smith, S. F. Aldred and Y. Fedorov, Functional profiling reveals critical role for miRNA in differentiation of human mesenchymal stem cells, *PLoS One* **4**: e5605 (2009).

40. Y. Haraguchi, T. Shimizu, M. Yamato and T. Okano, Concise review: cell therapy and tissue engineering for cardiovascular disease, *Stem Cells Transl Med* **1**: 136–141 (2012).

41. L. Hou, J. J. Kim, Y. J. Woo and N. F. Huang, Stem cell-based therapies to promote angiogenesis in ischemic cardiovascular disease, *Am J Physiol Heart Circ Physiol* **310**: H455–465 (2016).

42. J. G. Shake, P. J. Gruber, W. A. Baumgartner, G. Senechal, J. Meyers, J. M. Redmond, M. F. Pittenger and B. J. Martin, Mesenchymal stem cell implantation in a swine myocardial infarct model: engraftment and functional effects, *Ann Thorac Surg* **73**: 1919–1925; discussion 1926 (2002).

43. I. M. Barbash, P. Chouraqui, J. Baron, M. S. Feinberg, S. Etzion, A. Tessone, L. Miller, E. Guetta, D. Zipori, L. H. Kedes, R. A. Kloner and J. Leor, Systemic delivery of bone marrow-derived mesenchymal stem cells to the

infarcted myocardium: feasibility, cell migration, and body distribution, *Circulation* **108**: 863–868 (2003).

44. P. Locatelli, F. D. Olea, A. Hnatiuk, A. De Lorenzi, M. Cerda, C. S. Gimenez, D. Sepulveda, R. Laguens and A. Crottogini, Mesenchymal stromal cells overexpressing vascular endothelial growth factor in ovine myocardial infarction, *Gene Ther* **22**: 449–457 (2015).

45. W. Li, N. Ma, L. L. Ong, C. Nesselmann, C. Klopsch, Y. Ladilov, D. Furlani, C. Piechaczek, J. M. Moebius, K. Lutzow, A. Lendlein, C. Stamm, R. K, Li and G. Steinhoff, Bcl-2 engineered mscs inhibited apoptosis and improved heart function, *Stem Cells* **25**: 2118–2127 (2007).

46. N. F. Huang, J. Yu, R. Sievers, S. Li and R. J. Lee, Injectable biopolymers enhance angiogenesis after myocardial infarction, *Tissue Eng* **11**: 1860–1866 (2005).

47. K. L. Christman, H. H. Fok, R. E. Sievers, Q. Fang and R. J. Lee. Fibrin glue alone and skeletal myoblasts in a fibrin scaffold preserve cardiac function after myocardial infarction, *Tissue Eng* **10**: 403–409 (2004).

48. K. L. Christman, A. J. Vardanian, Q. Fang, R. E. Sievers, H. H. Fok and R. J. Lee, Injectable fibrin scaffold improves cell transplant survival, reduces infarct expansion, and induces neovasculature formation in ischemic myocardium, *J Am Coll Cardiol* **44**: 654–660 (2004).

49. N. Landa, L. Miller, M. S. Feinberg, R. Holbova, M. Shachar, I. Freeman, S. Cohen and J. Leor, Effect of injectable alginate implant on cardiac remodeling and function after recent and old infarcts in rat, *Circulation* **117**: 1388–1396 (2008).

50. N. F. Huang, A. Lam, Q. Fang, R. E. Sievers, S. Li and R. J. Lee, Bone marrow-derived mesenchymal stem cells in fibrin augment angiogenesis in the chronically infarcted myocardium, *Regen Med* **4**: 527–538 (2009).

51. J. P. Cooke and D. W. Losordo, Modulating the vascular response to limb ischemia: angiogenic and cell therapies, *Circ Res* **116**: 1561–1578 (2015).

52. T. Kinnaird, E. Stabile, M. S. Burnett, M. Shou, C. W. Lee, S. Barr, S. Fuchs and S. E. Epstein, Local delivery of marrow-derived stromal cells augments collateral perfusion through paracrine mechanisms, *Circulation* **109**: 1543–1549 (2004).

53. Y. Kang, C. Park, D. Kim, C. M. Seong, K. Kwon and C. Choi, Unsorted human adipose tissue-derived stem cells promote angiogenesis and myogenesis in murine ischemic hindlimb model, *Microvasc Res* **80**: 310–316 (2010).

54. Y. Kim, H. Kim, H. Cho, Y. Bae, K. Suh and J. Jung, Direct comparison of human mesenchymal stem cells derived from adipose tissues and bone marrow in mediating neovascularization in response to vascular ischemia, *Cell Physiol Biochem* **20**: 867–876 (2007).

55. A. El-Badawy, M. Amer, R. Abdelbaset, S. N. Sherif, M. Abo-Elela, Y. H. Ghallab, H. Abdelhamid, Y. Ismail and N. El-Badri, Adipose stem cells display higher regenerative capacities and more adaptable electro-kinetic properties compared to bone marrow-derived mesenchymal stromal cells, *Sci Rep* **6**: 37801 (2016).

56. C. K. Hashi, Y. Zhu, G. Y. Yang, W. L. Young, B. S. Hsiao, K. Wang, B. Chu and S. Li, Antithrombogenic property of bone marrow mesenchymal stem cells in nanofibrous vascular grafts, *Proc Natl Acad Sci USA* **104**: 11915–11920 (2007).

57. K. C. Kuo, R. Z. Lin, H. W. Tien, P. Y. Wu, Y. C. Li, J. M. Melero-Martin and Y. C. Chen, Bioengineering vascularized tissue constructs using an injectable cell-laden enzymatically crosslinked collagen hydrogel derived from dermal extracellular matrix, *Acta Biomater* **27**: 151–166 (2015).

58. M. Zhang, K. Wang, Z. Wang, B. Xing, Q. Zhao and D. Kong, Small-diameter tissue engineered vascular graft made of electrospun pcl/lecithin blend, *J Mater Sci Mater Med* **23**: 2639–2648 (2012).

59. A. Mirza, J. M. Hyvelin, G. Y. Rochefort, P. Lermusiaux, D. Antier, B. Awede, P. Bonnet, J. Domenech and V. Eder, Undifferentiated mesenchymal stem cells seeded on a vascular prosthesis contribute to the restoration of a physiologic vascular wall, *J Vasc Surg* **47**: 1313–1321 (2008).

60. J. T. Krawiec and D. A. Vorp, Adult stem cell-based tissue engineered blood vessels: a review, *Biomaterials* **33**: 3388–3400 (2012).

61. K. H. Kim, G. Blasco-Morente, N. Cuende and S. Arias-Santiago, Mesenchymal stromal cells: properties and role in management of cutaneous diseases, *J Eur Acad Dermatol Venereol* **31**: 414–423 (2016).

62. N. F. Huang, S. Zac-Verghese and S. Luke, Apoptosis in skin wound healing, *Wounds* **15**: 182–194 (2003).

63. D. Eisenbud, N. F. Huang, S. Luke and M. Silberklang, Skin substitutes and wound healing: current status and challenges, *Wounds* **16**: 2–17 (2004).

64. P. Liu, Z. Deng, S. Han, T. Liu, N. Wen, W. Lu, X. Geng, S. Huang and Y. Jin, Tissue-engineered skin containing mesenchymal stem cells improves burn wounds, *Artif Organs* **32**: 925–931 (2008).

65. E. Chung, V. Y. Rybalko, P. L. Hsieh, S. L. Leal, M. A. Samano, A. N. Willauer, R. S. Stowers, S. Natesan, D. O. Zamora, R. J. Christy and L. J. Sugg, Fibrin-based stem cell containing scaffold improves the dynamics of burn wound healing, *Wound Repair Regen* **24**: 810–819 (2016).

66. S. N. Redman, S. F. Oldfield and C. W. Archer, Current strategies for articular cartilage repair, *Eur Cell Mater* **9**: 23–32; discussion 23–32 (2005).

67. T. M. Simon and D. W. Jackson, Articular cartilage: injury pathways and treatment options, *Sports Med Arthrosc* **14**: 146–154 (2006).

68. D. Reissis, Q. O. Tang, N. C. Cooper, C. F. Carasco, Z. Gamie, A. Mantalaris and E. Tsiridis, Current clinical evidence for the use of mesenchymal stem cells in articular cartilage repair, *Expert Opin Biol Ther* **16**: 535–557 (2016).

69. S. M. Richardson, G. Kalamegam, P. N. Pushparaj, C. Matta, A. Memic, A. Khademhosseini, R. Mobasheri, F. L. Poletti, J. A. Hoyland and A. Mobasheri, Mesenchymal stem cells in regenerative medicine: focus on articular cartilage and intervertebral disc regeneration, *Methods* **99**: 69–80 (2016).

70. H. A. Awad, M. Q. Wickham, H. A. Leddy, J. M. Gimble and F. Guilak, Chondrogenic differentiation of adipose-derived adult stem cells in agarose, alginate, and gelatin scaffolds, *Biomaterials* **25**: 3211–3222 (2004).

71. W. Wang, N. He, C. Feng, V. Liu, L. Zhang, F. Wang, J. He, T. Zhu, S. Wang, W. Qiao, S. Li, G. Zhou, C. Dai and W. Cao, Human adipose-derived mesenchymal progenitor cells engraft into rabbit articular cartilage, *Int J Mol Sci* **16**: 12076–12091 (2015).

72. Z. Yang, Y. Shi, X. Wei, J. He, S. Yang, G. Dickson, J. Tang, J. Xiang, C. Song and G. Li, Fabrication and characterization of a novel acellular cartilage matrix scaffold for tissue engineering, *Tissue Eng Part C Methods* **16**: 865–876 (2010).

73. Q. Yang, J. Peng, Q. Guo, J. Huang, L. Zhang, J. Yao and F. Yang, S. Wang, W. Xu, A. Wang and S. Lu, A cartilage ecm-derived 3-d porous acellular matrix scaffold for *in vivo* cartilage tissue engineering with pkh26-labeled chondrogenic bone marrow-derived mesenchymal stem cells, *Biomaterials* **29**: 2378–2387 (2008).

74. D. L. Butler, N. Juncosa and M. R. Dressler, Functional efficacy of tendon repair processes, *Annu Rev Biomed Eng* **6**: 303–329 (2004).

75. B. A. Tucker, S. S. Karamsadkar, W. S. Khan and P. Pastides, The role of bone marrow derived mesenchymal stem cells in sports injuries, *J Stem Cells* **5**: 155–166 (2010).

76. S. Yokoya, Y. Mochizuki, K. Natsu, H. Omae, Y. Nagata and M. Ochi, Rotator cuff regeneration using a bioabsorbable material with bone marrow-derived mesenchymal stem cells in a rabbit model, *Am J Sports Med* **40**: 1259–1268 (2012).

77. H. Shen, R. H. Gelberman, M. J. Silva, S. E. Sakiyama-Elbert and S. Thomopoulos, Bmp12 induces tenogenic differentiation of adipose-derived stromal cells, *PLoS One* **8**: e77613 (2013).

78. A. Park, M. V. Hogan, G. S. Kesturu, R. James, G. Balian and A.B. Chhabra, Adipose-derived mesenchymal stem cells treated with growth differentiation factor-5 express tendon-specific markers, *Tissue Eng Part A* **16**: 2941–2951 (2010).

79. A. I. Goncalves, M. T. Rodrigues, S. J. Lee, A. Atala, J. J. Yoo, R. L. Reis and M. E. Gomes, Understanding the role of growth factors in modulating stem cell tenogenesis, *PLoS One* **8**: e83734 (2013).

80. J. Zhu, J. Li, B. Wang, W. J. Zhang, G. Zhou, Y. Cao and W. Liu, The regulation of phenotype of cultured tenocytes by microgrooved surface structure, *Biomaterials* **31**: 6952–6958 (2010).

81. V. Kishore, W. Bullock, X. Sun, W. S. Van Dyke and O. Akkus, Tenogenic differentiation of human mscs induced by the topography of electrochemically aligned collagen threads, *Biomaterials* **33**: 2137–2144 (2012).

82. K. L. Moffat, A. S. Kwei, J. P. Spalazzi, S. B. Doty, W. N. Levine and H. H. Lu, Novel nanofiber-based scaffold for rotator cuff repair and augmentation, *Tissue Eng Part A* **15**: 115–126 (2009).

83. G. Yang, R. C. Crawford and J. H. Wang, Proliferation and collagen production of human patellar tendon fibroblasts in response to cyclic uniaxial stretching in serum-free conditions, *J Biomech* **37**: 1543–1550 (2004).

84. C. K. Kuo and R. S. Tuan, Mechanoactive tenogenic differentiation of human mesenchymal stem cells, *Tissue Eng Part A* **14**: 1615–1627 (2008).

85. B. Xu, G. Song, Y. Ju, X. Li, Y. Song and S. Watanabe, Rhoa/rock, cytoskeletal dynamics, and focal adhesion kinase are required for mechanical stretch-induced tenogenic differentiation of human mesenchymal stem cells, *J Cell Physiol* **227**: 2722–2729 (2012).

86. T. Maeda, T. Sakabe, A. Sunaga, K. Sakai, A. L. Rivera, D. R. Keene, T. Sasaki, E. Stavnezer, J. Iannotti, R. Schweitzer, D. Ilic, H. Baskaran and T. Sakai, Conversion of mechanical force into tgf-beta-mediated biochemical signals, *Curr Biol* **21**: 933–941 (2011).

87. A. A. Mangi, N. Noiseux, D. Kong, H. He, M. Rezvani, J. S. Ingwall and V. J. Dzau, Mesenchymal stem cells modified with akt prevent remodeling and restore performance of infarcted hearts, *Nat Med* **9**: 1195–1201 (2003).

88. Y. Miyahara, N. Nagaya, M. Kataoka, B. Yanagawa, K. Tanaka, H. Hao, K. Ishino, H. Ishida, T. Shimizu, K. Kangawa, S. Sano, T. Okano, S. Kitamura and H. Mori, Monolayered mesenchymal stem cells repair scarred myocardium after myocardial infarction, *Nat Med* **12**: 459–465 (2006).

89. H. J. Wei, C. H. Chen, W. Y. Lee, I. Chiu, S. M. Hwang, W. W. Lin, C. C. Huang, Y. C. Yeh, Y, Chang and H. W. Sung, Bioengineered cardiac patch constructed from multilayered mesenchymal stem cells for myocardial repair, *Biomaterials* **29**: 3547–3556 (2008).

90. M. H. Moon, S. Y. Kim, Y. J. Kim, S. J. Kim, J. B. Lee, Y. C. Bae, S. M. Sung and J. S. Jung, Human adipose tissue-derived mesenchymal stem

cells improve postnatal neovascularization in a mouse model of hindlimb ischemia, *Cell Physiol Biochem* **17**: 279–290 (2006).

91. M. P. Brennan, A. Dardik, N. Hibino, J. D. Roh, G. N. Nelson, X. Papademitris, T. Shinoka and C. K. Breuer, Tissue-engineered vascular grafts demonstrate evidence of growth and development when implanted in a juvenile animal model, *Ann Surg* **248**: 370–377 (2008).

92. G. Di Rocco, M. G. Iachininoto, A. Tritarelli, S. Straino, A. Zacheo, A. Germani, F. Crea and M. C. Capogrossi, Myogenic potential of adipose-tissue-derived cells, *J Cell Sci* **119**: 2945–2952 (2006).

93. J. H. Lee and D. M. Kemp, Human adipose-derived stem cells display myogenic potential and perturbed function in hypoxic conditions, *Biochem Biophys Res Commun* **341**: 882–888 (2006).

94. M. Kim, Y. S. Choi, S. H. Yang, H. N. Hong, S. W. Cho, S. M. Cha, J. H. Pak, C. W. Kim, S. W. Kwon and C. J. Park, Muscle regeneration by adipose tissue-derived adult stem cells attached to injectable plga spheres, *Biochem Biophys Res Commun* **348**: 386–392 (2006).

95. C. H. Chen, Y. Chang, C. C. Wang, C. H. Huang, C. C. Huang, Y. C. Yeh, S. M. Hwang and H. W. Sung, Construction and characterization of fragmented mesenchymal-stem-cell sheets for intramuscular injection, *Biomaterials* **28**: 4643–4651 (2007).

96. F. Granero-Molto, J. A. Weis, M. I. Miga, B. Landis, T. J. Myers, L. O'Rear, L. Longobardi, E. D. Jansen, D. P. Mortlock and A. Spagnoli, Regenerative effects of transplanted mesenchymal stem cells in fracture healing, *Stem Cells* **27**: 1887–1898 (2009).

97. A. Hasharoni, Y. Zilberman, G. Turgeman, G. A. Helm, M. Liebergall and D. Gazit, Murine spinal fusion induced by engineered mesenchymal stem cells that conditionally express bone morphogenetic protein-2, *J Neurosurg Spine* **3**: 47–52 (2005).

98. H. Hattori, M. Sato, K. Masuoka, M. Ishihara, T. Kikuchi, T. Matsui, B. Takase, T. Ishizuka, M. Kikuchi and K. Fujikawa, Osteogenic potential of human adipose tissue-derived stromal cells as an alternative stem cell source, *Cells Tissues Organs* **178**: 2–12 (2004).

99. H. Hattori, K. Masuoka, M. Sato, M. Ishihara, T. Asazuma, B. Takase, M. Kikuchi and K. Nemoto, Bone formation using human adipose tissue-derived stromal cells and a biodegradable scaffold, *J Biomed Mater Res B* **76**: 230–239 (2006).

100. S. Sobajima, G. Vadala, A. Shimer, J. S. Kim, L. G. Gilbertson and J. D. Kang, Feasibility of a stem cell therapy for intervertebral disc degeneration, *Spine J* **8**: 888–896 (2008).

101. H. Koga, M. Shimaya, T. Muneta, A. Nimura, T. Morito, M. Hayashi, S. Suzuki, Y. J. Ju, T. Mochizuki and I. Sekiya, Local adherent technique for transplanting mesenchymal stem cells as a potential treatment of cartilage defect, *Arthritis Res Ther* **10**: R84 (2008).

102. M. Sasaki, R. Abe, Y. Fujita, S. Ando, D. Inokuma and H. Shimizu, Mesenchymal stem cells are recruited into wounded skin and contribute to wound repair by transdifferentiation into multiple skin cell type, *J Immunol* **180**: 2581–2587 (2008).

103. X. Fu, L. Fang, X. Li, B. Cheng B and Z. Sheng, Enhanced wound-healing quality with bone marrow mesenchymal stem cells autografting after skin injury, *Wound Repair Regen* **14**: 325–335 (2006).

104. A. M. Altman, N. Matthias, Y. Yan, Y. H. Song, X. Bai, E. S. Chiu, D. P. Slakey and E. U. Alt, Dermal matrix as a carrier for *in vivo* delivery of human adipose-derived stem cells, *Biomaterials* **29**: 1431–1442 (2008).

105. Z. Tan, Q. Zhao, P. Gong, Y. Wu, N. Wei, Q. Yuan, C. Wang, D. Liao and H. Tang, Research on promoting periodontal regeneration with human basic fibroblast growth factor-modified bone marrow mesenchymal stromal cell gene therapy, *Cytotherapy* **11**: 317–325 (2009).

106. M. Tobita, A. C. Uysal, R. Ogawa, H. Hyakusoku and H. Mizuno, Periodontal tissue regeneration with adipose-derived stem cells, *Tissue Eng Part A* **14**: 945–953 (2008).

107. X. Guo, Q. Zheng, I. Kulbatski, Q. Yuan, S. Yang, Z. Shao, H. Wang, B. Xiao, Z. Pan and S. Tang, Bone regeneration with active angiogenesis by basic fibroblast growth factor gene transfected mesenchymal stem cells seeded on porous beta-tcp ceramic scaffolds, *Biomed Mater* **1**: 93–99 (2006).

108. H. Li, F. Yan, L. Lei, Y. Li and Y. Xiao, Application of autologous cryopreserved bone marrow mesenchymal stem cells for periodontal regeneration in dogs, *Cells Tissues Organs* **190**: 94–101 (2009).

109. W. Gu, F. Zhang, Q. Xue, Z. Ma, P. Lu and B. Yu, Transplantation of bone marrow mesenchymal stem cells reduces lesion volume and induces axonal regrowth of injured spinal cord, *Neuropathology* **30**: 205–217 (2009).

110. Y. Ding, Q. Yan, J. W. Ruan, Y. Q. Zhang, W. J. Li, Y. J. Zhang, Y. Li, H. Dong and Y. S. Zeng, Electro-acupuncture promotes survival, differentiation of the bone marrow mesenchymal stem cells as well as functional recovery in the spinal cord-transected rats, *BMC Neurosci* **10**: 35 (2009).

111. J. Park, E. Lim, S. Back, H. Na, Y. Park and K. Sun, Nerve regeneration following spinal cord injury using matrix metalloproteinase-sensitive,

hyaluronic acid-based biomimetic hydrogel scaffold containing brain-derived neurotrophic factor, *J Biomed Mater Res A* **93**: 1091–1099 (2010).

112. L. He, Y. Zhang, C. Zeng, M. Ngiam, S. Liao, D. Quan, Y. Zeng, J. Lu and S. Ramakrishna, Manufacture of plga multiple-channel conduits with precise hierarchical pore architectures and *in vitro/vivo* evaluation for spinal cord injury, *Tissue Eng Part C Methods* **15**: 243–255 (2009).

2

DELIVERY VEHICLES FOR DEPLOYING MESENCHYMAL STEM CELLS IN TISSUE REPAIR

Ben P. Hung, Michael S. Friedman† and J. Kent Leach‡*

1. Introduction

Cell-based therapies are a promising alternative to organ transplantation. Numerous cell sources are under investigation for their efficacy in promoting or contributing to the repair and regeneration of human tissues. Over the past decade, mesenchymal stem cells (MSCs) have attracted much attention in the cell therapy and regenerative medicine fields. Unlike more mature cells, MSCs can be expanded through several passages without loss of differentiation potential, have multilineage potential,[1] possess potent anti-inflammatory and immunomodulatory properties,[2–4] and can be isolated from numerous tissues throughout the body. MSCs can be harvested from numerous tissue compartments of the postnatal organism, bone marrow-derived MSCs being the most commonly examined population, while adipose and umbilical cord-derived stem cells with similar potencies have also been studied.[5]

While the systemic injection of MSCs has shown some promise as a treatment for immune-mediated disease, their use in tissue regeneration applications has met with little clinical success. Recent work indicates that MSCs fail to engraft long-term and that any therapeutic benefit derived from MSCs is through transient tissue residence and secretion of trophic factors. However, many of these studies fail to integrate cellular therapy with tissue engineering approaches for deploying MSCs in the appropriate physiologic context with the relevant microenvironmental and mechanical cues. In this chapter, we will discuss some of the biomaterials that have been utilized to deploy MSCs in tissue repair and regeneration applications. We will highlight biomaterials for MSCs that may be delivered in a minimally invasive injectable formulation or utilize alternative implantable materials to promote tissue repair and regeneration (Fig. 1).

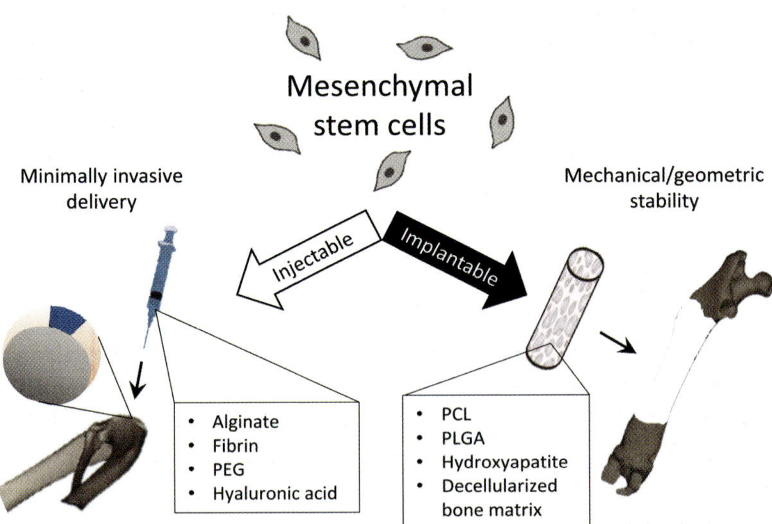

Fig. 1. Biomaterials are promising carriers of MSCs that may be injectable (*left*) or implantable (*right*). The biophysical properties of the material can guide cell function at the site of implantation.

2. Delivery of MSCs for Repairing Cardiovascular Tissues

Cardiovascular disease is the leading cause of death in the United States, and treatment of vascular disease accounts for a large proportion of health care expenditures. There is a vast shortage of viable cardiac tissues and graft segments available for transplantation due to disease and morbidity. Cell-based approaches to engineering cardiovascular structures represent an exciting alternative to traditional treatment options. In the paragraphs below, we will summarize several recent investigations that have deployed MSCs using various biomaterials to produce new blood vessels and repair damaged myocardium.

2.1 *Engineering long-lasting blood vessels*

The presence of a robust vasculature is critical to supply nutrients and circulating factors to developing and engineered tissues, as well as those tissues undergoing repair. The feasibility of engineering microvascular networks *in vivo* has been shown using human umbilical vein endothelial cells (HUVECs), human microvascular endothelial cells (HMVECs), and more recently using pluripotent cells differentiated from human embryonic stem cells or induced pluripotent stem cells. However, these leaky vessels are rapidly pruned and remodeled without stabilization by pericytes. MSCs are an ideal cell population to promote anastomosis and long-term stability due to their secretion of trophic factors, potential for differentiation toward the myogenic phenotype, and *in vivo* identity and function as pericytes.[6,7] (Fig. 2). The promise of MSCs has ushered in an exciting era of implanting co-cultures or accessory cells to form new tissues using injectable and implantable materials.

Hydrogels derived from natural or synthetic polymers are a preferred material for implanting cellular populations that facilitate neovascularization due to their mechanical properties and injectable capacity. Hydrogels composed of natural materials are advantageous, as they commonly degrade into easily metabolized degradation products. However, there is significant lot-to-lot variability during their preparation, potentially confounding their widespread use. Moreover, a substantial percentage of the patient population may have established immunogenicity to some animal-derived proteins (e.g. collagen), potentially stimulating an undesirable response to the cell

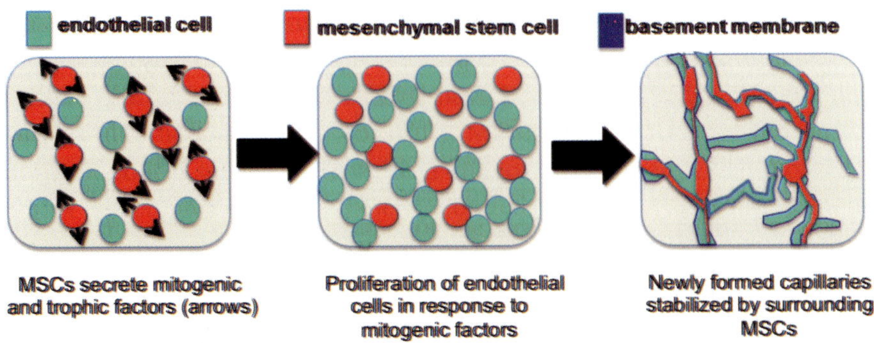

Fig. 2. The implantation of co-cultures of endothelial cells and mesenchymal stem cells for enhancing angiogenesis. MSCs secrete trophic and proangiogenic factors (*left*) that stimulate endothelial cell proliferation (*center*) and stabilize newly formed capillaries (*right*).

carrier. Synthetic polymers address the limitations associated with immunogenicity, availability, and variability, but may require the incorporation of binding domains to enable cell adhesion or advanced chemistries to attain the desirable degradation parameters. Regardless of the biomaterial, it is critical to utilize a substrate that enables long-term cell survival, remodeling of the gel over time, and the alignment of capillaries to facilitate anastomosis with the host vasculature.

Matrigel is a solubilized basement membrane preparation extracted from the Engelbreth-Holm-Swarm mouse sarcoma and is rich in extracellular matrix constituents such as laminin, collagen IV, and heparan sulfate proteoglycans. Bone marrow- or cord blood-derived MSCs, together with bone marrow-derived endothelial progenitor cells (EPCs), were entrapped in Matrigel and implanted into the dorsum of rats.[8] This co-culture yielded an extensive network of human blood vessels after one week and formed functional anastomoses with the existing vasculature. The implanted EPCs were restricted to the luminal aspect of the vessels, while mesenchymal progenitor cells were adjacent to lumens, confirming their role as perivascular cells. Importantly, the engineered vascular networks remained patent at four weeks *in vivo*. Similar results were found by deploying murine or human MSC-type populations, together with endothelial cells, in fibronectin-type I collagen gel implants.[9,10] Implantation of the endothelial-mesenchymal co-cultures consistently resulted in stable vessels that persisted *in vivo*, some for up to one year. However, those implants containing only endothelial cells quickly regressed. The implant material facilitated cell adhesion, proliferation, and arrangement into capillary vessels. Unlike implants formed from Matrigel or collagen that are limited by potential immunogenicity and safety concerns owing to the material source, hydrogels formed of fibrin can be produced from autologous blood samples. Mechanically robust fibrin hydrogels exhibited significant increases in vascular invasion when used to deploy MSCs due to the proteolytic degradation of the matrix by MSC-secreted metalloproteinases.[11] Synthetic polymers have also been used to implant MSCs with an eye toward neovascularization. Differentiated MSCs entrapped within cylinders formed of poly(ethylene glycol) diacrylate (PEGDA) enabled robust neovascularization and contributed to the formation of highly vascularized fibrous capsules representative of soft tissue.[12]

Blood vessels must be produced over several length scales, ranging from 3- to 8-μm capillaries to vascular grafts with diameters exceeding 6 mm. MSCs have been deployed on various biomaterials to assist in the formation of tissue-engineered vascular grafts (TEVGs) and support grafts that are endothelialized *in vivo*. Bone marrow-derived MSCs were embedded within the walls of nanofibrous mesh TEVGs produced from electrospun poly-L-lactic acid (PLLA) by seeding a flat film and rolling it around a mandrel to produce a vessel-like structure.[13] (Fig. 3a). Grafts implanted for up to 60 days in the common carotid artery of athymic rats facilitated efficient cell recruitment and organization, significant extracellular matrix (ECM) synthesis and self-assembly, and excellent long-term patency. Gong and Niklason seeded MSCs in mesh scaffolds formed of poly(glycolic acid) (PGA) and cells were cultured under optimized conditions in pulsatile perfusion bioreactors to induce differentiation toward the myogenic phenotype.[14] This graft was readily endothelialized when seeded and maintained in culture over eight weeks. After culture, MSCs expressed smooth muscle cell actin, a hallmark of the myogenic lineage, and MSC-seeded grafts possessed significantly more collagen content and elevated burst pressures compared to earlier protocols. These data suggest that MSCs provide a viable alternative to autologous smooth muscle cells to address the reduced production of vessel-strengthening collagen with

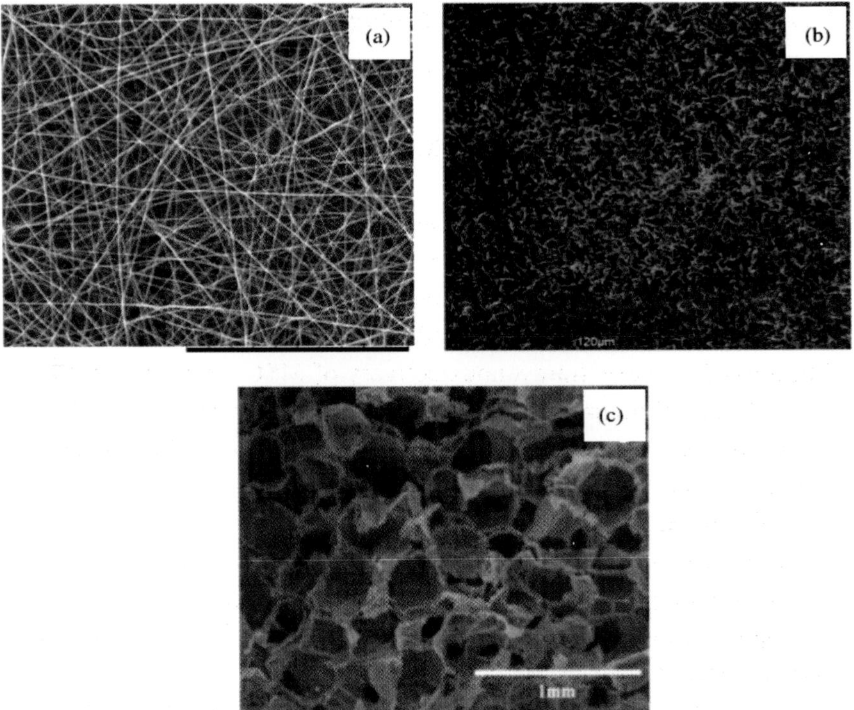

Fig. 3.   (a) Nanofiber mesh formed by electrospinning of poly(glycolic acid) (image courtesy of Randall Janairo and Song Li, University of California, Berkeley). (b) Fibrin gel observed with confocal reflectance microscopy (image courtesy of Ekaterina Kniazeva and Andrew Putnam, University of Michigan). (c) Polymeric sponge formed by gas foaming/particulate leaching of poly(lactide-co-glycolide).

age in TEVGs. Numerous other proteins and synthetic polymers have been electrospun into materials for bioresorbable vascular grafts.[15] MSCs have also been entrapped in natural materials and induced toward the myogenic phenotype in culture. Multipotent adult progenitor cells were incorporated in fibrin vascular molds, cultured in the presence of an optimized inductive cocktail for three weeks, and their responses to stimulation and mechanical load were characterized.[16] Cells within the molds aligned circumferentially as desired, expressed smooth muscle cell-specific genes and surface markers, and generated contractile force when chemically stimulated.

2.2 *Implanting MSCs to repair damaged myocardium*

Myocardial infarction (MI) and subsequent heart failure represent the main cause of death in industrialized nations. Left untreated, the loss of viable myocardium resulting from MI results in continued expansion of the initial infarct area, finally leading to heart failure. Since cardiomyocytes rarely proliferate or differentiate after injury, transplantation of stem cells is now increasingly recognized as an effective method to repair the infarcted myocardium.[17] The mechanism of MSC contribution toward myocardial repair is presently debated in the literature, as is the ideal route and timing of delivery. Intravenous delivery of MSCs is highly inefficient in light of the pulmonary first pass effect,[18] thereby necessitating the targeted delivery of these cells. Intramyocardial injection of MSCs has been reported to improve cardiac function and resting blood flow, and likely attains higher engraftment efficiency than systemic delivery. However, MSCs are stimulated by the properties of the surrounding diseased myocardium and may still migrate away from the injection site, thereby providing opportunities for instructing the behavior of these cells through materials-based delivery strategies.

Fibrin is a biodegradable, fast-gelling matrix, which easily entraps and enables the survival, adhesion, and proliferation of MSCs (Fig. 3b). The mechanical properties of this biopolymer can be tailored by modulating the fibrinogen concentration and thrombin concentration independently. However, there is a trade-off between mechanical properties and cell viability, with higher concentrations of each component facilitating rapid gelation of a high-strength product but failing to promote optimal cell viability and growth. To address these limitations, Zhang *et al.* modified fibrinogen with a benzotriazole carbonate derivative of polyethylene glycol (PEG) to increase the number of crosslinks with adjacent fibrin monomer molecules, thereby increasing mechanical properties while maintaining the viability of encapsulated cells.[19] PEGylation increased the storage modulus by nearly 40% without significantly increasing the gelation time. When MSCs were mixed into the gel, cells proliferated faster than MSCs grown on tissue culture plastic, migrated out of the gel when stimulated, and expressed cell surface markers indicative of differentiation toward the endothelial lineage in the absence of cytokine treatment. Liu *et al.* entrapped autologous porcine MSCs in a fibrin gel and applied the patch to the left ventricular (LV) anterior wall following MI.[20] After 21 days, LV contractile performance was improved in conjunction with increased neovascularization, likely as a function of improved MSC engraftment following migration from the gel and the localized production of trophic factors that stimulate host cell migration into the ischemic site.

MSCs have been implanted on the wall of ischemic myocardium using composite materials designed to capitalize on the benefits of each constituent. Poly-glycolide-co-caprolactone (PGCL) is a synthetic composite material that possesses elasticity, suggesting that it could be employed to engineer a patch for mechanically dynamic environments such as the heart. PGCL scaffolds were seeded with syngeneic bone marrow mononuclear cells and sutured onto the epicardial surface of rats seven days before induction of MI.[21] Animals treated with cell-seeded PGCL patches or PGCL alone exhibited significant reductions in LV remodeling associated with heart failure, while cell-seeded constructs were associated with increased neovascularization. These data suggest that PGCL contributed mechanical integrity to limit LV dilation, while the addition of cells stimulated repair of the damaged myocardium. The potential of MSC-seeded biopolymer composites to promote cardiac repair is still unknown, but this biomaterial represents a promising candidate that possesses elastic material properties and can promote the cardiomyogenic differentiation of MSCs.

3. Delivery Vehicles for Deploying Stem Cells in Dermal Regeneration

Chronic dermal wounds, including those associated with diabetes, affect 5.7 million patients and cost the United States health care system an estimated $20 billion annually.[22] Approximately 40,000 burn patients are admitted to US hospitals every year, and the use of skin grafting to treat burn patients represents a $1 billion market.[22] The use of autograft, allograft, and xenograft skin transplants by Hindus was described in Sanskrit texts and dates to around 3000–2500 BC. While such approaches are commonly used in current medical practice, extensive burns and diabetes-associated pathology limit the use of autografting in acute and chronic dermal wounds. Allografts can only be used as temporary cover due to immune rejection, and also carry the risk of disease transfer.[23] Single modality therapeutic approaches such as growth factor treatment using platelet-derived growth factor have met with little success, particularly in light of the complex pathophysiology of chronic and acute dermal wounds.[24] More recently, multifactorial approaches utilizing stem cells and tissue engineering strategies have been applied to treating acute and chronic dermal wounds. We will discuss some recent approaches that utilize bioengineered scaffolds to deploy keratinocyte stem cells, fibroblasts, MSCs, and adipose-derived stem cells (ASCs) to restore structure and function, reduce scarring, and improve the cosmetic appearance of skin.

3.1 *Cultured epithelial autograft*

One of the first cell and tissue engineering approaches developed to overcome some of the difficulties associated with autografting and allografting was the development of keratinocyte culture to generate cohesive sheets of stratified epithelium referred to as cultured epithelial autograft (CEA). Epicel® (Genzyme Biosurgery) is a commercialized CEA product that is generated in two to three weeks using bioreactor technology.[23] Several groups have also evaluated matrices such as fibrin, plasma, collagen, chitosan, hyaluronic acid, and polymers such as polyurethane, Teflon® film, and polyhydroxyethyl methacrylate for delivery of pre-confluent keratinocytes to wound sites.[25] However, CEAs are mechanically fragile, blister, and ulcerate due to the poor formation of a basement membrane and the absence of an underlying dermis. More recent approaches to dermal regeneration utilize multiple cell types in more complex, three-dimensional (3D) constructs to more closely mimic a complete dermal-epidermal structure for repair and regeneration of deeper wounds as described in Section 3.3.

3.2 *Dermal substitutes*

Dermal substitutes may be cellular or acellular, but lack an epidermal component. As opposed to CEAs, which can take weeks to culture, dermal substitutes are supplied as off-the-shelf products. Dermagraft® (Advanced BioHealing) is a dermal substitute consisting of metabolically active neonatal foreskin fibroblasts on a polyglactin mesh.[26] The foreskin fibroblasts deposit an ECM comprised of collagens, glycosaminoglycans, and growth factors onto the mesh that is then cryopreserved. Dermagraft® has been approved by the US Food and Drug Administration (FDA) for use in full-thickness diabetic foot ulcers. Alloderm® (LifeCell) is manufactured from donated skin that is decellularized to prevent an immune rejection response, and subsequently freeze-dried. Although it is only FDA-approved for breast reconstruction and hernia procedures, it has been used in research settings for treating full-thickness dermal wounds.[27]

Several dermal substitutes have also been described in a laboratory setting. In a pig model of dermal regeneration, autologous fibroblasts were deployed in a collagen gel containing alpha elastin hydrolysate.[28] Dermal fibroblasts were seeded onto this scaffold and cultured for ten days. The fibroblasts were applied to a full-thickness wound and covered with a split-skin mesh graft. The scaffolds seeded with the highest number of fibroblasts showed significantly improved healing relative to acellular constructs. In an *in vitro* model, HUVECs were overlaid onto a dermal layer generated by dermal fibroblasts. The dermal layer was shown to support the formation of vessel-like structures by HUVECs in a hepatocyte growth factor (HGF)-dependent manner.[29] In a similar model, MSCs were mixed with HUVECs and incorporated into the dermal layer.[30] The MSCs were shown to further stabilize and enhance the formation of these vessel-like structures. Recent efforts to apply these findings have utilized 3D bioprinting to place MSCs and HUVECs together in tightly controlled spatial layers, creating vascularized tissues that allow MSCs to support HUVECs.[31]

3.3 *Bilayered skin substitutes/living skin equivalents*

Living skin equivalents (LSEs) approximate the structure of skin, containing both an epidermal and dermal component in which different types of stem cells can be deployed. Similar to dermal equivalents, LSEs contain an array of tissue-engineered constructs for deploying stem cells that are either commercially available or described by individual laboratories. Apligraf® was the first bilayered, allogeneic skin substitute approved for treating chronic wounds. Apligraf® is derived from living neonatal foreskin fibroblasts seeded onto bovine type I collagen to generate a dermal layer. Neonatal foreskin keratinocytes are subsequently seeded onto the dermal layer to generate a functional epidermis. In clinical studies evaluating chronic wound healing, patients treated with Apligraf® were more likely to heal faster and more completely relative to the standard of care.[26]

Integra Bilayer Matrix Wound Dressing and Integra Dermal Regeneration Template are FDA-approved for the treatment of chronic wounds and burns, respectively. Both products have a dermal component comprised of crosslinked bovine tendon collagen, shark chondroitin 6-sulfate, and a pseudo-epidermal component comprised of silicon.[27] Neither product contains a cellular constituent, a native ECM, or the growth factors present in other bilayered skin substitutes, dermal substitutes, or CEAs. Relative to Alloderm® and Dermagraft®, Integra demonstrates very little construct remodeling.[27] The deployment of MSCs within Integra demonstrated significantly increased neovascularization and growth factor incorporation relative to the biomaterial without MSCs.[32]

Several laboratories have developed alternative delivery vehicles for deploying stem cells in dermal regeneration applications. An LSE comprised of autologous keratinocytes and fibroblasts deployed in clotted plasma gel have been successfully used for the treatment of burns. In an *in vitro* model, ASCs have been used in place of fibroblasts to generate a self-assembled dermal layer in the presence of ascorbate.[33] After merging three of these dermal layers, a keratinocyte layer was overlaid to yield a bilayered LSE. ASCs have also been differentiated *in vitro* to adipocytes and incorporated into a tri-layered LSE with an adipose containing fascia.[33]

3.4 *Other vehicles for deploying stem cells*

MSCs deployed in collagen gels generate a dermal-like substitute that is similar to collagen gels containing dermal fibroblasts.[34] Autologous MSCs deployed in fibrin gels promoted healing of large dermal wounds associated with skin cancer resection, as well as healing of chronic foot ulcers (duration of more than one year) associated with diabetes.[35] Platelet-rich plasma (PRP) has been used as a delivery vehicle for ASCs. In a porcine model of dermal wound healing, ASCs stimulated vessel formation, while PRP with ASCs also improved the cosmetic appearance of the dermal wounds.[36] As another example, MSCs were seeded onto freeze-dried collagen-glycosaminoglycan scaffolds, cultured for two days, and then raised to the air-liquid interface for two days to generate an epidermis.[37] In a porcine partial-thickness burn model, pigs treated with the MSC-seeded constructs exhibited significantly increased wound contraction and vascular density at four weeks.

4. Biomaterials for Implanting MSCs for Regenerating Osteochondral Tissues

MSCs are under intense investigation for use in cell-based approaches to repair skeletal defects. Although these cells can be readily induced to differentiate toward both the chondrogenic and osteoblastic lineages in culture, their direct participation in the formation of each tissue has been limited when delivered to the tissue site. The systemic or local injection of MSCs without an appropriate carrier or vehicle fails to provide the necessary structure and framework for cells to populate tissue defects. Furthermore, many cells migrate away from the defect site or undergo apoptosis or necrosis when placed in the harsh environment. To improve cell survival and participation in the repair of cartilage and bone defects, extensive efforts are ongoing to develop biomaterials that bridge the defect site and provide a platform for instructing the differentiation of these progenitor cells toward the desired phenotype. We will briefly review recent efforts in the development and application of biomaterials to implant MSCs for the formation and repair of cartilage and bone.

4.1 *Deploying MSCs for cartilage formation*

Cartilage is an avascular tissue with low cellularity and a limited capacity for self-repair, thereby making it an ideal candidate for cell-based approaches toward tissue engineering. MSCs have generated significant interest in cartilage formation as an alternative to autologous chondrocytes. Chondrogenesis with MSCs generally involves induction with transforming growth factor β (TGF-β) and a 3D culture environment (e.g.

Table 1. Commonly used biomaterials for implanting MSCs to repair bone and cartilage.

Biomaterial	Animal Model	Example References
Natural polymers		
Alginate	Rat, rabbit	59, 68
Collagen	Rabbit, rat	69
Fibrin	Equine	69
Hyaluronic acid	Rabbit	40
Silk fibroin	Rat	70
Synthetic polymers		
Poly(α-hydroxy esters)	Mouse, rabbit	56, 68
Poly(β-caprolactone)	Swine	44
Poly(ethylene oxide)	Mouse	71
Poly(propylene glycol) fumarate	Rabbit	72
Bioceramics		
Hydroxyapatite	Rat, rabbit, mouse	54, 56
Tricalcium phosphate	Sheep	45

micromass, cell pellet, seeded within biomaterials). With regard to the 3D culture environment, the degree of chondrogenesis is scaffold dependent,[38] thereby providing substantial motivation for careful biomaterial selection for the delivery of MSCs for cartilage repair. The application of biomaterial systems enables more extensive control over cell behavior than the popular cell pellet approach and provides a strategy for limiting cell migration away from the defect site. A host of natural and synthetic materials has been used for MSC chondrogenesis (Table 1), and these materials are broadly divided into substrates that are either injectable or implantable.

Hydrogels may be used as injectable scaffolds due to their ability to fill defects of any shape and size. Hydrogels have a high water content, support the transport of nutrients and waste, mimic many characteristics of the ECM, and have tailorable mechanical properties upon physical or chemical crosslinking. Furthermore, MSCs can be homogeneously suspended within these materials, where encapsulated cells generally retain a rounded morphology characteristic of a chondrocytic phenotype. Equine articular cartilage defects of the femoropatellar joint treated with fibrin-encapsulated MSCs exhibited enhanced tissue repair after 30 days characterized by greater tissue formation and increased presence of type II collagen, yet these improvements were lost compared to cell-free fibrin after eight months.[39] Liu *et al.* reported that MSCs transplanted in hyaluronic acid hydrogels resulted in firm, elastic cartilage that filled the entire defect after 12 weeks and implanted constructs were well integrated with host cartilage.[40] The encapsulation of MSCs in hydrogels formed of synthetic polymers releasing TGF-β exhibited increased type II collagen distribution and cartilage formation compared to cell-containing gels alone.[41] These findings suggest that chondrogenesis of MSCs *in situ* may require induction with TGF-β in order to realize the full potential of this approach. Indeed, previous studies using injectable alginate hydrogels for chondrogenic regeneration found that the inclusion of controlled-release TGF-β at 20 ng/mL enhanced regeneration compared to untreated rabbit controls.[42] These data were recently confirmed when using the same injectable alginate vehicle used to deliver MSCs for osteochondral regeneration in a rabbit model.[43]

Implantable materials such as sponges and meshes facilitate precise control over material morphological properties compared to hydrogels, and these substrates enable the generation of mechanically stable constructs *in vitro* for subsequent *in vivo* implantation. However, these materials suffer from difficulties with filling irregularly shaped defects and maintaining contact along the entire defect periphery, each of which facilitates host cell migration and inhibits the formation of weak fibrous tissue. Considering the responsiveness of MSCs to mechanical cues, whether from the substrate or mechanical loading, the properties of the implant will have a profound contribution toward the resulting phenotype upon implantation. In a recent study, circular poly(ε-caprolactone) (PCL) scaffolds cut from electrospun nanofibrous mats were seeded with MSCs and implanted into full-thickness defects in the femoral condyle of minipigs.[44] Six months after implantation, MSC-seeded scaffolds exhibited the most complete repair within the defects compared to materials seeded with autologous chondrocytes or cell-free scaffolds. Similarly, autologous ovine MSCs seeded on β-tricalcium phosphate (β-TCP) implants and implanted into full-thickness articular cartilage defects yielded hyaline-like tissue that was indistinguishable from the adjacent normal cartilage.[45] Defects treated with cell-free implants contained fiber-like tissue and exhibited incomplete repair after six months. Composite materials have also been studied for more precise matching to the different properties within cartilage tissue. For instance, silk fibroin was layered on composite silk fibroin/calcium phosphate and seeded with rabbit MSCs. When implanted in full-thickness knee defects, MSCs in the silk-only layer produced collagen II and glycosaminoglycans, while the layer with calcium phosphate integrated well with the calcified subchondral bone.[46]

4.2 *Deploying MSCs for bone formation and regeneration*

Bone autograft, commonly from the iliac crest, has been used for many years as the standard of care for bone regeneration applications such as the repair of non-union fractures and in ectopic ossification applications such as lumbar fusion. Intrinsic to autograft are the trinity of factors that is critical for optimal bone regeneration: an osteoconductive matrix to allow for new bone and vascular ingrowth, osteoinductive factors to induce new bone formation, and osteogenic cells to deposit and mineralize a bone matrix. The limitations of autograft, namely tissue availability, low fusion rates, and significant morbidity, have motivated the pursuit of alternative materials for bridging bone defects and stimulating bone repair. These materials include allogeneic cancellous bone chips, collagen sponges, synthetic polymer implants, and various hydrogels. Such materials have been described as alternatives to autograft and demonstrate osteoconductive properties; yet, for the most part, they lack osteoinductive factors as well as osteogenic cellular components.[47] In an effort to provide an osteogenic cellular constituent, several groups have combined stem cells from different sources with these materials. In this section, we will describe the materials that have been used in combination with different stem cells as alternatives to bone autograft in bone regeneration or ectopic/orthotopic ossification applications.

4.2.1 *Cancellous bone chips and demineralized bone matrix*

Bone allograft (cancellous bone chips) and demineralized bone matrix (DBM) are currently used to treat fractures and in lumbar fusion procedures.

They are sold as an off-the-shelf product that possesses osteoinductive activity elicited by endogenous growth factors that remain associated with the ECM. Most manufacturers offer cancellous bone chips as a decellularized product, but some offer metabolically active cellular cancellous products that are processed and cryopreserved in a manner that maintains the viability of endogenous stem cells (Trinity Evolution, Orthofix). For many years, bone marrow aspirate, containing a heterogeneous mixture of endogenous hematopoietic stem cells, MSCs, and EPCs, has been used in combination with cancellous bone chips and DBM to generate osteoconductive grafts with augmented osteogenic activity. Cancellous chips loaded with bone marrow were shown to enhance the healing of non-union fractures in the clinical setting.[48] In a posterior segmental lumbar fusion model in the dog, cancellous bone chips in combination with DBM and clotted bone marrow were shown to significantly improve the success rate.[47] Importantly, clotted bone marrow alone demonstrated no beneficial effect.

4.2.2 *Collagen sponges and collagen composites*

Collagen sponges, such as those found in the INFUSE Bone Graft (Medtronic), and collagen composites, such as Healos (collagen/ hydroxyapatite, Depuy), Mozaic, and Vitoss (collagen/β-TCP, Integra, and Orthovita, respectively), are osteconductive materials that lack both an intrinsic osteoinductive component as well as an osteogenic cellular component. Collagen sponges are highly compressible, while collagen-ceramic composites possess more robust mechanical properties. Both types of scaffolds incorporate the native cell adhesion domains associated with collagen. Recently, stem cells alone or in combination with growth factors have been added to these materials to enhance their performance in bone regeneration and ectopic/orthotopic ossification applications. In a rat model of lumbar fusion, bone marrow aspirate was shown to enhance lumbar fusion mediated by a BMP-2 soaked collagen sponge.[49] By contrast, platelet-rich plasma did not augment lumbar fusion when used with the INFUSE kit. In a similar model of lumbar fusion, ASCs transduced with an adenoviral BMP-2 vector and loaded into a collagen sponge demonstrated enhanced lumbar fusion relative to BMP-2 alone.[50] Interestingly, ASCs loaded in a collagen scaffold generated little orthotopic bone and failed to promote lumbar fusion. In an orthotopic, xenogeneic model of bone formation, undifferentiated ASCs and MSCs implanted in Healos persisted *in vivo*, while *in vitro* differentiated ASCs and MSCs did not.[51] In a similar model, *in vitro* differentiated ASCs loaded into a honeycombed collagen scaffold demonstrated increased bone formation relative to undifferentiated ASCs.[52]

4.2.3 *Calcium phosphate and hydroxyapatite ceramics*

Calcium phosphate ceramic scaffolds such as Copios bone void filler (calcium phosphate, dibasic, Zimmer) have been used in combination with several different stem cell sources to augment new bone formation. While ceramic scaffolds are osteoconductive, they lack osteoinductive factors and do not contain endogenous osteogenic cells. Macroporous biphasic calcium phosphate (MBCP) was used as a scaffold in a rat model of a radiation-induced defect of the long bones.[53] Bone marrow aspirate, MSCs, or ASCs were loaded into these scaffolds and the scaffolds were placed in the defect area. In this model, scaffolds loaded with bone marrow aspirate but not MSCs or ASCs demonstrated significantly greater bone formation than empty scaffolds, indicating that cultured cells were inferior to fresh isolates. Compared to the low specific surface area (SSA) of synthetic β-TCP, high SSA materials have been developed that demonstrate levels of osteogenic differentiation similar to low SSA materials. Calcium-deficient hydroxyapatite (CDHA) has an SSA of 20–80 m^2/g and belongs to the high SSA ceramic group.[54] In a rabbit tibial osteotomy model, CDHA scaffolds loaded with MSC or with PRP demonstrated significantly increased bone volume relative to empty scaffolds.[54] There was no added benefit when MSC and PRP were combined, suggesting that CDHA is enhancing bone formation through stimulating MSC differentiation as a function of substrate rigidity or composition.

4.2.4 *PLGA- and PCL-based scaffolds*

Poly(lactic-co-glycolic) acid (PLGA)-based and PCL-based scaffolds are minimally osteoconductive but do not have any osteoinductive properties. Both PLGA and PCL have good biocompatibility profiles and biodegrade at rates that are dependent on their composition. Different stem cell populations have been seeded on these scaffolds (Fig. 3c) to provide an osteogenic component for bone regeneration applications. In a rabbit critical-sized femoral defect model, porous PLGA scaffolds were loaded with the osteoinductive factor dihydroxy vitamin D3 (Calcitriol), MSCs, or Calcitriol with MSCs and placed in the defect.[55] Only scaffolds containing Calcitriol showed a bony union at nine weeks. Scaffolds incorporating MSCs had greater amounts of pre-mineralized matrix (osteoid) present, but Calcitriol was required for successful union. In a murine ectopic bone formation model, ASCs were loaded in a PLGA/hydroxyapatite composite in the presence or absence of BMP-2 and implanted subcutaneously.[56] Scaffolds with ASCs alone had negligible amounts of mineral, while scaffolds loaded with BMP-2 alone had significant amounts of mineral. ASC-loaded scaffolds eluting BMP-2 demonstrated the greatest amount of bone formation. In a similar model using porous, honeycombed PCL-TCP composite scaffolds, the osteogenic capacity of fetal bone marrow MSCs, umbilical cord MSCs (UC-MSCs), adult MSCs, and ASCs was compared.[57] The stem cells were loaded on scaffolds, differentiated with dexamethasone and ascorbate *in vitro*, and subsequently implanted. Scaffolds loaded with fetal MSCs and adult MSCs demonstrated the highest levels of bone formation, while adipose MSC-loaded scaffolds demonstrated the least. To overcome the limitations associated with the low osteoinductive nature of PCL and PCL-ceramic materials, an approach utilizing 3D printing of PCL with embedded native decellularized trabecular bone has recently been reported. Scaffolds containing the embedded matrix enhanced the mineralization of seeded ASCs for *in vitro* and in a murine critical-sized calvarial defect.[58]

4.2.5 *Hydrogels*

Hydrogels are appealing candidates as scaffolds given their similarity to native extracellular matrix and high degree of biocompatibility. Moreover, these materials enable highly efficient incorporation of cells and inductive factors to stimulate bone formation. In a rat critical-sized cranial defect model, ASCs alone or ASCs genetically modified with an adenovirus to overexpress BMP-2 were loaded into an alginate hydrogel and implanted into the defect.[59] Only scaffolds with BMP-2-expressing ASCs demonstrated closure of the critical-sized defect at 16 weeks. In a rat calvarial defect model, BMP-2, MSCs, or MSCs with BMP-2 were loaded into hyaluronic acid-based hydrogels and implanted in the defect.[60] While MSCs demonstrated poor adhesion to the scaffolds *in vitro*, scaffolds seeded with MSCs and BMP-2 demonstrated robust bone formation four weeks after implantation. Scaffolds seeded with MSCs alone showed slightly less bone formation, while scaffolds seeded with BMP-2 alone showed the least amount of bone formation. In an ectopic bone formation model, an MPEG-PCL *in situ* forming gel was used as a scaffold for implanting ASCs.[61] Only scaffolds loaded with ASCs demonstrated *in vivo* bone formation, while scaffolds loaded with ASCs and an osteogenic supplement had the highest levels of bone formation. Recently, hydrogels have been engineered using both thermal and chemical crosslinking as injectable and tunable systems for MSC delivery.[62] These materials facilitate the incorporation of hydroxyapatite nanoparticles[63] or pendant phosphate groups[64] for enhancing mineralization despite the relatively soft hydrogel substrate. Photocrosslinking has also been applied for functionalization and gelation of alginate hydrogels.[65] This system allows for the production of spatially defined concentration gradients of growth factor and adhesion ligands to instruct encapsulated MSCs.[66] Finally, the application of bio-orthogonal reactions, or click chemistry, enables a high degree of spatiotemporal control. Reactions stimulated by different wavelengths of light have been used to create highly tunable PEG hydrogels.[67]

5. Conclusions

MSCs have tremendous potential to contribute to the repair of human tissues. To capitalize on the promise of these multipotent cells to differentiate into many phenotypes, secrete tissue-inducing factors, and suppress inflammatory responses, new approaches for deploying these cells must be developed and optimized under conditions that approximate human physiology. Furthermore, the selection of the delivery vehicle enables control over the mechanical properties, degradation time, cell distribution, cytokine, growth factor, and morphogen gradients, and can even instruct cell fate. The examination of these systems under physiological conditions and in clinically relevant animal models will increase our understanding of how MSCs contribute to the development and repair of human tissues. Importantly, the results of these studies will provide new information on the efficacy of these materials to support the long-term survival and performance of MSCs in tissue repair and regeneration.

References

1. M. F. Pittenger, A. M. Mackay, S. C. Beck, R. K. Jaiswal, R. Douglas, J. D. Mosca, M. A. Moorman, D. W. Simonetti, S. Craig and D. R. Marshak, Multilineage potential of adult human mesenchymal stem cells, *Science* **284**: 143–147 (1999).

2. A. Uccelli, L. Moretta, V. Pistoia, Mesenchymal stem cells in health and disease, *Nat Rev Immunol* **8**: 726–736 (2008).

3. J. M. Ryan, F. P. Barry, J. M. Murphy and B. P. Mahon, Mesenchymal stem cells avoid allogeneic rejection, *J Inflamm (Lond)* **2**: 8 (2005).

4. K. H. Yoo, I. K. Jang, M. W. Lee, H. E. Kim, M. S. Yang, Y. Eom, J. E. Lee, Y. J. Kim, S. K. Yang, H. L. Jung, K. W. Sung, C. W. Kim and H. H. Koo, Comparison of immunomodulatory properties of mesenchymal stem cells derived from adult human tissues, *Cell Immunol* **259**: 150–156 (2009).

5. S. Kern, H. Eichler, J. Stoeve, H. Kluter and K. Bieback, Comparative analysis of mesenchymal stem cells from bone marrow, umbilical cord blood, or adipose tissue, *Stem Cells* **24**: 1294–1301 (2006).

6. T. Kinnaird, E. Stabile, M. S. Burnett, S. E. Epstein, Bone-marrow-derived cells for enhancing collateral development: mechanisms, animal data, and initial clinical experiences, *Circ Res* **95**: 354–363 (2004).

7. M. Crisan, S. Yap, L. Casteilla, C. W. Chen, M. Corselli, T. S. Park, G. Andriolo, B. Sun, B. Zheng, L. Zhang, C. Norotte, P. N. Teng, J. Traas, R. Schugar, B. M. Deasy, S. Badylak, H. J. Buhring, J. P. Giacobino, L. Lazzari, J. Huard and B. Peault, A perivascular origin for mesenchymal stem cells in multiple human organs, *Cell Stem Cell* **3**: 301–313 (2008).

8. J. M. Melero-Martin, M. E. De Obaldia, S. Y. Kang, Z. A. Khan, L. Yuan, P. Oettgen and J. Bischoff, Engineering robust and functional vascular networks *in vivo* with human adult and cord blood-derived progenitor cells, *Circ Res* **103**: 194–202 (2008).

9. P. Au, J. Tam, D. Fukumura, R. K. Jain, Bone marrow-derived mesenchymal stem cells facilitate engineering of long-lasting functional vasculature, *Blood* **111**: 4551–4558 (2008).

10. N. Koike, D. Fukumura, O. Gralla, P. Au, J. S. Schechner and R. K. Jain, Tissue engineering: creation of long-lasting blood vessels, *Nature* **428**: 138–139 (2004).

11. C. M. Ghajar, K. S. Blevins, C. C. Hughes, S. C. George and A. J. Putnam, Mesenchymal stem cells enhance angiogenesis in mechanically viable prevascularized tissues via early matrix metalloproteinase upregulation, *Tissue Eng* **12**: 2875–2888 (2006).

12. M. S. Stosich, B. Bastian, N. W. Marion, P. A. Clark, G. Reilly and J. J. Mao, Vascularized adipose tissue grafts from human mesenchymal stem cells with bioactive cues and microchannel conduits, *Tissue Eng* **13**: 2881–2890 (2007).

13. C. K. Hashi, Y. Zhu, G. Y. Yang, W. L. Young, B. S. Hsiao, K. Wang, B. Chu and S. Li, Antithrombogenic property of bone marrow mesenchymal stem cells in nanofibrous vascular grafts, *Proc Natl Acad Sci USA* **104**: 11915–11920 (2007).

14. Z. Gong and L. E. Niklason, Small-diameter human vessel wall engineered from bone marrow-derived mesenchymal stem cells (hMSCs), *FASEB J* **22**: 1635–1648 (2008).

15. S. A. Sell, M. J. McClure, K. Garg, P. S. Wolfe and G. L. Bowlin, Electrospinning of collagen/biopolymers for regenerative medicine and cardiovascular tissue engineering, *Adv Drug Deliv Rev* **61**: 1007–1019 (2009).

16. J. J. Ross, Z. G. Hong, B. Willenbring, L. P. Zeng, B. Isenberg, E. H. Lee, M. Reyes, S. A. Keirstead, E. K. Weir, R. T. Tranquillo and C. M. Verfaillie, Cytokine-induced differentiation of multipotent adult progenitor cells into functional smooth muscle cells, *J Clin Invest* **116**: 3139–3149 (2006).

17. V. F. Segers and R. T. Lee, Stem-cell therapy for cardiac disease, *Nature* **451**: 937–942 (2008).

18. U. M. Fischer, M. T. Harting, F. Jimenez, W. O. Monzon-Posadas, H. Xue, S. I. Savitz, G. A. Laine and C. S. Cox, Jr., Pulmonary passage is a major obstacle for intravenous stem cell delivery: the pulmonary first-pass effect, *Stem Cells Dev* **18**: 683–692 (2009).

19. G. Zhang, X. Wang, Z. Wang, J. Zhang and L. Suggs, A PEGylated fibrin patch for mesenchymal stem cell delivery, *Tissue Eng* **12**: 9–19 (2006).

20. J. Liu, Q. Hu, Z. Wang, C. Xu, X. Wang, G. Gong, A. Mansoor, J. Lee, M. Hou, L. Zeng, J. R. Zhang, M. Jerosch-Herold, T. Guo, R. J. Bache and J. Zhang, Autologous stem cell transplantation for myocardial repair, *Am J Physiol Heart Circ Physiol* **287**: H501–511 (2004).

21. H. Piao, J. S. Kwon, S. Piao, J. H. Sohn, Y. S. Lee, J. W. Bae, K. K. Hwang, D. W. Kim, O. Jeon, B. S. Kim, Y. B. Park and M. C. Cho, Effects of cardiac patches engineered with bone marrow-derived mononuclear cells and PGCL scaffolds in a rat myocardial infarction model, *Biomaterials* **28**: 641–649 (2007).

22. S. F. Miller, P. Q. Bessey, M. J. Schurr, S. M. Browning, J. C. Jeng, D. M. Caruso, M. Gomez, B. A. Latenser, C. W. Lentz, J. R. Saffle, R. J. Kagan, G. F. Purdue and J. A. Krichbaum, National Burn Repository 2005: a ten-year review, *J Burn Care Res* **27**: 411–436 (2006).

23. L. Macri and R. A. F. Clark, Tissue engineering for cutaneous wounds: selecting the proper time and space for growth factors, cells and the extracellular matrix, *Skin Pharmacol Physiol* **22**: 83–93 (2009).

24. V. Falanga, Wound healing and its impairment in the diabetic foot, *Lancet* **366**: 1736–1743 (2005).

25. B. S. Atiyeh and M. Costagliola, Cultured epithelial autograft (CEA) in burn treatment: three decades later, *Burns* **33**: 405–413 (2007).

26. M. Ehrenreich and Z. Ruszczak, Update on tissue-engineered biological dressings, *Tissue Eng* **12**: 2407–2424 (2006).

27. A. T. Truong, A. Kowal-Vern, B. A. Latenser, D. E. Wiley and R. J. Walter, Comparison of dermal substitutes in wound healing utilizing a nude mouse model, *J Burns Wounds* **4**: e4(2005).

28. E. N. Lamme, R. T. Van Leeuwen, K. Brandsma, J. Van Marle, E. Middelkoop, Higher numbers of autologous fibroblasts in an artificial dermal substitute improve tissue regeneration and modulate scar tissue formation, *J Pathol* **190**: 595–603 (2000).

29. J. M. Sorrell, M. A. Baber and A. I. Caplan, Human dermal fibroblast subpopulations; differential interactions with vascular endothelial cells in coculture: nonsoluble factors in the extracellular matrix influence interactions, *Wound Repair Regen*, **16**: 300–309 (2008).

30. J. M. Sorrell, M. A. Baber and A. I. Caplan, Influence of adult mesenchymal stem cells on *in vitro* vascular formation, *Tissue Eng Part A* **15**: 1751–1761 (2009).

31. D. B. Kolesky, K. A. Homan, M. A. Skylar-Scott and J. A. Lewis, Three-dimensional bioprinting of thick vascularized tissues, *Proc Natl Acad Sci USA* **113**: 3179–3184 (2016).

32. J. T. Egana, F. A. Fierro, S. Kruger, M. Bornhauser, R. Huss, S. Lavandero and H. G. Machens, Use of human mesenchymal cells to improve vascularization in a mouse model for scaffold-based dermal regeneration, *Tissue Eng Part A* **15**: 1191–1200 (2009).

33. V. Trottier, G. Marceau-Fortier, L. Germain, C. Vincent and J. Fradette, IFATS collection: using human adipose-derived stem/stromal cells for the production of new skin substitutes, *Stem Cells* **26**: 2713–2723 (2008).

34. F. Fioretti, C. Lebreton-DeCoster, F. Gueniche, M. Yousfi, P. Humbert, G. Godeau, K. Senni, A. Desmouliere and B. Coulomb, Human bone marrow-derived cells: an attractive source to populate dermal substitutes, *Wound Repair Regen* **16**: 87–94 (2008).

35. V. Falanga, S. Iwamoto, M. Chartier, T. Yufit, J. Butmarc, N. Kouttab, D. Shrayer and P. Carson, Autologous bone marrow-derived cultured mesenchymal stem cells delivered in a fibrin spray accelerate healing in murine and human cutaneous wounds, *Tissue Eng* **13**: 1299–1312 (2007).

36. M. W. Blanton, I. Hadad, B. H. Johnstone, J. A. Mund, P. I. Rogers, B. L. Eppley and K. L. March, Adipose stromal cells and platelet-rich plasma therapies synergistically increase revascularization during wound healing, *Plast Reconstr Surg* **123**: 56–64S (2009).

37. P. Liu, Z. Deng, S. Han, T. Liu, N. Wen, W. Lu, X. Geng, S. Huang and Y. Jin, Tissue-engineered skin containing mesenchymal stem cells improves burn wounds, *Artif Organs* **32**: 925–931 (2008).

38. R. M. Coleman, N. D. Case and R. E. Guldberg, Hydrogel effects on bone marrow stromal cell response to chondrogenic growth factors, *Biomaterials* **28**: 2077–2086 (2007).

39. M. M. Wilke, D. V. Nydam, A. J. Nixon, Enhanced early chondrogenesis in articular defects following arthroscopic mesenchymal stem cell implantation in an equine model, *J Orthop Res* **25**: 913–925 (2007).

40. Y. Liu, X. Z. Shu and G. D. Prestwich, Osteochondral defect repair with autologous bone marrow-derived mesenchymal stem cells in an injectable, in situ, cross-linked synthetic extracellular matrix, *Tissue Eng* **12**: 3405–3416 (2006).

41. K. H. Park, D. H. Lee and K. Na, Transplantation of poly(N-isopropylacrylamide-co-vinylimidazole) hydrogel constructs composed of rabbit chondrocytes and growth factor-loaded nanoparticles for neocartilage formation, *Biotechnol Lett* **31**: 337–346 (2009).

42. C. M. Mierisch, S. B. Cohen, L. C. Jordan, P. G. Robertson, G. Balian and D. R. Diduch, Transforming growth factor-beta in calcium alginate beads for the treatment of articular cartilage defects in the rabbit, *Arthroscopy* **18**: 892–900 (2002).

43. T. Igarashi, N. Iwasaki, Y. Kasahara and A. Minami, A cellular implantation system using an injectable ultra-purified alginate gel for repair of

osteochondral defects in a rabbit model, *J Biomed Mater Res A*, **94**: 844–855 (2010).

44. W. J. Li, H. Chiang, T. F. Kuo, H. S. Lee, C. C. Jiang and R. S. Tuan, Evaluation of articular cartilage repair using biodegradable nanofibrous scaffolds in a swine model: a pilot study, *J Tissue Eng Regen Med* **3**: 1–10 (2009).

45. X. M. Guo, C. Y. Wang, Y. F. Zhang, R. Y. Xia, M. Hu, C. M. Duan, Q. Zhao, L. Z. Dong, J. X. Lu and Y. Q. Song, Repair of large articular cartilage defects with implants of autologous mesenchymal stem cells seeded into beta-tricalcium phosphate in a sheep model, *Tissue Eng* **10**: 1818–1829 (2004).

46. L. P. Yan, J. Silva-Correia, M. B. Oliveira, C. Vilela, H. Pereira, R. A. Sousa, J. F. Mano, A. L. Oliveira, J. M. Oliveira and R. L. Reis, Bilayered silk/silk-nanoCaP scaffolds for osteochondral tissue engineering: *in vitro* and *in vivo* assessment of biological performance, *Acta Biomate*r **12**: 227–241 (2015).

47. G. F. Muschler, H. Nitto, Y. Matsukura, C. Boehm, A. Valdevit, H. Kambic, W. Davros, K. Powell and K. Easley, Spine fusion using cell matrix composites enriched in bone marrow-derived cells, *Clin Orthop Relat Res* **407**: 102–118 (2003).

48. A. Ateschrang, B. G. Ochs, M. Ludemann, K. Weise and D. Albrecht, Fibula and tibia fusion with cancellous allograft vitalised with autologous bone marrow: first results for infected tibial non-union, *Arch Orthop Trauma Surg* **129**: 97–104 (2009).

49. R. D. Rao, K. Gourab, V. B. Bagaria, V. B. Shidham, U. Metkar and B. C. Cooley, The effect of platelet-rich plasma and bone marrow on murine posterolateral lumbar spine arthrodesis with bone morphogenetic protein, *J Bone Joint Surg Am* **91**: 1199–1206 (2009).

50. W. K. Hsu, J. C. Wang, N. Q. Liu, L. Krenek, P. A. Zuk, M. H. Hedrick, P. Benhaim and J. R. Lieberman, Stem cells from human fat as cellular delivery vehicles in an athymic rat posterolateral spine fusion model, *J Bone Joint Surg Am* **90**: 1043–1052 (2008).

51. P. Niemeyer, J. Vohrer, H. Schmal, P. Kasten, J. Fellenberg, N. P. Suedkamp and A. T. Mehlhorn, Survival of human mesenchymal stromal cells from bone marrow and adipose tissue after xenogenic transplantation in immunocompetent mice, *Cytotherapy* **10**: 784–795 (2008).

52. N. Kakudo, A. Shimotsuma, S. Miyake, S. Kushida and K. Kusumoto, Bone tissue engineering using human adipose-derived stem cells and honeycomb collagen scaffold, *J Biomed Mater Res A* **84**: 191–197 (2008).

53. F. Espitalier, C. Vinatier, E. Lerouxel, J. Guicheux, P. Pilet, F. Moreau, G. Daculsi, P. Weiss and O. Malard, A comparison between bone reconstruction following the use of mesenchymal stem cells and total bone marrow in association with calcium phosphate scaffold in irradiated bone, Biomaterials **30**: 763–769 (2009).

54. P. Kasten, J. Vogel, F. Geiger, P. Niemeyer, R. Luginbuhl and K. Szalay, The effect of platelet-rich plasma on healing in critical-size long-bone defects, *Biomaterials* **29**: 3983–3992 (2008).

55. S. J. Yoon, K. S. Park, M. S. Kim, J. M. Rhee, G. Khang and H. B. Lee, Repair of diaphyseal bone defects with calcitriol-loaded PLGA scaffolds and marrow stromal cells, *Tissue Eng* **13**: 1125–1133 (2007).

56. O. Jeon, J. W. Rhie, I. K. Kwon, J. H. Kim, B. S. Kim and S. H. Lee, *In vivo* bone formation following transplantation of human adipose-derived stromal cells that are not differentiated osteogenically, *Tissue Eng Part A* **14**: 1285–1294 (2008).

57. Z. Y. Zhang, S. H. Teoh, M. S. Chong, J. T. Schantz, N. M. Fisk, M. A. Choolani and J. Chan, Superior osteogenic capacity for bone tissue engineering of fetal compared with perinatal and adult mesenchymal stem cells, *Stem Cells* **27**: 126–137 (2009).

58. B. P. Hung, B. A. Naved, E. L. Nyberg, M. Dias, C. A. Holmes, J. H. Elisseeff, A. H. Dorafshar and W. L. Grayson, Three-dimensional printing of bone extracellular matrix for craniofacial regeneration, *ACS Biomater Sci Eng* **2**: 1806–1816 (2016).

59. Y. Lin, W. Tang, L. Wu, W. Jing, X. Li, Y. Wu, L. Liu, J. Long and W. Tian, Bone regeneration by BMP-2 enhanced adipose stem cells loading on alginate gel, *Histochem Cell Biol* **129**: 203–210 (2008).

60. J. Kim, I. S. Kim, T. H. Cho, K. B. Lee, S. J. Hwang, G. Tae, I. Noh, S. H. Lee, Y. Park and K. Sun, Bone regeneration using hyaluronic acid-based hydrogel with bone morphogenic protein-2 and human mesenchymal stem cells, *Biomaterials* **28**: 1830–1837 (2007).

61. H. H. Ahn, K. S. Kim, J. H. Lee, J. Y. Lee, B. S. Kim, I. W. Lee, H. J. Chun, J. H. Kim, H. B. Lee and M. S. Kim, *In vivo* osteogenic differentiation of human adipose-derived stem cells in an injectable in situ-forming gel scaffold, *Tissue Eng Part A* **15**: 1821–1832 (2009).

62. T. N. Vo, A. K. Ekenseair, F. K. Kasper and A. G. Mikos, Synthesis, physicochemical characterization, and cytocompatibility of bioresorbable, dual-gelling injectable hydrogels, *Biomacromolecules* **15**: 132–142 (2014).

63. B. M. Watson, T. N. Vo, P. S. Engel and A. G. Mikos, Biodegradable, in situ-forming cell-laden hydrogel composites of hydroxyapatite nanoparticles for bone regeneration, *Ind Eng Chem Res* **54**: 10206–10211 (2015).

64. B. M. Watson, T. N. Vo, A. M. Tatara, S. R. Shah, D. W. Scott, P. S. Engel and A. G. Mikos, Biodegradable, phosphate-containing, dual-gelling macromers for cellular delivery in bone tissue engineering, *Biomaterials* **67**: 286–296 (2015).

65. O. Jeon, E. Alsberg, Photofunctionalization of alginate hydrogels to promote adhesion and proliferation of human mesenchymal stem cells, *Tissue Eng Part A* **19**: 1424–1432 (2013).

66. O. Jeon, D. S. Alt, S. W. Linderman, E. Alsberg, Biochemical and physical signal gradients in hydrogels to control stem cell behavior, *Adv Mater* **25**: 6366–6372 (2013).

67. C. A. DeForest and K. S. Anseth, Cytocompatible click-based hydrogels with dynamically tunable properties through orthogonal photoconjugation and photocleavage reactions, *Nat Chem* **3**: 925–931 (2011).

68. D. Hannouche, H. Terai, J. R. Fuchs, S. Terada, S. Zand, B. A. Nasseri, H. Petite, L. Sedel and J. P. Vacanti, Engineering of implantable cartilaginous structures from bone marrow-derived mesenchymal stem cells, *Tissue Eng* **13**: 87–99 (2007).

69. A. Dickhut, E. Gottwald, E. Steck, C. Heisel and W. Richter, Chondrogenesis of mesenchymal stem cells in gel-like biomaterials *in vitro* and *in vivo*, *Front Biosci* **13**: 4517–4528 (2008).

70. S. Hofmann, S. Knecht, R. Langer, D. L. Kaplan, G. Vunjak-Novakovic, H. P. Merkle and L. Meinel, Cartilage-like tissue engineering using silk scaffolds and mesenchymal stem cells, *Tissue Eng* **12**: 2729–2738 (2006).

71. C. N. Salinas, B. B. Cole, A. M. Kasko and K. S. Anseth, Chondrogenic differentiation potential of human mesenchymal stem cells photoencapsulated within poly(ethylene glycol)-arginine-glycine-aspartic acid-serine thiol-methacrylate mixed-mode networks, *Tissue Eng* **13**: 1025–1034 (2007).

72. H. Park, J. S. Temenoff, Y. Tabata, A. I. Caplan and A. G. Mikos, Injectable biodegradable hydrogel composites for rabbit marrow mesenchymal stem cell and growth factor delivery for cartilage tissue engineering, *Biomaterials* **28**: 3217–3227 (2007).

3

STEM CELLS FOR CARDIAC TISSUE ENGINEERING

Jennifer L. Young, Karen L. Christman[†] and Adam J. Engler[‡]*

1. Introduction

The myocardial wall post-myocardial infarction becomes ischemic, resulting in tremendous apoptosis which is counterbalanced by increased secretion of collagen to maintain wall integrity. This process creates a stiff, fibrotic scar that lacks the contractile functions of healthy myocardium and can even induce transdifferentiation into aberrant cell types. Several clinically relevant tissue engineering methods have been proposed using stem cells to either restore myocardial function or generate functional replacements to combat this problem, but each faces unique challenges associated with the design criteria of functional cardiac tissue. By examining *in vitro* studies of the native myocardial and stem cell niches, here we provide a succinct review of the design criteria that should be associated with the field of cardiac tissue engineering and stem cells. We also discuss how the mixed results of current stem cell-based clinical trials may be the result of methods that perhaps do not always fit these criteria and remodel the stiff scar. By shifting the focus to how the native environment programs cells, we are hopeful that future therapies designed around these goals may dramatically improve patient outcomes.

2. Cell Therapies for Myocardial Infarction and Heart Failure

As a leading cause of death in the United States, congestive heart failure (CHF) post-myocardial infarction (MI) has incited the need to develop novel techniques that prevent muscle wall scarring and ventricular dilation, symptoms which contribute to impaired heart function. CHF is typically induced from an MI, in which a major coronary vessel becomes occluded, preventing appropriate oxygen exchange in the muscle downstream of this blockage. Ensuing cell death in the myocardium is counterbalanced by increased secretion of collagen, which forms a fibrotic scar to maintain wall integrity. As a consequence of this, ventricle contraction is impaired and results in negative left ventricular (LV) remodeling and dilation of the heart.[1] CHF-induced impairment results in a high morbidity rate as the body has limited capacity to replace damaged tissue and restore function. Although recent studies have reported on the ability of the heart to repair itself after injury via progenitor cardiac stem cells (CSCs) and homing of hematopoietic stem cells (HSCs),[2–4] intrinsic cardiac regeneration is limited by extremely slow cell turnover in the myocardium,[5–7] and turnover only further slows in older patients where CHF predominates. Over the past two decades, novel strategies have taken many routes and used many cell types to treat MI and CHF. Here we present a succinct review of three common regenerative medicine and tissue engineering strategies: cellular cardiomyoplasty, cardiac patches, and injectable scaffolds (Fig. 1). Though numerous variations have been attempted for these techniques, we will limit our discussion to the clinically relevant cell types used, their mechanistic characterization *in vitro*, and subsequent *in vivo* results.

Before presenting these methodologies, it is perhaps helpful to categorize the clinically used cell types today. Since its introduction in the 1990s, cellular cardiomyoplasty, where somatic cells are injected in liquid

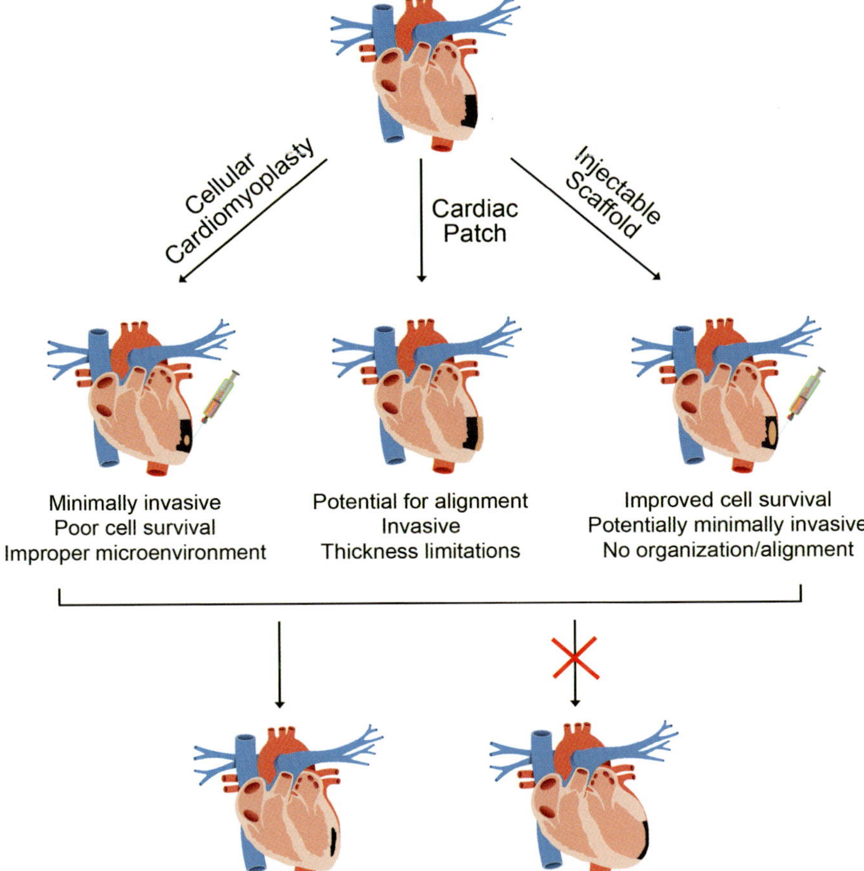

Fig. 1. Approaches to cell-based therapy in the heart. Following a myocardial infarction, cells in liquid solutions of saline or cell culture medium (cellular cardiomyoplasty), and in combination with a biomaterial scaffold (cardiac patch and injectable scaffold) have been explored to prevent progressive LV remodeling that leads to LV dilation and heart failure. Each technique has its own pros and cons as listed here, yet all approaches have had some success in animal models.

solutions of saline or cell culture medium into the post-MI heart wall, has gained much attention as one of the first cell-based methods proposed for repairing, replacing and restoring function to the injured heart. Early studies in animal infarct models showed implanted somatic cells survive, prevent further ventricular dilation, and confer some level of improved myocardial output.[8–10] Despite some promising results in animal models, trials with autologous somatic cells using various injection methods, while not showing adverse patient effects, have produced mixed results for improvement of global contractility,[11–14] although a recent meta-analysis of intracoronary transplantation trials for acute myocardial infarction showed no improvement in LV function from these interventions.[15] As the authors warn, however, these results must be taken with caution, as many studies were excluded from the analysis, and the influence of experimental parameters, e.g. cell type injected, was not taken into account. The studies that have shown improvement in cardiac function have yet to identify the precise mechanisms involved, but hypotheses are mostly related to transdifferentiation of injected cells, promotion of new blood vessel formation, or paracrine effects.[14] One significant shift in the technique to better determine its efficacy has been to use adult stem cells, as they are regarded as the optimal source since they are easily obtained, readily expanded *in vitro*, non-immunogenic, and can be an autologous cell source.

HSCs are bone marrow-derived cells that can regenerate myeloid and lymphoid lineage cells of the blood, though there is some evidence that these cells are multipotent and become non-blood derived lineages.[16,17] Their ability to differentiate into cardiomyocytes *in vitro* and in animal models with infarcted myocardium, however, has proven less than successful.[16,18,19] Mesenchymal stem cells (MSCs), named after their ability to differentiate into mesodermal lineages such as adipose, muscle, bone, etc.,[20] are also multipotent stem cells which can be easily isolated from the bone marrow where they are relatively abundant.[21] MSCs can also be derived from adipose tissue, dental pulp and Wharton's jelly, and have been shown to exhibit tissue derivation-dependent cell behavior and differentiation potential, which are important to consider when designing MSC-based therapies.[22] These cells lack several cell-surface markers enabling them to be immune privileged,[23] and in culture, these cells have been shown to differentiate into beating cardiomyocytes by the addition of the demethylating agent, 5-azacytidine.[24] It should be noted that though cells produced by this method exhibit some genotypic and phenotypic characteristics of adult cardiomyocytes, they lack the presence of some contractile elements.

With this supporting evidence, human MSCs were once thought to be the most clinically relevant source for myocardial cell therapy. Some studies have reported the ability of MSCs to marginally restore function and structure to animal infarct models by differentiating into cardiomyocytes and inducing angiogenesis,[25] though it was later demonstrated that improved cardiac function was likely due to paracrine effects of the transplanted cells[19,26] and not functional restoration. Others suggest that implanted MSCs fused with resident cardiomyocytes and thus do not actually differentiate into functioning cardiac muscle cells.[18,27] Increasing infarct cellularity, which increases wall thickness and thus reduces wall stress, is also postulated to result in the observed functional effects.[28] Although the mechanisms underlying improvements in heart function have yet to be identified, the results of cellular cardiomyoplasty in animal models have prompted several human clinical trials. However, both skeletal myoblasts and human MSCs were injected into the myocardium post-MI, and little to no improvement was observed;[29–31] thus it appears that injection of cells alone into injured myocardium may not be sufficient to restore function and structure due to some mechanism blocking their regenerative capacity. In fact, evidence, which we will discuss next, appears to indicate that MSCs in the infarct differentiate not into muscle but into cells that form small calcified lesions typical of MSC-derived osteoblasts[32] (Fig. 2c). Human induced pluripotent stem cells (hiPSCs) have been increasingly explored as a promising cell source in cellular cardiomyoplasty due to their capacity for infinite expansion and ability to efficiently differentiate into cardiomyocytes,[33,34] although concerns of tumor formation still exist.

3. Cellular Cardiomyoplasty Revisited: The Influence of *In Vitro* Mechanics

Cells are highly responsive to their surroundings, and when presented with a disease microenvironment, their function is likely different from what is observed in a healthy setting. Healthy tissue exists in a balance where forces, integrins, and the surrounding fibrous scaffold called the extracellular matrix (ECM) reside in a state of "dynamic reciprocity." In such a state, altered cells programmed by a damaged environment may not be able to sufficiently regenerate tissue function or may even worsen the cell niche.[35] The inconclusive results from MSCs used in myocardial infarction treatment could be explained in part by improper epigenetic regulation as dictated by the aberrant surrounding ECM. Indeed, as cells necrose in the heart wall, fibroblasts secrete a substantial amount of collagen to form a scar in the infarct area. This is initially a compensatory mechanism to resist wall deformation and rupture.[36] However, as noted by Berry and co-workers, this fibrotic scar is three- to four-fold stiffer than normal muscle, as measured using an atomic force microscope (AFM),[37] and is accompanied by ventricular remodeling.

Resistance to deformation from a force, i.e. the "stiffness" or, more formally, the elasticity (E, measured in Pascals, Pa), provided by the ECM for cells to contract against, has an important role in development and function. Tissue-specific matrices have distinct mechanical properties that help to direct developing cells to a specific lineage.[20] It has been demonstrated that MSCs plated onto collagen I-coated polyacrylamide (PA) gels that mimic the elasticity of the brain (0.1–1 kPa), muscle (~ 8–17 kPa), and bone (>30 kPa), give rise to neurocytes, myocytes, and osteocytes, respectively[38,39] (Fig. 2a). Such responses are also seen with fibronectin, but not with laminin- and collagen IV-coated gels,[39] suggesting that cells are highly responsive to their physical environment when ligating appropriate integrins.[40] If placed into an abnormally rigid environment *in vivo*, e.g. one that is three- to four-fold more stiff than normal myocardium and full of collagen I,[37] such results would imply that MSCs would respond by becoming osteoblast-like rather than muscle-like (Fig. 2b). Though there are a variety of soluble factors that may attenuate this process, much of that is likely absent in this infarcted niche.

Such observations also highlight the glaring differences between the *in vivo* environment, *in vitro* mimics (e.g. PA gels), and traditional myocyte cultures from the past several decades, i.e. cells grown on collagen-coated rigid glass and tissue culture polystyrene. On the rigid substrate, three-step myofibrillogenesis[41] is halted at the initial pre-myofibril step in isolated myotubes, as shown by Griffin and co-workers.[42] Instead of mature fibrils, they find that large stress fibers appear and myotubes are overly adherent to the rigid substrate.[42] Conversely, many labs have studied myofibrillogenesis in culture, with cells grown either (1) in syncytia, where clustered cells

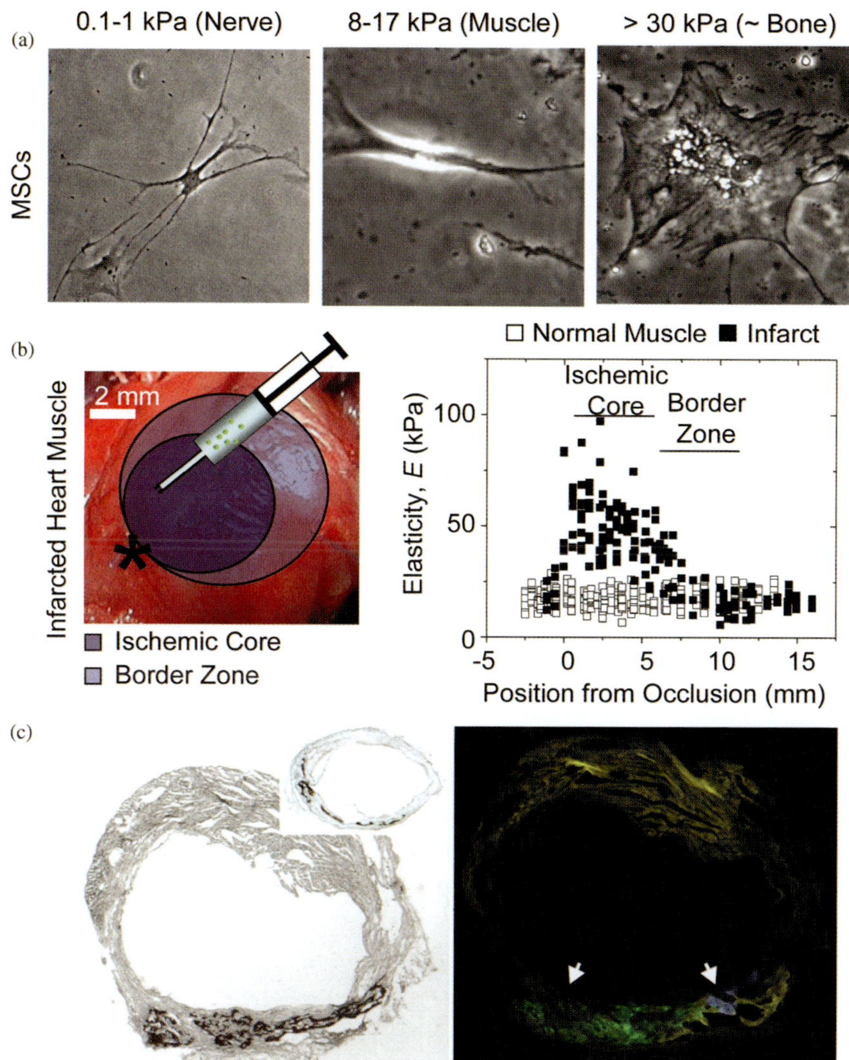

Fig. 2. Cellular cardiomyoplasty. (a) MSCs grown on hydrogels of defined stiffness develop cell morphologies reflective of neural, myocyte, and osteocyte lineages. (b) MSCs were injected into infarcted hearts (left; with inner ischemic and outer border zones highlighted in shades of blue) where the elasticity of the wall was significantly changed downstream of the occlusion site as shown in the right panel using atomic force microscopy. (c) Histological analysis of the heart wall shows that 29 days after injecting 10^5 MSCs, hearts showed massive calcifications (black deposits; left image) that originated from injected EGFP + MSCs.

can contract against one another rather than the rigid matrix, or (2) on glass coated with thicker collagen layers, effectively making the surface more compliant. For example, when cultured on collagen gel-coated glass, with elasticity closer to muscle than rigid substrates,[43] pre-cardiac mesoderm was able to undergo myofibrillogenesis[41] and could beat rhythmically in similar collagen gels.[44] Detailed analysis of cardiomyocyte behavior as a function of matrix stiffness has shown that mature[45] and neonatal myocytes[46] require a compliant niche that is as soft or softer than healthy muscle for sustained contraction and proper expression of sarcoplasmic/endoplasmic reticular calcium ATPase, calcium storage and transients, and traction stresses. Yet until recently, the relatively low rigidity of the *in vivo* environment for the myocardium has largely been underappreciated.

When returning to the infarct example, it is thus not surprising that somatic cells as well as stem cells are regulated by the mechanical context of their environment. While the highly ischemic region in the center of the infarct may lack sufficient blood supply for a cell therapy alone to work, stiffness within the border zone, which is only 50–200% higher than normal,[37] likely prevents MSCs from restoring function there as well since no MSC-derived cardiomyocytes were found. As illustrated in Fig. 2, MSC-treated infarcted hearts sometimes show limited recovery[37] and abnormal differentiation,[32] agreeing with *in vitro* data demonstrating that MSCs become osteogenic on gels with E of ~30–50 kPa.[38,39] Surprisingly, these lesions are found throughout the infarct, indicating that at least modest cell engraftment occurred through a dominant fraction of both the highly ischemic necrotic core and the modestly ischemic border zone.

However, clinical studies using marrow-derived stem cells, presumably with the same abnormal differentiation characteristics, and other somatic cells appear to demonstrate that there is still some improvement in global function as cells either are directly injected or home to the injury site; observed improvement includes 6–9% increase in LV ejection fraction, reduced end-systolic LV volumes, and enhanced perfusion in the infarcted area four to six months after cell transplantation.[11–13,47] Such discrepancies between studies directly measuring elasticity and implicating problems with differentiation versus those that demonstrate modest functional improvement may likely be the result of differences in variables such as time-post infarct for cell injection, injection cell number, cell preparation protocols, spatial distribution of injections, etc. With differences in time post-MI for cell injection in particular, cells seeing a pre-formed rigid scar versus a matrix that is actively remodeling could present environments that are either already rigid or one that is still soft but actively remodeling, respectively. Indeed, cells injected up to three days post-MI versus seven days post-MI showed reduced retention and functional benefits, indicating that timing of injected cells is important.[48]

Though more appropriate cell types for this therapy may exist, such as more recently discovered cardiac stem cells, these data imply that it is not possible to rely solely on direct injection of cells to remodel the matrix as it provides too few cues in the proper direction to promote the formation of appropriate cardiomyocyte lineages. Instead, two alternate approaches have been proposed which involve delivering cells in epicardial sheets or in injectable scaffolds.

4. Tissue Engineering Approach: Utilizing Biomaterial Scaffolds

While there has been some success in animal models with cellular cardiomyoplasty, and numerous clinical trials are currently ongoing, this technique is plagued by limited cell retention and transplant survival.[49,50] Injection into an ischemic environment and anoikis (cell death from the absence of cell-matrix interactions) are two major factors attributed to poor cell transplant survival. To overcome both of these challenges, tissue engineering approaches are currently being examined. In these strategies, cells are given a temporary extracellular matrix through the use of a biomaterial scaffold, unlike the typical cellular cardiomyoplasty approach where cells are delivered in only a liquid solution. Biomaterial approaches also have potential for growth factor delivery, which can further promote cell survival and reduce the ischemic environment.

4.1 *Cardiac patches*

Myocardial tissue engineering can be categorized into *in vitro* engineering and injectable approaches (Fig. 1).[51,52] *In vitro* engineered myocardial tissue, involving the classical tissue engineering approach of seeding cells on a scaffold and culturing the construct prior to implantation, was the first to be examined in the myocardium.[53,54] In addition to utilizing pre-formed scaffolds, soluble scaffolds have been pre-mixed with cells and shaped into the appropriate construct. Moreover, temperature-responsive polymer-coated substrates have been employed to generate cell monolayers or sheets, which retain their secreted matrices.[55,56] These can be subsequently combined to form a cardiac patch over the infarcted region. By providing an extracellular matrix for transplanted cells, cardiac patches have been shown to increase cell survival, induce neovascularization, attenuate negative LV remodeling, and preserve cardiac function in pre-clinical models.[52] While improvements in contractility are theoretically possible with stem cell-derived cardiomyocytes, such effects to date are likely through paracrine mechanisms.

Clinical results with this approach are sparse to date, yet results from a non-randomized Phase I clinical trial with a cell-seeded collagen scaffold are encouraging, at least in terms of safety and feasibility.[57] Mononuclear bone marrow cells isolated from the patient were seeded on a porous collagen matrix ($7 \times 5 \times 0.6$ cm) in the operating room and then sutured onto the epicardium following a single off-pump coronary artery bypass graft surgery (OP-CABG). In chronic infarcts, with an average age of approximately eight months, this cardiac patch increased wall thickness, limited negative LV remodeling, and improved diastolic function; however, patients also received injections of the bone marrow cells in autologous serum into the infarct and border zone. Additionally, pre-clinical trials have begun for a cardiac patch made of hiPSC-cardiomyocytes (CMs) and human pericytes entrapped in a fibrin gel. hiPSC-CM survival, proliferation, and subsequent improvement in infarct size and cardiac function have been reported in rat models of MI after transplantation; however, some later-stage metrics, i.e. LV thinning and fractional shortening, showed no benefit from the patch, indicating that experimental parameters are still in need of optimization.[58]

While *in vitro* engineering of a patch to cover up damaged myocardial tissue is a viable strategy for cardiac repair, there are limitations with this approach. For instance, as with other vascularized tissues, the development of sizable constructs *in vitro* is a major challenge. Secondly, these patches can only be applied to the epicardial surface and thus do not directly treat the infarct. Implantation of a patch would also require an invasive surgical procedure, unlike injectable, percutaneous approaches. However, there are some distinct advantages that this *in vitro* approach offers. For instance, organization and alignment of transplanted cells can be uniquely achieved.[59,60] While this is likely not critical for cells that are functioning strictly through paracrine effects, it would be important for cell types that can undergo cardiomyogenesis.

One approach to cardiac patch design includes the combination of pro-survival and angiogenic growth factors, and prevascularization on the omentum, which is a blood vessel-enriched membrane.[61] This approach improved subsequent patch integration with host myocardium, leading to preservation of LV volume and ejection fraction, and has potential for the use with other cell types. A caveat with this study, however, was that LV volume and cardiac function were not statistically different than acellular patches also prevascularized on the omentum. More sophisticated biomimetic scaffolds are also providing potential advances in the field.

Cardiac patches composed of nanoelectronics for online monitoring of cardiac function post-implantation have recently been developed.[62] In addition, these embedded electrodes can be used to synchronize cardiac patch contraction, as shown by electrical pacing of non-contracting portions of the scaffold, which may assist in the integration with host tissue and eventual improvement in function. It has yet to be studied, however, how such patches function *in vivo*, as well as their longevity in living systems. Moving away from synthetic approaches, protocols for decellularizing myocardium[63,64] have opened up the possibility for creating a cardiac patch with a scaffold that contains many of the biochemical and mechanical cues that match original myocardial ECM composition and potentially its mechanical properties, i.e. 10–20 kPa.[37] To that end, synthetic scaffolds have also been engineered to match the anisotropic properties of healthy myocardium for both the right ventricular myocardium and the left ventricular myocardium.[65,66]

4.2 *Injectable scaffolds*

Given the push toward more minimally invasive surgeries that require less recovery time and reduce the chances of infection, injectable approaches to myocardial tissue engineering, which could be delivered minimally invasively through a catheter, are particularly attractive. Moreover, an injectable therapy could be delivered throughout the infarct wall and border zone, and not solely to the epicardium as with cardiac patches.

Injectable scaffolds for *in situ* myocardial tissue engineering can be utilized as either acellular or cellular treatments.[52] In the first approach, material is injected into the infarct alone, and can serve to increase cell migration into the infarct area, including neovascularization, to thicken and support the LV wall, or both. As a cell therapy, injectable scaffolds can be employed to increase cell transplant survival over the typical cellular cardiomyoplasty approach. An acellular approach may reach the clinic sooner since it has the potential to be an off-the-shelf treatment without the added complications that cells bring, including appropriate source, the need for *in vitro* expansion, or potential disease transmission. However, cellular therapies are currently in clinical trials, despite poor cell survival as with the typical cellular cardiomyoplasty technique. Injectable scaffolds have the potential to increase cell transplant survival, with potential functional benefit, and as such are often viewed as an improvement over previously discussed methods.

With the knowledge of these pros and cons, it is critical to define design criteria associated with an ideal injectable scaffold for cell delivery in the myocardium. First, an injectable myocardial scaffold should promote neovascularization to reduce the ischemic environment, promote cell adhesion, survival, and maturation in the case of progenitor or stem cell delivery, and can be injected through a catheter. In terms of cell adhesion and survival, material choice and composition are especially important and should likely mimic both biochemical composition, which includes a complex mixture of proteins and polysaccharides, and mechanical properties of the native myocardial ECM which they are attempting to replace. To date, several materials have been examined for injectable approaches to myocardial tissue engineering, including fibrin,[67,68] collagen,[69] Matrigel,[70–72] alginate,[73,74] self-assembling peptides,[52] and synthetic materials,[75] among others.

Collagen has been utilized to promote neovascularization,[69] thicken the infarct wall, and improve LV geometry,[76] and deliver cells into the myocardium;[77] however, neutralized collagen gels rapidly at even room temperature. Matrigel, both alone[70,71] and in combination with collagen,[72] has been shown to facilitate cell transplantation. Yet, clinical translation of Matrigel is very unlikely since it is a matrix derived from mouse sarcoma, and is known to promote tumor growth *in vitro* and *in vitro*. Alginate, which is a polysaccharide derived from seaweed, has been utilized mostly for thickening and supporting the LV wall;[73,74] however, as a regenerative scaffold, alginate is likely not the best choice since it is known to have poor cell adhesion,[78] although it can be modified to contain cell adhesion peptides.[79] Self-assembling peptides can form nanofibrous networks inside the myocardium, which promote cell and vessel ingrowth,[80] yet these matrices were not capable of improving cell transplant survival.[81] Synthetic materials are an attractive source for tissue engineering applications because their properties can be easily tuned to their desired application. In small animal models, however, synthetic materials have not shown great improvement in overall cardiac function post-MI, likely due to the fact that they lack bioactivity.[75,82] Of all of these materials, the only neutralized material to be injected via catheter is alginate, which has shown success in large animal models,[83] but has failed to show improvement thus far in clinical trials.[84] Taken all together, there is still much to be accomplished in the design of a clinically relevant, minimally invasive, injectable material for MI treatment. Furthermore, the importance of tissue regeneration in positive functional outcomes must not be ignored.

The ECM is known to play a role in almost every cell process including attachment, differentiation, morphogenesis, and whether a cell proliferates or apoptosis,[20,37–39,42–46] and thus the typical goal of tissue engineering scaffolds is to mimic the *in vivo* ECM for a particular tissue, mainly because of the importance of cell-matrix interactions. An injectable, *in situ* gelling scaffold derived from decellularized ventricular myocardium, which retains many of the original biochemical cues of native cardiac ECM, has been tested *in vivo*, both in small and large animal models.[85–87] In the larger porcine model, this material was injected via an endocardial, percutaneous approach, which is the preferred delivery method for cell therapy in the myocardium.[88] These studies demonstrated improvements in both global cardiac function as well as regional cardiac function, thereby reinforcing the importance of native ECM cues in tissue repair. While the biochemical aspects of the myocardial ECM have been effectively mimicked, no injectable materials have been designed to mimic the mechanical properties, whose importance have been highlighted above and likely play an important role in determining transplanted cell fate.

4.3 *Material selection*

Considering the key role of cell-ECM interactions in affecting cell behavior, and the tissue specificity of the ECM, a scaffold that possesses the appropriate ventricular ECM cues would be ideal for myocardial tissue engineering. However, few materials for cardiac tissue engineering have involved this design approach. Neither naturally derived or synthetic materials possess all of the necessary attributes for a scaffold; whether for a cardiac patch or an injectable scaffold, each has its pros and cons.

Protein and natural-based hydrogels are especially applicable in tissue engineering strategies due to their ability to better mimic the *in vivo* microenvironment by exhibiting familiar ECM components and structural organization. Cells can thus better respond to the signals provided by familiar components in their surrounding matrix. Additionally, these molecules are generally biocompatible, thereby reducing concerns over potential toxicity. Synthetic biomaterials, on the other hand, can provide better mechanical properties and can be more easily manipulated than natural polymers.

Currently no ideal materials exist in terms of mimicking both biochemical and mechanical properties of the myocardium. Advancement in this field will likely be achieved by the rational design of materials to appropriately mimic the desired properties of the native ECM. This will likely change depending on the intended mechanism of therapy: structural versus paracrine versus regeneration. Furthermore, designing materials with clinical translation in mind is a necessity. For example, although a material may be injectable via syringe, it does not necessarily translate to a minimally invasive approach. Design criteria must therefore include translation to man from the very beginning with small animal studies.

References

1. M. S. Figueroa and J. I. Peters, Congestive heart failure: diagnosis, pathophysiology, therapy, and implications for respiratory care, *Respir Care* **51**: 403–412 (2006).

2. M. A. Laflamme, D. Myerson, J. E. Saffitz and C. E. Murry, Evidence for cardiomyocyte repopulation by extracardiac progenitors in transplanted human hearts, *Circ Res* **90**: 634–640 (2002).

3. A. Leri, J. Kajstura and P. Anversa, Cardiac stem cells and mechanisms of myocardial regeneration, *Physiol Rev* **85**: 1373–1416 (2005).

4. K. Urbanek, D. Torella, F. Sheikh, A. De Angelis, D. Nurzynska, F. Silvestri, C. A. Beltrami, R. Bussani, A. P. Beltrami, F. Quaini, R. Bolli, A. Leri, J. Kajstura and P. Anversa, Myocardial regeneration by activation of multipotent cardiac stem cells in ischemic heart failure, *Proc Natl Acad Sci USA* **102**: 8692–8697 (2005).

5. O. Bergmann, R. D. Bhardwaj, S. Bernard, S. Zdunek, F. Barnabe-Heider, S. Walsh, J. Zupicich, K. Alkass, B. A. Buchholz, H. Druid, S. Jovinge and J. Frisen, Evidence for cardiomyocyte renewal in humans, *Science* **324**: 98–102 (2009).

6. S. E. Senyo, M. L. Steinhauser, C. L. Pizzimenti, V. K. Yang, L. Cai, M. Wang, T. D. Wu, J. L. Guerquin-Kern, C.P. Lechene and R. T. Lee, Mammalian heart renewal by pre-existing cardiomyocytes, *Nature* **493**: 433–436 (2012).

7. E. R. Porrello, A. I. Mahmoud, E. Simpson, J. A. Hill, J. A. Richardson, E. N. Olson and H. A. Sadek, Transient regenerative potential of the neonatal mouse heart, *Science* **331**: 1078–1080 (2011).

8. R. C. Chiu, A. Zibaitis and R. L. Kao, Cellular cardiomyoplasty: myocardial regeneration with satellite cell implantation, *Ann Thorac Surg* **60**: 12–18 (1995).

9. R. K. Li, Z. Q. Jia, R. D. Weisel, D. A. Mickle, J. Zhang, M. K. Mohabeer, V. Rao and J. Ivanov, Cardiomyocyte transplantation improves heart function, *Ann Thorac Surg* **62**: 654–660; discussion 660–651 (1996).

10. D. A. Taylor, B. Z. Atkins, P. Hungspreugs, T. R. Jones, M. C. Reedy, K. A. Hutcheson, D. D. Glower and W. E. Kraus, Regenerating functional myocardium: improved performance after skeletal myoblast transplantation, *Nat Med* **4**: 929–933 (1998).

11. K. Lunde, S. Solheim, S. Aakhus, H. Arnesen, M. Abdelnoor and K. Forfang, Autologous stem cell transplantation in acute myocardial infarction: the ASTAMI randomized controlled trial: intracoronary transplantation

of autologous mononuclear bone marrow cells, study design and safety aspects, *Scand Cardiovasc J* **39**: 150–158 (2005).

12. J. C. Chachques, C. Acar, J. Herreros, J. C. Trainini, F. Prosper, N. D'Attellis, J. N. Fabiani and A. F. Carpentier, Cellular cardiomyoplasty: clinical application, *Ann Thorac Surg* **77**: 1121–1130 (2004).

13. S. Janssens, C. Dubois, J. Bogaert, K. Theunissen, C. Deroose, W. Desmet, M. Kalantzi, L. Herbots, P. Sinnaeve, J. Dens, J. Maertens, F. Rademakers, S. Dymarkowski, O. Gheysens, J. Van Cleemput, G. Bormans, J. Nuyts, A. Belmans, L. Mortelmans, M. Boogaerts and F. Van de Werf, Autologous bone marrow-derived stem-cell transfer in patients with ST-segment elevation myocardial infarction: double-blind, randomised controlled trial, *Lancet* **367**: 113–121 (2006).

14. S. K. Sanganalmath and R. Bolli, Cell therapy for heart failure, *Circ Res* **113**: 810–834 (2013).

15. M. Gyongyosi, W. Wojakowski, P. Lemarchand, K. Lunde, M. Tendera, J. Bartunek, E. Marban, B. Assmus, T. D. Henry, J. H. Traverse, L. A. Moye, D. Surder, R. Corti, H. Huikuri, J. Miettinen, J. Wohrle, S. Obradovic, J. Roncalli, K. Malliaras, E. Pokushalov, A. Romanov, J. Kastrup, M. W. Bergmann, D. E. Atsma, A. Diederichsen, I. Edes, I. Benedek, T. Benedek, H. Pejkov, N. Nyolczas, N. Pavo, J. Bergler-Klein, I. J. Pavo, C. Sylven, S. Berti, E. P. Navarese, G. Maurer and ACCRUE investigators, Meta-Analysis of Cell-based CaRdiac stUdiEs (ACCRUE) in patients with acute myocardial infarction based on individual patient data. *Circ Res* **116**: 1346–1360 (2015).

16. C. E. Murry, M. H. Soonpaa, H. Reinecke, H. Nakajima, H. O. Nakajima, M. Rubart, K. B. S. Pasumarthi, J. Ismail Virag, S. H. Bartelmez, V. Poppa, G. Bradford, J. D. Dowell, D. A. Williams and L. J. Field, Haematopoietic stem cells do not transdifferentiate into cardiac myocytes in myocardial infarcts, *Nature* **428**: 664–668 (2004).

17. A. Xynos, P. Corbella, N. Belmonte, R. Zini, R. Manfredini and G. Ferrari, bone marrow-derived hematopoietic cells undergo myogenic differentiation following a pax-7 independent pathway, *Stem Cells* **28**: 965–973 (2010).

18. J. M. Nygren, S. Jovinge, M. Breitbach, P. Sawen, W. Roll, J. Hescheler, J. Taneera, B. K. Fleischmann and S. E. Jacobsen, Bone marrow-derived hematopoietic cells generate cardiomyocytes at a low frequency through cell fusion, but not transdifferentiation, *Nat Med* **10**: 494–501 (2004).

19. R. Uemura, M. Xu, N. Ahmad and M. Ashraf, Bone marrow stem cells prevent left ventricular remodeling of ischemic heart through paracrine signaling, *Circ Res* **98**: 1414–1421 (2006).

20. D. E. Discher, P. Janmey and Y. L. Wang, Tissue cells feel and respond to the stiffness of their substrate, *Science* **310**: 1139–1143 (2005).

21. A. I. Caplan, Mesenchymal stem cells, *J Orthop Res* **9**: 641–650 (1991).

22. N. H. Abu Kasim, V. Govindasamy, N. Gnanasegaran, S. Musa, P. J. Pradeep, T. C. Srijaya and Z. A. C. A. Aziz, Unique molecular signatures influencing the biological function and fate of post-natal stem cells isolated from different sources, *J Tissue Eng Regen Med* **9**: E252–266 (2015).

23. K. Le Blanc, L. Tammik, B. Sundberg, S. E. Haynesworth and O. Ringden, Mesenchymal stem cells inhibit and stimulate mixed lymphocyte cultures and mitogenic responses independently of the major histocompatibility complex, *Scand J Immunol* **57**: 11–20 (2003).

24. S. Makino, K. Fukuda, S. Miyoshi, F. Konishi, H. Kodama, J. Pan, M. Sano, T. Takahashi, S. Hori, H. Abe, J. Hata, A. Umezawa and S. Ogawa, Cardiomyocytes can be generated from marrow stromal cells *in vitro*, *J Clin Invest* **103**: 697–705 (1999).

25. D. Orlic, J. Kajstura, S. Chimenti, I. Jakoniuk, S. M. Anderson, B. Li, J. Pickel, R. McKay, B. Nadal-Ginard, D. M. Bodine, A. Leri and P. Anversa, Bone marrow cells regenerate infarcted myocardium, *Nature* **410**: 701–705 (2001).

26. M. Gnecchi, H. He, O. D. Liang, L. G. Melo, F. Morello, H. Mu, N. Noiseux, L. Zhang, R. E. Pratt, J. S. Ingwall and V. J. Dzau, Paracrine action accounts for marked protection of ischemic heart by Akt-modified mesenchymal stem cells, *Nat Med* **11**: 367–368 (2005).

27. M. Alvarez-Dolado, R. Pardal, J. M. Garcia-Verdugo, J. R. Fike, H. O. Lee, K. Pfeffer, C. Lois, S. J. Morrison and A. Alvarez-Buylla, Fusion of bone-marrow-derived cells with Purkinje neurons, cardiomyocytes and hepatocytes, *Nature* **425**: 968–973 (2003).

28. M. A. Laflamme, S. Zbinden, S. E. Epstein and C. E. Murry, Cell-based therapy for myocardial ischemia and infarction: pathophysiological mechanisms, *Annu Rev Pathol* **2**: 307–339 (2007).

29. S. Fazel, L. Chen, R. D. Weisel, D. Angoulvant, C. Seneviratne, A. Fazel, P. Cheung, J. Lam, P. W. Fedak, T. M. Yau and R.K. Li, Cell transplantation preserves cardiac function after infarction by infarct stabilization: augmentation by stem cell factor. *J Thorac Cardiovasc Surg* **130**: 1310 (2005).

30. M. B. Britten, N. D. Abolmaali, B. Assmus, R. Lehmann, J. Honold, J. Schmitt, T. J. Vogl, H. Martin, V. Schachinger, S. Dimmeler and A.M. Zeiher, Infarct remodeling after intracoronary progenitor cell treatment in patients with acute myocardial infarction (TOPCARE-AMI): mechanistic insights from serial contrast-enhanced magnetic resonance imaging, *Circulation* **108**: 2212–2218 (2003).

31. G. P. Meyer, K. C. Wollert, J. Lotz, J. Steffens, P. Lippolt, S. Fichtner, H. Hecker, A. Schaefer, L. Arseniev, B. Hertenstein, A. Ganser and H. Drexler, Intracoronary bone marrow cell transfer after myocardial infarction: eighteen months' follow-up data from the randomized, controlled BOOST (BOne marrow transfer to enhance ST-elevation infarct regeneration) trial. *Circulation* **113**: 1287–1294 (2006).

32. M. Breitbach, T. Bostani, W. Roell, Y. Xia, O. Dewald, J. M. Nygren, J. W. Fries, K. Tiemann, H. Bohlen, J. Hescheler, A. Welz, W. Bloch, S. E. Jacobsen and B. K. Fleischmann, Potential risks of bone marrow cell transplantation into infarcted hearts, *Blood* **110**: 1362–1369 (2007).

33. L. Zwi, O. Caspi, G. Arbel, I. Huber, A. Gepstein, I. H. Park and L. Gepstein, Cardiomyocyte differentiation of human induced pluripotent stem cells, *Circulation* **120**: 1513–1523 (2009).

34. H. Masumoto, T. Ikuno, M. Takeda, H. Fukushima, A. Marui, S. Katayama, T. Shimizu, T. Ikeda, T. Okano, R. Sakata and J. K. Yamashita, Human iPS cell-engineered cardiac tissue sheets with cardiomyocytes and vascular cells for cardiac regeneration, *Sci Rep* **4**: 6716 (2014).

35. A. J. Engler, P. O. Humbert, B. Wehrle-Haller and V. M. Weaver, Multiscale modeling of form and function, *Science* **324**: 208–212 (2009).

36. D. L. Mann, Mechanisms and models in heart failure: a combinatorial approach, *Circulation* **100**: 999–1008 (1999).

37. M. F. Berry, A. J. Engler, Y. J. Woo, T. J. Pirolli, L. T. Bish, V. Jayasankar, K. J. Morine, T. J. Gardner, D. E. Discher and H. L. Sweeney, Mesenchymal stem cell injection after myocardial infarction improves myocardial compliance, *Am J Physiol Heart Circ Physiol* **290**: H2196–2203 (2006).

38. A. J. Engler, S. Sen, H. L. Sweeney and D. E. Discher, Matrix elasticity directs stem cell lineage specification, *Cell* **126**(4): 677–689 (2006).

39. A. S. Rowlands, P. A. George and J. J. Cooper-White, Directing osteogenic and myogenic differentiation of MSCs: interplay of stiffness and adhesive ligand presentation, *Am J Physiol Cell Physiol* **295**: C1037–1044 (2008).

40. J. E. Frith, R. J. Mills, J. E. Hudson and J. J. Cooper-White, Tailored integrin–extracellular matrix interactions to direct human mesenchymal stem cell differentiation, *Stem Cells Dev* **21**: 2442–2456 (2012).

41. A. Du, J. M. Sanger, K. K. Linask and J. W. Sanger, Myofibrillogenesis in the first cardiomyocytes formed from isolated quail precardiac mesoderm, *Dev Biol* **257**: 382–394 (2003).

42. M. A. Griffin, S. Sen, H. L. Sweeney and D. E. Discher, Adhesion-contractile balance in myocyte differentiation, *J Cell Sci* **117**: 5855–5863 (2004).

43. M. T. Sheu, J. C. Huang, G. C. Yeh and H. O. Ho, Characterization of collagen gel solutions and collagen matrices for cell culture, *Biomaterials* **22**: 1713–1719 (2001).

44. T. Eschenhagen, C. Fink, U. Remmers, H. Scholz, J. Wattchow, J. Weil, W. Zimmermann, H. II. Dohmen, H. Schafer, N. Bishopric, T. Wakatsuki and E. L. Elson, Three-dimensional reconstitution of embryonic cardiomyocytes in a collagen matrix: a new heart muscle model system. *FASEB J* **11**: 683–694 (1997).

45. A. J. Engler, C. Carag-Krieger, C. P. Johnson, M. Raab, H. Y. Tang, D. W. Speicher, J. W. Sanger, J. M. Sanger and D. E. Discher, Embryonic cardiomyocytes beat best on a matrix with heart-like elasticity: scar-like rigidity inhibits beating, *J Cell Sci* **121**: 3794–3802 (2008).

46. J. G. Jacot, A. D. McCulloch and J. H. Omens, Substrate stiffness affects the functional maturation of neonatal rat ventricular myocytes, *Biophys J* **95**: 3479–3487 (2008).

47. V. Schachinger, T. Tonn, S. Dimmeler and A. M. Zeiher, Bone-marrow-derived progenitor cell therapy in need of proof of concept: design of the REPAIR-AMI trial, *Nat Clin Pract Cardiovasc Med* **3**: S23–28 (2006).

48. J. S. Nakamuta, M. E. Danoviz, F. L. N. Marques, L. dos Santos, C. Becker, G. A. Gonçalves, P. F. Vassallo, I. T. Schettert, P. J. F. Tucci and J. E. Krieger, cell therapy attenuates cardiac dysfunction post myocardial infarction: effect of timing, routes of injection and a fibrin scaffold, *PLoS ONE* **4**: e6005 (2009).

49. J. Muller-Ehmsen, P. Whittaker, R. A. Kloner, J. S. Dow, T. Sakoda, T. I. Long, P. W. Laird and L. Kedes, Survival and development of neonatal rat cardiomyocytes transplanted into adult myocardium, *J Mol Cell Cardiol* **34**: 107–116 (2002).

50. H. Reinecke and C. E. Murry, Taking the death toll after cardiomyocyte grafting: a reminder of the importance of quantitative biology, *J Mol Cell Cardiol* **34**: 251–253 (2002).

51. W. H. Zimmermann and T. Eschenhagen, Cardiac tissue engineering for replacement therapy, *Heart Fail Rev* **8**: 259–269 (2003).

52. K. L. Christman and R. J. Lee, Biomaterials for the treatment of myocardial infarction, *J Am Coll Cardiol* **48**: 907–913 (2006).

53. R. K. Li, Z. Q. Jia, R. D. Weisel, D. A. Mickle, A. Choi and T. M. Yau, Survival and function of bioengineered cardiac grafts, *Circulation* **100**: 1163–1169 (1999).

54. J. Leor, S. Aboulafia-Etzion, A. Dar, L. Shapiro, I. M. Barbash, A. Battler, Y. Granot and S. Cohen, Bioengineered cardiac grafts: a new approach to repair the infarcted myocardium? *Circulation* **102**: 11156–11161 (2000).

55. Y. Miyahara, N. Nagaya, M. Kataoka, B. Yanagawa, K. Tanaka, H. Hao, K. Ishino, H. Ishida, T. Shimizu, K. Kangawa, S. Sano, T. Okano, S. Kitamura and H. Mori, Monolayered mesenchymal stem cells repair scarred myocardium after myocardial infarction, *Nat Med* **12**: 459–465 (2006).

56. H. Sekine, T. Shimizu, K. Hobo, S. Sekiya, J. Yang, M. Yamato, H. Kurosawa, E. Kobayashi and T. Okano, Endothelial cell coculture within tissue-engineered cardiomyocyte sheets enhances neovascularization and improves cardiac function of ischemic hearts, *Circulation* **118**: S145–152 (2008).

57. J. C. Chachques, J. C. Trainini, N. Lago, M. Cortes-Morichetti, O. Schussler and A. Carpentier, Myocardial assistance by grafting a new bio-artificial upgraded myocardium (MAGNUM trial): clinical feasibility study, *Ann Thorac Surg* **85**: 901–908 (2008).

58. J. S. Wendel, L. Ye, R. Tao, J. Zhang, J. Zhang, T. J. Kamp and R. T. Tranquillo, Functional effects of a tissue-engineered cardiac patch from human induced pluripotent stem cell-derived cardiomyocytes in a rat infarct model, *Stem Cells Transl Med* **4**: 1324–1332 (2015).

59. N. Bursac, Y. Loo, K. Leong and L. Tung, Novel anisotropic engineered cardiac tissues: studies of electrical propagation, *Biochem Biophys Res Commun* **361**: 847–853 (2007).

60. M. Radisic, H. Park, H. Shing, T. Consi, F. J. Schoen, R. Langer, L. E. Freed and G. Vunjak-Novakovic, Functional assembly of engineered myocardium by electrical stimulation of cardiac myocytes cultured on scaffolds, *Proc Natl Acad Sci USA* **101**: 18129–18134 (2004).

61. T. Dvir, A. Kedem, E. Ruvinov, O. Levy, I. Freeman, N. Landa, R. Holbova, M. S. Feinberg, S. Dror, Y. Etzion, J. Leor and S. Cohen, Prevascularization of cardiac patch on the omentum improves its therapeutic outcome, *Proc Natl Acad Sci USA* **106**: 14990–14995 (2009).

62. R. Feiner, L. Engel, S. Fleischer, M. Malki, I. Gal, A. Shapira, Y. Shacham-Diamand and T. Dvir, Engineered hybrid cardiac patches with multifunctional electronics for online monitoring and regulation of tissue function, *Nat Mater* **15**: 679–685 (2016).

63. H. C. Ott, T. S. Matthiesen, S. K. Goh, L. D. Black, S. M. Kren, T. I. Netoff and D. A. Taylor, Perfusion-decellularized matrix: using nature's platform to engineer a bioartificial heart, *Nat Med* **14**: 213–221 (2008).

64. J. M. Wainwright, C. A. Czajka, U. B. Patel, D. O. Freytes, K. Tobita, T. W. Gilbert and S. F. Badylak, Preparation of cardiac extracellular matrix from an intact porcine heart, *Tissue Eng Part C Methods* **16**: 525–532 (2010).

65. G. C. Engelmayr, Jr., M. Cheng, C. J. Bettinger, J. T. Borenstein, R. Langer and L. E. Freed, Accordion-like honeycombs for tissue engineering of cardiac anisotropy, *Nat Mater* **7**: 1003–1010 (2008).

66. A. D'Amore, T. Yoshizumi, S. K. Luketich, M. T. Wolf, X. Gu, M. Cammarata, R. Hoff, S. F. Badylak and W.R. Wagner, Bi-layered polyurethane: extracellular matrix cardiac patch improves ischemic ventricular wall remodeling in a rat model, *Biomaterials* **107**: 1–14 (2016).

67. K. L. Christman, H. H. Fok, R. E. Sievers, Q. Fang and R. J. Lee, Fibrin glue alone and skeletal myoblasts in a fibrin scaffold preserve cardiac function after myocardial infarction, *Tissue Eng* **10**: 403–409 (2004).

68. K. L. Christman, A. J. Vardanian, Q. Fang, R. E. Sievers, H. H. Fok and R. J. Lee, Injectable fibrin scaffold improves cell transplant survival, reduces infarct expansion, and induces neovasculature formation in ischemic myocardium, *J Am Coll Cardiol* **44**: 654–660 (2004).

69. N. F. Huang, J. Yu, R. Sievers, S. Li and R. J. Lee, Injectable biopolymers enhance angiogenesis after myocardial infarction, *Tissue Eng* **11**: 1860–1866 (2005).

70. M. A. Laflamme, K. Y. Chen, A. V. Naumova, V. Muskheli, J. A. Fugate, S. K. Dupras, H. Reinecke, C. H. Xu, M. Hassanipour, S. Police, C. O'Sullivan, L. Collins, Y. H. Chen, E. Minami, E. A. Gill, S. Ueno, C. Yuan, J. Gold and C. E. Murry, Cardiomyocytes derived from human embryonic stem cells in pro-survival factors enhance function of infarcted rat hearts, *Nat Biotechnol* **25**: 1015–1024 (2007).

71. T. Kofidis, J. L. de Bruin, G. Hoyt, D. R. Lebl, M. Tanaka, T. Yamane, C. P. Chang and R. C. Robbins, Injectable bioartificial myocardial tissue for large-scale intramural cell transfer and functional recovery of injured heart muscle, *J Thorac Cardiovasc Surg* **128**: 571–578 (2004).

72. P. Zhang, H. Zhang, H. Wang, Y. Wei and S. Hu, Artificial matrix helps neonatal cardiomyocytes restore injured myocardium in rats, *Artif Organs* **30**: 86–93 (2006).

73. J. Yu, K. L. Christman, E. Chin, R. E. Sievers, M. Saeed and R. J. Lee, Restoration of left ventricular geometry and improvement of left ventricular function in a rodent model of chronic ischemic cardiomyopathy, *J Thorac Cardiovasc Surg* **137**: 180–187 (2009).

74. R. Mukherjee, J. A. Zavadzkas, S. M. Saunders, J. E. McLean, L. B. Jeffords, C. Beck, R. E. Stroud, A. M. Leone, C. N. Koval, W. T. Rivers, S. Basu, A. Sheehy, G. Michal and F. G. Spinale, Targeted myocardial microinjections of a biocomposite material reduces infarct expansion in pigs, *Ann Thorac Surg* **86**: 1268–1276 (2008).

75. X. Y. Li, T. Wang, X. J. Jiang, T. Lin, D. Q. Wu, X. Z. Zhang, E. Okello, H. X. Xu and M. J. Yuan, Injectable hydrogel helps bone marrow-derived mononuclear cells restore infarcted myocardium, *Cardiology* **115**: 194–199 (2010).

76. W. Dai, L. E. Wold, J. S. Dow and R. A. Kloner, Thickening of the infarcted wall by collagen injection improves left ventricular function in rats: a novel approach to preserve cardiac function after myocardial infarction, *J Am Coll Cardiol* **46**: 714–719 (2005).

77. C. A. Thompson, B. A. Nasseri, J. Makower, S. Houser, M. McGarry, T. Lamson, I. Pomerantseva, J. Y. Chang, H. K. Gold, J. P. Vacanti and S. N. Oesterle, Percutaneous transvenous cellular cardiomyoplasty, A novel nonsurgical approach for myocardial cell transplantation, *J Am Coll Cardiol* **41**: 1964–1971 (2003).

78. T. W. Chung, J. Yang, T. Akaike, K. Y. Cho, J. W. Nah, S. I. Kim and C. S. Cho, Preparation of alginate/galactosylated chitosan scaffold for hepatocyte attachment, *Biomaterials* **23**: 2827–2834 (2002).

79. J. Yu, Y. Gu, K. T. Du, S. Mihardja, R. E. Sievers and R. J. Lee, The effect of injected RGD modified alginate on angiogenesis and left ventricular function in a chronic rat infarct model, *Biomaterials* **30**: 751–756 (2009).

80. M. E. Davis, J. P. Motion, D. A. Narmoneva, T. Takahashi, D. Hakuno, R. D. Kamm, S. Zhang and R. T. Lee, Injectable self-assembling peptide nanofibers create intramyocardial microenvironments for endothelial cells, *Circulation* **111**: 442–450 (2005).

81. G. Dubois, V. F. Segers, V. Bellamy, L. Sabbah, S. Peyrard, P. Bruneval, A. A. Hagege, R. T. Lee and P. Menasche, Self-assembling peptide nanofibers and skeletal myoblast transplantation in infarcted myocardium, *J Biomed Mater Res B Appl Biomater* **87**: 222–228 (2008).

82. S. Dobner, D. Bezuidenhout, P. Govender, P. Zilla and N. Davies, A synthetic non-degradable polyethylene glycol hydrogel retards adverse post-infarct left ventricular remodeling, *Journal of Cardiac Failure* **15**: 629–636 (2009).

83. J. Leor, S. Tuvia, V. Guetta, F. Manczur, D. Castel, U. Willenz, O. Petnehazy, N. Landa, M. S. Feinberg, E. Konen, O. Goitein, O. Tsur-Gang, M. Shaul, L. Klapper and S. Cohen, Intracoronary injection of *in situ* forming alginate hydrogel reverses left ventricular remodeling after myocardial infarction in swine, *J Am Coll Cardiol* **54**: 1014–1023 (2009).

84. S. V. Rao, U. Zeymer, P. S. Douglas, H. Al-Khalidi, J. A. White, J. Liu, H. Levy, V. Guetta, C. M. Gibson, J. F. Tanguay, P. Vermeersch, J. Roncalli, J. D. Kasprzak, T. D. Henry, N. Frey, O. Kracoff, J. H. Traverse, D. P.

Chew, J. Lopez-Sendon, R. Heyrman and M. W. Krucoff, Bioabsorbable intracoronary matrix for prevention of ventricular remodeling after myocardial infarction, *J Am Coll Cardiol* **68**: 715–723 (2016).

85. J. M. Singelyn, J. A. DeQuach, S. B. Seif-Naraghi, R. B. Littlefield, P. J. Schup-Magoffin and K. L. Christman, Naturally derived myocardial matrix as an injectable scaffold for cardiac tissue engineering, *Biomaterials* **30**: 5409–5416 (2009).

86. J. M. Singelyn, P. Sundaramurthy, T. D. Johnson, P. J. Schup-Magoffin, D. P. Hu, D. M. Faulk, J. Wang, K. M. Mayle, K. Bartels, M. Salvatore, A. M. Kinsey, A. N. DeMaria, N. Dib and K. L. Christman, Catheter-deliverable hydrogel derived from decellularized ventricular extracellular matrix increases endogenous cardiomyocytes and preserves cardiac function post-myocardial infarction. *J Am Coll Cardiol* **59**: 751–763 (2012).

87. S. B. Seif-Naraghi, J. M. Singelyn, M. A. Salvatore, K. G. Osborn, J. J. Wang, U. Sampat, O. L. Kwan, G. M. Strachan, J. Wong, P. J. Schup-Magoffin, R. L. Braden, K. Bartels, J. A. DeQuach, M. Preul, A. M. Kinsey, A. N. DeMaria, N. Dib and K. L. Christman, Safety and efficacy of an injectable extracellular matrix hydrogel for treating myocardial infarction, *Sci Transl Med* **5**: 173ra125 (2013).

88. R. J. Laham, N. A. Chronos, M. Pike, M. E. Leimbach, J. E. Udelson, J. D. Pearlman, R. I. Pettigrew, M. J. Whitehouse, C. Yoshizawa and M. Simons, Intracoronary basic fibroblast growth factor (FGF-2) in patients with severe ischemic heart disease: results of a phase I open-label dose escalation study, *J Am Coll Cardiol* **36**: 2132–2139 (2000).

4

ENGINEERED MECHANICAL FACTORS TO MATURE PLURIPOTENT STEM CELL-DERIVED CARDIOMYOCYTES

Alexandre J. S. Ribeiro, Robin E. Wilson[†] and Beth L. Pruitt[‡]*

1. Introduction

The ability to transform human somatic cells into human pluripotent stem cells (hPSCs) presents high potential to differentiate different cell types specific to human tissues,[1] including cardiomyocytes (CMs). CMs are the muscle cells of the heart whose beating activity governs heart function. Several protocols can differentiate hPSCs into populations of beating CMs (hPSC-CMs), which are homologous to primary CMs in the expression of several CM-specific proteins found in sarcomeres, ion channels, and other functional structures.[2] hPSC-CMs for biomedical research have strong potential for modeling heart disease, developing regenerative therapies, and studying fundamental heart biology. Studies using hPSC-CMs derived from patients with genetic cardiac disorders have demonstrated that these cells can even recapitulate many of phenotypic abnormalities of diseases.[3–5] However, all of these applications have been hindered by the immature shape and function of hPSC-CMs.[2] This chapter focuses on strategies to mature the shape and the organization of sarcomeres along myofibrils in hPSC-CMs with extracellular mechanical cues by engineering cell adhesion, substrate topography, substrate stiffness, and stretching. Long, rectangular cell shapes and alignment of myofibrils are key features of mature CMs that can be recapitulated by using these engineering approaches. We discuss details of methodological approaches, the effects of the different engineering approaches on hPSC-CM shape and myofibril organization, and potential structural mechanisms thought to drive myofibril alignment through tension-driven mechanisms. We also provide application examples for these maturation strategies in the engineering of cardiac tissues derived from hPSC-CMs.

2. hPSC-CMs Lack a Mature Shape and Myofibril Organization

The shape of single primary mature CMs is anisotropic and can be roughly approximated as a rectangle,[6] where the sarcomeres that compose myofibrils are registered along the minor axis of the cell and distributed in parallel along the cell major axis (Fig. 1a). Shape does not vary or remodel in mature CMs. Instead, CM shape is constant after neonatal heart development and relates to beating function.[7] During development, several morphological changes occur in CMs at different length scales that are known to improve their cardiomyogenic function.[7,8] Specifically, CMs elongate within the myocardium, lateral membrane invaginations (transverse or t-tubules) form, and dense intercalated discs develop longitudinally between cells. Therefore, the level of anisotropy observed in mature CMs also occurs at the tissue level in the myocardium.[8] Without engineering approaches, these morphological properties are absent in hPSC-CMs.[2] In comparison to mature CMs, hPSC-CMs exhibit morphological immaturity and disarrayed organization of myofibrils characterized by (1) more than ten times smaller size; (2) fewer mitochondria and irregularly distributed mitochondria; (3) different electrophysiological parameters and calcium cycling; (4) less total force generation; (5) different gap junction distributions; and (6) different expression of myofibril protein isoforms, t-tubule proteins, β-receptors and cardiac-specific calcium channels.[2] This absence of morphological maturity and associated differences in function and gene expression relative to mature CMs are also observed in immature fetal CMs and neonatal CMs (Fig. 1b). As detailed in the next section, delivering tunable extracellular cues to hPSC-CMs and other types of immature CMs can enhance their cardiomyogenic maturity. Conditioning the mechanics of the extracellular environment relies on several engineering approaches that aim to enhance anisotropy of cell shape or alignment of myofibrils (Fig. 1c).

3. Engineering Cardiomyocyte Shape to Align Myofibrils and Mature Their Function

To be useful models of cardiac activity, hPSC-CMs must have the same morphological and functional characteristics of primary mature adult CMs. Currently, hPSC-CMs substantially differ from adult CMs and instead display characteristics more similar to neonatal CMs.[2] In terms of mechanical function, hPSC-CMs generate less contractile force as both single cells[9] and engineered tissues.[10] hPSC-CMs also lack mature electrophysiology[11] and calcium cycling,[12] which are critical for stimulating contraction. The expression of β-adrenergic receptors, G protein-coupled receptors that regulate cardiac contractile function,[13] is low in hPSC-CMs,[14] which also affects the maturity of these cells. However, an increase in the expression of β-adrenergic receptors and improved activation of their signaling pathways occurs in hPSC-CMs at late stages of differentiation (day 90).[15] In addition, higher expression and signaling of β-adrenergic receptors are coincident with other maturity markers.[15] Forcing a rectangular shape on hPSC-CMs with microfabricated substrates further increases β-adrenergic receptor expression and signaling, and further improves the maturity of functional phenotypes associated to β-adrenergic receptor pathways.[15] These cells do not naturally assume the same morphology as adult CMs,[16] unless submitted to engineered mechanical cues. Mature CMs have a rectangular shape with a longitudinal area around 2000 μm^2 and an aspect ratio of length over width in the

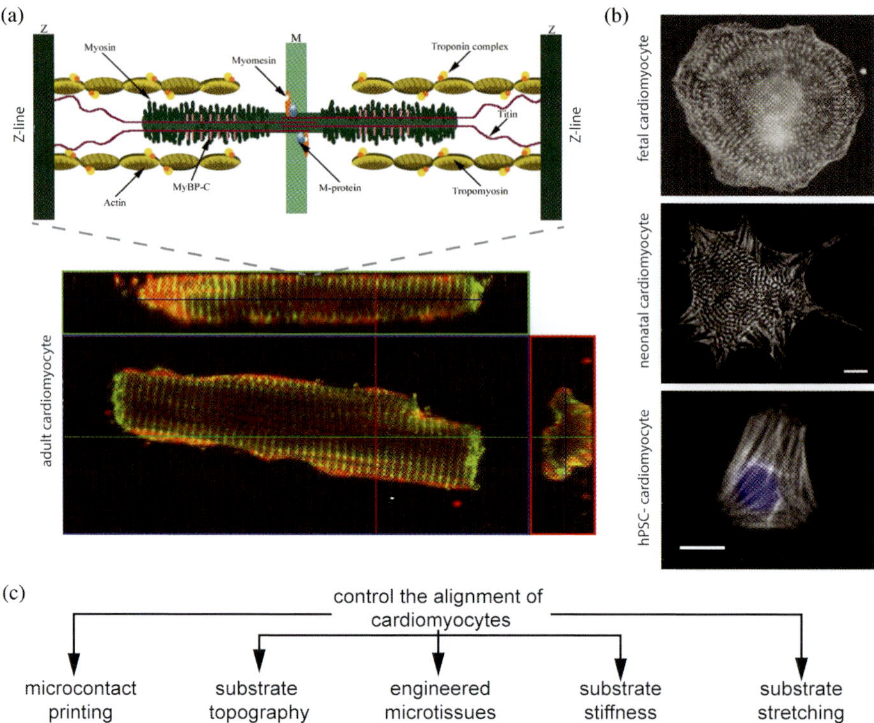

Fig. 1. Mechanical cues enhance the maturity of hPSC-CM shape and myofibril organization. (a) Top: molecular model of the sarcomere. Sarcomere is the basic contractile unit of CMs and consists of well-organized protein-based structures between Z-lines that support the contractility that primarily results from the motor activity of myosin thick filaments interacting with actin thin filaments. Bottom: single adult mature CM with labeled Z-lines (green), which are parallel along the cell main axis. Figures reproduced with permission.[126,127] (b) Sarcomeres (grey) labeled in single fetal-, neonatal- and hPSC-CMs. Blue is nucleus. Sarcomeres along myofibrils are organized in disarray in the three different types of CMs. Figures reproduced with permission.[23,120,123] (c) Different engineering approaches to elongate the shape of sarcomeres and align the orientation of myofibrils along the main axis of cell elongation. Images reproduced with permission.

range of 5:1 to 7:1.[2,17,18] In addition to the rectangular shape of CMs, alignment of myofibrils along the cell major axis is a hallmark of maturation of differentiated CMs (Fig. 1).[2,19] Myofibrils are comprised of a series of sarcomeres and thus alignment is necessary for proper CM contractility.

Sarcomeres are the contractile units of muscle cells containing myosin thick filaments that interact with actin thin filaments to enable muscle contractions.[20] Sarcomeres contract uniaxially along each myofibril and are delimited by Z-lines (Fig. 1a). In addition to Z-lines, electron microscopy and tissue staining also reveal the existence of three different regions within sarcomeres: I-band, A-band, and M-line regions.[20] Isolated immature CMs, such as fetal or neonatal CMs, tend to present a random intracellular distribution of myofibrils; such disarray is also observed in hPSC-CMs (Fig. 1b). Culturing hPSC-CMs for months in specialized extracellular matrices can induce myofibril alignment in some cardiomyocytes.[21] However, strategies to engineer the extracellular environment (tuning substrate stiffness and forcing hPSC-CMs to assume rectangular shapes) can accelerate intracellular alignment of myofibrils because cell shape and myofibril alignment are interrelated. Several strategies have been developed to align myofibrils in hPSC-CMs (Fig. 1c). Alignment of myofibrils leads to increased maturation of hPSC-CMs,[22–24] as assayed by electrophysiology, calcium signaling, contractility, and gene expression. Engineering approaches to control the shape of hPSC-CMs and thus alignment of their myofibrils (Fig. 1c) are detailed in the following sections.

3.1 *Substrate micropatterning*

The biological roles of cell geometry regulated with substrate micropatterning have been studied for more than 15 years in different cell types.[22,25–29] Micropatterning techniques can be used to engineer single hPSC-CMs,[22,23,30] as well as microtissue structures.[31]

In general, substrate micropatterning is used to control the geometry of extracellular matrix proteins, which limits cellular adhesion and thus controls cell geometry. The shape of hPSC-CMs will match the shape of protein micropatterns, and such micropatterned cells present higher maturity when seeded on patterns with a rectangular shape.[22,23,30] Two primary approaches have been developed for patterning proteins on surfaces: (1) microcontact printing and (2) lift-off techniques.

In microcontact printing, protein patterns are transferred to surfaces via contact of microstamps "inked" with extracellular matrix protein(s).

Microstamps are usually made of polydimethylsiloxane (PDMS) molded from master wafers fabricated using soft lithography techniques.[32] Selecting the type of mask depends on the length scale of the desired features.[33] These features are designed with CAD software and printed onto a mask, which is then used to selectively expose a photoresist to UV light. For mammalian single cells the desired features are often 10–200 μm.[32] Transparency masks printed on high-resolution printers can readily reproduce these features at 50,000 dpi with 5-μm resolution. However, resolution of features can be further reduced to submicron length scales with chrome-on-quartz masks.[34] A master mold for microstamps is obtained after spin-coating, exposing, curing, and developing a photoresist.[32] This master is then used to mold the PDMS microstamps. After peeling microstamps from the master mold, solutions of extracellular matrix proteins are incubated on the surface of PDMS microstamps. Surfaces with high surface energy, charged surfaces, and surfaces with chemical functions that bind proteins have been micropatterned with microcontact printing.[32]

In lift-off techniques, protein micropatterns are directly created on surfaces for cell culture with lithography approaches.[35] A sacrificial photoresist material is initially spin coated onto the substrate. This sacrificial surface layer is then exposed to light through a mask, and uncured regions of the surface are developed and washed away. A simple approach to protein patterning with lift-off then uses incubation of the exposed surface area in a passivation agent.[36] After passivation, the sacrificial photoresist is removed to expose the unpassivated patterned area. Physisorbtion of extracellular matrix proteins then fills these newly exposed patterned areas.

Cells will preferentially attach to patterned regions when seeded on surfaces because of their affinity to extracellular matrix proteins on patterns. Adhesion of hPSC-CMs is optimal when Matrigel is patterned,[23] but fibronectin[30] and laminin[37] can also be used as extracellular matrices for culturing hPSC-CMs. Many cells can also secrete extracellular matrix proteins, and cell culture media can contain components that induce cell adhesion to the unpatterned regions of surfaces. Surface passivation approaches are used to avoid unspecific cell adhesion to unpatterned regions by decreasing non-specific interactions with proteins that bind cells.[38–40] Bovine serum albumin, Pluronic-F127, and poly(ethylene glycol) (PEG) and poly(ethylene oxide) PEO derivatives are some of the most used macromolecules to passivate surfaces from non-specific protein interactions.

Polyacrylamide is often used as a substrate for patterning cells because substrate stiffness can be tuned, and cell generated forces can be measured with this material.[29,41] In association with cell shape, substrate stiffness affects the function and maturity of hPSC-CMs through a tension-driven mechanism.[41] To fabricate a gel, polyacrylamide is polymerized under protein-patterned glass surfaces; this process promotes transfer of protein patterns onto the surface of polyacrylamide gels. Patterned glass surfaces with microcontact printing[29] or lift-off techniques[42] can be used for micropatterning of polyacrylamide. Since the steric hindrance of polyacrylamide surfaces prevents protein adhesion, these surfaces do not require any passivation agent after protein micropatterning. PDMS surfaces can also be micropatterned with proteins to control cell shape.[43] Several methods allow tunability of PDMS stiffness at the kPa range,[43,44] allowing it to be an alternative to polyacrylamide. However, as opposed to polyacrylamide, passivation agents are required to ensure specificity of cell adhesion to patterns on PDMS surfaces.[45]

These techniques rely on controlling cell adhesion to the substrate by having regions for cell adhesion functionalized with extracellular proteins and the remaining regions passivated to avoid cell adhesion. However, other approaches to induce myofibril alignment can also be developed where substrate surfaces are homogeneously distributed with extracellular matrix. Such approaches consist of tuning the topography of the substrate or stretching substrates with specialized devices.

3.2 *Substrate anisotropic topography and substrate stretching*

CMs within the myocardium are oriented relative to each other along their major axes due to cell and tissue polarity that occurs during development.[46,47] Defects in CM polarity and orientation of CMs relate to cardiac defects.[48] In fact, the most successful strategies for the development of engineered tissues for heart regeneration involve inducing the orientation and polarization of CMs in culture along a main axis of orientation.[49]

As reviewed by Tamiello and colleagues,[50] orientation and elongation of adherent mammalian cells *in vitro* can generally be induced with substrate topography of the substrate or by stretching of a deformable substrate. Substrate topography consists of the three-dimensional shape of the substrate surface in contact with cells. Substrate topography is anisotropic when its surface and solid features on the surface align along a specific orientation, normally termed the degree of alignment. Stretching of a two-dimensional substrate can also have different orientations. Stretching of 2D substrates can be biaxial, e.g. a substrate region is stretched equally in the plane with concomitant thinning out of plane, or it can be anisotropic, e.g. uniaxial stretching in the plane along a principal direction.

Different cell types respond differently to anisotropic topography or uniaxial stretch by aligning along a major axis at different angles relative to the angle of topographic anisotropy or of substrate stretching.[50,51] However, the proposed mechanisms that trigger remodeling of cell shape and alignment under these cues involve induced formation of cell adhesions at the edges of oriented cells and the intracellular orientation of actin stress fibers. Proteins involved in mechanotransduction pathways affect the ability of cells to polarize along an axis of orientation, such as focal adhesion proteins,[52,53] proteins of the cytoskeleton,[54,55] and proteins that connect the cytoskeleton to the nucleoskeleton.[55–57] As adherent cells,[50] hPSC-CMs orient along a main axis due to topographic anisotropies[15] and substrate stretching.[58] In addition, cells can respond to substrate topographic anisotropies from the nanometer to the micron length scales that can be engineered with techniques reviewed by Nikkhah and colleagues.[59] These strategies to engineer substrate anisotropies on substrates for cell culture and approaches to stretch hPSC-CMs in culture are summarized below.

Microfabrication is the most reliable and common approach to systematically produce cell culture substrates with ordered topographies at different length scales. The main challenge of microfabrication techniques is to produce features at submicron and nano scales. Microfabrication relies on photolithography to produce stiff features on surfaces composed of silicon or silicon oxide. To microfabricate features using photolithography, a photoresist polymer is first coated onto a surface. The coated surface is then exposed to light filtered through a mask, which induces selective exposure of the surface to light and selective crosslinking or degradation of the photoresist. Uncrosslinked photoresist is developed away from selected areas controlled by the mask design. Therefore, the wavelength of light limits the sizes of these features and the production of submicron features with this technique is not possible due to diffraction of light. X-rays and electron beams produce shorter wavelengths to overcome this limitation. However, electron beam lithography is time consuming and more expensive than simpler approaches to generate features at the nanoscale that induce CM orientation along a main axis: electrospining,[60] nanowrinkles,[61] and nanotubes.[62] Hard features are also often used as molds to generate topographical features on softer materials for cell culture such as PDMS.[15]

For stretching adherent mammalian cells, the following parameters should be considered in stretching experiments: frequency of stretching, magnitude of stretching, direction of stretching, and stretching times. Static stretch consists of applying stretch and maintaining the level of stretch. Stretch can also be applied periodically and dynamically as cyclic stretch. Several devices have been built to stretch cells in 2D or 3D culture systems under cyclic or static stretch with uniaxial or isotropic stretch profiles.[63–65] Stretchable silicone is the most common material for stretching cells cultured on 2D surfaces. Hydrogel-based substrates for cell culture can be bound to silicone surfaces to undergo stretch.[66] Stretchable 3D culture systems for CM culture typically consist of engineered heart tissues (EHTs) immobilized between two poles that can be micromanipulated to apply stretch. Such platforms may also sense the beating forces collectively exerted by CMs within EHTs.[67] Uniaxial stretch induces alignment and maturation of CMs.[68] In addition, static or cyclic stretching can mature CMs in culture.[69,70] However, cyclic stretch has also been shown to induce hyperplasia and higher passive stress in cultures of CMs isolated from chicken embryos.[71] In this study, the optimal magnitude of stretching varied between CMs isolated at different days of development, which suggests that different stretch parameters must be used with CMs at different stages of development and maturation. The magnitude of stretch is defined by the strain value and is calculated as the length of deformation of a region along the axis of stretching divided by the length of the stretched region in the absence of applied loads. Strain is usually presented as a percentage. A local +5% strain means that stretching caused deformations in a region resulting in dimensions that are 5% longer than those in an unloaded state. Cyclically stretching mouse embryonic stem cell derived CMs at 10% strain increased the expression of cardiac-specific genes and improved the synchronicity of beating between cells in culture.[72] In this study, stretching was applied at a frequency of 1 Hz. The frequency of stretching in cyclic stretch is of high relevance because it may translate to the frequency of beating *in vivo*. Another study has shown different effects in mouse embryonic stem cell-derived CMs of 10% stretch at different frequencies (1, 2, and 3Hz).[73] The electrophysiology, contractility, and expression of sarcomere proteins differ between human and murine CMs, as well as their beating activity in heart development and maturation.[74–79] However, the same ranges of stretching parameters that mature mouse stem cell-derived CMs also induce maturation of hPSC-CMs in association with the alignment of myofibrils in cells on 2D culture systems[72] and in EHTs.[80] Unfortunately, stretching protocols vary considerably among studies and are hard to compare. Labs often use custom stretching devices, differing surface chemistries, and stretching regimes that vary between studies. These differences may underlie the high variability in reported cellular responses to stretch. Thus, optimization of stretching parameters for specific cells and chemistries is required before using stretch for maturing the myofibril organization of hPSC-CMs and enhancing CM maturity.

3.3 *Substrate stiffness*

Substrate stiffness is defined by its rigidity or by how deformable it is under an externally applied load. As with other adherent cells, hPSC-CMs exert loads on substrates upon attachment and substrates deform as a function of the applied forces and of the deformability of the substrate.[41] The extent of deformation of the substrate in cell biology has been shown to affect the biology of adherent cells at different levels.[81,82] For example, substrate stiffness can drive the commitment of stem cell differentiation toward a specific lineage,[83,84] but the mechanotransduction mechanisms that drive changes in cell biology due to substrate stiffness still require elucidation. It is known that cytoskeletal proteins that physically interconnect the nucleus to cell adhesions and specific signaling pathways mediate the effects of extracellular mechanics in cell biology.[85,86] We now focus on studies that use substrate stiffness to mature the shape, myofibril alignment, and function of CMs.

Substrate stiffness has been shown to drive the differentiation of CMs from murine iPSCs.[87] Differentiation was accelerated in cells cultured on substrates with a stiffness ranging from 0.6 kPa to 50 kPa relative to cells on rigid tissue culture plastic. In this study, soft materials of tunable stiffness were made of polyacrylamide hydrogels. On these soft substrates, differentiated CMs expressed higher levels of sarcomeric proteins and cardiac-specific L-type calcium channels. Another study with differentiating hPSC-CMs on polyacrylamide showed that optimal differentiation occurs around substrate stiffness of 50 kPa and not at lower stiffnesses.[88] It is generally known that specific cell types perform better their physiological roles when their stiffness matches the stiffness of their surrounding substrate;[81,89] in other words, when cell stiffness matches tissue stiffness *in vivo*. The organization of subcellular structures, such as myofibrils and the remaining cytoskeleton, set the stiffness of cells,[90] suggesting that substrate stiffness regulates these structures. These are the same structures that also set the contractile beating of CMs.[91] The stiffness of myocardial tissues varies during heart development.[70,92] Therefore, variations in tissue stiffness during heart development may regulate both the cell stiffness and contractility of CMs. Tissue elasticity of the embryonic heart increases 1–2 kPa per day. In neonatal rat hearts, stiffness ranges from 1 to10 kPa and stiffness ranges from 10 to 70 kPa in the adult rat.[70,92] These changes in tissue stiffness during development and maturation may also play a role in the biological changes that occur in CMs during the different stages of heart development. Majkut and colleagues [91,92] have suggested mechanical mechanisms for the effects of tissue stiffness in heart development, where cell shape and aligned organization of sarcomeres within myofibrils are tuned by the elasticity of the extracellular environment to drive the maturation of the heart. In addition, the alignment of sarcomeres agreed with mathematical models of force-induced registration between sarcomeres in neighboring myofibrils.[93] The authors also showed that maturity in electrophysiology and excitation contraction coupling were dependent on extracellular elasticity. Several studies experimentally and theoretically showed that alignment of myofibrils and registry of sarcomeres in CMs match their beating activity and vary with substrate stiffness.[94] CM shape is then driven by the organization of myofibrils and sarcomere registry in a tension-dependent mechanism regulated by extracellular mechanical cues. Recent work with live single hPSC-CMs with labeled myofibrils supports the idea that substrate stiffness regulates their contractility and maturity through a tension-driven mechanism that sets myofibril organization in relation to cell shape.[23]

Polyacrylamide substrates are most frequently used for testing the roles of extracellular stiffness in the biology of adherent cells because this material is easily available in laboratories that study cell biology, and its stiffness is easily tunable to match the physiological stiffness of mammalian tissues.[95–97] However, other naturally occurring materials, synthetic hydrogels, and PDMS-based devices can be developed as cell substrates with tunable stiffness.[41,96,98,99] Functionalization of the surface of these substrates with bioactive proteins that bind cell surface receptors is necessary to ensure stable cell attachment.[41]

3.4 *Engineered heart tissues*

EHTs composed of hPSC-CMs serve the experimental purpose of measuring properties that are not easily measurable in tissue culture dishes. For instance, forces from collective cell contractions can be measured in EHTs.[100] In addition, EHTs enable a number of modes of experimental stimulation and manipulation including electrical stimulation, applied stretching, variations of extracellular matrix composition, and the presence of other cell types in hPSC-CM biology and microtissue function.[69,101,102] EHTs mainly consist of tissue constructs immobilized between two posts. CMs within EHTs have a more mature long shape and have myofibrils aligned along the main tissue axis. This organization of cell orientation and myofibril alignment enables uniaxial tissue contractions that better mimic the beating of myocardial tissue. Immobilization of single CMs and other multicellular CM-based systems between posts also drive a more physiological contractile function, cellular morphology, and myofibril alignment within CMs.[103–105] Microtissues develop under tension between posts, which may also drive the formation of aligned myofibrils within CMs.

Other approaches based on microfabricating on-a-chip devices for cell culture can be successfully implemented to engineer microtissues containing hPSC-CMs and drive their structural and functional maturation.[79,106,107] These approaches develop surfaces with different methods and designs that confine cells, forcing them to remodel their shape to collectively orient. In the same way that single cell confinement to wells, ridges, or patterns force cells to align,[50] these systems demonstrate that adding multiple cells to confined surfaces also forces them to self-organize and rearrange their distribution and shape to establish an equilibrium of forces between cells. The contractile activity and the intracellular organization of myofibrils seem to play strong roles in the organization of CMs within these microtissues and in forcing them to assume an elongated shape. However, cues from the extracellular matrix and of neighboring cells also contribute to the maturation of hPSC-CMs within microtissues.[69] In addition, electrical chronic pacing of microtissues can also significantly mature hPSC-CMs in terms of mechanical output, electrophysiology, calcium signaling, cell shape, and alignment of myofibrils.[102] The greatest advantage of microtissues relative to other platforms for testing cell function is the ability to miniaturize these systems for drug testing or screening cardiac function under different cues.[108] However, intercellular variability within differentiated populations of hPSC-CMs may lead to increased variability of functional results in microtissues. A future application of engineered microtissues is to use them as bioreactors to produce single mature cells. Dissociation of functional microtissues into single cells has already been shown to provide populations of matured hPSC-CMs in terms of myofibril organization, cell shape, and *t*-tubules. After separation from non-mature cells and isolation, these cells can be powerful platforms to study heart development, physiology, and disease in a dish.

Altogether, we show that protein micropatterning, stretching, substrate stiffness, and microtissue culture approaches (Fig. 1c) affect cell shape and induce alignment of myofibrils in association with the enhancement of cardiomyogenic maturity of hPSC-CMs. The mechanisms that link cell shape to other cardiomyogenic functional phenotypes are still poorly understood, mainly for hPSC-CMs. A tension-driven mechanism in hPSC-CMs mediated by extracellular mechanical cues and intracellular contractile forces may drive myofibril alignment, improved electrophysiology, mitochondria organization, and calcium signaling.[23] These results agree with other studies on the effects of mechanics in CM shape and sarcomere organization, where the machinery supporting sarcomere contractility seems to respond to these cues.[91,92,94] A mature cell shape and alignment of myofibrils also increase the expression of caveolin 3,[15] a precursor of t-tubule formation. This protein has a cell membrane activity known to be affected by mechanical cues.[109] In addition to caveolin 3, the expressions of beta-adrenergic receptors also increased in cells with more mature shape and myofibril alignment.[15] These receptors are markers of mature CM cardiomyogenic function.[14] Overall, cell shape and myofibril organization seem to be key upstream properties of maturation of hPSC-CMs and they are intimately related to each other. In the next section, we elaborate on potential mechanisms that link cell shape to organization of myofibrils.

4. Mechanical Mechanisms that Mature Alignment of Myofibrils and Cell Shape hPSC-CMs

The formation of actomyosin stress fibers shares similar fundamental mechanisms even when performing a variety of mechanical and structural roles in different cell types.[110] Both myosin and actin in their monomeric state form filaments that interact with each other and form contractile structures. These structures develop into stress fibers upon interaction with other stress fiber-associated proteins, such as α-actinin, Arp2/3, tropomyosin, mDia2, focal adhesion proteins, and others, depending on cell type. Actin polymerizes linearly into filaments that arrange as bipolar arrays in stress fibers. Myosin also assembles into bipolar bundles. The intercalation of bipolar actin with bipolar myosin bundles tends to also occur along bipolar arrays of variable persistence length, depending on the participation of other proteins in the formation of actomyosin structures. The force triggering the motor activity of stress fibers is produced by ATP-driven interactions of the motor domain of myosin along actin filaments.

The motor activity of stress fibers requires a bipolar organization of acto-myosin stress fibers. Therefore, the magnitude of contractile forces increases with the persistence length of stress fibers.[111] Cell elongated shapes increase the length of stress fibers, which explains why cells with rectangular shapes have the potential to generate stronger contractile forces.[112] The contractility and organization of non-muscle actomyosin fibers have structural and functional similarities with their myofibril iso-forms.[110,113,114] Such similarities justify the common alignment of actomyosin stress fibers along the main cell axis of many different adherent cell types with rectangular shapes[28,115,116] because of their bipolar formation and periodic intercalation of actin thin filaments, myosin thick filaments, and clusters of other actomyosin-associated proteins.[117–119]

Muscle-specific actin, myosin, and associated proteins in CMs assemble into well-organized bipolar structures that originate sarcomeres distributed in series along myofibrils. This mechanism is termed myofibrillogenesis and is initiated in CMs during embryonic development after an increased expression of muscle-specific cytoskeleton proteins.[120–122] Later in development, as myofibrils mature in CMs, sarcomeres in neighboring myofibrils register with each other by establishing physical interactions of Z-lines.[94]

Myofibrillogenesis and maturation of myofibrils *in vitro* first occur near the cytoplasmic membrane,[120–122] suggesting that interactions between cell membrane structures and pre-myofibrils trigger myofibrillogenesis. Myofibrils replace non-muscle cytoskeletal actomyosin structures as they form toward the inner cytoplasmic regions. Therefore, the cytoplasmic membrane may be a key player in the maturation of myofibrils and sarcomeres in hPSC-CMs with rectangular shapes *in vitro*. This would also contribute to the enhancement of functional maturity derived from structural maturation.[15,23] Pre-myofibrils align along the cell's major axis in CMs with an elongated shape, and myofibrillogenesis in these cells leads to well-aligned and mature myofibrils. However, if engineered cell shape can induce myofibril maturation, alignment of myofibrils can also drive an elongated cellular shape of CMs. The intracellular tension along myofibrils set by extracellular and intracellular mechanics and the contractile activity of sarcomeres is known to align and mature myofibrils.[23,91,92] This tension-induced myofibril alignment has been shown to force CMs to

    A. J. S. Ribeiro, R. E. Wilson and B. L. Pruitt

assume a mature elongated shape. Setting both extracellular stiffness and cell shape to physiological levels can in fact enhance the maturity of hPSC-CMs.[23] Therefore the mechanical mechanisms that induce structural maturation in hPSC-CMs cultured on rectangular patterns, substrates of anisotropic topography, engineered microtissues, substrates of physiological stiffness, and under stretching (Fig. 1c) rely on a combination of factors that force myofibrils to align or the cell to assume an elongated shape. These factors can be extracellular mechanical cues, the cellular contractile activity, and restriction of cell shape with engineered surfaces.

5. Conclusions

This chapter describes how mechanical cues can be delivered to hPSC-CMs with engineered devices to induce elongated cell shapes and the alignment of myofibrils along the cell main axis. Devices used for this purpose and device fabrication were described, taking into consideration experimental requirements for culture of hPSC-CM and how mechanical stimulation *in vitro* may match what is known to physiologically occur *in vivo*. A more mature rectangular cell shape with associated myofibril alignment can be achieved with tunable mechanical cues. These cues include micropatterning of extracellular matrix proteins to spatially restrict cell adhesion to specific locations of the substrate surface, microfabrication of anisotropic topographies on the substrate, uniaxially stretching cells, tuning the stiffness of the extracellular environment, and engineering more complex microtissues. A more mature cardiomyogenic function seems to follow from a more mature shape and myofibril alignment. Specifically, magnitudes of contractile force, electrophysiology, and calcium signaling are some of the phenotypes that have been identified to be more mature in hPSC-CMs with matured shape and myofibril alignment. Intracellular tension and activity of the cell contractile machinery play strong roles in the functional maturity of these matured cells. The mechanisms that regulate myofibril assembly and cell shape under mechanical stimulation seem to be identical to what has been generally observed in the formation of actin stress fibers in adherent cells. Bundles of actin fibers interact with myosin bundles and form aligned structures that later originate myofibrils. However, the cell membrane seems to play a role in the formation of the first myofibrils, suggesting that cell shape and myofibril alignment are intimately related in the development of cardiomyocytes. These platforms for engineering hPSC-CM shape and structure have high potential to be used for producing more mature cell populations that are better models to study heart disease or heart function in a dish, as well as to produce cells that can better perform in regenerative approaches.

Acknowledgements

The authors would like to thank Ngan Huang, and acknowledge support from the American Heart Association — postdoctoral fellowship 14POST18360018, National Science Foundation — Graduate Research Fellowship Program, Stanford Graduate Fellowship in Science and Engineering, Child Health Research Institute — Stanford University School of Medicine, ChEM-H Stanford University — Postdocs at the Interface Seed Grant, National Science Foundation — Grant MIKS-1136790.

References

1. S. Yamanaka, Strategies and new developments in the generation of patient-specific pluripotent stem cells, *Cell Stem Cell* **1**: 39–49 (2007).
2. X. Yang, L. Pabon and C. E. Murry, Engineering adolescence: maturation of human pluripotent stem cell-derived cardiomyocytes, *Circ Res* **114**: 511–523 (2014).
3. I. Itzhaki, L. Maizels, I. Huber, L. Zwi-Dantsis, O. Caspi, A. Winterstern, O. Feldman, A. Gepstein, G. Arbel, H. Hammerman, M. Boulos and L. Gepstein, Modelling the long QT syndrome with induced pluripotent stem cells, *Nature* **471**: 225–229 (2011).
4. N. Sun, M. Yazawa, J. Liu, L. Han, V. Sanchez-Freire, O. J. Abilez, E. G. Navarrete, S. Hu, L. Wang, A. Lee, A. Pavlovic, S. Lin, R. Chen, R. J. Hajjar, M. P. Snyder, R. E. Dolmetsch, M. J. Butte, E. A. Ashley, M. T. Longaker, R. C. Robbins and J. C. Wu, Patient-specific induced pluripotent stem cells as a model for familial dilated cardiomyopathy, *Sci Transl Med* **4**: 130ra47 (2012).
5. F. Lan, A. S. Lee, P. Liang, V. Sanchez-Freire, P. K. Nguyen, L. Wang, L. Han, M. Yen, Y. Wang, N. Sun, O. J. Abilez, S. Hu, A. D. Ebert, E. G. Navarrete, C. S. Simmons, M. Wheeler, B. Pruitt, R. Lewis, Y. Yamaguchi, E. A. Ashley, D. M. Bers, R. C. Robbins, M. T. Longaker and J. C. Wu,

Abnormal calcium handling properties underlie familial hypertrophic car-diomyopathy pathology in patient-specific induced pluripotent stem cells, *Cell Stem Cell* **12**: 101–113 (2013).

6. N. J. Severs, A. F. Bruce, E. Dupont and S. Rothery, Remodelling of gap junctions and connexin expression in diseased myocardium, *Cardiovasc Res* **80**: 9–19 (2008).

7. C. F. Kline and P. J. Mohler, Evolving form to fit function: cardiomyocyte intercalated disc and transverse-tubule membranes, *Curr Top Membr* **72**: 121–158 (2013).

8. M. L. McCain and K. K. Parker, Mechanotransduction: the role of mechanical stress, myocyte shape, and cytoskeletal architecture on cardiac function, *Pflugers Arch* **462**: 89–104 (2011).

9. H. Kita-Matsuo, M. Barcova, N. Prigozhina, N. Salomonis, K. Wei, J. G. Jacot, B. Nelson, S. Spiering, R. Haverslag, C. Kim, M. Talantova, R. Bajpai, D. Calzolari, A. Terskikh, A. D. McCulloch, J. H. Price, B. R. Conklin, H. S. V. Chen and M. Mercola, Lentiviral vectors and protocols for creation of stable hesc lines for fluorescent tracking and drug resistance selection of cardiomyocytes, *PLoS One* **4**: (2009).

10. J. Xi, M. Khalil, N. Shishechian, T. Hannes, K. Pfannkuche, H. Liang, A. Fatima, M. Haustein, F. Suhr, W. Bloch, M. Reppel, T. Saric, M. Wernig, R. Janisch, K. Brockmeier, J. Hescheler and F. Pillekamp, Comparison of contractile behavior of native murine ventricular tissue and cardiomyocytes derived from embryonic or induced pluripotent stem cells, *FASEB J* **24**: 2739–2751 (2010).

11. L. Sartiani, E. Bettiol, F. Stillitano, A. Mugelli, E. Cerbai and M. E. Jaconi, Developmental changes in cardiomyocytes differentiated from human embryonic stem cells: a molecular and electrophysiological approach, *Stem Cells* **25**: 1136–1144 (2007).

12. J. Satin, I. Itzhaki, S. Rapoport, E. A. Schroder, L. Izu, G. Arbel, R. Beyar, C. W. Balke, J. Schiller and L. Gepstein, Calcium handling in human embryonic stem cell-derived cardiomyocytes, *Stem Cells* **26**: 1961–1972 (2008).

13. A. Madamanchi, β-Adrenergic receptor signaling in cardiac function and heart failure, *Mcgill J Med* **10**: 99–104 (2007).

14. C. Robertson, D. D. Tran and S. C. George, Concise review: maturation phases of human pluripotent stem cell-derived cardiomyocytes, *Stem Cells* **31**: 829–837 (2013).

15. G. Jung, G. Fajardo, A. J. Ribeiro, K. B. Kooiker, M. Coronado, M. Zhao, D. Q. Hu, S. Reddy, K. Kodo, K. Sriram, P. A. Insel, J. C. Wu, B. L. Pruitt and D.

Bernstein, Time-dependent evolution of functional *vs.* remodeling signaling in induced pluripotent stem cell-derived cardiomyocytes and induced maturation with biomechanical stimulation, *FASEB J* (2015) [Epub ahead of print].

16. M. Gherghiceanu, L. Barad, A. Novak, I. Reiter, J. Itskovitz-Eldor, O. Binah and L. M. Popescu, Cardiomyocytes derived from human embryonic and induced pluripotent stem cells: comparative ultrastructure, *J Cell Mol Med* **15**: 2539–2551 (2011).

17. S. D. Bird, P. A. Doevendans, M. A. van Rooijen, A. Brutel de la Riviere, R. J. Hassink, R. Passier and C. L. Mummery, The human adult cardiomyocyte phenotype, *Cardiovasc Res* **58**: 423–434 (2003).

18. L. H. Opie (ed.), *Heart Physiology: From Cell to Circulation.* (Lippincott Williams & Wilkins, Philadelphia, PA, London, 2004).

19. F. B. Bedada, M. Wheelwright and J. M. Metzger, Maturation status of sarcomere structure and function in human iPSC-derived cardiac myocytes, *Biochim Biophys Acta* **1863**: 1829–1838 (2016).

20. K. A. Clark, A. S. McElhinny, M. C. Beckerle and C. C. Gregorio, Striated muscle cytoarchitecture: an intricate web of form and function, *Annu Rev Cell Dev Biol* **18**: 637–706 (2002).

21. S. D. Lundy, W. Z. Zhu, M. Regnier and M. A. Laflamme, Structural and functional maturation of cardiomyocytes derived from human pluripotent stem cells, *Stem Cells Dev* **22**: 1991–2002 (2013).

22. M. C. Ribeiro, L. G. Tertoolen, J. A. Guadix, M. Bellin, G. Kosmidis, C. D'Aniello, J. Monshouwer-Kloots, M. J. Goumans, Y. L. Wang, A. W. Feinberg, C. L. Mummery and R. Passier, Functional maturation of human pluripotent stem cell derived cardiomyocytes *in vitro*: correlation between contraction force and electrophysiology, *Biomaterials* **51**: 138–150 (2015).

23. A. J. Ribeiro, Y. S. Ang, J. D. Fu, R. N. Rivas, T. M. Mohamed, G. C. Higgs, D. Srivastava and B. L. Pruitt, Contractility of single cardiomyocytes differentiated from pluripotent stem cells depends on physiological shape and substrate stiffness, *Proc Natl Acad Sci USA* **112**: 12705–12710 (2015).

24. J. D. Kijlstra, D. Hu, N. Mittal, E. Kausel, P. van der Meer, A. Garakani and I. J. Domian, Integrated analysis of contractile kinetics, force generation, and electrical activity in single human stem cell-derived cardiomyocytes, *Stem Cell Reports* **5**: 1226–1238 (2015).

25. C. S. Chen, M. Mrksich, S. Huang, G. M. Whitesides and D. E. Ingber, Geometric control of cell life and death, *Science* **276**: 1425–1428 (1997).

26. R. McBeath, D. M. Pirone, C. M. Nelson, K. Bhadriraju and C. S. Chen, Cell shape, cytoskeletal tension, and RhoA regulate stem cell lineage commitment, *Dev Cell* **6**: 483–495 (2004).

27. K. K. Parker, J. Tan, C. S. Chen and L. Tung, Myofibrillar architecture in engineered cardiac myocytes, *Circ Res* **103**: 340–342 (2008).

28. P. Roca-Cusachs, J. Alcaraz, R. Sunyer, J. Samitier, R. Farre and D. Navajas, Micropatterning of single endothelial cell shape reveals a tight coupling between nuclear volume in G1 and proliferation, *Biophys J* **94**: 4984–4995 (2008).

29. A. D. Rape, W. H. Guo and Y. L. Wang, The regulation of traction force in relation to cell shape and focal adhesions, *Biomaterials* **32**: 2043–2051 (2011).

30. G. Wang, M. L. McCain, L. Yang, A. He, F. S. Pasqualini, A. Agarwal, H. Yuan, D. Jiang, D. Zhang, L. Zangi, J. Geva, A. E. Roberts, Q. Ma, J. Ding, J. Chen, D. Z. Wang, K. Li, J. Wang, R. J. Wanders, W. Kulik, F. M. Vaz, M. A. Laflamme, C. E. Murry, K. R. Chien, R. I. Kelley, G. M. Church, K. K. Parker and W. T. Pu, Modeling the mitochondrial cardiomyopathy of Barth syndrome with induced pluripotent stem cell and heart-on-chip technologies, *Nat Med* **20**: 616–623 (2014).

31. M. R. Salick, B. N. Napiwocki, J. Sha, G. T. Knight, S. A. Chindhy, T. J. Kamp, R. S. Ashton and W. C. Crone, Micropattern width dependent sarcomere development in human ESC-derived cardiomyocytes, *Biomaterials* **35**: 4454–4464 (2014).

32. R. S. Kane, S. Takayama, E. Ostuni, D. E. Ingber and G. M. Whitesides, Patterning proteins and cells using soft lithography, *Biomaterials* **20**: 2363–2376 (1999).

33. B. A. Grzybowski, R. Haag, N. Bowden and G. M. Whitesides, Generation of micrometer-sized patterns for microanalytical applications using a laser direct-write method and microcontact printing, *Anal Chem* **70**: 4645–4652 (1998).

34. K. L. Christman, M. V. Requa, V. D. Enriquez-Rios, S. C. Ward, K. A. Bradley, K. L. Turner and H. D. Maynard, Submicron streptavidin patterns for protein assembly, *Langmuir* **22**: 7444–7450 (2006).

35. H. Sorribas, C. Padeste and L. Tiefenauer, Photolithographic generation of protein micropatterns for neuron culture applications, *Biomaterials* **23**: 893–900 (2002).

36. J. Moller, T. Luhmann, M. Chabria, H. Hall and V. Vogel, Macrophages lift off surface-bound bacteria using a filopodium-lamellipodium hook-and-shovel mechanism, *Sci Rep* **3**: 2884 (2013).

37. P. W. Burridge, E. Matsa, P. Shukla, Z. C. Lin, J. M. Churko, A. D. Ebert, F. Lan, S. Diecke, B. Huber, N. M. Mordwinkin, J. R. Plews, O. J. Abilez, B. Cui, J. D. Gold and J. C. Wu, Chemically defined generation of human cardiomyocytes, *Nat Methods* **11**: 855–860 (2014).

38. M. Thery and M. Piel, Adhesive micropatterns for cells: a microcontact printing protocol, *Cold Spring Harb Protoc* **2009**: pdb prot5255 (2009).

39. H. Chen, L. Yuan, W. Song, Z. Wu and D. Li, Biocompatible polymer materials: Role of protein-surface interactions, *Prog Polym Sci* **33**: 1059–1087 (2008).

40. J. W. Lussi, D. Falconnet, J. A. Hubbell, M. Textor and G. Csucs, Pattern stability under cell culture conditions: a comparative study of patterning methods based on PLL-g-PEG background passivation, *Biomaterials* **27**: 2534–2541 (2006).

41. A. J. Ribeiro, A. K. Denisin, R. E. Wilson and B. L. Pruitt, For whom the cells pull: hydrogel and micropost devices for measuring traction forces, *Methods* **94**: 51–64 (2016).

42. J. Y. Sim, J. Moeller, K. C. Hart, D. Ramallo, V. Vogel, A. R. Dunn, W. J. Nelson and B. L. Pruitt, Spatial distribution of cell-cell and cell-ECM adhesions regulates force balance while maintaining E-cadherin molecular tension in cell pairs, *Mol Biol Cell* **26**: 2456–2465 (2015).

43. R. N. Palchesko, L. Zhang, Y. Sun and A. W. Feinberg, Development of polydimethylsiloxane substrates with tunable elastic modulus to study cell mechanobiology in muscle and nerve, *PLoS One* **7**: e51499 (2012).

44. J. Schaefer and R. T. Tranquillo, Tissue contraction force microscopy for optimization of engineered cardiac tissue, *Tissue Eng Part C Methods*, **22**: 76–83 (2016).

45. N. Carpi and M. Piel, Stretching micropatterned cells on a PDMS membrane, *J Vis Exp*, e51193 (2014).

46. S. M. Meilhac, R. G. Kelly, D. Rocancourt, S. Eloy-Trinquet, J. F. Nicolas and M. E. Buckingham, A retrospective clonal analysis of the myocardium reveals two phases of clonal growth in the developing mouse heart, *Development* **130**: 3877–3889 (2003).

47. P. S. Jouk, Y. Usson, G. Michalowicz and L. Grossi, Three-dimensional cartography of the pattern of the myofibres in the second trimester fetal human heart, *Anat Embryol (Berl)* **202**: 103–118 (2000).

48. H. M. Phillips, H. J. Rhee, J. N. Murdoch, V. Hildreth, J. D. Peat, R. H. Anderson, A. J. Copp, B. Chaudhry and D. J. Henderson, Disruption of planar cell polarity signaling results in congenital heart defects and cardio-myopathy attributable to early cardiomyocyte disorganization, *Circ Res* **101**: 137–145 (2007).

49. K. L. Coulombe, V. K. Bajpai, S. T. Andreadis and C. E. Murry, Heart regeneration with engineered myocardial tissue, *Annu Rev Biomed Eng* **16**: 1–28 (2014).

50. C. Tamiello, A. C. Buskermolen, F. T. Baaijens, J. V. Broers and C. C. Bouten, Heading in the right direction: understanding cellular orientation responses to complex biophysical environments, *Cell Mol Bioeng* **9**: 1–26 (2015).

51. C. Sears and R. Kaunas, The many ways adherent cells respond to applied stretch, *J Biomech* (2015) [Epub ahead of print].

52. H. Ngu, Y. Feng, L. Lu, S. J. Oswald, G. D. Longmore and F. C. Yin, Effect of focal adhesion proteins on endothelial cell adhesion, motility and orientation response to cyclic strain, *Ann Biomed Eng* **38**: 208–222 (2010).

53. M. Ventre, C. F. Natale, C. Rianna and P. A. Netti, Topographic cell instructive patterns to control cell adhesion, polarization and migration, *J R Soc Interface* **11**: 20140687 (2014).

54. T. Vignaud, L. Blanchoin and M. Thery, Directed cytoskeleton self-organization, *Trends Cell Biol* **22**: 671–682 (2012).

55. D. Tremblay, L. Andrzejewski, A. Leclerc and A. E. Pelling, Actin and microtubules play distinct roles in governing the anisotropic deformation of cell nuclei in response to substrate strain, *Cytoskeleton (Hoboken)* **70**: 837–848 (2013).

56. W. Chang, S. Antoku, C. Ostlund, H. J. Worman and G. G. Gundersen, Linker of nucleoskeleton and cytoskeleton (LINC) complex-mediated actin-dependent nuclear positioning orients centrosomes in migrating myoblasts, *Nucleus* **6**: 77–88 (2015).

57. L. E. McNamara, R. Burchmore, M. O. Riehle, P. Herzyk, M. J. Biggs, C. D. Wilkinson, A. S. Curtis and M. J. Dalby, The role of microtopography in cellular mechanotransduction, *Biomaterials* **33**: 2835–2847 (2012).

58. A. Mihic, J. Li, Y. Miyagi, M. Gagliardi, S. H. Li, J. Zu, R. D. Weisel, G. Keller and R. K. Li, The effect of cyclic stretch on maturation and 3D tissue formation of human embryonic stem cell-derived cardiomyocytes, *Biomaterials* **35**: 2798–2808 (2014).

59. M. Nikkhah, F. Edalat, S. Manoucheri and A. Khademhosseini, Engineering microscale topographies to control the cell-substrate interface, *Biomaterials* **33**: 5230–5246 (2012).

60. M. Khan, Y. Xu, S. Hua, J. Johnson, A. Belevych, P. M. L. Janssen, S. Gyorke, J. Guan and M. G. Angelos, Evaluation of changes in morphology and function of human induced pluripotent stem cell derived cardiomyocytes (hipsc-cms) cultured on an aligned-nanofiber cardiac patch, *PLoS One* **10**: e0126338 (2015).

61. J. I. Luna, J. Ciriza J., M. E. Garcia-Ojeda, M. Kong, A. Herren, D. K. Lieu, R. A. Li, C. C. Fowlkes, M. Khine and K. E. McCloskey, Multiscale biomimetic topography for the alignment of neonatal and embryonic stem cell-derived heart cells, *Tissue Eng Part C Methods* **17**: 579–588 (2011).

62. M. Kharaziha, S. R. Shin, M. Nikkhah, S. N. Topkaya, N. Masoumi, N. Annabi, M. R. Dokmeci and A. Khademhosseini, Tough and flexible CNT-polymeric hybrid scaffolds for engineering cardiac constructs, *Biomaterials* **35**: 7346–7354 (2014).

63. C. S. Simmons, B. C. Petzold and B. L. Pruitt, Microsystems for biomimetic stimulation of cardiac cells, *Lab Chip* **12**: 3235–3248 (2012).

64. B. D. Riehl, J. H. Park, I. K. Kwon and J. Y. Lim, Mechanical stretching for tissue engineering: two-dimensional and three-dimensional constructs, *Tissue Eng Part B Rev* **18**: 288–300 (2012).

65. A. Shradhanjali, B. Riehl, I. Kwon and J. Lim, Cardiomyocyte stretching for regenerative medicine and hypertrophy study, *Tissue Eng Regen Med* **12**: 398–409 (2015).

66. C. S. Simmons, A. J. Ribeiro and B. L. Pruitt, Formation of composite polyacrylamide and silicone substrates for independent control of stiffness and strain, *Lab Chip* **13**: 646–649 (2013).

67. W. H. Zimmermann, K. Schneiderbanger, P. Schubert, M. Didie, F. Munzel, J. F. Heubach, S. Kostin, W. L. Neuhuber and T. Eschenhagen, Tissue engineering of a differentiated cardiac muscle construct, *Circ Res* **90**: 223–230 (2002).

68. Y. Qi, Z. Li, C. W. Kong, N. L. Tang, Y. Huang, R. A. Li and X. Yao, Uniaxial cyclic stretch stimulates TRPV4 to induce realignment of human embryonic stem cell-derived cardiomyocytes, *J Mol Cell Cardiol* **87**: 65–73 (2015).

69. G. Kensah, A. Roa Lara, J. Dahlmann, R. Zweigerdt, K. Schwanke, J. Hegermann, D. Skvorc, A. Gawol, A. Azizian, S. Wagner, L. S. Maier, A. Krause, G. Drager, M. Ochs, A. Haverich, I. Gruh and U. Martin, Murine and human pluripotent stem cell-derived cardiac bodies form contractile myocardial tissue *in vitro*, *Eur Heart J* **34**: 1134–1146 (2013).

70. M. Tallawi, R. Rai, A. R. Boccaccini and K. E. Aifantis, Effect of substrate mechanics on cardiomyocyte maturation and growth, *Tissue Eng Part B Rev* **21**: 157–165 (2015).

71. K. Tobita, L. J. Liu, A. M. Janczewski, J. P. Tinney, J. M. Nonemaker, S. Augustine, D. B. Stolz, S. G. Shroff and B. B. Keller, Engineered early

embryonic cardiac tissue retains proliferative and contractile properties of developing embryonic myocardium, *Am J Physiol Heart Circ Physiol* **291**: H1829–1837 (2006).

72. S. J. Gwak, S. H. Bhang, I. K. Kim, S. S. Kim, S. W. Cho, O. Jeon, K. J. Yoo, A. J. Putnam and B. S. Kim, The effect of cyclic strain on embryonic stem cell-derived cardiomyocytes, *Biomaterials* **29**: 844–856 (2008).

73. V. F. Shimko and W. C. Claycomb, Effect of mechanical loading on three-dimensional cultures of embryonic stem cell-derived cardiomyocytes, *Tissue Eng Part A* **14**: 49–58 (2008).

74. D. Sedmera, R. Kockova, F. Vostarek and E. Raddatz, Arrhythmias in the developing heart, *Acta Physiol (Oxf)* **213**: 303–320 (2015).

75. M. Krenz, S. Sadayappan, H. E. Osinska, J. A. Henry, S. Beck, D. M. Warshaw and J. Robbins, Distribution and structure-function relationship of myosin heavy chain isoforms in the adult mouse heart, *J Biol Chem* **282**: 24057–24064 (2007).

76. P. J. Reiser, M. A. Portman, X. H. Ning and C. Schomisch Moravec, Human cardiac myosin heavy chain isoforms in fetal and failing adult atria and ventricles, *Am J Physiol Heart Circ Physiol* **280**: H1814–1820 (2001).

77. I. N. Sabir, M. J. Killeen, A. A. Grace and C. L. Huang, Ventricular arrhythmogenesis: insights from murine models, *Prog Biophys Mol Biol* **98**: 208–218 (2008).

78. L. Wang, Z. P. Feng, C. S. Kondo, R. S. Sheldon and H. J. Duff, Developmental changes in the delayed rectifier K+ channels in mouse heart, *Circ Res* **79**: 79–85 (1996).

79. S. S. Nunes, J. W. Miklas, J. Liu, R. Aschar-Sobbi, Y. Xiao, B. Zhang, J. Jiang, S. Masse, M. Gagliardi, A. Hsieh, N. Thavandiran, M. A. Laflamme, K. Nanthakumar, G. J. Gross, P. H. Backx, G. Keller and M. Radisic, Biowire: a platform for maturation of human pluripotent stem cell-derived cardiomyocytes, *Nat Methods* **10**: 781–787 (2013).

80. N. L. Tulloch, V. Muskheli, M. V. Razumova, F. S. Korte, M. Regnier, K. D. Hauch, L. Pabon, H. Reinecke and C. E. Murry, Growth of engineered human myocardium with mechanical loading and vascular coculture, *Circ Res* **109**: 47–59 (2011).

81. D. E. Discher, P. Janmey and Y. L. Wang, Tissue cells feel and respond to the stiffness of their substrate, *Science* **310**: 1139–1143 (2005).

82. A. L. Zajac and D. E. Discher, Cell differentiation through tissue elasticity-coupled, myosin-driven remodeling, *Curr Opin Cell Biol* **20**: 609–615 (2008).

83. A. J. Engler, S. Sen, H. L. Sweeney and D. E. Discher, Matrix elasticity directs stem cell lineage specification, *Cell* **126**: 677–689 (2006).

84. N. Huebsch, P. R. Arany, A. S. Mao, D. Shvartsman, O. A. Ali, S. A. Bencherif, J. Rivera-Feliciano and D. J. Mooney, Harnessing traction-mediated manipulation of the cell/matrix interface to control stem-cell fate, *Nat Mater* **9**: 518–526 (2010).

85. K. N. Dahl, A. J. Ribeiro and J. Lammerding, Nuclear shape, mechanics, and mechanotransduction, *Circ Res* **102**: 1307–1318 (2008).

86. Z. Jahed, H. Shams, M. Mehrbod and M. R. Mofrad, Mechanotransduction pathways linking the extracellular matrix to the nucleus, *Int Rev Cell Mol Biol* **310**: 171–220 (2014).

87. L. Macri-Pellizzeri, B. Pelacho, A. Sancho, O. Iglesias-Garcia, A. M. Simon-Yarza, M. Soriano-Navarro, S. Gonzalez-Granero, J. M. Garcia-Verdugo, E. M. De-Juan-Pardo and F. Prosper, Substrate stiffness and composition specifically direct differentiation of induced pluripotent stem cells, *Tissue Eng Part A* **21**: 1633–1641 (2015).

88. L. B. Hazeltine, M. G. Badur, X. Lian, A. Das, W. Han and S. P. Palecek, Temporal impact of substrate mechanics on differentiation of human embryonic stem cells to cardiomyocytes, *Acta Biomater* **10**: 604–612 (2014).

89. F. Chowdhury, S. Na, D. Li, Y. C. Poh, T. S. Tanaka, F. Wang and N. Wang, Material properties of the cell dictate stress-induced spreading and differentiation in embryonic stem cells, *Nat Mater* **9**: 82–88 (2010).

90. D. A. Fletcher and R. D. Mullins, Cell mechanics and the cytoskeleton, *Nature* **463**: 485–492 (2010).

91. S. Majkut, P. C. Dingal and D. E. Discher, Stress sensitivity and mechanotransduction during heart development, *Curr Biol* **24**: R495–501 (2014).

92. S. Majkut, T. Idema, J. Swift, C. Krieger, A. Liu and D. E. Discher, Heart-specific stiffening in early embryos parallels matrix and myosin expression to optimize beating, *Curr Biol* **23**: 2434–2439 (2013).

93. B. M. Friedrich, E. Fischer-Friedrich, N. S. Gov and S. A. Safran, Sarcomeric pattern formation by actin cluster coalescence, *PLoS Comput Biol* **8**: e1002544 (2012).

94. K. Dasbiswas, S. Majkut, D. E. Discher and S. A. Safran, Substrate stiffness-modulated registry phase correlations in cardiomyocytes map structural order to coherent beating, *Nat Commun* **6**: 6085 (2015).

95. M. Dembo and Y. L. Wang, Stresses at the cell-to-substrate interface during locomotion of fibroblasts, *Biophys J* **76**: 2307–2316 (1999).

96. C. Storm, J. J. Pastore, F. C. MacKintosh, T. C. Lubensky and P. A. Janmey, Nonlinear elasticity in biological gels, *Nature* **435**: 191–194 (2005).

97. S. Kumar, Cellular mechanotransduction: stiffness does matter, *Nat Mater* **13**: 918–920 (2014).

98. A. M. Kloxin, C. J. Kloxin, C. N. Bowman and K. S. Anseth, Mechanical properties of cellularly responsive hydrogels and their experimental determination, *Adv Mater* **22**: 3484–3894 (2010).

99. Q. Xing, K. Yates, C. Vogt, Z. Qian, M. C. Frost and F. Zhao, Increasing mechanical strength of gelatin hydrogels by divalent metal ion removal, *Sci Rep* **4**: 4706 (2014).

100. C. S. Sondergaard, G. Mathews, L. Wang, A. Jeffreys, A. Sahota, M. Wood, C. M. Ripplinger and M. S. Si, Contractile and electrophysiologic characterization of optimized self-organizing engineered heart tissue, *Ann Thorac Surg* **94**: 1241–1248; discussion 1249 (2012).

101. W. H. Zimmermann, I. Melnychenko and T. Eschenhagen, Engineered heart tissue for regeneration of diseased hearts, *Biomaterials* **25**: 1639–1647 (2004).

102. M. N. Hirt, J. Boeddinghaus, A. Mitchell, S. Schaaf, C. Bornchen, C. Muller, H. Schulz, N. Hubner, J. Stenzig, A. Stoehr, C. Neuber, A. Eder, P. K. Luther, A. Hansen and T. Eschenhagen, Functional improvement and maturation of rat and human engineered heart tissue by chronic electrical stimulation, *J Mol Cell Cardiol* **74**: 151–161 (2014).

103. R. E. Taylor, K. Kim, N. Sun, S. J. Park, J. Y. Sim, G. Fajardo, D. Bernstein., J. C. Wu and B. L. Pruitt, Sacrificial layer technique for axial force post assay of immature cardiomyocytes, *Biomed Microdevices* **15**: 171–181 (2013).

104. T. Boudou, W. R. Legant, A. Mu, M. A. Borochin, N. Thavandiran, M. Radisic, P. W. Zandstra, J. A. Epstein, K. B. Margulies and C. S. Chen, A microfabricated platform to measure and manipulate the mechanics of engineered cardiac microtissues, *Tissue Eng Part A* **18**: 910–919 (2012).

105. G. Vunjak Novakovic, T. Eschenhagen and C. Mummery, Myocardial tissue engineering: *in vitro* models, *Cold Spring Harb Perspect Med* **4** (2014).

106. A. Agarwal, J. A. Goss, A. Cho, M. L. McCain and K. K. Parker, Microfluidic heart on a chip for higher throughput pharmacological studies, *Lab Chip* **13**: 3599–3608 (2013).

107. Z. Ma, S. Koo, M. A. Finnegan, P. Loskill, N. Huebsch, N. C. Marks, B. R. Conklin, C. P. Grigoropoulos and K. E. Healy, Three-dimensional filamentous human diseased cardiac tissue model, *Biomaterials* **35**: 1367–1377 (2014).

108. S. N. Bhatia and D. E. Ingber, Microfluidic organs-on-chips, *Nat Biotechnol* **32**: 760–772 (2014).

109. P. Nassoy and C. Lamaze, Stressing caveolae new role in cell mechanics, *Trends Cell Biol* **22**: 381–389 (2012).

110. S. Tojkander, G. Gateva and P. Lappalainen, Actin stress fibers: assembly, dynamics and biological roles, *J Cell Sci* **125**: 1855–1864 (2012).

111. W. Nie, M. T. Wei, D. H. Ou-Yang, S. S. Jedlicka and D. Vavylonis, Formation of contractile networks and fibers in the medial cell cortex through myosin-II turnover, contraction, and stress-stabilization, *Cytoskeleton (Hoboken)* **72**: 29–46 (2015).

112. M. Murrell, P. W. Oakes, M. Lenz and M. L. Gardel, Forcing cells into shape: the mechanics of actomyosin contractility, *Nat Rev Mol Cell Biol* **16**: 486–498 (2015).

113. L. J. Peterson, Z. Rajfur, A. S. Maddox, C. D. Freel, Y. Chen, M. Edlund, C. Otey and K. Burridge, Simultaneous stretching and contraction of stress fibers *in vivo*, *Mol Biol Cell* **15**: 3497–3508 (2004).

114. S. Ono, Dynamic regulation of sarcomeric actin filaments in striated muscle, *Cytoskeleton (Hoboken)* **67**: 677–692 (2010).

115. P. W. Oakes, S. Banerjee, M. C. Marchetti and M. L. Gardel, Geometry regulates traction stresses in adherent cells, *Biophys J* **107**: 825–833 (2014).

116. D. Wang, W. Zheng, Y. Xie, P. Gong, F. Zhao, B. Yuan, W. Ma, Y. Cui, W. Liu, Y. Sun, M. Piel, W. Zhang and X. Jiang, Tissue-specific mechanical and geometrical control of cell viability and actin cytoskeleton alignment, *Sci Rep* **4**: 6160 (2014).

117. G. Langanger, M. Moeremans, G. Daneels, A. Sobieszek, M. De Brabander and J. De Mey, The molecular organization of myosin in stress fibers of cultured cells, *J Cell Biol* **102**: 200–209 (1986).

118. L. P. Cramer, M. Siebert and T. J. Mitchison, Identification of novel graded polarity actin filament bundles in locomoting heart fibroblasts: implications for the generation of motile force, *J Cell Biol* **136**: 1287–1305 (1997).

119. T. Butt, T. Mufti, A. Humayun, P. B. Rosenthal, S. Khan, S. Khan and J. E. Molloy, Myosin motors drive long range alignment of actin filaments, *J Biol Chem* **285**: 4964–4974 (2010).

120. D. Rhee, J. M. Sanger and J. W. Sanger, The premyofibril: evidence for its role in myofibrillogenesis, *Cell Motil Cytoskeleton* **28**: 1–24 (1994).

121. S. M. LoRusso, D. Rhee, J. M. Sanger and J. W. Sanger, Premyofibrils in spreading adult cardiomyocytes in tissue culture: evidence for reexpression of the embryonic program for myofibrillogenesis in adult cells, *Cell Motil Cytoskeleton* **37**: 183–198 (1997).

122. G. A. Dabiri, K. K. Turnacioglu, J. M. Sanger and J. W. Sanger, Myofibrillogenesis visualized in living embryonic cardiomyocytes, *Proc Natl Acad Sci USA* **94**: 9493–9498 (1997).

123. M. A. Bray, S. P. Sheehy and K. K. Parker, Sarcomere alignment is regulated by myocyte shape, *Cell Motil Cytoskeleton* **65**: 641–651 (2008).

124. A. Chopra, E. Tabdanov, H. Patel, P. A. Janmey and J. Y. Kresh, Cardiac myocyte remodeling mediated by N-cadherin-dependent mechanosensing, *Am J Physiol Heart Circ Physiol* **300**: H1252–1266 (2011).

125. R. Zhu, A. Blazeski, E. Poon, K. D. Costa, L. Tung and K. R. Boheler, Physical developmental cues for the maturation of human pluripotent stem cell-derived cardiomyocytes, *Stem Cell Res Ther* **5**: 117 (2014).

5

CARDIOVASCULAR SYSTEM: STEM CELLS IN TISSUE-ENGINEERED BLOOD VESSELS

Rajendra Sawh-Martinez, Edward McGillicuddy,
Gustavo Villalona, Yong-Ung Lee, Toshiharu Shin'oka
and Christopher K. Breuer

1. Introduction

Cardiovascular disease (CVD) is the leading cause of death in the world. Ischemic heart disease, cerebrovascular disease, and peripheral vascular disease account for 13 million deaths annually.[1] Additionally, congenital heart disease, which represents the most common birth defect affecting nearly 1% of all live births, remains a leading cause of death in the newborn period. In the United States alone, the American Heart Association estimates the cost of treating CVD in 2009 to be US$475.3 billion.[2]

For end-stage cardiovascular disease in adults, or in children with congenital cardiovascular anomalies, surgical reconstruction remains the treatment of choice. Unfortunately, there is a relative paucity of autologous tissue available for reconstructive surgeries, necessitating the use of synthetic materials. Biomaterials available in the cardiovascular surgeon's armamentarium include autologous vessels and prosthetic materials. Autologous tissue is superior to prosthetic materials such as Goretex®, as prosthetics are more prone to thrombosis, infectious complications, neointimal hyperplasia, and accelerated atherosclerosis.[3–6] In an adult patient undergoing multiple cardiac or peripheral vascular procedures, however, suitable autologous tissue may be scarce due to the diffuse nature of atherosclerosis. Additionally, in children undergoing surgery for complex congenital heart defects, repairs may be extra-anatomic and require more autologous tissue than is available. Clearly, there is demand for improved, biocompatible vascular conduits in these populations.

The great promise and hope of vascular tissue engineering is the development of biological substitutes that restore, maintain, or improve tissue function. The origins of vascular tissue engineering can be traced back to Alexis Carrel who received the Nobel Prize in medicine for his early work in vascular biology. This established the basic principles that provided the foundation for vascular surgery. Later in his career while working in collaboration with Charles Lindbergh at the Rockefeller Institute, some of the early pioneering work in cell culture was performed. In their manuscript entitled "The culture of whole organs," Carrel and Lindbergh postulated that cell culture would one day be used to grow entire organs.[7]

However, it was not until Weinberg and Bell's seminal paper in 1986 that Lindberg and Carrel's prediction became a reality for vascular tissue.[8] This search led to many advances in our basic understanding of tissue interactions, including the critical importance of cellular and matrix interactions, the modulation of cellular recruitment, growth and differentiation, as well as a deeper appreciation for the challenges that are faced in attempting to apply these discoveries clinically. In this chapter we aim to guide the reader through the major milestones of vascular tissue engineering while highlighting the importance of continued scientific investigations aimed at reducing the tremendous burden of CVD.

1.1 *Historical overview*

The first report of the use of synthetic vascular grafts dates back to 1952 when Voorhees *et al.* used Vinyon N cloth tubes as arterial interpositions in dogs.[9–12] Prior to this report, scientists were focused on using native arteries as conduits.[9,11] In the ensuing years, other synthetic materials aimed at passively transporting blood with minimal reaction were developed, leading to the clinical success of polyethylene terephthalate (Dacron) and expanded polytetrafluoroethylene (ePTFE).

Major achievements in creating an ideal conduit have driven the continued search for the best strategy to create fully functioning blood vessel substitutes. In 1986, Weinberg and Bell's landmark publication described the use of bovine cells with rat collagen gel to create an artificial blood vessel model. In 1998, Shinoka *et al.* first described the use of a biodegradable synthetic scaffold with ovine cells, demonstrating long-term autologous implantation in a low-pressure pulmonary artery system.[13] Also in 1998, L'Heureux *et al.* published their creation of an engineered graft that used human cells and was tested as a short-term xenogeneic implant in a high-pressure arterial bypass model.[14] In 1999, Nicklason *et al.* and Shum-Tim *et al.* described the creation of an artery grown *in vitro* using autologous porcine/ovine cells on a biodegradable polymer.[15,16] In their studies, Niklason *et al.* used small-diameter grafts, and studied their vessels under arterial pressures in the short term. Shum-Tim *et al.* developed large-diameter grafts, studied under lower pressures with long-term evaluation of function.

In 2001, Shin'oka *et al.* reported the first clinical use of a tissue-engineered vascular construct. The vascular graft was implanted in a four-year-old girl to reconstruct an occluded pulmonary artery, after a prior Fontan procedure.[17] In another major milestone for vascular tissue engineering, in 2007 L'Heureux *et al.* used their sheet-based tissue-engineered blood vessel clinically as replacements for failing arteriovenous shunts.[18] Despite these tremendous achievements, the field has yet to report the clinical use of a tissue-engineered vascular construct as a fully arterial interposition in humans. Further, the mainstay therapy for vascular reconstruction continues to be autologous arteries and veins. In the following pages we will detail these early successes, discuss the applications of vascular tissue engineering in clinical trials, and describe the frontiers of our technology.

2. Critical Elements of an Artificial Blood Vessel

Synthetic vascular grafts were developed for patients with insufficient autologous tissue for bypass grafting. These grafts have reasonable safety profiles, satisfactory surgical handling characteristics (i.e. suture retention), and are available "off-the-shelf" for use as large-caliber bypass grafts. Currently available synthetic vascular conduits, however, have several significant limitations. These grafts have no growth potential, thus many pediatric patients will "outgrow" the graft and require re-operation. Re-do operations are associated with an increased risk of complications and death. Bioprosthetic materials such as gluteraldehyde-fixed xeno- or allo- grafts used for grafting are prone to ectopic calcification, resulting in poor durability. Both prosthetic and bioprosthetic materials are prone to infection, putting the patient at risk for sepsis, graft rupture, distal septic emboli, as well as re-operation for explantation of graft material. Lastly, the use of synthetic grafts as small-caliber bypass grafts is limited as they are prone to thrombosis and neointimal hyperplasia, likely from a lack of biocompatibility and an inability to repair and remodel.

Tissue-engineered vascular conduits address the shortcomings of currently available vascular conduits. The ideal tissue-engineered graft possesses excellent surgical handling characteristics. An experienced surgeon must be able to handle the graft, modify it as necessary for the patient's anatomy, perform anastomoses using standard surgical technique and instrumentation, and obtain hemostasis immediately following implantation. Because of the morbidity and mortality associated with longer operative and anesthetic time, the surgeon must perform these anastomoses in a timely fashion. Optimally, tissue-engineered grafts would be available "off-the shelf" (similar to prosthetic grafts), and require minimal manipulation other than seeding the cell on the substrate the day of surgery.

In addition to satisfactory surgical handling techniques, the ideal tissue-engineered vascular graft is biocompatible. The polymerized scaffold or decellularized matrix should degrade over time, leaving intact vascular neotissue that provides the structural integrity for the conduit. The degradation of scaffold materials results in a completely biocompatible structure that is not prone to infection or ectopic calcification, and does not require immunosuppression. Finally, the ideal tissue-engineered graft possesses the intrinsic ability to grow with the patient, obviating the need for re-operations in the pediatric population.

The most important characteristic of a tissue-engineered vascular graft is, however, recapitulation of vessel form and function. Put simply, grafts implanted in the arterial and venous circulation should mirror as closely as possible the native artery and vein. The ideal arterial interposition graft possesses a functional, confluent, non-thrombogenic endothelium and a thick smooth muscle-laden tunica media that can accommodate mean arterial pressure and constrict appropriately to ensure perfusion to end-organs during low flow states. Thus, the intrinsic mechanical strength of the seeded scaffold should be higher and the degradation time longer in arterial grafts than in venous grafts where mechanical integrity is less critical but adequate compliance is necessary.

3. Approaches to Creating Tissue-Engineered Blood Vessels (TEBVs)

3.1 The first artificial artery: Weinberg and Bell

The first reported successful construction of a tissue-engineered vascular graft (TEVG) came from Weinberg and Bell in 1986. Prior work had established the use of synthetic materials as vascular conduits, albeit not for small-diameter vessels (<6 mm diameter), and autologous tissue had become the mainstay of vascular repair. Building on previous reports of partial vascular constructs, including *in vitro* growth of endothelial cells and models of the vascular wall in mock circulatory loops, this seminal paper reported culturing bovine vascular cells to seed a collagen matrix in a tubular mold.[8,19]

Specifically, bovine smooth muscle cells were combined with collagen using a casting culture medium to create a tubular lattice. After a week of culture, a synthetic Dacron sleeve was placed on the outer surface of the construct, and seeded with fibroblasts to create a neo-adventitia. After another two weeks, the grafts were seeded with endothelial cells and left in a rotational culture (1 rev/min) for one week. This procedure provided a template for the creation of a tubular cellular structure that demonstrated extracellular matrix deposition of collagen, alignment of smooth muscle cells, and a confluent endothelial cell layer, in a construct that had burst pressures of up to 323 mmHg when reinforced with added layers of

Dacron. This led the authors to describe the use of their graft as a model for the study of the biological properties of blood vessels, rather than as a potentially clinically useful construct. Despite these shortcomings, Weinberg and Bell laid the foundation and provided a roadmap for scientists to develop artificial blood vessels, as their publication marked the birth of vascular tissue engineering.

3.2 *First clinical use of a TEBV*

Congenital cardiac anomalies, a diverse spectrum of defects, result in significant perinatal morbidity and mortality. Untreated single ventricle anomalies are associated with 70% mortality in the first year of life. The therapy of choice is surgical reconstruction. Without surgery, survival of this cohort into adulthood is rare. Initially described in 1971, the Fontan operation separates pulmonic and systemic blood flow; subsequently decreasing the incidence of chronic hypoxia and high output cardiac failure. This suite of procedures requires a synthetic graft, commonly polytetrafluoroethylene (PTFE, or Gore-Tex®). As mentioned above, PTFE has several limitations, including potential for infection, thrombosis, and ectopic calcification.[20–22] Furthermore, PTFE does not grow, resulting in patients "outgrowing" the graft and requiring re-operation, or suffering complications related to intentional graft over-sizing.

The fabrication and seeding of biodegradable polymer scaffolds for use in humans was an outgrowth of work performed in large animal models. Precursors to the clinical study included the creation of a tissue- engineered heart valve by Zund and others.[23–26] This valve construct consisted of biodegradable polyglycolic acid fibers serially seeded with fibroblasts and endothelial cells. These tissue-engineered valvular constructs were implanted in the pulmonary valve position in juvenile lambs. Functional performance of the grafts was satisfactory on ultrasound, and histological analysis revealed appropriate cellular architecture. These results were validated by a larger study with acellular valve controls, in which the seeded scaffolds demonstrated superior functionality.[27]

The natural extension of tissue-engineered heart valves was the creation of tissue-engineered vascular conduits. These tubular constructs could be fabricated from the same scaffold materials and cellular elements, with the additional advantage of less complex biomechanical constraints.[16,28,29]

Although both tissue-engineered heart valves and blood vessels demonstrated mechanical integrity and vascular neotissue formation *in vivo*, the original seeding models relied on time-consuming expansion of endothelial cells and smooth muscle cells *in vitro*. This made experiments time consuming and vulnerable to culture contamination, in addition to limiting the practicality of human use as each patient would require multiple procedures.[30] Attention was turned to other sources of seeded cells, with a focus on cells that could be procured on the day of surgical implantation of tissue-engineered grafts. Seeking an abundant cell source that did not require *ex vivo* tissue culture expansion, Matsumura and colleagues seeded polymer scaffolds with autologous bone marrow and implanted these constructs into the inferior vena cava of dogs.[31,32] Not only did these grafts remain patent, histological analysis revealed cell populations elaborating vascular endothelial growth factor (VEGF) and expressing endothelial or smooth muscle cell markers.[31]

The first reported use of a tissue-engineered graft in humans occurred in 1999, in a four-year-old girl with a single right ventricle and pulmonary atresia.[17] At the age of three she had undergone pulmonary artery angioplasty and the Fontan procedure. Subsequent angiography revealed total occlusion of the right intermediate pulmonary artery. A 2-cm segment of peripheral vein was explanted and cells expanded *ex vivo*. Cells were seeded onto a caprolactone–polylactic acid copolymer scaffold, reinforced with woven polyglycolic acid. The occluded pulmonary artery was successfully reconstructed with the tissue-engineered graft. No postoperative complications were reported. On follow-up angiography, the transplanted vessel was patent with no signs of aneurismal dilation.

Promising large animal study data and the first successful human application of a tissue-engineered graft provided the impetus for a larger clinical study. From May 2000 to December 2004, 25 patients (mean age 5.5 years) with single ventricle physiology were implanted with tissue-engineered grafts at Tokyo Women's Medical University. Extra cardiac total cavopulmonary connection (EC TCPC) conduits (a connection between the vena cava and the pulmonary arteries) for surgical correction of single ventricle physiology represented an ideal hemodynamic starting point for tissue-engineered grafts in humans. In this system, high flow rates minimize thrombotic risks while relatively low pressure minimizes wall tension on the conduit. This pilot study evaluated two types of scaffolds, PCLA-PGA ($n = 11$) and PCLA-PLA ($n = 14$). Autologous bone marrow (5 mm/Kg BW), aspirated from the patient's ilium under general anesthesia, was used to seed the polymer scaffold and the construct was incubated for two hours in media prior to implantation. All 25 patients survived to hospital discharge.[33] Duplex ultrasonography, computed tomography, magnetic resonance angiography, and cineangiography were utilized for graft surveillance (Fig. 1). All grafts remained patent, and no aneurismal dilation was detected with any imaging modality.[34]

Of note, six patients developed silent graft stenosis, and four underwent successful balloon angioplasty. Four patients died during follow-up (mean follow-up 5.8 years). The causes of death, however, were not related to tissue-engineered graft dysfunction; all four patients had imag-

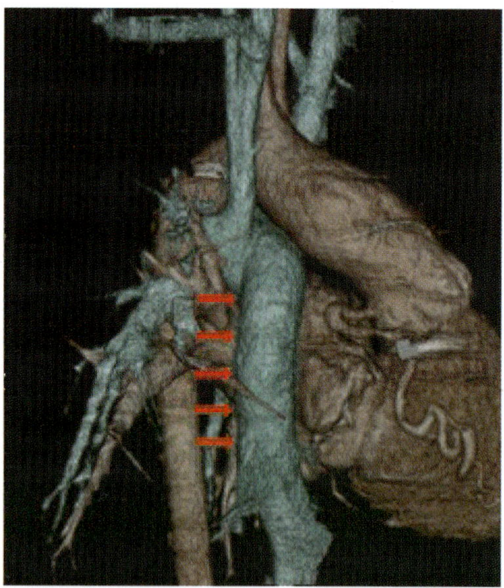

Fig. 1.   Three-dimensional computed tomography reconstruction of *in vivo* TEBV, tissue-engineered vascular graft in a 13-year-old with single ventricle physiology. The graft connects the superior vena cava to the pulmonary arteries. Three-dimensional reconstructed computed tomography of the heart, great vessels, and tissue-engineered vascular graft demonstrate a widely patent graft with no evidence of stenosis, thrombosis, or aneurismal dilation.

ing demonstrating a widely patent graft prior to death. Of the 21 surviving patients, 18 were classified as New York Heart Association (NYHA) Class I (no impairment of physical activity) and three were classified as NYHA Class II (mild impairment of physical activity).

Although the Tokyo Women's study remains a landmark in tissue engineering, the data represent a single institution's experience and patients were not randomized. Additionally, although autologous bone marrow cells remain an attractive cell source for tissue-engineered graft seeding, no histological specimens exist from this clinical trial (autopsies are not usually performed in Japan, and most patients are still alive). In order to address the problem of TEVG stenosis and to investigate the cellular and molecular mechanisms underlying this pathological process, the same group developed a murine model of TEBV.[35,36] Using this model system, it was discovered that cell seeding reduces the incidence of the development of TEVG stenosis. However, in contrast to the conventional assumption that the cells seeded onto the tissue-engineering scaffold directly contribute to vascular neotissue formation, the seeded cells rapidly disappeared from the TEVG, suggesting that they exert their effect *via* a paracrine mechanism.[37,38] Moreover, Hibino and colleagues showed that neovessel formation is an immune-mediated regenerative process, and that the neovessel arises from ingrowth of the vascular cells (EC and SMC) from the neighboring vessel wall along the luminal surface of the scaffold. It is also shown that normal macrophage infiltration is essential for vascular neotissue formation, but when this process is excessive it leads to the development of TEVG stenosis.[38,39] The pharmacological intervention, either systemically or locally, to prevent stenosis remains to be investigated.

3.3 *Sheet-based tissue engineering*

In the continuing search for an ideal conduit to repair damaged human vessels, L'Heureux *et al.* employed a new methodology termed tissue engineering by self-assembly (TESA) to produce human vessels *in vitro*. TESA channels the ability of cells of mesenchymal origin to secrete and assemble large amounts of extracellular matrix (ECM) in various geometries. Sheet-based tissue engineering (SBTE), a variation of TESA, employs sheets of living cells and the natural ECM they synthesized to produce tubular structures and create vessel-like structures. In developing SBTE, L'Heureux *et al.* sought to overcome the need to use synthetic material for structural support of biological grafts as was the case for Weinberg and Bell.[8,14,40] In this approach only cellular components were employed in the creation of their artificial blood vessel.

The first TEVG produced by SBTE contained only human skin fibroblasts cultured from dermal specimens removed during reductive breast surgery; smooth muscle cells and endothelial cells were isolated from newborn umbilical veins or adult sapheneous veins. Nonetheless, it had a burst pressure of 2594 ± 501 mmHg.[14,41] Cell sheets were produced in culture medium supplemented with 50 μg/ml of sodium ascorbate, peeled off from culture flasks and tubularized around inert tubular support cylinders. After a maturation period of at least eight weeks, the sheets become cohesive tubes that can be seeded with endothelial cells. The endothelial layer of the constructs expressed von Willenbrand factor immunohisto- chemical staining and demonstrated ac-LDL uptake; smooth muscle cells stained positively for alpha-smooth muscle actin and desmin. Fibroblasts did not stain for smooth muscle cell markers, but expressed vimentin and synthesized elastin, which organized in circular arrays. The ECM contained laminin, fibronectin, chondroitin sulfates, and various collagen subtypes. Further research demonstrated that these artificial blood vessels retained concentration-dependent contractile properties, responsive to various vasoconstricting (histamine, bradykinin, ATP) and vasodilating (sodium nitroprusside, SIN-1, forskolin) agents with and without inhibitory agents.[41]

In the following years, the model was simplified to exclude smooth muscle cells and was shown to be feasible using skin fibroblasts and venous endothelium isolated from elderly patients with CVD. Pre-clinical studies in immunodeficient nude rats, evaluated for up to 225 days, demonstrated patency rates of 85% without aneurism as abdominal aorta interpositional grafts. To study the *in vivo* development of these grafts in a more representative biomechanical environment, the TEBVs were implanted in immunosuppressed cynomolgus primates. Explants at six and eight weeks demonstrated patent vessels with no signs of luminal narrowing or aneurismal formation. Alpha-actin positive smooth muscle cells and proteoglycan expression were also observed in both models.[41,42]

In 2007, L'Heureux, McAllister and de la Fuente reported the second clinical use of vascular tissue-engineered constructs when they described the implantation of their TEVG in three patients undergoing hemodialysis treatment (Fig. 2).[18] These dialysis patients had a history of previously failed access grafts. After 24 patient-months of follow-up for this cohort, only one of the grafts used for dialysis access had a thrombogenic failure, secondary to low postoperative flow rate and moderate dilatation of the graft. McAllister *et al.* then expanded their cohort to ten patients in total, drawing from two centers, in Argentina and in Poland.[43]

Three out of the ten implanted grafts failed within the safety phase of their study, and one patient required surgical re-intervention to maintain patency at 11 months. These results are consistent with the expected failure rates with native hemodialysis access grafts (fistula) in such high-risk patient populations. Overall, after six months, the artificial grafts implanted as hemodialysis access grafts had a 60% primary patency rate which is superior to that of the standard of care synthetic graft (ePTFE).

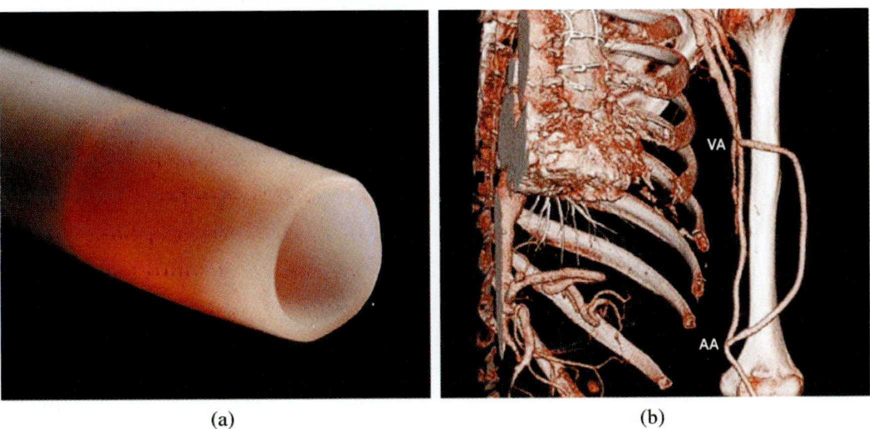

(a)　　　　　　　　　　　　　　(b)

Fig. 2.   The tissue-engineered blood vessel preoperatively (a), at three months after implantation (b), computed tomographic angiography. VA: venous anastomosis; AA: arterial anastomosis. Reproduced with permission from N. L'Heureux, T. N. McAllister and L. M. de la Fuente, Tissue-engineered blood vessel for adult arterial revascularization, *N Engl J Med* **357**: 1451–1453 (2007). Copyright © 2007 Massachusetts Medical Society. All rights reserved.

3.4 *Stem cells for vascular tissue engineering*

During the past decade, significant attention has been turned to the use of stem cells in TEVGs. Stem cells and endothelial progenitor cells (EPCs) can differentiate into vascular lineages and thus have the potential to repair vascular systems. Despite their obvious potential in clinical practice, there still remain many controversies regarding how EPCs actually enhance endothelial repair and neovascularization.[44] Additionally, because of the limited expansion ability of EPCs, expansion of sufficient EPC populations for therapeutic angiogenesis remains a significant impediment for most investigators. On the other hand, embryonic stem cells (ESCs) have an extensive self-renewal activity and can be expanded without limit, thus ESC-derived endothelial cells may be feasible as a novel cell source for therapeutic angiogenesis.[45,46]

Nourse and colleagues used VEGF to induce differentiation of functional endothelium from human embryonic stem cells.[47] Continuous VEGF treatment of embryonic stem cells resulted in a four- to five-fold enrichment of CD31(+) cells but did not increase endothelial proliferation rates, suggesting a primary effect on differentiation. CD31(+) cells purified from differentiating embyroid bodies upregulated ICAM-1 and VCAM-1 in response to TNF-α, confirming their ability to function as endothelial cells. Collagen gel constructs containing the human ESC-derived endothelial cells and implanted into infarcted nude rat hearts formed dense networks of patent vessels filled with host blood cells. Thus Nourse *et al.* demonstrated the ability of human ESC-derived endothelial cells to facilitate neovascularization of tissue-engineered constructs.

In addition to driving ESCs into an endothelial lineage, investigators have turned to embryonic stem cell-derived smooth muscle cells as a potential cell source in cardiovascular tissue engineering.[48,49] NADPH oxidase (Nox4) overexpression in ESC culture resulted in increased smooth muscle cell marker production, whereas knockdown of Nox4 induced a decrease in production. Moreover, Nox4 was demonstrated to drive smooth muscle differentiation through generation of H_2O_2.[50] Thus Nox4 expression maintains differentiation status and functional features of stem cell-derived smooth muscle cells, highlighting its impact on vessel formation *in vivo*. It is clear that ESCs and endothelial progenitor cells

(EPCs) will be important in vascular tissue engineering in the future, and endothelial cell and smooth muscle cells derived from ESC culture will require further characterization.

Autologous, readily available, bone marrow mononuclear cells (BMC) in both human and animal models promote angiogenesis and early endothelialization, and contain multipotent cells that can also form part of the growing neovessel.[51] In trying to develop a TEVG, Matsumana *et al.* reported initial success with the use of harvested vessel wall cells after isolation and culture. They found that this was a time-consuming process that required previous hospitalization for vessel harvesting, and that in half of the cases they were not able to obtain sufficient cells on the day of surgery. As a result, this group turned to BMCs as the cell source for their constructs, removing the impediments of long-term cell culture and contamination, and more importantly, providing sufficient cells for the seeding of biodegradable scaffolds.[33,52–54] In the first human clinical trial with long-term follow-up,[34] 25 TEVGs were implanted using autologous bone marrow-derived mononuclear cells with no evidence of aneurysm formation, graft rupture, graft infection, or ectopic calcification; four patients had graft stenosis and underwent successful percutaneous angioplasty. These initial findings demonstrated the feasibility and safety of this technique and provided evidence for the use of BMCs for the construction of TEVG.

4. Conclusion

The history of reconstructive medicine and surgery dates back centuries as physicians and scientists have continually sought to restore function to damaged tissue. Vascular tissue engineering began its path toward clinical utility with the construction of the first blood vessel model by Weinberg and Bell. In the ensuing years, several major breakthroughs in biological vascular graft production have led this young field into its first successful clinical applications. Vigorous research continues, as these clinical trials are in their early phase; it is a matter of time before they are attempted in the United States under the guidance of the Food and Drug Administration (FDA).

In the last 30 years we have witnessed the success, and the various pros and cons of the methodologies used to create neovessels. The first generation of tissue-engineered arterial grafts represents a significant step forward, but several challenges remain to be addressed. Approaches that rely on bioreactors require significant culture time, preclude "off-the-shelf" availability, and place the conduit at risk for contamination. Also, the cell source remains challenging, as adult human SMCs have limited number of passages *in vitro*, precluding the development of a well-developed tunica media. Finally, standardized imaging algorithms for implanted grafts will provide a more structured analysis of graft patency, dilation, and/or stenosis.

Undoubtedly, new approaches and techniques have yet to be developed. Modern technology in polymer fabrication or assembly may yield a more ideal scaffolding material. Advances in cell culture techniques may allow for faster production of artificial tissue. Ultimately, a deeper understanding of the signaling cascades involved in cellular interactions in neotissue development will be critical to the construction of third-generation grafts. Modification of the genetic and molecular constitution of cells may lead to the development of tissue with selective maturation properties. Employing particle release technology, grafts may be constructed that would elute growth factors, cytokines or other molecular signals that may recruit host cell development onto scaffolding material, doing away with the need for cell seeding altogether.

As this technology develops, successful clinical applicability will be paramount to future implementation. The focus in the field remains on responsible development of modern tools to combat diseases that affect a large portion of the population. Potential widespread use for TEBVs will be dependent on patient safety, and functionality of TEBVs that is as least equal to currently used synthetic grafts. Through landmark accomplishments and continued clinical achievements, we are nearing the promise of artificially constructing neotissues that replicate the function of the human body.

Acknowledgements

The authors would like to thank Nicolas L'Heureux and the editors for their careful attention and helpful critiques.

References

1. World Health Organization, *Global Burden of Disease* (2004).

2. American Heart Association, *Cardiovascular Disease Cost* (2009).

3. Y. J. Hong, G. S. Mintz, S. W. Kim, S. Y. Lee, S. Y. Kim, T. Okabe, A. D. Pichard, L. F. Satler, R. Waksman, K. M. Kent, W. O. Suddath and N. J. Weissman, Disease progression in nonintervened saphenous vein graft segments: a serial intravascular ultrasound analysis, *J Am Coll Cardiol* **53**: 1257–1264 (2009).

4. P. G. Malvindi, S. Jacob, A. Kallikourdis and N. Vitale, What is the patency of the gastroepiploic artery when used for coronary artery bypass grafting? *Interact Cardiovasc Thorac Surg* **6**: 397–402 (2007).

5. N. D. Desai and S. E. Fremes, Radial artery conduit for coronary revascularization: as good as an internal thoracic artery? *Curr Opin Cardiol* **22**: 534–540 (2007).

6. M. R. Dashwood and A. Loesch, The saphenous vein as a bypass conduit: the potential role of vascular nerves in graft performance, *Curr Vasc Pharmacol* **7**: 47–57 (2009).

7. A. Carrel and C. A. Lindbergh, The culture of whole organs, *Science* **81**: 621–623 (1935).

8. C. B. Weinberg and E. Bell, A blood vessel model constructed from collagen and cultured vascular cells, *Science* **231**: 397–400 (1986).

9. A. D. Callow, Arterial homografts. *Eur J Vasc Endovasc Surg* **12**: 272–281 (1996).

10. A. H. Blakemore and A. B. Voorhees, Jr., The use of tubes constructed from vinyon N cloth in bridging arterial defects; experimental and clinical, *Ann Surg* **140**: 324–334 (1954).

11. A. B. Voorhees, Jr., A. Jaretzki, 3rd and A. H. Blakemore, The use of tubes constructed from vinyon "N" cloth in bridging arterial defects, *Ann Surg* **135**: 332–336 (1952).

12. R. P. Lanza, R. S. Langer and J. P. Vacanti, *Principles of Tissue Engineering* (Elsevier Academic Press, Burlington, MA, 2007).

13. T. Shinoka, D. Shum-Tim, P. X. Ma, R. E. Tanel, N. Isogai, R. Langer, J. P. Vacanti and J. E. Mayer, Jr., Creation of viable pulmonary artery autografts through tissue engineering, *J Thorac Cardiovasc Surg* **115**: 536–545; discussion 545–536 (1998).

14. N. L'heureux, S. Pâquet, R. Labbé, L. Germain and F. A. Auger, A completely biological tissue-engineered human blood vessel, *FASEB J* **12**: 47–56 (1998).

15. L. E. Niklason, J. Gao, W. M. Abbott, K. K. Hirschi, S. Houser, R. Marini and R. Langer, Functional arteries grown *in vitro, Science* **284**: 489–493 (1999).

16. D. Shum-Tim, U. Stock, J. Hrkach, T. Shinoka, J. Lien, M. A. Moses, A. Stamp, G. Taylor, A. M. Moran, W. Landis, R. Langer, J. P. Vacanti and J. E. Mayer, Jr., Tissue engineering of autologous aorta using a new biodegradable polymer, *Ann Thorac Surg* **68**: 2298–2304; discussion 2305 (1999).

17. T. Shin'oka, Y. Imai and Y. Ikada, Transplantation of a tissue-engineered pulmonary artery, *N Engl J Med* **344**: 532–533 (2001).

18. N. L'Heureux, T. N. McAllister and L. M. de la Fuente, Tissue-engineered blood vessel for adult arterial revascularization, *N Engl J Med* **357**: 1451–1453 (2007).

19. N. L'Heureux, N. Dusserre, A. Marini, S. Garrido, L. de la Fuente and T. McAllister, Technology insight: the evolution of tissue-engineered vascular grafts — from research to clinical practice. *Nat Clin Pract Cardiovasc Med* **4**: 389–395 (2007).

20. S. Sarkar, T. Schmitz-Rixen, G. Hamilton and A. M. Seifalian, Achieving the ideal properties for vascular bypass grafts using a tissue engineered approach: a review, *Med Biol Eng Comput* **45**: 327–336 (2007).

21. H. P. Gildein, A. Ahmadi, F. Fontan and R. Mocellin, Special problems in Fontan-type operations for complex cardiac lesions, *Int J Cardiol* **29**: 21–28 (1990).

22. M. P. Brennan, A. Dardik, N. Hibino, J. D. Roh, G. N. Nelson, X. Papademitris, T. Shinoka and C. K. Breuer, Tissue-engineered vascular grafts demonstrate evidence of growth and development when implanted in a juvenile animal model, *Ann Surg* **248**: 370–377 (2008).

23. G. Zund, C. K. Breuer, T. Shinoka, P. X. Ma, R. Langer, J. E. Mayer and J. P. Vacanti, The *in vitro* construction of a tissue engineered bioprosthetic heart valve, *Eur J Cardiothorac Surg* **11**: 493–497 (1997).

24. T. Shinoka, C. K. Breuer, R. E. Tanel, G. Zund, T. Miura, P. X. Ma, R. Langer, J. P. Vacanti and J. E. Mayer, Jr., Tissue engineering heart valves: valve leaflet replacement study in a lamb model, *Ann Thorac Surg* **60**: S513–516 (1995).

25. C. K. Breuer, T. Shin'oka, R. E. Tanel, G. Zund, D. J. Mooney, P. X. Ma, T. Miura, S. Colan, R. Langer, J. E. Mayer and J. P. Vacanti, Tissue engineering lamb heart valve leaflets, *Biotechnol Bioeng* **50**: 562–567 (1996).

26. T. Shinoka, D. Shum-Tim, P. X. Ma, R. E. Tanel, R. Langer, J. P. Vacanti and J. E. Mayer, Jr., Tissue-engineered heart valve leaflets: does cell origin affect outcome? *Circulation* **96**: II-102–107 (1997).

27.   T. Shinoka, P. X. Ma, D. Shum-Tim, C. K. Breuer, R. A. Cusick, G. Zund, R. Langer, J. P. Vacanti and J. E. Mayer, Jr., Tissue-engineered heart valves: autologous valve leaflet replacement study in a lamb model, *Circulation* **94**: II-164–168 (1996).

28.   C. K. Breuer, B. A. Mettler, T. Anthony, V. L. Sales, F. J. Schoen and J. E. Mayer, Application of tissue-engineering principles toward the development of a semilunar heart valve substitute, *Tissue Eng* **10**: 1725–1736 (2004).

29.   R. M. Nerem, Tissue engineering a blood vessel substitute: the role of biomechanics. *Yonsei Med J*, **41**: 735–739 (2000).

30.   G. M. Riha, P. H. Lin, A. B. Lumsden, Q. Yao and C. Chen, Review: application of stem cells for vascular tissue engineering, *Tissue Eng* **11**: 1535–1552 (2005).

31.   G. Matsumura, S. Miyagawa-Tomita, T. Shin'oka, Y. Ikada and H. Kurosawa, First evidence that bone marrow cells contribute to the construction of tissue-engineered vascular autografts *in vivo*, *Circulation* **108**: 1729–1734 (2003).

32.   M. Watanabe, T. Shin'oka, S. Tohyama, N. Hibino, T. Konuma, G. Matsumura, Y. Kosaka, T. Ishida, Y. Imai, M. Yamakawa, Y. Ikada and S. Morita, Tissue-engineered vascular autograft: inferior vena cava replacement in a dog model, *Tissue Eng* **7**: 429–439 (2001).

33.   T. Shin'oka, G. Matsumura, N. Hibino, Y. Naito, M. Watanabe, T. Konuma, T. Sakamoto, M. Nagatsu and H. Kurosawa, Midterm clinical result of tissue-engineered vascular autografts seeded with autologous bone marrow cells, *J Thorac Cardiovasc Surg* **129**: 1330–1338 (2005).

34.   N. Hibino, E. McGillicuddy, G. Matsumura, Y. Ichihara, Y. Naito, C. Breuer and T. Shinoka, Late-term results of tissue-engineered vascular grafts in humans, *J Thorac Cardiovasc Surg* **139**: 431–436; e432 (2010).

35.   R. I. Lopez-Soler, M. P. Brennan, A. Goyal, Y. Wang, P. Fong, G. Tellides, A. Sinusas, A. Dardik and C. Breuer, Development of a mouse model for evaluation of small diameter vascular grafts, *J Surg Res* **139**: 1–6 (2007).

36.   J. D. Roh, G. N. Nelson, M. P. Brennan, T. L. Mirensky, T. Yi, T. F. Hazlett, G. Tellides, A. J. Sinusas, J. S. Pober, W. M. Saltzman, T. R. Kyriakides and C. K. Breuer, Small-diameter biodegradable scaffolds for functional vascular tissue engineering in the mouse model, *Biomaterials* **29**: 1454–1463 (2008).

37.   J. D. Roh, R. Sawh-Martinez, M. P. Brennan, S. M. Jay, L. Devine, D. A. Rao, T. Yi, T. L. Mirensky, A. Nalbandian, B. Udelsman, N. Hibino, T. Shinoka, W. M. Saltzman, E. Snyder, T. R. Kyriakides, J. S. Pober and C.

K. Breuer, Tissue-engineered vascular grafts transform into mature blood vessels via an inflammation-mediated process of vascular remodeling, *Proc Natl Acad Sci USA* **107**: 4669–4674 (2010).

38. N. Hibino, T. Yi, D. R. Duncan, A. Rathore, E. Dean, Y. Naito, A. Dardik, T. Kyriakides, J. Madri, J. S. Pober, T. Shinoka and C. K. Breuer, A critical role for macrophages in neovessel formation and the development of stenosis in tissue-engineered vascular grafts, *FASEB J* **25**: 4253–4263 (2011).

39. N. Hibino, G. Villalona, N. Pietris, D. R. Duncan, A. Schoffner, J. D. Roh, T. Yi, L. W. Dobrucki, D. Mejias, R. Sawh-Martinez, J. K. Harrington, A. Sinusas, D. S. Krause, T. Kyriakides, W. M. Saltzman, J. S. Pober, T. Shin'oka and C. K. Breuer, Tissue-engineered vascular grafts form neovessels that arise from regeneration of the adjacent blood vessel, *FASEB J* **25**: 2731–2739 (2011).

40. N. L'Heureux, L. Germain, R. Labbe and F. A. Auger, *In vitro* construction of a human blood vessel from cultured vascular cells: a morphologic study, *J Vasc Surg* **17**: 499–509 (1993).

41. N. L'Heureux, J. C. Stoclet, F. A. Auger, G. J. Lagaud, L. Germain and R. Andriantsitohaina, A human tissue-engineered vascular media: a new model for pharmacological studies of contractile responses, *FASEB J* **15**: 515–524 (2001).

42. N. L'Heureux, N. Dusserre, G. Konig, B. Victor, P. Keire, T. N. Wight, N. A. Chronos, A. E. Kyles, C. R. Gregory, G. Hoyt, R. C. Robbins and T. N. McAllister, Human tissue-engineered blood vessels for adult arterial revascularization, *Nat Med* **12**: 361–365 (2006).

43. T. N. McAllister, M. Maruszewski, S. A. Garrido, W. Wystrychowski, N. Dusserre, A. Marini, K. Zagalski, A. Fiorillo, H. Avila, X. Manglano, J. Antonelli, A. Kocher, M. Zembala, L. Cierpka, L. M. de la Fuente and N. L'Heureux, Effectiveness of haemodialysis access with an autologous tissue-engineered vascular graft: a multicentre cohort study, *Lancet* **373**: 1440–1446 (2009).

44. D. Hanjaya-Putra and S. Gerecht, Vascular engineering using human embryonic stem cells, *Biotechnol Prog* **25**: 2–9 (2009).

45. M. D. Cuenca-Lopez, P. Zamora-Navas, J. M. Garcia-Herrera, M. Godino, J. M. Lopez-Puertas, E. Guerado, J. Becerra and J. A. Andrades, Adult stem cells applied to tissue engineering and regenerative medicine, *Cell Mol Biol(Noisy-le-grand)* **54**: 40–51 (2008).

46. K. Yamahara and H. Itoh, Potential use of endothelial progenitor cells for regeneration of the vasculature, *Ther Adv Cardiovasc Dis* **3**: 17–27 (2009).

47. M. B. Nourse, D. E. Halpin, M. Scatena, D. J. Mortisen, N. L. Tulloch, K. D. Hauch, B. Torok-Storb, B. D. Ratner, L. Pabon and C. E. Murry, VEGF induces differentiation of functional endothelium from human embryonic stem cells: implications for tissue engineering, *Arterioscler Thromb Vasc Biol* **30**: 80–89 (2010).

48. L. J. Harris, H. Abdollahi, P. Zhang, S. McIlhenny, T. N. Tulenko and P. J. DiMuzio, Differentiation of adult stem cells into smooth muscle for vascular tissue engineering, *J Surg Res* **168**: 306–314 (2011).

49. A. Noghero, F. Bussolino and A. Gualandris, Role of the microenvironment in the specification of endothelial progenitors derived from embryonic stem cells, *Microvasc Res* **79**: 178–183 (2010).

50. Q. Xiao, Z. Luo, A. E. Pepe, A. Margariti, L. Zeng and Q. Xu, Embryonic stem cell differentiation into smooth muscle cells is mediated by Nox4-produced H_2O_2, *Am J Physiol Cell Physiol* **296**: C711–723 (2009).

51. G. Matsumura, N. Hibino, Y. Ikada, H. Kurosawa and T. Shin'oka, Successful application of tissue engineered vascular autografts: clinical experience, *Biomaterials* **24**: 2303–2308 (2003).

52. Y. Noishiki, Y. Tomizawa, Y. Yamane and A. Matsumoto, Autocrine angiogenic vascular prosthesis with bone marrow transplantation, *Nat Med* **2**: 90–93 (1996).

53. S. Shintani, T. Murohara, H. Ikeda, T. Ueno, T. Honma, A. Katoh, K. Sasaki, T. Shimada, Y. Oike and T. Imaizumi, Mobilization of endothelial progenitor cells in patients with acute myocardial infarction, *Circulation* **103**: 2776–2779 (2001).

54. R. McKay, Stem cells: hype and hope, *Nature* **406**: 361–364 (2000).

6

STEM CELL-DERIVED ENDOTHELIAL CELLS FOR CARDIOVASCULAR REGENERATION

Luqia Hou and Ngan F. Huang

1. Introduction

Owing to greater knowledge in stem cell biology and advances in regenerative medicine, stem cell-based therapies are being clinically applied toward treatment of cardiovascular diseases.[1-3] Endothelial cells (ECs) derived from stem cells have been applied in a number of transplantation studies in which they have shown significant therapeutic improvement in diseased animal models.[4,5] With the rapid development of reprogramming approaches, induced pluripotent stem cells (iPSCs) are regarded as one of the most promising cell sources that can provide unlimited patient-specific therapeutic cells, including iPSC-derived ECs (iPSC-ECs), for regenerative medicine and tissue engineering applications.[6] Recently, transdifferentiation approaches have been developed, which further accelerate the process of generating therapeutic ECs.[7–9] This chapter will summarize the progress in generating ECs from pluripotent and adult stem cells for vascular repair and regenerative medicine. In addition, the advantages and limitations of current approaches to generate therapeutic cells will be discussed, as well as the challenges in applying these approaches in a clinical setting.

2. Cell Sources and Characterization of Stem Cell-Derived ECs

ECs can be generated from different cell sources and characterized based on known phenotypic and functional markers (Table 1). ECs phenotypically express cluster of differentiation 31 (CD31), vascular endothelial cadherin (VE-cadherin), von Willebrand Factor (vWF), and endothelial nitric oxide synthase (eNOS). CD31, also known as platelet/endothelial cell adhesion molecule-1 (PECAM-1), is a transmembrane glycoprotein that can be found on the surface of platelets and at EC intercellular junctions.[10] Previous studies have suggested that CD31 is critical for endothelial cell-cell adhesion, as antibodies against CD31 inhibit ECs from forming a confluent monolayer *in vitro*.[11] Another important EC marker is VE-cadherin, also known as CD144, which belongs to the cadherin superfamily and can be found at endothelial adherens junctions.[12] As a multimeric glycoprotein present in blood plasma, in vessel walls, in ECs, vWF plays a critical role in platelet activation during vascular injury by interacting with a number of subendothelial matrix components and membrane receptors.[13] Another intracellular marker is eNOS, the endothelial-specific isoform of nitric oxide synthase that produces nitric oxide, which mediates vascular relaxation and regulates blood pressure and blood flow.[14] Besides phenotypic markers, ECs are functionally well characterized to be able to uptake acetylated low-density lipoprotein (acLDL) and form an endothelial tube network when cultured *in vitro* on matrigel. A more definitive proof of endothelial phenotype is inosculation, which is the ability of cells to connect with the existing endothelial network.[15]

So far, a number of different types of stem cells have been tested in animal studies, including embryonic stem cells (ESCs), iPSCs, adult stem cells, and bone marrow stem cells. It is important to understand the

Table 1. Cell source, differentiation, and characterizations of stem cell-derived ECs.

Cell Source	Differentiation Type	Feeder Cell	Co-Culture /Growth Factor	*In vitro* Characterization	Ref. #
hESC (H9)	EB formation			RT-PCR: CD31; VE-cad; CD34; FLK1; TIE-2; GATA-2 and 3; CD133 IF: CD31; VE-cad; vWF Ac-LDL uptake; tube formation on matrigel	[18]
hESC (H1, H1.1, H9.2)	EB formation		Mouse bone marrow stromal cell line S17 or mouse yolk-sac EC line C166	Flow cytometry: CD34, CD31 Magnetic selection: CD34 RT-PCR: TAL-1, GATA-2	[51]
hESC (H1, H7, H9; MA01, MA03, MA40, MA09)	EB formation		2 steps (EB for 3.5 days; EB single cells for 4–6 days in methycellulose	hES-BCs: Wright-Giemsa staining; IF with CD235a, D13, and CD45; hES-BCs-ECs: Ac-LDL uptake, IF of vWF and CD31	[117]
hESC (H9)	EB formation			Flow cytometry of CD31, Ac-LDL uptake assay, matrigel assay	[118]
hESC (H9, H1)	EB formation			Flow cytometry of CD45, CD34, CD31, c-Kit, CD33, CD13, CD41, AC133, Glycophorin A, CD105, VE-cadherin. IF with VE-cadherin, CD31, CD45	[119]

(*Continued*)

Table 1. *(Continued)*

Cell Source	Differentiation Type	Feeder Cell	Co-Culture /Growth Factor	*In vitro* Characterization	Ref. #
hESC	2D Culture	MEF		Flow cytometry of CD34; IF of VE-cad, CD31; Ac-LDL uptake, tube formation on matrigel	[120]
mESC	2D culture	Collagen IV	VEGF	Flow cytometry: FLK1 Immunohischemistry: VE-cad, CD34, endoglin Ac-LDL uptake	[75]
miPSC	2D culture	Collagen IV	VEGF	RT-PCR IF: CD31, VE-cad, SM22a, Calponin	[27]
mESC/iPSC	2D culture	Gelatin	OP9/angiopoietin-1	Flow cytometry of CD31, CD144, CD45, CD41, CD117; IF with CD144, CD41; matrigel tube formation	[63]

Abbreviations: acetylated low density lipoprotein (acLDL); embryonic body (EB); endothelial cell (EC); human embryonic stem cell (hESC); human embryonic stem cell-derived blast cells (hES-BCs); immunofluorescence (IF); mouse embryonic stem cell (mESC); mouse induced pluripotent stem cell (miPSC); two dimensional (2D); mouse embryonic fibroblasts (MEF).

phenotype and characterization of these cells, in order to utilize them for cell therapy and tissue engineering. Moreover, it has been reported that ECs from various stem cell sources exhibit heterogeneity in their cell surface marker expression as well as their functional properties.[16] This heterogeneity adds an additional level of complexity in characterizing the identity of stem cell-derived ECs. Therefore, it is critical to recognize the benefits and limitations associated with using different sources of ECs.

2.1 *ESCs*

ESCs can be obtained from the inner cell mass of a blastocyst.[17] These cells have unlimited renewal capacity and can differentiate into all three germ layers, namely the ectoderm, endoderm, and mesoderm. In 2002, Levenberg *et al.* were among the first to derive ECs from human ESCs and characterize these cells *in vitro* and *in vivo*.[18] When human ESCs were removed from mouse embryonic fibroblast (MEF) feeder cells and subjected to suspension culture conditions, they formed embryoid bodies (EBs) that promoted spontaneous differentiation. With the help of endothelial cell growth factors, the expression of EC genes increased from day 5 to days 13–15.[18] Human ESC-derived ECs (ESC-ECs) can be isolated by flow cytometry using one or a combination of several endothelial cell surface markers, including CD31 and VE-cadherin. When cultured *in vitro* on matrigel, these cells can form tube-like structures, similar to human umbilical vein endothelial cells (HUVECs).[18] Recently, detailed protocols of isolation, differentiation, and characterization of human ESC-ECs have been developed, where more than 80% endothelial-like/smooth muscle-like cells were generated.[19] However, the prominent limitations of ESCs include teratoma formation *in vivo* and the ethical issues associated with destruction of embryos.

Human ESC-ECs have been transplanted into different ischemic cardiovascular disease models.[4,20,21] These studies have shown improvement in cardiac function and neovascular network formation from two to eight weeks after cell delivery. In an earlier study by Levenberg and colleagues, mouse blood cells were successfully detected in transplanted human vessels, which indicated that transplanted ESC-ECs were able to form functional microvessels that integrate with host vasculature *in vivo*.[18] In a more recent study using a mouse model of myocardial infarction (MI), human ESC-ECs that were delivered by intramyocardial injection showed increased capillary density in the peri-infarct area after six weeks.[22] Although large animal data are still lacking, the positive results in rodent disease models point to the potential benefit of ESC-ECs for treating peripheral artery disease (PAD), ischemic heart disease and other cardiovascular diseases.

Although ESC-ECs are not currently tested in clinical trials, other ESC derivatives are currently under investigation. Advanced Cell Technology's clinical trial to treat Stargardt's Macular Dystrophy using human ESC-derived retinal pigment epithelial cells has shown positive safety results, although with only a few patients. Prior to this trial, Geron's clinical trial involving human ESC-derived oligodendrocyte progenitors for treatment of spinal cord injury was suspended due to concerns of cyst formation, and ultimately terminated because of business reasons.

2.2 *iPSCs*

Less than one decade ago, the successful generation of iPSCs from mouse and human fibroblasts became a milestone for stem cell biology.[6,23] This tremendous achievement by Shinya Yamanaka's group paved the way for stem cell research and regenerative medicine. In 2006, Yamanaka and colleagues first described the reprogramming approach using four transcriptional factor genes (OCT3/4, C-MYC, NANOG, and SOX2) to generate pluripotent stem cells from mouse fibroblasts.[23] James Thomson's laboratory showed similar results using a different set of factor genes (OCT3/4, SOX2, NANOG, and LIN28).[24] These cells are derived from the somatic cells and can obviate immune rejection when later transplanted back into the same patient. More importantly, iPSCs avoid the destruction of embryos and bypass the ethical issues that human ESCs encountered. In 2012, Dr. Yamanaka was awarded the Nobel Prize in Physiology for Medicine for his contribution in discovering the approach of converting somatic cells to pluripotent cells. In the last decade, iPSCs have been widely used in different research fields, including cell therapy, disease modeling, drug screening, and tissue engineering.[25] Functionally, iPSCs behave similarly to ESCs in terms of their self-renewal capacity and capability of generating all three germ layers. Different cell lineages, including ECs, have been successfully derived from these cells and been characterized *in vitro* and *in vivo*. Patient-specific disease models have also been generated in order to study the mechanism of cardiovascular diseases. For example, Moretti *et al.* developed the first patient-specific Long QT syndrome (LQT) model from iPSC-derived cardiomyocytes.[26] ECs were also derived from iPSCs, first from Yamanaka's group.[27] These iPSC-ECs had similar phenotype as the primary ECs. In 2011, Rufaihah *et al.* tested the effect of iPSC-EC delivery in the ischemic hindlimb of SCID mice.[5] The results indicated an increase in the capillaries in the ischemic limb area, as well as a higher degree of blood perfusion. Although there are a number of hurdles to overcome before iPSC-ECs can be used in the clinical setting, such as overcoming genetic and epigenetic instability as well as scaling up a large number of highly purified cells, iPSCs remain a promising cell source for EC cell therapy.

2.3 *Adult stem cells*

Although pluripotent stem cells have recently become a popular research field for vascular regeneration, adult stem cells are among the earliest cells that have been studied in the pre-clinical setting, and now are being investigated in a number of clinical trials. As multipotent stem cells, these cells are partially cell fate-committed in that they generally give rise to cells belonging to one germ layer. Because these cells can be obtained autologously, there is no immunological barrier. Furthermore, they do not require embryo destruction and do not appear to form teratomas. Therefore, adult stem cells can serve as a good candidate for cell therapy. Various adult stem cells, including endothelial progenitor cells (EPCs), mesenchymal stem cells, and multipotent adult progenitor cells, have been investigated intensively in the past 20 years.

Among the adult stem cells, EPCs are a rising star for vascular regeneration. They were first introduced in 1997 as a group of cells derived from the bone marrow that can differentiate into ECs and form blood vessels.[28] Isner's group pioneered the characterization of EPCs, and provided strong evidence that these cells were involved in angiogenesis in animal models of ischemia. These cells were characterized by the expression of both hematopoietic cell markers (CD133 or CD34) and endothelial marker vascular endothelial growth factor receptor-2 (VEGFR-2, also known as KDR or FLK1). Data from various studies have suggested that EPCs may consist of different cell populations since the cell surface markers are heterogeneously expressed.[29–32] Furthermore, no single marker is found to be exclusively expressed by EPCs. For example, CD34 is a single-transmembrane sialomucin that can be identified in a number of mesoderm derivatives, including blood, ECs, fibroblasts, and cancer stem cells. Therefore, it remains a challenge to distinguish EPCs from circulating ECs and blood cells or divide them into subtypes. Nevertheless, EPCs are a promising target for therapeutic neovascularization. In 1999, Takahashi *et al.* demonstrated that ischemia and exogenous cytokines were able to stimulate EPC mobilization and promote neovascularization in both mice and rabbits.[33] Later, Vasa *et al.* examined the relationship between EPCs and the risk factors of coronary artery disease.[34] They found that the number and migratory activity of EPCs were inversely associated with atherosclerotic risk factors, which were scored considering age, gender, medical history, and left ventricular ejection fraction (LVEF). Further investigation of the prognostic values indicated that an increased EPC level is correlated with reduced cardiovascular-related deaths.[35] Animal studies also support the idea that EPCs play an important role in the circulation. When human EPCs were transplanted into athymic nude mice with hindlimb ischemia, not only was limb perfusion recovery significantly improved, but the capillary density was higher, along with a reduced rate of limb loss.[36]

Adult stem cells have been tested in clinical trials and have shown therapeutic benefit. In 2002, Tateishi-Yuyama *et al.* performed a pilot study in 52 patients with unilateral/bilateral leg ischemia.[37] They found that injection of bone marrow mononuclear cells (BMMNCs) significantly improved ankle-branchial index (ABI), rest pain, and pain-free walking time at four weeks, compared to the injection of peripheral blood mononuclear cells (PBMNCs). A number of reported clinical trials followed and suggested modest improvement in ABI, transcutaneous oxygen level, rest pain, or pain-free walking distance. However, due to the fact that most of the trials were performed with a small number of patients, and a short follow-up time period, it is still not clear how long-term cell therapy affected these patients. Other ongoing clinical trials using adult stem cells for treatment of cardiovascular diseases such as myocardial infarction and stroke include REPAIR-AMI, BOOST trial, and IACT.[38–41] Although most clinical trials using adult stem cells have shown positive results, the mechanism of these beneficial therapeutic effects is still unclear and awaiting further studies to verify the clinical benefit.

2.4 *Comparison between multipotent and pluripotent stem cells*

The main advantages of multipotent adult stem cells include the ease of isolation, high expansion ratio, ability to be derived autologously, and the lack of teratoma formation. The drawbacks include the rare presence of these cells, and fairly long expansion time to obtain large quantity *in vitro*. Although certain markers can be used to distinguish different progenitor cells, selection of specific adult stem cell subpopulations still remains difficult due to the fact that they share a number of common surface markers with other stem cells. In addition, there is no general consensus in the characteristics and phenotypic markers of certain stem cell populations, such as EPCs and mesenchymal stem cells.

On the other hand, pluripotent stem cells are advantageous because they are theoretically able to provide an unlimited number of cells. Compared to multipotent adult stem cells, pluripotent stem cells have greater plasticity and able to differentiate into cells in all three germ layers. Since the generation of the first human ESC line in 1998, pluripotent stem cells have attracted tremendous attention as a therapeutic cell source.[17] However, a major concern of using ESCs is the ethical issue surrounding the destruction of embryos. With the generation of iPSCs, pluripotent stem cells remain a promising cell source for regenerative medicine. However, the molecular signaling pathways and microenvironment conditions for optimal differentiation of ECs are still not well understood. Despite the fact that vascular cells can be obtained from pluripotent stem cells, researchers will need to significantly improve the efficiency and purity of the stem cell-derived ECs in order to accelerate the progress toward clinical application.

3. Co-Culture System in Endothelial Differentiation

The co-culture system is defined by directly culturing stem cells with other cell types in order to induce lineage-specific differentiation. It mimics the natural environment of embryonic development by providing inductive signaling cues such as mechanical stimulation, activation of signaling pathways, and releasing of soluble and diffusable molecular factors. Many studies have shown that the co-culture system can efficiently promote differentiation of stem cells into specific lineages. For example, dopaminergic neurons have been obtained from differentiation when ESCs were co-cultured with Sertoli cells;[42] retinal cells can be generated from ESCs co-culturing with ESC-derived pigmented cells; and cardiomyocyte differentiation can be promoted by co-culturing ESCs with bone marrow stromal cells.[43]

3.1 *OP9 stromal cells*

It has been demonstrated that hematopoietic differentiation can be significantly enhanced when mouse/human ESCs were co-cultured with OP9 mouse stromal cell line.[44,45] In 2005, Vodyanik *et al.* reported a yield of 20% $CD34^+$ cells with high purity (>95%) using the OP9 co-culture approach for ESC differentiation.[45] In 2009, Choi *et al.* investigated the effect of OP9 cells in hematopoietic and endothelial differentiation with seven different iPSC lines, and they compared the efficiency with five different ESC lines.[46] The results suggested that human iPSCs co-cultured with OP9 cells were able to generate both hematopoietic progenitor cells ($CD34^+/CD43^+$) and endothelial cells ($CD31^+/CD43^-$), similar to human ESCs. In addition, Kono *et al.* successfully obtained lymphatic ECs from ESCs on OP9 stromal cells but not on type IV collagen.[47] These results indicated the importance of OP9 stromal cells in mimicking the differentiation environment of endothelium development. Although the mechanism remains unclear, a number of soluble factors secreted by OP9 cells could contribute to the differentiation process, including vascular endothelial growth factor-C (VEGF-C), angiopoietin-1, interferon-γ, interleukin-3, and interleukin-6.[47,48] More recently, a population of "small cells" in adult murine bone marrow was studied by Ratajczak *et al.*[49] The authors found that only after co-culturing these cells with OP9 stromal cells, can hematopoietic differentiation be achieved.

Although the mechanism is not yet elucidated, OP9 cells are believed to augment the proliferation of hemogenic precursors and at the same time inhibit their apoptosis.[50]

3.2 *Other cell types in co-culture*

A number of other cell types have also been utilized in EC differentiation. Murine bone marrow cell line S17 was used to generate hematopoietic precursor cells from human ESCs by Kaufman *et al.*[51] A number of cell surface markers (i.e. CD34) and hematopoietic transcription factors (i.e. TAL-1, LMO-2, GATA-2) were identified in these cells. M2-10B4 stromal cells were also used in co-culture to generate vascular progenitor cells from human ESCs.[52] On the other hand, Vodyanik *et al.* suggested that MS-5 stromal cells were able to promote differentiation of CD34$^+$ cells into lymphoid and myeloid lineages in the presence of several stem cell factors.[45] The mechanism of different co-culture systems promoting EC differentiation is under exploration.

4. Endothelial Transdifferentiation from Somatic Cells As an Alternative Approach

Besides differentiation from stem cells, a recent paradigm is direct reprogramming of somatic cells into other cell lineages in a process known as transdifferentiation. Transdifferentiation eliminates the intermediate step of reprograming somatic cells into iPSCs before then differentiating them into the cell type of interest, which reduces the risk of teratoma formation. In 2010, Szabo *et al.* first reported that blood progenitor cells can be generated with direct conversion from human fibroblasts.[53] One year later in 2011, neurons were obtained directly from *C. elegans* germ cells,[54] as well as from mouse and human fibroblasts.[55] Around the same time, cardiomyocytes were directly reprogrammed from fibroblasts using either defined transcription factors (GATA4, MEF2C, and TBX5) or developmental cues such as small molecules following the conventional reprogramming approach.[56,57]

These early successes in transdifferentiation from stem cells into neurons and cardiomyocytes encouraged researchers in the vascular field to find a shortcut to generate ECs in a similar manner. In 2012, Rafii's group first reported that amniotic cells can be reprogrammed into ECs with transforming growth factor-β inhibition and expression of three members of the E-twenty six (ETS) family of transcription factors (ETV2, FLI1, and ERG1).[58] These amniotic cell-derived ECs behave similarly to adult ECs genetically and functionally. Moreover, they were able to form perfused vessels *in vivo* when engrafted and transplanted into animal models. Around the same time, Margariti *et al.* reported success in direct reprogramming of human fibroblasts into ECs.[7] In their study, fibroblasts turned into partial-iPSCs (PiPS) when treated with the four Yamanaka transcription factors (OCT4, SOX2, KLF4, and cMYC) for only four days. These PiPS did not form tumors *in vivo* and were able to further differentiate into ECs (PiPS-ECs). Characterization of these cells showed that PiPS-ECs improved blood flow recovery in the hindlimb ischemic model. More recently in 2013, Kurian *et al.* published their results in turning human fibroblasts into $CD34^+$ vascular progenitor cells that could differentiate into ECs and smooth muscle cells.[9] They obtained ECs through a three-step protocol, each of which takes eight days. First, the same four Yamanaka factors were used to induce fibroblasts into a plastic reprogrammable state. Next, a mesodermal induction medium was used to generate $CD34^+$ progenitor cells. Finally, endothelial differentiation could be induced in EC expansion media. This approach has been validated with both lentiviral constructs and non-integrative approaches. ECs from these progenitor cells were able to form vessels and improve blood circulation after 17 days of implantation in a Matrigel plug. In another study, Li *et al.* demonstrated that only OCT4 and KLF4 were required to convert fibroblasts into ECs.[8] A 28-day protocol was developed to generate ECs with relatively high purity. The success of avoiding using the oncogene cMYC makes the transdifferentiation procedure safer for clinical application.

Overall, transdifferentiation paved a way for quicker and safer generation of patient-specific ECs for regenerative medicine. Although ECs generated from transdifferentiation have been characterized and their functional phenotypes are similar to normal ECs *in vitro* and *in vivo*, the slight genetic and functional difference between these cells and mature adult ECs need to be carefully investigated before using them in clinical studies.

5. Role of the Microenvironment

The microenvironment surrounding the cells plays an important role in regulating stem cell proliferation, migration, and differentiation. Stem cell fate commitment is a key concern for generating a sufficient amount of specific cell types for regenerative medicine. A number of factors have been suggested to be involved in the differentiation process to different cell lineages. These microenvironment factors include soluble factors (cytokines and growth factors), extracellular matrix proteins, cell-cell interactions, substrate rigidity, mechanical stimulation, and two dimensional (2D) vs. three-dimensional (3D) architecture.[59]

5.1 *Differentiation of human stem cells using growth factors*

Among the soluble factors that induce EC differentiation, VEGF has been suggested to be the most potent inducer, and is able to increase differentiation of EC in a dose-dependent manner.[60] VEGF induces endothelial differentiation by acting with its two receptors (FLT1 and FLK1), activating the Rho/ROCK signaling pathway, and enhance nuclear translocation of myocardin-related transcription factor-A (MRTF-A).[61] VEGF has been shown important at late stages of differentiation, while bone morphogenetic protein-4 has been shown to play an important role toward the beginning of differentiation to specify mesodermal lineages through the Smad1/5/8 pathway.[62] Another important soluble factor is angiopoietin-1. In 2011, Joo *et al.* studied the effect of angiopoietin-1 in EC differentiation.[63] They showed the percentage of EC (CD31$^+$/CD144$^+$) doubled in mouse ESC when differentiation was conducted with the OP9 co-culture system in presence of angiopoietin-1. This increase was also observed in mouse iPSCs and human ESC. Angiopoietin-1 acts on endothelial fate commitment through Tie2 signaling pathways, which is well known for its effects on angiogenesis.[64] In addition, with BIO, a GSK-3 inhibitor, in the culture system, endothelial differentiation has also been shown to increase significantly as VE-cadherin$^+$ and FLK1$^+$ cells were detected at a higher density.[65] More recently, an angiogenic peptide, TB4, has been identified to promote capillary growth through an unknown mechanism, which may involve activation of VEGF.[66] In 2012, Lippmann *et al.* established a monolayer culture system on Matrigel substrate and showed >60% efficiency in generating CD31[+]/GLUT-1[+] ECs.[67]

5.2 *Role of ECMs on endothelial differentiation*

5.2.1 *Naturally derived ECMs*

It has become increasingly recognized that the extracellular matrix (ECM) is not merely a scaffolding material that structurally supports cells, but also can functionally regulate cell behavior and phenotype through a number of biological molecules. Stem cell-ECM interaction is mainly mediated through heterodimeric transmembrane adhesion receptors known as integrins. Two major components of ECM, collagens (27 collagen types and nine families) and cell adhesive glycoproteins (laminin and fibronectin), have been suggested to promote EC differentiation. Other components of ECM, including proteoglycans and elastins, are mainly involved in the maintenance of mechanical stability.

Collagens and glycoproteins are able to bind to different types of integrin receptors and generate instructive signals to influence cell behavior. The role of integrins in early development of the embryo was first studied in transgenic mice models. In 1995, Stephens *et al.* reported that β1-integrin knockout mice had severe defects in early development of embryos, with inner cell mass failure and collapsed blastocoeles.[68] Also in mice models, laminins and fibronectin have been suggested to be critical in the process of embryogenesis.[69] These early studies indicated that ECM proteins could be important in controlling the stem cell differentiation process.

Different substrates of cell culture can be involved in the balance between pluripotency and differentiation of stem cells. When ESCs are plated on type I or type IV collagen, pluripotency is maintained. However, when they are cultured on laminin- or fibronectin-coated dishes, differentiation is induced.[70] A variety of laminin subtypes have been suggested to differentially influence the differentiation process. For example, laminin-322 can induce osteogenic differentiation, while laminin-111 preferably stimulates neural differentiation.[71,72] On the other hand, fibronectin has been shown to upregulate integrin $\alpha5\beta1$ expression in ESCs to promote meso-endodermal lineages such as skeletal lineages.[70,73]

5.2.2 *ECMs induce EC differentiation*

A number of ECMs have been shown to induce EC differentiation. Collagen IV has been suggested to support ESCs differentiate into hemopoietic cells via proximal lateral mesodermal stage *in vitro*, whereas gelatin, fibronectin, and collagen I support mesodermal differentiation with less efficiency.[74] Collagen IV has been used to induce ESC-derived FLK1[+] progenitor cells *in vitro* to explore the mechanisms of further specification down hematopoietic or endothelial lineage. Yamashita *et al.* successfully purified and differentiated mouse ESC-derived FLK1[+] cells obtained from collagen IV-coated dishes into ECs and mural cells. These cells organized into vascular tubes in three dimensions in culture.[75] Besides mouse ESCs, human ECs have also been successfully differentiated from ESCs on collagen IV-treated dishes *in vitro*.[76] In addition, fibronectin was also shown to increase EC differentiation. Wijelath *et al.* observed a five-fold increase in EC colonies when human EPCs were cultured in the presence of both fibronectin and VEGF.[77] It has been suggested that $\alpha5\beta1$ integrin was involved in the enhancement of VEGF activity. Similar results were obtained by Kanayasu-Toyoda *et al.* when they isolated CD133[+] cells from periphery blood on fibronectin-coated dishes.[78] Moreover, other ECMs, such as gelatin, were also used in combination with VEGF to obtain ECs from human and mouse ESC differentiation.[18,79]

Besides inducing endothelial differentiation based on the chemical composition, ECMs can also promote endothelial differentiation by concurrently delivering inductive growth factors. Hydrogels with growth factors showed greater angiogenesis when delivered within engineered tissue constructs. For example, Phelps *et al.* developed protease-degradable polyethylene glycol (PEG)-based hydrogel matrices to deliver VEGF in order to stimulate vasculature growth in an *in vivo* setting of murine hindlimb ischemia.[80] They showed that the efficiency was higher when VEGF was immobilized to the hydrogel, than when compared to soluble VEGF. In addition, *in vivo* experiments suggested that there was a significant increase in the number of regenerated vessels using immobilized VEGF. Naturally derived materials have also been considered in EC tissue engineering after chemical modification. Zhang *et al.* has studied PEGylated fibrin patches and utilized it to promote EC differentiation from MSCs.[81] The authors showed a positive effect on myocardial repair with increased cell viability. On the other hand, Krenning *et al.* focused on searching new cell types for vascular regeneration using biomaterial. They reported that CD14[+] monocytes can be differentiated into EC-like cells on three different biodegradable biomaterials.[82] Their results suggested that CD14[+] cells can serve as a suitable cell source because they are more abundant in peripheral blood than CD34[+] EPCs, and they have the potential to differentiate into ECs in polymer films. Recently, Ferreira and colleagues developed a bioactive 3D hydrogel composed of a tethered RGD peptide and microencapsulated VEGF that can be used to enhance vascular differentiation.[83,84] Encapsulation of human ESCs in hydrogels with RGD peptide and VEGF led to a 20-fold increase in cells that expressing FLK1 compared to when the cells were allowed to spontaneously differentiate as EBs. These studies demonstrate that ECMs can be multifunctional in providing differentiation cues to stem cells.

5.3 *Physical and mechanical regulation of ECMs on differentiation*

The mechanical properties of ECM are also important in ESC differentiation, and are extensively reviewed elsewhere.[85] The stiffness/rigidity of the culture substrate can manipulate cell fate as well. Engler *et al.* built the fundamental framework to correlate the rigidity of the culture substrate to stem cell differentiation capacity. Their results indicated that stem cells differentiated into specific lineages on substrate rigidities that match the native stiffness of the target cell lineages. For example, neurogenic differentiation can be induced by using substrates with stiffness similar to brain tissue, whereas stiffer substrates resulted in osteogenic differentiation.[86] Moreover, when the rigidity of cell substrate matched that of myocardium (12 to 16 kPa), cardiosphere-derived cells (CDCs) were demonstrated to have an enhanced endothelial differentiation *in vitro*, as well as increased cell survival and vascular integration *in vivo*.[87] Further, matrix elasticity showed profound effect on rhythmical contractile properties and intracellular protein structures of isolated embryonic cardiomyocytes.[88]

6. Pre-clinical Testing of Therapeutic ECs

Experimental models of tissue ischemia have been developed and widely used to evaluate the functional effect of EC transplantation in improving neovascularization (Table 2). Common disease models are hindlimb ischemia, myocardial infarction, and stroke. In hindlimb ischemia, the femoral artery that supplies blood flow to the hindlimb is ligated and excised, significantly reducing blood flow to the leg.[89] To experimentally induce myocardial infarction (MI), the left anterior descending (LAD) coronary artery is ligated to restrict blood flow to the left ventricle.[90] In the stroke model, occlusion of the middle cerebral artery is commonly used to generate focal ischemia and produce brain infarction.[91] Although the results of EC transplantation in small animal models show therapeutic benefit, large animal models will likely be required in order for EC therapy to advance into clinical testing.

6.1 *PAD*

Among the animal models of PAD, the most popular and widely used one is the hindlimb ischemia model. The hindlimb ischemia model has been used to evaluate the functional capacity of stem cell-derived ECs. EPCs were first isolated from human blood by Asahara and colleagues and then delivered into mouse and rabbit ischemic hindlimbs. The EPCs showed integration with host capillary vessel walls and formed capillaries in the ischemic limb.[28] In 2001, Ikenaga *et al.* reported that implantation of autologous bone marrow cells into the ischemia limb of rats restored blood flow and improved deteriorated exercise capacity.[92] Recently, in order to overcome the problems of low cell number, cell survival, and implant efficiency in transplantation, Suuronen *et al.* developed an injectable collagen matrix to deliver CD133[+] progenitor cells into the rat hindlimb ischemia model.[93] A significant improvement in vascular net-

Table 2. Application of stem cell-derived ECs for cardiovascular repair.

Disease Model	Cell Type	Injection Approach	Follow-up Time	Outcome	Ref. #
Mouse/rabbit HLI	Putative progenitor ECs	Inject into tail vein	1 to 6 weeks	Integration to capillary vessel walls, arranged into capillaries by 6 weeks; incorporation of cells in capillaries and small arteries; cells localized to neovascular zones	[28]
Rat HLI	Bone marrow cells	Inject percutaneously at six points into gastrocnemius muscle	2 to 4 weeks	Angiogenesis; restored blood flow; increased exercise capacity	[92]
Rat HLI	$CD133^+$ progenitor cells	Injectable collagen-based matrix	2 weeks	$CD133^+$ cells retained better; cells incorporated into vascular structures' increased intramuscular arteriole and capillary density	[93]
Mouse HLI	Human ESC-ECs	Intramuscular injection into ischemic limbs	4 weeks	Increased limb salvage; increased blood perfusion; increased capillary and arteriole densities	[94]
Mouse HLI	Human ESC-ECs and hESC-SMC	Inject into right femoral artery	1 to 6 weeks	Increased blood flow and capillary density. Cells incorporated into host circulating vessels	[95]
Mouse HLI	Mouse ESC; ESC-ECs	Intramuscular, intrafemoral artery, intrafemoral vein injections	2 weeks	ESC-ECs localized to the ischemic limb; engraftment of ESC-ECs into vasculature; improved limb perfusion and neovascularization	[4]
Mouse HLI	Human iPSC-ECs	Intramuscular injection at day 0 and day 7	2 weeks	Increased blood flow and capillaries	[5]

Mouse MI	Mouse ESC-ECs	Intramyocardial injection	8 weeks	Echocardiogram shows improved LV function; increased capillaries and venules in infacted zones	[96]
Mouse MI	Porcine iPSC-ECs	Intramyocardial injection	4 weeks	Echocardiogram shows improved LV function; MRI shows increased ejection fraction; proangiogenic and antiapoptotic factors released from piPSC-ECs	[97]
Porcine MI	Human ESC-ECs and SMC	Fibrin 3D porous scaffold biomatrix with hESC-ECs and SMCs seeded injected	1 to 4 weeks	MRI shows improved LV function; significant engraftment of hESC-ECs	[98]
Porcine MI	Human iPSCs	Intramyocardial injection	15 weeks	hiPSCs can be visualized up to 15 weeks; hiPSC-ECs contribute to vascularization	[99]
Mouse stroke	EPCs	Infused intravenously via the tail vein	2 weeks	Neovascularization in ischemic zone; endogenous neurogenesis	[100]
Mouse stroke	Bone marrow-derived ECs and neurons	Intravenously into retro-orbital sinus	1 to 2 weeks	Cells incorporated into the vasculature in ischemic zone	[101]
Mouse/rat stroke	Human iPSCs	Intracerebral transplantation	1 week to 4 months	Recovery of forepaw movements; grafted cells exhibited elctrophysiological properties of mature neurons and received synaptic input from host neurons	[102]

Abbreviations: endothelial cells (ECs); endothelial progenitor cells (EPCs); hindlimb ischemia (HLI); myocardial infarction (MI); human embryonic stem cell-derived endothelial cells (hESC-ECs); induced pluripotent stem cell-derived endothelial cells (iPSC-ECs).

work restoration was found in this study. ECs differentiated from human ESCs and iPSCs were also examined in limb ischemia models. In 2007, Cho *et al.* injected human ESC-ECs intramuscularly into ischemic limbs of mouse, and followed them for 4 weeks.[94] An increase in limb salvage and blood perfusion were reported, along with improvement in capillary and arteriole densities in treated animals. Yamahara *et al.* investigated the effect of combined transplantation of ESC-ECs and ESC-derived smooth muscle cells into mice with ischemic hindlimbs.[95] Their data suggested that there were synergistic effects in terms of neovascularization, as well as improvement in blood flow when both cell types were transplanted, compared to when only one cell type was delivered. Huang *et al.* utilized bioluminescence imaging (BLI) to track murine ESC-ECs delivered to the ischemia hindlimb by intramuscular, intrafemoral artery and intrafemoral vein injection.[4] They showed that ESC-ECs localized in ischemic limb using all three modalities, but the systemically delivered cells resulted in greater improvement in limb perfusion and neovascularization when compared to local delivery. More recently, Rufaihah *et al.* examined the therapeutic efficacy of human iPSC-ECs in the mouse hindlimb ischemia model.[5] They reported an increase in capillary density in ischemic limb and improvement in blood perfusion when iPSC-ECs were injected. This further improved the feasibility of using human iPSC-ECs in developing novel cell therapy for PAD patients.

6.2 *MI*

ECs derived from stem cells were also delivered and tested in a number of MI models. In 2007, Li *et al.* first injected murine ESC-ECs into mice heart after acute MI was induced by LAD ligation.[96] Using BLI, the authors tracked these implanted cells for up to eight weeks. In the ESC-EC injection group, echocardiogram and CD31 immunostaining results indicated that there was a higher density of capillaries in the infarcted zones accompanied with functional improvement based on the fractional shortening (FS) measurement of left ventricles. Recently, Gu and colleagues generated ECs from porcine iPSCs and transplanted them into the mouse model of MI.[97] They showed that the transplantation of iPSC-ECs significantly improved FS by up to 5% and LFEV by 3.5% on average. More importantly, using quantitative protein assays and single-cell PCR, they identified the proangiogenic and antiapoptotic factors released from iPSC-ECs that are believed to be responsible for the neo-vascularization and the survival of cardiomyocytes in the ischemic region. A similar transplantation procedure was also applied to large animal models of MI such as pigs. In 2011, Xiong *et al.* utilized a biomatrix scaffold to deliver ECs and smooth muscle cells derived from human ESCs into a porcine model of MI.[98] BLI was used to confirm the engraftment of ESC-derived cells. Left ventricular (LV) function has been shown significantly improved based on cardiac magnetic resonance imaging results. More recently, Templin *et al.* evaluated transplantation of human iPSCs in a pig model of MI.[99] Using an *in vivo* imaging technique, the authors tracked these transgenic human iPSC lines for up to 15 weeks. They reported that ECs derived from these transplanted iPSCs contributed to vascularization. However, current studies were not able to determine if these improvements in cardiac function were the result of differentiation of stem cells into cardiomyocytes, or purely due to paracrine factors secreted by them in the infarcted region.

6.3 *Stroke*

Stem cell therapy has also been tested in stroke models. Focal ischemia is normally induced by middle cerebral artery occlusion in animals by cauterization, clips, or threads. Since neurogenesis and angiogenesis are closely coupled processes, researchers evaluated the therapeutic effect of ECs upon transplantation into stroke models. Taguchi *et al.* reported that systemic administration of EPCs into the mice model of stroke favored neuronal regeneration by an enhanced migration of neuronal progenitor cells to the ischemic zone.[100] Bone marrow-derived ECs and neurons were also suggested to be involved in the recovery after stoke when they were transplanted into a mice model.[101] Postnatal vasculogenesis has been observed in this study in cerebral infarcted brain after transplantation of bone marrow stem cells. More recently, Oki *et al.* demonstrated that human iPSC transplantation can improve recovery in stroke-injured brain along with increased VEGF levels that enhance endogenous plasticity.[102] This data demonstrate the importance of angiogenesis in the treatment of patients suffering from stroke.

7. Role of Prevascularization in Engineered Heart Tissues

Vasculature is the key in providing efficient nutrition supply when bioengineered tissues exceed the thickness that permits diffusion. Stem cell-derived ECs may provide a means to vascularize the tissue construct, as well as to interact with other cell types to regulate cell proliferation, survival, and migration. In recent studies to remuscularize damaged heart tissue with adult or pluripotent stem cells, constructs consisting of only cardiomyocytes survived poorly after transplantation. In contrast, when ECs and stromal cells are included in the construct, these cells organized into endothelial networks and integrated to host circulation.[103] These results are supported by other studies that show that prevascularized constructs can improve cell survival and function after transplantation.[104,105] These studies together suggest that the interaction of multiple cell types can have additive or synergistic effects in improving the survival or function of engineered tissues, presumably by cell-cell interactions or paracrine factors.

8. Clinical Relevance

Stem cells have enabled researchers to generate therapeutic cells for cell transplantation, patient-specific disease modeling, drug screening, and construction of artificial vessels and cardiac patches. Development of therapeutic approaches to stimulate or mimic angiogenesis is promising in treating patients with cardiovascular diseases. Clinical trials with PAD patients normally involve intramuscular administration of cells in the scale of $1–10 \times 10^7$ to multiple sites in the calf and thigh locations of patients. BMMNCs and PBMNCs were commonly used to treat PAD patients with either critical limb ischemia or only symptomatic PAD.[106] So far relatively small numbers of patients have been included in these clinical trials, and the follow-up time period is short. However, the efficacy is significant when considering the endpoint measurement such as ABI, transcutaneous oxygen level, laser Doppler perfusion, or pain-free walking distance and time in the case of PAD patients. Stem cell therapy was also applied in clinical trials with patients suffering from MI or stroke. Bone marrow stem/progenitor cells (BMSC) were commonly used in a number of randomized controlled trials worldwide.[107–110] Among the clinical trials, some selected subpopulations with specific surface markers, such as CD133[+],[111] or CD34[+]/CXCR4[+].[112] Improvement in LVEF was reported over short- or long-term follow-up, as well as reduction in LV end systolic and diastolic volumes and infarct size in certain cases.[113] Meanwhile, a number of Phase I and Phase II clinical trials of stem cell therapy to treat stroke have been completed or are ongoing. Mesenchymal stem cells and BMMNC were used in both acute and chronic stroke trials, with successes in reducing infarct volume or improvements in National Institutes of Health Stroke Scale scores.[114–116] Although small clinical trials suggested a modest benefit by directly injecting adult progenitor cells into the ischemic muscle, large-scale tissue constructs such as a biological conduit made from stem cell-derived EC might be more efficient in improving functions of injured tissues.

9. Challenges and Future Directions

There are a number of challenges that need to be overcome to improve stem cell-based therapy. First is the source of cells and the generally low-efficient isolation approach. With the current *in vitro* cell differentiation, culture, and passaging techniques, the ability to generate adequate numbers of clinical-grade ECs (billions of cells) for regenerative medicine is still far from ideal. The detailed mechanism of EC differentiation is largely unknown. We are on our way to understand the differentiation pathways involving growth factors and cytokines, as well as the ECM microenvironment to optimize the EC differentiation from pluripotent stem cells. Another concern is contamination by xenoproteins from animal serum or animal feeder cells. Animal serum and feeder cells would increase the risk of graft rejection and pathogenic transmission. Defined medium and human fibroblasts may provide an alternative solution to resolve this problem. Obviating teratoma formation will require better quality control to obtain pure and fully committed cells for transplantation. In addition, it is unclear whether venous, arterial, and lymphatic EC subtypes will have differential therapeutic effects, and their epigenetic stability will need to be fully characterized. Although there are challenges in using stem cell-derived ECs in regenerative medicine, the future of stem cell therapy in cardiovascular tissue regeneration remains bright.

Acknowledgements

This work was supported in part by grants to NFH from the US National Institutes of Health (R00HL098688, R01HL127113, and R21EB0202-35-01), Merit Review Award (1I01BX002310) from the Department of Veterans Affairs Biomedical Laboratory Research and Development, the Stanford Chemistry Engineering and Medicine for Human Health, and the Stanford Cardiovascular Institute.

References

1. N. F. Huang, R. J. Lee and S. Li, Chemical and physical regulation of stem cells and progenitor cells: potential for cardiovascular tissue engineering, *Tissue Eng* **13**: 1809–1823 (2007).

2. N. F. Huang, B. Patlolla, O. Abilez, H. Sharma, J. Rajadas, R. E. Beygui, C. K. Zarins and J. P. Cooke, A matrix micropatterning platform for cell localization and stem cell fate determination, *Acta Biomater* **6**: 4614–4621 (2010).

3. W. T. Wong, N. F. Huang, C. M. Botham, N. Sayed and J. P. Cooke, Endothelial cells derived from nuclear reprogramming, *Circ Res* **111**: 1363–1375 (2012).

4. N. F. Huang, H. Niiyama, C. Peter, A. De, Y. Natkunam, F. Fleissner, Z. Li, M. D. Rollins, J. C. Wu, S. S. Gambhir and J. P. Cooke, Embryonic stem cell-derived endothelial cells engraft into the ischemic hindlimb and restore perfusion, *Arterioscler Thromb Vasc Biol* **30**: 984–991 (2010).

5. A. J. Rufaihah, N. F. Huang, S. Jame, J. C. Lee, H. N. Nguyen, B. Byers, A. De, J. Okogbaa, M. Rollins, R. Reijo-Pera, S. S. Gambhir and J. P. Cooke, Endothelial cells derived from human iPSCS increase capillary density and improve perfusion in a mouse model of peripheral arterial disease, *Arterioscler Thromb Vasc Biol* **31**: e72–79 (2011).

6. K. Takahashi, K. Tanabe, M. Ohnuki, M. Narita, T. Ichisaka, K. Tomoda and S. Yamanaka, Induction of pluripotent stem cells from adult human fibroblasts by defined factors, *Cell* **131**: 861–872 (2007).

7. A. Margariti, B. Winkler, E. Karamariti, A. Zampetaki, T. N. Tsai, D. Baban, J. Ragoussis, Y. Huang, J. D. Han, L. Zeng, Y. Hu and Q. Xu, Direct reprogramming of fibroblasts into endothelial cells capable of angiogenesis and reendothelialization in tissue-engineered vessels, *Proc Natl Acad Sci USA* **109**: 13793–13798 (2012).

8. J. Li, N. F. Huang, J. Zou, T. J. Laurent, J. C. Lee, J. Okogbaa, J. P. Cooke and S. Ding, Conversion of human fibroblasts to functional endothelial cells by defined factors, *Arterioscler Thromb Vasc Biol* **33**:1366–1275 (2013).

9. L. Kurian, I. Sancho-Martinez, E. Nivet, A. Aguirre, K. Moon, C. Pendaries, C. Volle-Challier, F. Bono, J. M. Herbert, J. Pulecio, Y. Xia, M. Li, N. Montserrat, S. Ruiz, I. Dubova, C. Rodriguez, A. M. Denli, F. S. Boscolo, R. D. Thiagarajan, F. H. Gage, J. F. Loring, L. C. Laurent and J. C. Izpisua Belmonte, Conversion of human fibroblasts to angioblast-like progenitor cells, *Nat Methods* **10**: 77–83 (2013).

10. S. M. Albelda, W. A. Muller, C. A. Buck and P. J. Newman, Molecular and cellular properties of PECAM-1 (endoCAM/CD31): a novel vascular cell-cell adhesion molecule, *J Cell Biol* **114**: 1059–1068 (1991).

11. S. M. Albelda, P. D. Oliver, L. H. Romer and C. A. Buck, EndoCAM: a novel endothelial cell-cell adhesion molecule, *J Cell Biol* **110**: 1227–1237 (1990).

12. M. G. Lampugnani, M. Resnati, M. Raiteri, R. Pigott, A. Pisacane, G. Houen, L. P. Ruco and E. Dejana, A novel endothelial-specific membrane protein is a marker of cell-cell contacts, *J Cell Biol* **118**: 1511–1522 (1992).

13. Z. M. Ruggeri, von Willebrand factor, *J Clin Invest* **99**: 559–564 (1997).

14. R. M. Palmer, A. G. Ferrige and S. Moncada, Nitric oxide release accounts for the biological activity of endothelium-derived relaxing factor, *Nature* **327**: 524–526 (1987).

15. K. K. Hirschi, D. A. Ingram and M. C. Yoder, Assessing identity, phenotype, and fate of endothelial progenitor cells, *Arterioscler Thromb Vasc Biol* **28**: 1584–1595 (2008).

16. A. J. Rufaihah, N. F. Huang, J. Kim, J. Herold, K. S. Volz, T. S. Park, J. C. Lee, E. T. Zambidis, R. Reijo-Pera and J. P. Cooke, Human induced pluripotent stem cell-derived endothelial cells exhibit functional heterogeneity, *Am J Transl Res* **5**: 21–35 (2013).

17. J. A. Thomson, J. Itskovitz-Eldor, S. S. Shapiro, M. A. Waknitz, J. J. Swiergiel, V. S. Marshall and J. M. Jones, Embryonic stem cell lines derived from human blastocysts, *Science* **282**: 1145–1147 (1998).

18. S. Levenberg, J. S. Golub, M. Amit, J. Itskovitz-Eldor and R. Langer, Endothelial cells derived from human embryonic stem cells, *Proc Natl Acad Sci USA*, **99**: 4391–4396 (2002).

19. S. Levenberg, L. S. Ferreira, L. Chen-Konak, T. P. Kraehenbuehl and R. Langer, Isolation, differentiation and characterization of vascular cells

derived from human embryonic stem cells, *Nat Protoc* **5**: 1115–1126 (2010).

20. A. Lesman, M. Habib, O. Caspi, A. Gepstein, G. Arbel, S. Levenberg and L. Gepstein, Transplantation of a tissue-engineered human vascularized cardiac muscle, *Tissue Eng Part A* **16**: 115–125 (2010).

21. Z. Li, K. D. Wilson, B. Smith, D. L. Kraft, F. Jia, M. Huang, X. Xie, R. C. Robbins, S. S. Gambhir, I. L. Weissman and J. C. Wu, Functional and transcriptional characterization of human embryonic stem cell-derived endothelial cells for treatment of myocardial infarction, *PLoS One* **4**: e8443 (2009).

22. J. Yu, N. F. Huang, K. D. Wilson, J. B. Velotta, M. Huang, Z. Li, A. Lee, R. C. Robbins, J. P. Cooke and J. C. Wu, nAChRs mediate human embryonic stem cell-derived endothelial cells: proliferation, apoptosis, and angiogenesis, *PLoS One* **4**: e7040 (2009).

23. K. Takahashi and S. Yamanaka, Induction of pluripotent stem cells from mouse embryonic and adult fibroblast cultures by defined factors, *Cell* **126**: 663–676 (2006).

24. J. Yu, M. A. Vodyanik, K. Smuga-Otto, J. Antosiewicz-Bourget, J. L. Frane, S. Tian, J. Nie, G. A. Jonsdottir, V. Ruotti, R. Stewart, I. I. Slukvin and J. A. Thomson, Induced pluripotent stem cell lines derived from human somatic cells, *Science* **318**: 1917–1920 (2007).

25. E. T. Pashuck and M. M. Stevens, Designing regenerative biomaterial therapies for the clinic, *Sci Transl Med* **4**: 160sr164 (2012).

26. A. Moretti, M. Bellin, A. Welling, C. B. Jung, J. T. Lam, L. Bott-Flugel, T. Dorn, A. Goedel, C. Hohnke, F. Hofmann, M. Seyfarth, D. Sinnecker, A. Schomig and K. L. Laugwitz, Patient-specific induced pluripotent stem-cell models for long-QT syndrome, *N Engl J Med* **363**: 1397–1409 (2010).

27. G. Narazaki, H. Uosaki, M. Teranishi, K. Okita, B. Kim, S. Matsuoka, S. Yamanaka and J. K. Yamashita, Directed and systematic differentiation of cardiovascular cells from mouse induced pluripotent stem cells, *Circulation* **118**: 498–506 (2008).

28. T. Asahara, T. Murohara, A. Sullivan, M. Silver, R. van der Zee, T. Li, B. Witzenbichler, G. Schatteman and J. M. Isner, Isolation of putative progenitor endothelial cells for angiogenesis, *Science* **275**: 964–967 (1997).

29. J. K. McKenney, S. W. Weiss and A. L. Folpe, CD31 expression in intratumoral macrophages: a potential diagnostic pitfall, *Am J Surg Pathol* **25**: 1167–1173 (2001).

30. M. Miettinen, A. E. Lindenmayer and A. Chaubal, Endothelial cell markers CD31, CD34, and BNH9 antibody to H- and Y-antigens: evaluation of their

specificity and sensitivity in the diagnosis of vascular tumors and comparison with von Willebrand factor, *Mod Pathol* **7**: 82–90 (1994).

31. B. L. Ziegler, M. Valtieri, G. A. Porada, R. De Maria, R. Muller, B. Masella, M. Gabbianelli, I. Casella, E. Pelosi, T. Bock, E. D. Zanjani and C. Peschle, KDR receptor: a key marker defining hematopoietic stem cells, *Science* **285**: 1553–1558 (1999).

32. A. Y. Khakoo and T. Finkel, Endothelial progenitor cells, *Annu Rev Med* **56**: 79–101 (2005).

33. T. Takahashi, C. Kalka, H. Masuda, D. Chen, M. Silver, M. Kearney, M. Magner, J. M. Isner and T. Asahara, Ischemia- and cytokine-induced mobilization of bone marrow-derived endothelial progenitor cells for neovascularization, *Nat Med* **5**: 434–438 (1999).

34. M. Vasa, S. Fichtlscherer, A. Aicher, K. Adler, C. Urbich, H. Martin, A. M. Zeiher and S. Dimmeler, Number and migratory activity of circulating endothelial progenitor cells inversely correlate with risk factors for coronary artery disease, *Circ Res* **89**: E1–7 (2001).

35. N. Werner, S. Kosiol, T. Schiegl, P. Ahlers, K. Walenta, A. Link, M. Bohm and G. Nickenig, Circulating endothelial progenitor cells and cardio-vascular outcomes, *N Engl J Med* **353**: 999–1007 (2005).

36. C. Kalka, H. Masuda, T. Takahashi, W. M. Kalka-Moll, M. Silver, M. Kearney, T. Li, J. M. Isner and T. Asahara, Transplantation of *ex vivo* expanded endothelial progenitor cells for therapeutic neovascularization, *Proc Natl Acad Sci USA* **97**: 3422–3427 (2000).

37. E. Tateishi-Yuyama, H. Matsubara, T. Murohara, U. Ikeda, S. Shintani, H. Masaki, K. Amano, Y. Kishimoto, K. Yoshimoto, H. Akashi, K. Shimada, T. Iwasaka, T. Imaizumi and Therapeutic Angiogenesis using Cell Transplantation (TACT) Study Investigators, Therapeutic angiogenesis for patients with limb ischaemia by autologous transplantation of bone-marrow cells: a pilot study and a randomised controlled trial, *Lancet* **360**: 427–435 (2002).

38. V. Schachinger, S. Erbs, A. Elsasser, W. Haberbosch, R. Hambrecht, H. Holschermann, J. Yu, R. Corti, D. G. Mathey, C. W. Hamm, T. Suselbeck, N. Werner, J. Haase, J. Neuzner, A. Germing, B. Mark, B. Assmus, T. Tonn, S. Dimmeler, A. M. Zeiher and R.A. Investigators, Improved clinical outcome after intracoronary administration of bone-marrow-derived progenitor cells in acute myocardial infarction: final 1-year results of the REPAIR-AMI trial, *Eur Heart J* **27**: 2775–2783 (2006).

39. V. Schachinger, S. Erbs, A. Elsasser, W. Haberbosch, R. Hambrecht, H. Holschermann, J. Yu, R. Corti, D. G. Mathey, C. W. Hamm, T. Suselbeck,

B. Assmus, T. Tonn, S. Dimmeler, A. M. Zeiher and R.A. Investigators, Intracoronary bone marrow-derived progenitor cells in acute myocardial infarction, *N Engl J Med* **355**: 1210–1221 (2006).

40. K. C. Wollert, G. P. Meyer, J. Lotz, S. Ringes-Lichtenberg, P. Lippolt, C. Breidenbach, S. Fichtner, T. Korte, B. Hornig, D. Messinger, L. Arseniev, B. Hertenstein, A. Ganser and H. Drexler, Intracoronary autologous bone-marrow cell transfer after myocardial infarction: the BOOST randomised controlled clinical trial, *Lancet* **364**: 141–148 (2004).

41. B. E. Strauer, M. Brehm, T. Zeus, T. Bartsch, C. Schannwell, C. Antke, R. V. Sorg, G. Kogler, P. Wernet, H. W. Muller and M. Kostering, Regeneration of human infarcted heart muscle by intracoronary autologous bone marrow cell transplantation in chronic coronary artery disease: the IACT Study, *J Am Coll Cardiol* **46**: 1651–1658 (2005).

42. S. H. Lee, N. Lumelsky, L. Studer, J. M. Auerbach and R. D. McKay, Efficient generation of midbrain and hindbrain neurons from mouse embryonic stem cells, *Nat Biotechnol* **18**: 675–679 (2000).

43. F. M. Yue, S. Shirasawa, H. Ichikawa, S. Yoshie, A. Mogi, S. Masuda, M. Nagai, T. Yokohama, T. Daihachiro and K. Sasaki, Induce Differentiation of Embryonic Stem Cells by Co-Culture system, *INTECH* (2013).

44. T. Nakano, H. Kodama and T. Honjo, *In vitro* development of primitive and definitive erythrocytes from different precursors, *Science* **272**: 722–724 (1996).

45. M. A. Vodyanik, J. A. Bork, J. A. Thomson and I. I. Slukvin, Human embryonic stem cell-derived CD34+ cells: efficient production in the coculture with OP9 stromal cells and analysis of lymphohematopoietic potential, *Blood* **105**: 617–626 (2005).

46. K. D. Choi, J. Yu, K. Smuga-Otto, G. Salvagiotto, W. Rehrauer, M. Vodyanik, J. Thomson and I. Slukvin, Hematopoietic and endothelial differentiation of human induced pluripotent stem cells, *Stem Cells* **27**: 559–567 (2009).

47. T. Kono, H. Kubo, C. Shimazu, Y. Ueda, M. Takahashi, K. Yanagi, N. Fujita, T. Tsuruo, H. Wada and J. K. Yamashita, Differentiation of lymphatic endothelial cells from embryonic stem cells on OP9 stromal cells, *Arterioscler Thromb Vasc Biol* **26**: 2070–2076 (2006).

48. M. Groger, R. Loewe, W. Holnthoner, R. Embacher, M. Pillinger, G. S. Herron, K. Wolff and P. Petzelbauer, IL-3 induces expression of lymphatic markers Prox-1 and podoplanin in human endothelial cells, *J Immunol* **173**: 7161–7169 (2004).

49. J. Ratajczak, M. Wysoczynski, E. Zuba-Surma, W. Wan, M. Kucia, M. C. Yoder and M. Z. Ratajczak, Adult murine bone marrow-derived very small

embryonic-like stem cells differentiate into the hematopoietic lineage after coculture over OP9 stromal cells, *Exp Hematol* **39**: 225–237 (2011).

50. J. Ji, K. Vijayaragavan, M. Bosse, P. Menendez, K. Weisel and M. Bhatia, OP9 stroma augments survival of hematopoietic precursors and progenitors during hematopoietic differentiation from human embryonic stem cells, *Stem Cells* **26**: 2485–2495 (2008).

51. D. S. Kaufman, E. T. Hanson, R. L. Lewis, R. Auerbach and J. A. Thomson, Hematopoietic colony-forming cells derived from human embryonic stem cells, *Proc Natl Acad Sci USA* **98**: 10716–10721 (2001).

52. K. L. Hill, P. Obrtlikova, D. F. Alvarez, J. A. King, S. A. Keirstead, J. R. Allred and D. S. Kaufman, Human embryonic stem cell-derived vascular progenitor cells capable of endothelial and smooth muscle cell function, *Exp Hematol* **38**: 246–257 (2010).

53. E. Szabo, S. Rampalli, R. M. Risueno, A. Schnerch, R. Mitchell, A. Fiebig-Comyn, M. Levadoux-Martin and M. Bhatia, Direct conversion of human fibroblasts to multilineage blood progenitors, *Nature* **468**: 521–526 (2010).

54. B. Tursun, T. Patel, P. Kratsios and O. Hobert, Direct conversion of C. elegans germ cells into specific neuron types, *Science* **331**: 304–308 (2011).

55. E. Y. Son, J. K. Ichida, B. J. Wainger, J. S. Toma, V. F. Rafuse, C. J. Woolf and K. Eggan, Conversion of mouse and human fibroblasts into functional spinal motor neurons, *Cell Stem Cell* **9**: 205–218 (2011).

56. M. Ieda, J. D. Fu, P. Delgado-Olguin, V. Vedantham, Y. Hayashi, B. G. Bruneau and D. Srivastava, Direct reprogramming of fibroblasts into functional cardiomyocytes by defined factors, *Cell* **142**: 375–386 (2010).

57. J. A. Efe, S. Hilcove, J. Kim, H. Zhou, K. Ouyang, G. Wang, J. Chen and S. Ding, Conversion of mouse fibroblasts into cardiomyocytes using a direct reprogramming strategy, *Nat Cell Biol* **13**: 215–222 (2011).

58. M. Ginsberg, D. James, B. S. Ding, D. Nolan, F. Geng, J. M. Butler, W. Schachterle, V. R. Pulijaal, S. Mathew, S. T. Chasen, J. Xiang, Z. Rosenwaks, K. Shido, O. Elemento, S. Y. Rabbany and S. Rafii, Efficient direct reprogramming of mature amniotic cells into endothelial cells by ETS factors and TGFbeta suppression, *Cell* **151**: 559–575 (2012).

59. C. H. Wang, T. M. Wang, T. H. Young, Y. K. Lai and M. L. Yen, The critical role of ECM proteins within the human MSC niche in endothelial differentiation, *Biomaterials* **34**: 4223–4234 (2013).

60. F. Lanner, M. Sohl and F. Farnebo, Functional arterial and venous fate is determined by graded VEGF signaling and notch status during embryonic

stem cell differentiation, *Arterioscler Thromb Vasc Biol* **27**: 487–493 (2007).

61. N. Wang, R. Zhang, S. J. Wang, C. L. Zhang, L. B. Mao, C. Y. Zhuang, Y. Y. Tang, X. G. Luo, H. Zhou and T. C. Zhang, Vascular endothelial growth factor stimulates endothelial differentiation from mesenchymal stem cells via Rho/myocardin-related transcription factor-A signaling pathway, *Int J Biochem Cell Biol* **45**: 1447–1456 (2013).

62. H. Bai, Y. Gao, M. Arzigian, D. M. Wojchowski, W. S. Wu and Z. Z. Wang, BMP4 regulates vascular progenitor development in human embryonic stem cells through a Smad-dependent pathway, *J Cell Biochem* **109**: 363–374 (2010).

63. H. J. Joo, H. Kim, S. W. Park, H. J. Cho, H. S. Kim, D. S. Lim, H. M. Chung, I. Kim, Y. M. Han and G. Y. Koh, Angiopoietin-1 promotes endothelial differentiation from embryonic stem cells and induced pluripotent stem cells, *Blood* **118**: 2094–2104 (2011).

64. H. G. Augustin, G. Y. Koh, G. Thurston and K. Alitalo, Control of vascular morphogenesis and homeostasis through the angiopoietin-Tie system, *Nat Rev Mol Cell Biol* **10**: 165–177 (2009).

65. R. Tatsumi, Y. Suzuki, T. Sumi, M. Sone, H. Suemori and N. Nakatsuji, Simple and highly efficient method for production of endothelial cells from human embryonic stem cells, *Cell Transplant* **20**: 1423–1430 (2011).

66. L. L. Chiu, M. Montgomery, Y. Liang, H. Liu and M. Radisic, Perfusable branching microvessel bed for vascularization of engineered tissues, *Proc Natl Acad Sci USA* **109**: E3414–3423 (2012).

67. E. S. Lippmann, S. M. Azarin, J. E. Kay, R. A. Nessler, H. K. Wilson, A. Al-Ahmad, S. P. Palecek and E. V. Shusta, Derivation of blood-brain barrier endothelial cells from human pluripotent stem cells, *Nat Biotechnol* **30**: 783–791 (2012).

68. L. E. Stephens, A. E. Sutherland, I. V. Klimanskaya, A. Andrieux, J. Meneses, R. A. Pedersen and C. H. Damsky, Deletion of beta 1 integrins in mice results in inner cell mass failure and peri-implantation lethality, *Genes Dev* **9**: 1883–1895 (1995).

69. D. R. Armant, H. A. Kaplan and W. J. Lennarz, Fibronectin and laminin promote *in vitro* attachment and outgrowth of mouse blastocysts, *Dev Biol* **116**: 519–523 (1986).

70. Y. Hayashi, M. K. Furue, T. Okamoto, K. Ohnuma, Y. Myoishi, Y. Fukuhara, T. Abe, J. D. Sato, R. Hata and M. Asashima, Integrins regulate mouse embryonic stem cell self-renewal, *Stem Cells* **25**: 3005–3015 (2007).

71. R. F. Klees, R. M. Salasznyk, S. Vandenberg, K. Bennett and G. E. Plopper, Laminin-5 activates extracellular matrix production and osteogenic gene focusing in human mesenchymal stem cells, *Matrix Biol* **26**: 106–114 (2007).

72. S. Mruthyunjaya, R. Manchanda, R. Godbole, R. Pujari, A. Shiras and P. Shastry, Laminin-1 induces neurite outgrowth in human mesenchymal stem cells in serum/differentiation factors-free conditions through activation of FAK-MEK/ERK signaling pathways, *Biochem Biophys Res Commun* **391**: 43–48 (2010).

73. P. Pimton, S. Sarkar, N. Sheth, A. Perets, C. Marcinkiewicz, P. Lazarovici and P. I. Lelkes, Fibronectin-mediated upregulation of alpha5beta1 integrin and cell adhesion during differentiation of mouse embryonic stem cells, *Cell Adh Migr* **5**: 73–82 (2011).

74. S. I. Nishikawa, S. Nishikawa, M. Hirashima, N. Matsuyoshi and H. Kodama, Progressive lineage analysis by cell sorting and culture identifies FLK1+VE-cadherin+ cells at a diverging point of endothelial and hemopoietic lineages, *Development* **125**: 1747–1757 (1998).

75. J. Yamashita, H. Itoh, M. Hirashima, M. Ogawa, S. Nishikawa, T. Yurugi, M. Naito, K. Nakao and S. Nishikawa, Flk1-positive cells derived from embryonic stem cells serve as vascular progenitors, *Nature* **408**: 92–96 (2000).

76. S. Gerecht-Nir, A. Ziskind, S. Cohen and J. Itskovitz-Eldor, Human embryonic stem cells as an *in vitro* model for human vascular development and the induction of vascular differentiation, *Lab Invest* **83**: 1811–1820 (2003).

77. E. S. Wijelath, S. Rahman, J. Murray, Y. Patel, G. Savidge and M. Sobel, Fibronectin promotes VEGF-induced CD34 cell differentiation into endothelial cells, *J Vasc Surg* **39**: 655–660 (2004).

78. T. Kanayasu-Toyoda, T. Yamaguchi, T. Oshizawa and T. Hayakawa, CD31 (PECAM-1)-bright cells derived from AC133-positive cells in human peripheral blood as endothelial-precursor cells, *J Cell Physiol* **195**: 119–129 (2003).

79. S. Marchetti, C. Gimond, K. Iljin, C. Bourcier, K. Alitalo, J. Pouyssegur and G. Pages, Endothelial cells genetically selected from differentiating mouse embryonic stem cells incorporate at sites of neovascularization *in vivo*, *J Cell Sci* **115**: 2075–2085 (2002).

80. E. A. Phelps, N. Landazuri, P. M. Thule, W. R. Taylor and A. J. Garcia, Bioartificial matrices for therapeutic vascularization, *Proc Natl Acad Sci USA* **107**: 3323–3328 (2010).

81. G. Zhang, X. Wang, Z. Wang, J. Zhang and L. Suggs, A PEGylated fibrin patch for mesenchymal stem cell delivery, *Tissue Eng* **12**: 9–19 (2006).

82. G. Krenning, P. Y. Dankers, D. Jovanovic, M. J. van Luyn and M. C. Harmsen, Efficient differentiation of CD14+ monocytic cells into endothelial cells on degradable biomaterials, *Biomaterials* **28**: 1470–1479 (2007).

83. L. S. Ferreira, S. Gerecht, J. Fuller, H. F. Shieh, G. Vunjak-Novakovic and R. Langer, Bioactive hydrogel scaffolds for controllable vascular differentiation of human embryonic stem cells, *Biomaterials* **28**: 2706–2717 (2007).

84. S. Gerecht, L. S. Ferreira and R. Langer, Vascular differentiation of human embryonic stem cells in bioactive hydrogel-based scaffolds, *Methods Mol Biol* **584**: 333–354 (2010).

85. F. Guilak, D. M. Cohen, B. T. Estes, J. M. Gimble, W. Liedtke and C. S. Chen, Control of stem cell fate by physical interactions with the extracellular matrix, *Cell Stem Cell* **5**: 17–26 (2009).

86. A. J. Engler, S. Sen, H. L. Sweeney and D. E. Discher, Matrix elasticity directs stem cell lineage specification, *Cell* **126**: 677–689 (2006).

87. Kshitiz, M. E. Hubbi, E. H. Ahn, J. Downey, J. Afzal, D. H. Kim, S. Rey, C. Chang, A. Kundu, G. L. Semenza, R. M. Abraham and A. Levchenko, Matrix rigidity controls endothelial differentiation and morphogenesis of cardiac precursors, *Sci Signal* **5**: ra41 (2012).

88. A. J. Engler, C. Carag-Krieger, C. P. Johnson, M. Raab, H. Y. Tang, D. W. Speicher, J. W. Sanger, J. M. Sanger and D. E. Discher, Embryonic cardiomyocytes beat best on a matrix with heart-like elasticity: scar-like rigidity inhibits beating, *J Cell Sci* **121**: 3794–3802 (2008).

89. H. Niiyama, N. F. Huang, M. D. Rollins and J. P. Cooke, Murine model of hindlimb ischemia, *J Vis Exp* (2009).

90. L. Gepstein, A. Goldin, J. Lessick, G. Hayam, S. Shpun, Y. Schwartz, G. Hakim, R. Shofty, A. Turgeman, D. Kirshenbaum and S. A. Ben-Haim, Electromechanical characterization of chronic myocardial infarction in the canine coronary occlusion model, *Circulation* **98**: 2055–2064 (1998).

91. A. J. Hunter, A. R. Green and A. J. Cross, Animal models of acute ischaemic stroke: can they predict clinically successful neuroprotective drugs? *Trends Pharmacol Sci* **16**: 123–128 (1995).

92. S. Ikenaga, K. Hamano, M. Nishida, T. Kobayashi, T. S. Li, S. Kobayashi, M. Matsuzaki, N. Zempo and K. Esato, Autologous bone marrow implantation induced angiogenesis and improved deteriorated exercise capacity in a rat ischemic hindlimb model, *J Surg Res* **96**: 277–283 (2001).

93. E. J. Suuronen, J. P. Veinot, S. Wong, V. Kapila, J. Price, M. Griffith, T. G. Mesana and M. Ruel, Tissue-engineered injectable collagen-based matrices for improved cell delivery and vascularization of ischemic tissue using

CD133[+] progenitors expanded from the peripheral blood, *Circulation* **114**: I138–144 (2006).

94. S. W. Cho, S. H. Moon, S. H. Lee, S. W. Kang, J. Kim, J. M. Lim, H. S. Kim, B. S. Kim and H. M. Chung, Improvement of postnatal neovascularization by human embryonic stem cell derived endothelial-like cell transplantation in a mouse model of hindlimb ischemia, *Circulation* **116**: 2409–2419 (2007).

95. K. Yamahara, M. Sone, H. Itoh, J. K. Yamashita, T. Yurugi-Kobayashi, K. Homma, T. H. Chao, K. Miyashita, K. Park, N. Oyamada, N. Sawada, D. Taura, Y. Fukunaga, N. Tamura and K. Nakao, Augmentation of neovascularization [corrected] in hindlimb ischemia by combined transplantation of human embryonic stem cells-derived endothelial and mural cells, *PLoS One* **3**: e1666 (2008).

96. Z. Li, J. C. Wu, A. Y. Sheikh, D. Kraft, F. Cao, X. Xie, M. Patel, S. S. Gambhir, R. C. Robbins, J. P. Cooke and J. C. Wu, Differentiation, survival, and function of embryonic stem cell derived endothelial cells for ischemic heart disease, *Circulation* **116**: I46–54 (2007).

97. M. Gu, P. K. Nguyen, A. S. Lee, D. Xu, S. Hu, J. R. Plews, L. Han, B. C. Huber, W. H. Lee, Y. Gong, P. E. de Almeida, J. Lyons, F. Ikeno, C. Pacharinsak, A. J. Connolly, S. S. Gambhir, R. C. Robbins, M. T. Longaker and J. C. Wu, Microfluidic single-cell analysis shows that porcine induced pluripotent stem cell-derived endothelial cells improve myocardial function by paracrine activation, *Circ Res* **111**: 882–893 (2012).

98. Q. Xiong, K. L. Hill, Q. Li, P. Suntharalingam, A. Mansoor, X. Wang, M. N. Jameel, P. Zhang, C. Swingen, D. S. Kaufman and J. Zhang, A fibrin patch-based enhanced delivery of human embryonic stem cell-derived vascular cell transplantation in a porcine model of postinfarction left ventricular remodeling, *Stem Cells* **29**: 367–375 (2011).

99. C. Templin, R. Zweigerdt, K. Schwanke, R. Olmer, J. R. Ghadri, M. Y. Emmert, E. Muller, S. M. Kuest, S. Cohrs, R. Schibli, P. Kronen, M. Hilbe, A. Reinisch, D. Strunk, A. Haverich, S. Hoerstrup, T. F. Luscher, P. A. Kaufmann, U. Landmesser and U. Martin, Transplantation and tracking of human-induced pluripotent stem cells in a pig model of myocardial infarction: assessment of cell survival, engraftment, and distribution by hybrid single photon emission computed tomography/computed tomography of sodium iodide symporter transgene expression, *Circulation* **126**: 430–439 (2012).

100. A. Taguchi, T. Soma, H. Tanaka, T. Kanda, H. Nishimura, H. Yoshikawa, Y. Tsukamoto, H. Iso, Y. Fujimori, D. M. Stern, H. Naritomi and

T. Matsuyama, Administration of CD34+ cells after stroke enhances neurogenesis via angiogenesis in a mouse model, *J Clin Invest* **114**: 330–338 (2004).

101. D. C. Hess, W. D. Hill, A. Martin-Studdard, J. Carroll, J. Brailer and J. Carothers, Bone marrow as a source of endothelial cells and NeuN-expressing cells after stroke, *Stroke* **33**: 1362–1368 (2002).

102. K. Oki, J. Tatarishvili, J. Wood, P. Koch, S. Wattananit, Y. Mine, E. Monni, D. Tornero, H. Ahlenius, J. Ladewig, O. Brustle, O. Lindvall and Z. Kokaia, Human-induced pluripotent stem cells form functional neurons and improve recovery after grafting in stroke-damaged brain, *Stem Cells* **30**: 1120–1133 (2012).

103. O. Caspi, A. Lesman, Y. Basevitch, A. Gepstein, G. Arbel, I. H. Habib, L. Gepstein and S. Levenberg, Tissue engineering of vascularized cardiac muscle from human embryonic stem cells, *Circ Res* **100**: 263–272 (2007).

104. K. R. Stevens, K. L. Kreutziger, S. K. Dupras, F. S. Korte, M. Regnier, V. Muskheli, M. B. Nourse, K. Bendixen, H. Reinecke and C. E. Murry, Physiological function and transplantation of scaffold-free and vascularized human cardiac muscle tissue, *Proc Natl Acad Sci USA*, **106**: 16568–16573 (2009).

105. T. Dvir, A. Kedem, E. Ruvinov, O. Levy, I. Freeman, N. Landa, R. Holbova, M. S. Feinberg, S. Dror, Y. Etzion, J. Leor and S. Cohen, Prevascularization of cardiac patch on the omentum improves its therapeutic outcome, *Proc Natl Acad Sci USA* **106**: 14990–14995 (2009).

106. S. Matoba and H. Matsubara, Therapeutic angiogenesis for peripheral artery diseases by autologous bone marrow cell transplantation, *Curr Pharm Des* **15**: 2769–2777 (2009).

107. F. Cao, D. Sun, C. Li, K. Narsinh, L. Zhao, X. Li, X. Feng, J. Zhang, Y. Duan, J. Wang, D. Liu and H. Wang, Long-term myocardial functional improvement after autologous bone marrow mononuclear cells transplantation in patients with ST-segment elevation myocardial infarction: 4 years follow-up, *Eur Heart J* **30**: 1986–1994 (2009).

108. J. Ge, Y. Li, J. Qian, J. Shi, Q. Wang, Y. Niu, B. Fan, X. Liu, S. Zhang, A. Sun and Y. Zou, Efficacy of emergent transcatheter transplantation of stem cells for treatment of acute myocardial infarction (TCT-STAMI), *Heart* **92**: 1764–1767 (2006).

109. S. Grajek, M. Popiel, L. Gil, P. Breborowicz, M. Lesiak, R. Czepczynski, K. Sawinski, E. Straburzynska-Migaj, A. Araszkiewicz, A. Czyz, M. Kozlowska-Skrzypczak and M. Komarnicki, Influence of bone marrow stem cells on left ventricle perfusion and ejection fraction in patients with acute myocardial infarction of anterior wall: randomized clinical trial: impact of bone

marrow stem cell intracoronary infusion on improvement of microcirculation, *Eur Heart J* **31**: 691–702 (2010).

110. R. C. Huang, K. Yao, Y. Z. Zou, L. Ge, J. Y. Qian, J. Yang, S. Yang, Y. H. Niu, Y. L. Li, Y. Q. Zhang, F. Zhang, S. K. Xu, S. H. Zhang, A. J. Sun and J. B. Ge, [Long term follow-up on emergent intracoronary autologous bone marrow mononuclear cell transplantation for acute inferior-wall myocardial infarction], *Zhonghua Yi Xue Za Zhi* **86**: 1107–1110 (2006).

111. A. A. Quyyumi, E. K. Waller, J. Murrow, F. Esteves, J. Galt, J. Oshinski, S. Lerakis, S. Sher, D. Vaughan, E. Perin, J. Willerson, D. Kereiakes, B. J. Gersh, D. Gregory, A. Werner, T. Moss, W. S. Chan, R. Preti and A. L. Pecora, CD34(+) cell infusion after ST elevation myocardial infarction is associated with improved perfusion and is dose dependent, *Am Heart J* **161**: 98–105 (2011).

112. M. Tendera, W. Wojakowski, W. Ruzyllo, L. Chojnowska, C. Kepka, W. Tracz, P. Musialek, W. Piwowarska, J. Nessler, P. Buszman, S. Grajek, P. Breborowicz, M. Majka, M. Z. Ratajczak and R. Investigators, Intracoronary infusion of bone marrow-derived selected CD34+CXCR4+ cells and non-selected mononuclear cells in patients with acute STEMI and reduced left ventricular ejection fraction: results of randomized, multicentre Myocardial Regeneration by Intracoronary Infusion of Selected Population of Stem Cells in Acute Myocardial Infarction (REGENT) Trial, *Eur Heart J* **30**: 1313–1321 (2009).

113. D. M. Clifford, S. A. Fisher, S. J. Brunskill, C. Doree, A. Mathur, S. Watt and E. Martin-Rendon, Stem cell treatment for acute myocardial infarction, *Cochrane Database Syst Rev* **2**: CD006536 (2012).

114. G. Aptisa, F. Benavente, V. Sanz-Nebot, E. Chirila and J. Barbosa, Evaluation of migration behaviour of therapeutic peptide hormones in capillary electrophoresis using polybrene-coated capillaries, *Anal Bioanal Chem* **396**: 1571–1579 (2010).

115. O. Honmou, K. Houkin, T. Matsunaga, Y. Niitsu, S. Ishiai, R. Onodera, S. G. Waxman and J. D. Kocsis, Intravenous administration of auto serum-expanded autologous mesenchymal stem cells in stroke, *Brain* **134**: 1790–1807 (2011).

116. S. I. Savitz, V. Misra, M. Kasam, H. Juneja, C. S. Cox, Jr., S. Alderman, I. Aisiku, S. Kar, A. Gee and J. C. Grotta, Intravenous autologous bone marrow mononuclear cells for ischemic stroke, *Ann Neurol* **70**: 59–69 (2011).

117. S. J. Lu, Q. Feng, S. Caballero, Y. Chen, M. A. Moore, M. B. Grant and R. Lanza, Generation of functional hemangioblasts from human embryonic stem cells, *Nat Methods* **4**: 501–509 (2007).

118. Z. Li, Y. Suzuki, M. Huang, F. Cao, X. Xie, A. J. Connolly, P. C. Yang and J. C. Wu, Comparison of reporter gene and iron particle labeling for tracking fate of human embryonic stem cells and differentiated endothelial cells in living subjects, *Stem Cells* **26**: 864–873 (2008).

119. L. Wang, L. Li, F. Shojaei, K. Levac, C. Cerdan, P. Menendez, T. Martin, A. Rouleau and M. Bhatia, Endothelial and hematopoietic cell fate of human embryonic stem cells originates from primitive endothelium with hemangioblastic properties, *Immunity* **21**: 31–41 (2004).

120. Z. Z. Wang, P. Au, T. Chen, Y. Shao, L. M. Daheron, H. Bai, M. Arzigian, D. Fukumura, R. K. Jain and D. T. Scadden, Endothelial cells derived from human embryonic stem cells form durable blood vessels *in vivo*, *Nat Biotechnol* **25**: 317–318 (2007).

7

ANGIOGENIC CYTOKINES
IN THE TREATMENT OF ISCHEMIC
HEART DISEASE

Michael J. Paulsen and Y. Joseph Woo

1. Introduction

Approximately 15.5 million American adults suffer from coronary artery disease, a number that is expected to increase over 18% by 2030.[1] Over the past several decades, the mortality rate for coronary artery disease has fallen significantly, largely due to the development of effective treatments for acute coronary syndrome, along with improvements in primary and secondary prevention.[1] Unfortunately, many of these survivors will eventually succumb to heart failure, as our ability to reverse or prevent the development of ischemic cardiomyopathy has not been nearly as successful. Treatments are often initiated late in the course of the disease, and revascularization with either coronary artery bypass grafting (CABG) or percutaneous coronary intervention (PCI) is only possible in 60–80% of

161

patients.[2] Revascularization techniques such as CABG and PCI focus largely on macrovascular disease. Even after successful revascularization, a deficit in microvascular perfusion remains, and this deficit likely contributes to the adverse left ventricular remodeling and eventual dilated cardiomyopathy that results despite revascularization.[3] While this is a shortcoming in our current treatment of ischemic heart disease, many have recognized it as an enormous opportunity to direct research on future therapies. One particular therapeutic option is the use of cytokines that signal the body to develop new blood vessels.

2. Mechanisms of Blood Vessel Formation

While often used interchangeably, there are three mechanisms for the formation of vasculature in adults: angiogenesis, vasculogenesis, and arteriogenesis. Arteriogenesis is the mechanism behind collateralization, whereby arterioles remodel into larger, muscular-walled collateral arteries. In terms of growing "new" blood vessels, angiogenesis and vasculogenesis are the primary mechanisms of interest, and are quite similar. The main difference between these two mechanisms is that angiogenesis is the creation of blood vessels from already existing vessels. This occurs through the activation of differentiated endothelial cells (EC) within the walls of blood vessels; once activated, ECs migrate through the basement membrane and then bloom into new vasculature. Vasculogenesis, on the other hand, is the *de novo* formation of new blood vessels from which none existed previously. In vasculogenesis, endothelial progenitor cells (EPCs) are recruited from the bone marrow by signaling molecules, where they then differentiate into ECs and form new vasculature.[4] These signaling molecules are termed angiogenic cytokines. Recall that a cytokine is typically a small protein that acts through receptors found on cell surfaces. Angiogenic cytokines are simply cytokines that stimulate the formation or remodeling of blood vessels, either through angiogenesis, vasculogenesis, or arteriogenesis. Many angiogenic cytokines have been discovered, including vascular endothelial growth factor (VEGF), fibroblast growth factor (FGF), placental growth factor (PGF), hepatocyte growth factor (HGF), and stromal cell-derived factor-1α (SDF-1). While many of these molecules are expressed *in vivo* as a response to stimuli, such as ischemia or even changes in fluid shear stresses experienced by vessels, they can also be delivered intentionally as treatments to induce angiogenesis and vasculogenesis.[5,6]

3. Angiogenic Cytokines with Therapeutic Potential

Angiogenic cytokines have tremendous therapeutic potential. Two established methods of delivering angiogenic cytokines exist, either as protein therapy or as gene therapy. In protein therapy, a recombinant protein is delivered via intramyocardial or intracoronary injection. The effects are local and short-lived. This approach results in few side effects, but also less therapeutic effect because the effector molecules are not present in the tissue for an extended period of time. Gene therapy involves encoding the protein of interest on plasmid DNA and then delivering the gene using a vector, typically a modified virus. Gene therapy causes the desired protein to be expressed for a more prolonged period of time, which increases the therapeutic effect. Unfortunately, this approach also increases the risk of side effects such as overexpression. Too much angiogenesis could result in equally as morbid and as mortal pathologies, such as cancers or macular degeneration. Gene therapy can also result in severe and potentially lethal immune responses. Numerous angiogenic cytokines have been used in pre-clinical and clinical studies.

3.1 *Vascular endothelial growth factor (VEGF)*

VEGF, a heparin-binding glycoprotein, is one of the first-discovered and perhaps most widely studied angiogenic cytokines. There are five isoforms of VEGF (denoted VEGF-A through VEGF-E). The number of amino acids making up the ligand is used to further delineate the various isoforms. All of the VEGF isoforms seem to promote angiogenesis, and play important roles within embryological development, tissue ischemia, and neoplasia.[7,8] VEGF has three receptors: VEGFR-1, VEGFR-2, and VEGFR-3. The receptor involved in the pro-angiogenic properties of the cytokine is VEGFR-2 (human KDR/mouse Flk-1).[9,10] Binding of VEGF to VEGFR-2 on endothelial cells results in the production of nitric oxide (NO) through the activation of AKT.[11] During times of hypoxia, NO further enhances VEGF production, creating a positive feedback loop.[11] An example of this can be seen after a myocardial infarction, where elevations in circulating VEGF levels can be detected for four to five months following infarction.[12] Other pro-angiogenic effects of VEGF include the enhancement of EC cell survival, mobilization of EPCs, induction of EPC mitosis, enhancement of vasculature permeability and vasodilation, as well as regulation of coagulation and fibrinolysis.[11]

In pre-clinical animal studies of ischemic heart disease, VEGF was found to safely stimulate angiogenesis, limit infarct size, and prevent or reduce ventricular remodeling, which improved cardiac function compared to controls.[13–15] This finding paved the way for clinical trials of VEGF treatment, which is by far the most heavily studied of all angiogenic cytokines in human trials. A multitude of phase I trials showed that cytokine therapy with VEGF was safe and promoted symptomatic relief, collateralization, and enhanced perfusion with reduced ischemia.[16–19] Initial phase II studies were somewhat encouraging, with symptomatic benefits and trends toward better function. However, more rigorous later phase II studies, such as the NORTHERN trial, did not demonstrate differences between treatment and control groups.[20–25] A more recent trial used a temporal control model of patients with severe coronary artery disease deemed inoperable for revascularization; these patients underwent a minimum of six months of maximal medical therapy followed by injection of high-dose plasmid VEGF. This trial showed a transient improvement in myocardial perfusion on three-month perfusion scans, though at one year, the perfusion levels had returned to baseline.[26] Given the pathological progression of ischemic heart disease, the fact that improvement was noted following a trial of medical therapy is encouraging, although the effects appeared to be transient as reported in the phase II clinical trials. Overcoming this limitation will be an important opportunity moving forward, which we will discuss later in this chapter.

3.2 *Fibroblast growth factor (FGF)*

Fibroblast growth factors comprise a large family of heparin-binding proteins that includes 23 various isoforms (FGF-1 through FGF-23). The isoform most heavily involved in angiogenesis and in other critical myocardial cellular functions is FGF-2 (also known as basic FGF or bFGF). FGF-2 plays critical roles within cardiovascular cells throughout development and into adulthood. During embryogenesis, FGF-2 directs cells within the mesoderm to initiate cardiogenesis, and later, guides the differentiation of cardiac stem cells.[27–29] Other actions of FGF-2 include stimulating EC proliferation, enhancing EC survival, and altering the adhesion, migration, and motility of ECs. FGF-2 also plays critical roles in collateralization, angiogenesis, and vascuologenesis.[30–32] Like VEGF, FGF is also upregulated following ischemic injury or during hypoxia. FGF and VEGF have synergistic effects with one another, likely involving common mechanistic pathways.[31]

In small and large animal studies, FGF therapy caused angiogenesis and enhanced perfusion, which ultimately led to functional benefits in terms of cardiac performance following ischemic insults to the heart.[33–40] Phase I trials of FGF therapy demonstrated that the treatment was safe and overall well tolerated. These trials included both recombinant protein therapy as well as gene therapy using a viral vector.[41,42] When used as an adjunct to CABG, FGF was shown in some phase I trials to induce angiogenesis, reduce symptoms, and enhance myocardial perfusion. For example, a study by Ruel *et al.* showed that CABG patients receiving FGF often had disappearance of areas with reversible perfusion deficits, and in some cases, regression of fixed perfusion defects as well; these effects appeared to be sustained at long-term follow-up.[43] Several other early clinical studies of FGF also showed promising results in terms of improved symptoms and trends toward enhanced myocardial perfusion.[44–46] In inoperable patients with ischemic heart disease, phase I studies demonstrated improvements in symptoms and perfusion, but a larger and more robust phase II study demonstrated no benefits aside from a transient reduction in chest pain.[47–49]

3.3 *Placental growth factor (PGF)*

PGF is structurally very similar to VEGF; however, it binds specifically to the VEGFR-1 receptor.[50] Discovered initially in placental tissue, hence its name, PGF exerts actions in several other tissues, including the heart and lung.[51] Despite its name, PGF does not have a significant role in angiogenesis during development, but rather plays an important role following ischemic or hypoxic insults to the myocardium.[52] Like FGF, PGF also has a synergistic relationship with VEGF, which is not surprising given their structural and mechanistic similarities. PGF also plays an important role in mediating EPC proliferation and differentiation, guiding vasculogenesis directly through the VEGFR-2 receptor found on the surface of EPCs.[53,54] Studies in small animals have demonstrated that PGF effectively induces angiogenesis, reduces ventricular remodeling, and helps preserve cardiac function.[55–57] The use of PGF therapy in humans for ischemic heart disease has not been extensively studied, however.

3.4 *Hepatocyte growth factor (HGF)*

HGF, also a heparin-binding glycoprotein, binds to the hepatocyte growth factor receptor (HGFR, also known as c-MET). HGF was discovered originally in hepatocytes and plays a major role in the regenerative properties of the liver, as well as in the prevention of fibrosis and cirrhosis through an anti-apoptotic mechanism.[58,59] HGF has subsequently been discovered to exert effects on other tissues, such as the myocardium, where it likely plays a role in mitigating ischemic damage, as HGF levels are elevated following myocardial infarction.[60,61] Sweeney and colleagues experimented with the use of HGF therapy using an adenovirus vector in a small animal model of myocardial infarction and found that angiogenesis was increased and apoptosis decreased, preserving left ventricular form and cardiac function.[62,63] Two phase I trials have taken place using an adenovirus vector to deliver HGF during CABG procedures; the trials showed that HGF was safe, and improved myocardial perfusion in areas injected with HGF in a dose-dependent fashion.[64,65]

3.5 *Stromal cell-derived factor-1α (SDF-1)*

SDF-1 (CXCL12), a chemokine that binds to the CXCR4 receptor, is an attractive target for use in therapeutic angiogenesis. The CXCR4 receptor is found in high concentrations on the surface of EPCs, and SDF-1 stimulates the migration of EPCs and other hematopoietic stem cells to participate in hematopoiesis, cardiogenesis, and vasculogenesis during embryotic development.[66] In adults, SDF-1 recruits EPCs from the bone marrow to sites in which SDF-1 is highly concentrated, where it then stimulates angiogenesis and vasculogenesis in addition to limiting ventricular remodeling and fibrosis.[67,68] The mechanism that fibrosis and left ventricular remodeling is mitigated by SDF-1 appears to be through the preservation of tissue elasticity following infarction as well as inducing reinforcement of weakened scar tissue.[69] SDF-1 also appears to play an important role in hypoxia. For example, a recent study found that exercise triggers SDF-1 release, which then mobilizes EPCs.[70] The effects of SDF-1 are concentration-dependent, and conditions of myocardial ischemia appear to increase serum SDF-1 levels.[71,72]

The most significant factor in the ability of SDF-1 to exert its regenerative effects, however, is the ability of SDF-1 to bind the CXCR4 receptor on target EPCs and other stem cells. This conclusion was made after early studies demonstrated that SDF-1 was only effective when administered along with transplanted EPCs.[67] It was postulated that SDF-1 was only effective if an adequate number of circulating EPCs were present. Our lab hypothesized that treatment with exogenous EPCs could be replaced by SDF-1-mediated activation of native circulating EPCs. To test this hypothesis, in a small animal model of ischemic heart disease, we administered granulocyte-monocyte colony-stimulating factor (GM-CSF) alone, SDF-1 alone, and GM-CSF with SDF-1. We found that the group receiving combined therapy had an increased concentration of EPCs, along with increased vasculogenesis, and developed less ventricular remodeling which was associated with increased myocardial function.[73–75] Further studies, however, suggested that other factors were involved in determining the efficacy of SDF-1. Abbott *et al.* showed that SDF-1 only resulted in EPC homing following ischemic injury.[76] This mechanism was further elucidated in a study by Tang *et al.* where it was shown that SDF-1 homing required that EPCs express the CXCR4 receptor, which is upregulated during times of hypoxia.[77] The current hypothesis is that peak levels of SDF do not temporally align with peak upregulation of CXCR4 on EPCs. SDF-1 is upregulated almost immediately after the onset of ischemia and is rapidly broken down. By the time CXCR4 upregulation peaks at 96 hours, SDF-1 is no longer present in sufficient quantities.[77–80] Several methods to overcome this temporal mismatch, ranging from gene therapy with viral and non-viral vectors to the inhibition of SDF-1 break-down, have been attempted with some success in increasing angiogenesis, enhancing cardiac function, and reducing ventricular remodeling and fibrosis.[81–85] Other strategies, such as engineering more efficient synthetic cytokine analogs, using scaffold technology, and designing sustained release formulations with hydrogels, have also been developed. We will elaborate on some of the technological and engineering aspects of these technologies in the following section.

In terms of clinical translation, non-viral gene transfer using a plasmid expressing SDF-1 has been most heavily studied. In a phase I trial, patients with symptomatic ischemic cardiomyopathy were treated with various endomyocardial doses of SDF-1-encoded plasmid. The treatment was found to be safe and all treated patients experienced improvements in six-minute walk distance.[86] Those who received higher doses were also found to have improvements in quality-of-life scores. Notably, these improvements were found to be sustained after 12 months.[86] The encouraging results of this study prompted the development of the STOP-HF trial, a phase II study of SDF-1 therapy in patients with chronic ischemic heart failure.[87] In the STOP-HF trial, patients with ischemic cardiomyopathy received treatments of low-dose SDF-1, high-dose SDF-1, or placebo. Although the treatment was found to be safe without any severe adverse events, no significant improvement in left ventricular function was found at 12 months. For patients with very severe cardiomyopathy, measured by an ejection fraction less than 26%, a small improvement in EF and reduction in N-terminal pro-brain natriuretic peptide was noted, in comparison to the placebo arm.[87] Additional trials studying repeated administration are proceeding.

4. Moving Forward: Overcoming Limitations with Bioengineering

Perhaps one of the greatest limitations of cytokine therapy encountered thus far is that its effects are transient and there is minimal persistence of the therapeutic agents shortly after administration. Currently, many active avenues of research into delivery vectors and biomaterials that can provide sustained delivery of cytokine therapy are underway.

One potential reason for the short-lived effects of angiogenic cytokines is the small size of these proteins. As a result, they are rapidly broken down by various proteases that are often present at sites of tissue injury, such as in the ischemic myocardium.[88,89] To address this limitation, our lab developed a small peptide analog that could emulate the function of full-size cytokines but without many of the drawbacks, such as large molecular size, rapid degradation, and excessive cost associated with the manufacturing of recombinant human cytokines. Using computational protein design, we spliced the N-terminus and C-terminus of SDF-1, deleting much of the central region of the protein, to create a functionally active engineered SDF-1 analog (ESA).[90,91] As predicted, ESA surpassed the effects of native SDF-1 in augmenting EPC migration, resulting in improved left ventricular function, along with geometric and mechanical improvements following myocardial infarction in small animal models.[92–94] Pre-clinical studies of ESA in large mammals also demonstrate benefits in terms of left ventricular function preservation, enhanced angiogenesis, improved EPC migration, and decreased infarct size.[95] An engineered HGF analog has also been developed that affords similar advantages over recombinant HGF.[96]

Incorporating angiogenic cytokines into various biodegradable materials, such as hydrogels or scaffolds, also has the potential to enhance or prolong the therapeutic effect of cytokine therapy. By protecting the cytokines from degradation and diffusion, biomaterials provide sustained release of therapeutic cytokines. To investigate the efficacy of hydrogels to enhance the function of angiogenic cytokines, a study compared an FGF hydrogel to FGF in phosphate-buffered saline in a murine model of myocardial infarction.[97] This study found that hydrogels did indeed appear to augment efficacy of cytokine therapy, with significant improvements noted in left ventricular ejection fraction and vasculogenesis with attenuation of infarct size and fibrotic remodeling.[97] Studies of HGF incorporated into a gelatin hydrogel demonstrated that the controlled release of HGF produced a long-term survival benefit in a rat model of heart failure.[98] Our lab encapsulated ESA in a hyaluronic acid hydrogel and demonstrated that this approach provided up to four weeks of sustained ESA release, as measured by fluorescently tagged ESA, along with tracking of EPC chemotaxis.[79] As a result, the benefits of ESA were more pronounced in a murine model of myocardial infarction with preservation of ventricular function through increased vasculogenesis, conserved ventricular geometry, enhanced contractility, and augmented ejection fraction, in comparison to the control group.[79]

An important consideration in utilizing hydrogels to treat ischemic cardiomyopathy is the delivery modality. To optimally exert effects on heart tissue, injection into the myocardium is necessary. Ideally, percutaneous catheter delivery enables a minimally invasive approach to deliver the hydrogel treatment locally. Multiple materials have been studied as hydrogels, such as alginate, polyester, hyaluronic acid, temperature-/pH-sensitive polymers, polyethylene glycol/fibrinogen, and others.[99–105] However, it is unknown which kind of material results in the ideal hydrogel for sustained release of angiogenic cytokines. More recently, shear-thinning hydrogels have been studied, which show great promise in allowing for precise, percutaneous delivery.[106]

As described earlier in this chapter, many of the cytokines used in angiogenic therapy are created by the body naturally in response to hypoxia or tissue damage. Unfortunately, cytokines often do not accumulate in sufficient time or quantity to entirely offset injury and prevent the development of clinically significant pathology.[107,108] This is particularly true in the case of large and acute insults, such as myocardial infarction. An interesting approach to enhance cytokine delivery during times of need is to create an "on demand" cytokine delivery vector that takes advantage of transcription factors, such as p53, that are released during times of cellular stress.[109] In this approach, an adeno-associated virus vector expressing VEGF was enhanced with a synthetic p53-responsive promoter.[109] As cellular stress results in p53 accumulation, VEGF expression is also increased, which was shown to result in improved vasculogenesis, decreased fibrosis, and ultimately increased ventricular function in rats subjected to ventricular pressure overload.[109]

The transient persistence of cytokines in target tissues, coupled with a single gene or protein therapy, may not provide sustained clinical benefits, especially in complex diseases. Accordingly, hydrogels can be used to deliver multiple growth factors, which can even be released sequentially by combining hydrogels with differing molecular weights and degradation rates.[110] The sustained release of multiple angiogenic cytokines can also be achieved through the use of therapeutic cell constructs, which play an important role in tissue healing and angiogenesis through the release and modulation of various cytokines, such as VEGF and HGF. Our laboratory created a three-dimensional human fibroblast scaffold and tested its effects in a murine model of myocardial infarction, where it was shown to limit ischemic damage and preserve ventricular function.[111] Another option to enhance therapeutic benefit is to combine angiogenic cytokine therapy with other treatment modalities to create a synergistic effect. Since low cell survival is a critical limitation of cell therapy, combining cytokine therapy with cell therapy may result in more profound functional effects, compared to cell therapy in isolation.[112] Taking this a step further, Chung *et al.*[113] delivered cardiac stem cells in a VEGF scaffold that not only provided the cells with an environment in which to grow, but also provided sustained release of VEGF. When tested in a rat model, this combined approach resulted in significantly more angiogenesis, cardiomyogenesis, functional benefit, and ventricular geometry preservation, in comparison to either treatments alone or controls. We seeded a tissue-engineered extracellular matrix (ECM) scaffold with EPCs pre-primed with SDF-1 over areas of infarcted rat myocardium after left anterior descending artery ligation. The ECM matrix alone and ECM matrix with non-primed EPCs were also studied in addition to a control group.[114] The group receiving combination therapy with EPCs primed with SDF-1 demonstrated superior preservation of ventricular function, enhanced vasculogenesis, and decreased post-infarction fibrotic scar size, suggesting a synergistic effect of cell therapy combined with cytokine therapy.[114]

5. Conclusion

Angiogenic cytokine therapy holds tremendous potential in the treatment of ischemic heart disease. Perfecting delivery mechanisms that provide sustained and targeted release of therapeutic cytokines is an important next step. It is unknown if single protein or single gene cytokine therapy alone will provide enough therapeutic benefit to effectively treat a multifactorial disease such as ischemic cardiomyopathy. As we increase our understanding of the complex pathophysiologic mechanisms involved in the natural history of coronary disease and ischemic cardiomyopathy, combining or sequentially delivering various targeted genes and/or cytokines will likely yield further benefit. Perhaps the most exciting application of angiogenic cytokines is the increasingly important role they will play as adjunctive agents to other regenerative therapies. In cell therapy, angiogenic cytokines may hold the key to promoting survival and integration of stem cells. Harnessing the body's own machinery and unlocking the regenerative potential that resides inside of our cells is within reach. We are fortunate to be witness and part of the revolution that is currently taking place within the field of medicine.

References

1. D. Mozaffarian, E. J. Benjamin, A. S. Go, D. K. Arnett, M. J. Blaha, M. Cushman, S. de Ferranti, J. P. Despres, H. J. Fullerton, V. J. Howard, M. D. Huffman, S. E. Judd, B. M. Kissela, D. T. Lackland, J. H. Lichtman, L. D. Lisabeth, S. Liu, R. H. Mackey, D. B. Matchar, D. K. McGuire, E. R. Mohler, 3rd, C. S. Moy, P. Muntner, M. E. Mussolino, K. Nasir, R. W. Neumar, G. Nichol, L. Palaniappan, D. K. Pandey, M. J. Reeves, C. J. Rodriguez, P. D. Sorlie, J. Stein, A. Towfighi, T. N. Turan, S. S. Virani, J. Z. Willey, D. Woo, R. W. Yeh, M. B. Turner; American Heart Association Statistics Committee and Stroke Statistics Subcommittee, Heart disease and stroke statistics — 2015 update: a report from the American Heart Association, *Circulation* **131**: e29–322 (2015).
2. M. Boodhwani, N. R. Sodha, R. J. Laham and F. W. Sellke, The future of therapeutic myocardial angiogenesis, *Shock* **26**: 332–341 (2006).
3. A. Araszkiewicz, S. Grajek, M. Lesiak, M. Prech, M. Pyda, M. Janus and A. Cieslinski, Effect of impaired myocardial reperfusion on left ventricular remodeling in patients with anterior wall acute myocardial infarction treated with primary coronary intervention, *Am J Cardiol* **98**: 725–728 (2006).
4. Y. S. Yoon, I. A. Johnson, J. S. Park, L. Diaz and D. W. Losordo, Therapeutic myocardial angiogenesis with vascular endothelial growth factors, *Mol Cell Biochem* **264**: 63–74 (2004).
5. P. F. Davies, K. A. Barbee, M. V. Volin, A. Robotewskyj, J. Chen, L. Joseph, M. L. Griem, M. N. Wernick, E. Jacobs, D. C. Polacek, N. dePaola and A. I. Barakat, Spatial relationships in early signaling

events of flow-mediated endothelial mechanotransduction, *Annu Rev Physiol* **59**: 527–549 (1997).

6. J. N. Topper and M. A. Gimbrone, Jr., Blood flow and vascular gene expression: fluid shear stress as a modulator of endothelial phenotype, *Mol Med Today* **5**: 40–46 (1999).

7. S. Takeshita, T. Isshiki and T. Sato, Increased expression of direct gene transfer into skeletal muscles observed after acute ischemic injury in rats, *Lab Invest* **74**: 1061–1065 (1996).

8. T. Asahara, T. Takahashi, H. Masuda, C. Kalka, D. Chen, H. Iwaguro, Y. Inai, M. Silver and J. M. Isner, VEGF contributes to postnatal neovascularization by mobilizing bone marrow-derived endothelial progenitor cells, *EMBO J* **18**: 3964–3972 (1999).

9. N. Ferrara, H. P. Gerber and J. LeCouter, The biology of VEGF and its receptors, *Nat Med* **9**: 669–676 (2003).

10. J. Kroll and J. Waltenberger, The vascular endothelial growth factor receptor KDR activates multiple signal transduction pathways in porcine aortic endothelial cells, *J Biol Chem* **272**: 32521–32527 (1997).

11. E. Toyota, T. Matsunaga and W. M. Chilian, Myocardial angiogenesis, *Mol Cell Biochem* **264**: 35–44 (2004).

12. K. W. Lee, G. Y. Lip and A. D. Blann, Plasma angiopoietin-1, angiopoietin-2, angiopoietin receptor tie-2, and vascular endothelial growth factor levels in acute coronary syndromes, *Circulation* **110**: 2355–2360 (2004).

13. P. Atluri and Y. J. Woo, Pro-angiogenic cytokines as cardiovascular therapeutics: assessing the potential, *BioDrugs* **22**: 209–222 (2008).

14. C. Kim, R. K. Li, G. Li, Y. Zhang, R. D. Weisel and T. M. Yau, Effects of cell-based angiogenic gene therapy at 6 months: persistent angiogenesis and absence of oncogenicity, *Ann Thorac Surg* **83**: 640–646 (2007).

15. G. Vera Janavel, A. Crottogini, P. Cabeza Meckert, L. Cuniberti, A. Mele, M. Papouchado, N. Fernandez, A. Bercovich, M. Criscuolo, C. Melo and R. Laguens, Plasmid-mediated VEGF gene transfer induces cardiomyogenesis and reduces myocardial infarct size in sheep, *Gene Ther* **13**: 1133–1142 (2006).

16. S. Fuchs, N. Dib, B. M. Cohen, P. Okubagzi, E. B. Diethrich, A. Campbell, J. Macko, P. D. Kessler, H. S. Rasmussen, S. E. Epstein and R. Kornowski, A randomized, double-blind, placebo-controlled, multicenter, pilot study of the safety and feasibility of catheter-based intramyocardial injection of AdVEGF121 in patients with refractory advanced coronary artery disease, *Catheter Cardiovasc Interv* **68**: 372–378 (2006).

17. T. D. Henry, K. Rocha-Singh, J. M. Isner, D. J. Kereiakes, F. J. Giordano, M. Simons, D. W. Losordo, R. C. Hendel, R. O. Bonow, S. M. Eppler, T. F. Zioncheck, E. B. Holmgren and E. R. McCluskey, Intracoronary administration of recombinant human vascular endothelial growth factor to patients with coronary artery disease, *Am Heart J* **142**: 872–880 (2001).

18. T. K. Rosengart, L. Y. Lee, S. R. Patel, T. A. Sanborn, M. Parikh, G. W. Bergman, R. Hachamovitch, M. Szulc, P. D. Kligfield, P. M. Okin, R. T. Hahn, R. B. Devereux, M. R. Post, N. R. Hackett, T. Foster, T. M. Grasso, M. L. Lesser, O. W. Isom and R. G. Crystal, Angiogenesis gene therapy: phase I assessment of direct intramyocardial administration of an adenovirus vector expressing VEGF121 cDNA to individuals with clinically significant severe coronary artery disease, *Circulation* **100**: 468–474 (1999).

19. J. F. Symes, D. W. Losordo, P. R. Vale, K. G. Lathi, D. D. Esakof, M. Mayskiy and J. M. Isner, Gene therapy with vascular endothelial growth factor for inoperable coronary artery disease, *Ann Thorac Surg* **68**: 830–836; discussion 836–837 (1999).

20. T. D. Henry, B. H. Annex, G. R. McKendall, M. A. Azrin, J. J. Lopez, F. J. Giordano, P. K. Shah, J. T. Willerson, R. L. Benza, D. S. Berman, C. M. Gibson, A. Bajamonde, A. C. Rundle, J. Fine, E. R. McCluskey and V. Investigators, The VIVA trial: vascular endothelial growth factor in ischemia for vascular angiogenesis, *Circulation* **107**: 1359–1365 (2003).

21. J. Kastrup, E. Jorgensen, A. Ruck, K. Tagil, D. Glogar, W. Ruzyllo, H. E. Botker, D. Dudek, V. Drvota, B. Hesse, L. Thuesen, P. Blomberg, M. Gyongyosi, C. Sylven and G. Euroinject One, Direct intramyocardial plasmid vascular endothelial growth factor-A165 gene therapy in patients with stable severe angina pectoris, a randomized double-blind placebo-controlled study: the Euroinject One trial, *J Am Coll Cardiol* **45**: 982–988 (2005).

22. D. W. Losordo, P. R. Vale, R. C. Hendel, C. E. Milliken, F. D. Fortuin, N. Cummings, R. A. Schatz, T. Asahara, J. M. Isner and R. E. Kuntz, Phase 1/2 placebo-controlled, double-blind, dose-escalating trial of myocardial vascular endothelial growth factor 2 gene transfer by catheter delivery in patients with chronic myocardial ischemia, *Circulation* **105**: 2012–2018 (2002).

23. R. S. Ripa, E. Jorgensen, Y. Wang, J. J. Thune, J. C. Nilsson, L. Sondergaard, H. E. Johnsen, L. Kober, P. Grande and J. Kastrup, Stem cell mobilization induced by subcutaneous granulocyte-colony stimulating factor to improve cardiac regeneration after acute ST-elevation myocardial infarction: result of the double-blind, randomized, placebo-controlled stem cells in myocardial infarction (STEMMI) trial, *Circulation* **113**: 1983–1992 (2006).

24. D. J. Stewart, J. D. Hilton, J. M. Arnold, J. Gregoire, A. Rivard, S. L. Archer, F. Charbonneau, E. Cohen, M. Curtis, C. E. Buller, F. O.

Mendelsohn, N. Dib, P. Page, J. Ducas, S. Plante, J. Sullivan, J. Macko, C. Rasmussen, P. D. Kessler and H. S. Rasmussen, Angiogenic gene therapy in patients with nonrevascularizable ischemic heart disease: a phase 2 randomized, controlled trial of AdVEGF(121) (AdVEGF121) versus maximum medical treatment, *Gene Ther* **13**: 1503–1511 (2006).

25. D. J. Stewart, M. J. Kutryk, D. Fitchett, M. Freeman, N. Camack, Y. Su, A. Della Siega, L. Bilodeau, J. R. Burton, G. Proulx, S. Radhakrishnan and N. T. Investigators, VEGF gene therapy fails to improve perfusion of ischemic myocardium in patients with advanced coronary disease: results of the NORTHERN trial, *Mol Ther* **17**: 1109–1115 (2009).

26. I. I. Giusti, C. G. Rodrigues, F. B. Salles, R. T. Sant'Anna, B. Eibel, S. W. Han, E. Ludwig, G. Grossman, P. R. Prates, J. R. Sant'Anna, G. F. Filho, M. M. Markoski, I. A. Nesralla, N. B. Nardi and R. A. Kalil, High doses of vascular endothelial growth factor 165 safely, but transiently, improve myocardial perfusion in no-option ischemic disease, *Hum Gene Ther Methods* **24**: 298–306 (2013).

27. N. Rosenblatt-Velin, M. G. Lepore, C. Cartoni, F. Beermann and T. Pedrazzini, FGF-2 controls the differentiation of resident cardiac precursors into functional cardiomyocytes, *J Clin Invest* **115**: 1724–1733 (2005).

28. M. J. Solloway and R. P. Harvey, Molecular pathways in myocardial development: a stem cell perspective, *Cardiovasc Res* **58**: 264–277 (2003).

29. Y. Sugi and J. Lough, Activin-A and FGF-2 mimic the inductive effects of anterior endoderm on terminal cardiac myogenesis *in vitro*, *Dev Biol* **168**: 567–574 (1995).

30. P. E. Burger, S. Coetzee, W. L. McKeehan, M. Kan, P. Cook, Y. Fan, T. Suda, R. P. Hebbel, N. Novitzky, W. A. Muller and E. L. Wilson, Fibroblast growth factor receptor-1 is expressed by endothelial progenitor cells, *Blood* **100**: 3527–3535 (2002).

31. K. A. Detillieux, F. Sheikh, E. Kardami and P. A. Cattini, Biological activities of fibroblast growth factor-2 in the adult myocardium, *Cardiovasc Res* **57**: 8–19 (2003).

32. I. Kashiwakura and T. A. Takahashi, Basic fibroblast growth factor-stimulated *ex vivo* expansion of haematopoietic progenitor cells from human placental and umbilical cord blood, *Br J Haematol* **122**: 479–488 (2003).

33. A. Battler, M. Scheinowitz, A. Bor, D. Hasdai, Z. Vered, E. Di Segni, N. Varda-Bloom, D. Nass, S. Engelberg, M. Eldar, M. Belkin and N. Savion, Intracoronary injection of basic fibroblast growth factor enhances angiogenesis in infarcted swine myocardium, *J Am Coll Cardiol* **22**: 2001–2006 (1993).

34. B. Fernandez, A. Buehler, S. Wolfram, S. Kostin, G. Espanion, W. M. Franz, H. Niemann, P. A. Doevendans, W. Schaper and R. Zimmermann, Transgenic

myocardial overexpression of fibroblast growth factor-1 increases coronary artery density and branching, *Circ Res* **87**: 207–213 (2000).

35. F. J. Giordano, P. Ping, M. D. McKirnan, S. Nozaki, A. N. DeMaria, W. H. Dillmann, O. Mathieu-Costello and H. K. Hammond, Intracoronary gene transfer of fibroblast growth factor-5 increases blood flow and contractile function in an ischemic region of the heart, *Nat Med* **2**: 534–539 (1996).

36. K. Harada, W. Grossman, M. Friedman, E. R. Edelman, P. V. Prasad, C. S. Keighley, W. J. Manning, F. W. Sellke and M. Simons, Basic fibroblast growth factor improves myocardial function in chronically ischemic porcine hearts, *J Clin Invest* **94**: 623–630 (1994).

37. G. C. Hughes, S. S. Biswas, B. Yin, R. E. Coleman, T. R. DeGrado, C. K. Landolfo, J. E. Lowe, B. H. Annex and K. P. Landolfo, Therapeutic angiogenesis in chronically ischemic porcine myocardium: comparative effects of bFGF and VEGF, *Ann Thorac Surg* **77**: 812–818 (2004).

38. R. J. Laham, M. Rezaee, M. Post, X. Xu and F. W. Sellke, Intrapericardial administration of basic fibroblast growth factor: myocardial and tissue distribution and comparison with intracoronary and intravenous administration, *Catheter Cardiovasc Interv* **58**: 375–381 (2003).

39. Y. Sakakibara, Toward surgical angiogenesis using slow-released basic fibroblast growth factor, *Eur J Cardiothorac Surg* **24**: 105–112 (2003).

40. K. Sato, R. J. Laham, J. D. Pearlman, D. Novicki, F. W. Sellke, M. Simons and M. J. Post, Efficacy of intracoronary versus intravenous FGF-2 in a pig model of chronic myocardial ischemia, *Ann Thorac Surg* **70**: 2113–2118 (2000).

41. C. L. Grines, M. W. Watkins, G. Helmer, W. Penny, J. Brinker, J. D. Marmur, A. West, J. J. Rade, P. Marrott, H. K. Hammond and R. L. Engler, Angiogenic Gene Therapy (AGENT) trial in patients with stable angina pectoris, *Circulation* **105**: 1291–1297 (2002).

42. E. F. Unger, L. Goncalves, S. E. Epstein, E. Y. Chew, C. B. Trapnell, R. O. Cannon, 3rd and A. A. Quyyumi, Effects of a single intracoronary injection of basic fibroblast growth factor in stable angina pectoris, *Am J Cardiol* **85**: 1414–1419 (2000).

43. M. Ruel, R. J. Laham, J. A. Parker, M. J. Post, J. A. Ware, M. Simons and F. W. Sellke, Long-term effects of surgical angiogenic therapy with fibroblast growth factor 2 protein, *J Thorac Cardiovasc Surg* **124**: 28–34 (2002).

44. C. L. Grines, M. W. Watkins, J. J. Mahmarian, A. E. Iskandrian, J. J. Rade, P. Marrott, C. Pratt, N. Kleiman and G. T. S. G. Angiogene, A randomized, double-blind, placebo-controlled trial of Ad5FGF-4 gene therapy and its effect on myocardial perfusion in patients with stable angina, *J Am Coll Cardiol* **42**: 1339–1347 (2003).

45. R. J. Laham, F. W. Sellke, E. R. Edelman, J. D. Pearlman, J. A. Ware, D. L. Brown, J. P. Gold and M. Simons, Local perivascular delivery of basic fibroblast growth factor in patients undergoing coronary bypass surgery: results of a phase I randomized, double-blind, placebo-controlled trial, *Circulation* **100**: 1865–1871 (1999).

46. B. Schumacher, T. Stegmann and P. Pecher, The stimulation of neoangiogenesis in the ischemic human heart by the growth factor FGF: first clinical results, *J Cardiovasc Surg (Torino)* **39**: 783–789 (1998).

47. R. J. Laham, N. A. Chronos, M. Pike, M. E. Leimbach, J. E. Udelson, J. D. Pearlman, R. I. Pettigrew, M. J. Whitehouse, C. Yoshizawa and M. Simons, Intracoronary basic fibroblast growth factor (FGF-2) in patients with severe ischemic heart disease: results of a phase I open-label dose escalation study, *J Am Coll Cardiol* **36**: 2132–2139 (2000).

48. M. Simons, B. H. Annex, R. J. Laham, N. Kleiman, T. Henry, H. Dauerman, J. E. Udelson, E. V. Gervino, M. Pike, M. J. Whitehouse, T. Moon and N. A. Chronos, Pharmacological treatment of coronary artery disease with recombinant fibroblast growth factor-2: double-blind, randomized, controlled clinical trial, *Circulation* **105**: 788–793 (2002).

49. J. E. Udelson, V. Dilsizian, R. J. Laham, N. Chronos, J. Vansant, M. Blais, J. R. Galt, M. Pike, C. Yoshizawa and M. Simons, Therapeutic angiogenesis with recombinant fibroblast growth factor-2 improves stress and rest myocardial perfusion abnormalities in patients with severe symptomatic chronic coronary artery disease, *Circulation* **102**: 1605–1610 (2000).

50. S. Iyer, D. D. Leonidas, G. J. Swaminathan, D. Maglione, M. Battisti, M. Tucci, M. G. Persico and K. R. Acharya, The crystal structure of human placenta growth factor-1 (PlGF-1), an angiogenic protein, at 2.0 A resolution, *J Biol Chem* **276**: 12153–12161 (2001).

51. M. Autiero, A. Luttun, M. Tjwa and P. Carmeliet, Placental growth factor and its receptor, vascular endothelial growth factor receptor-1: novel targets for stimulation of ischemic tissue revascularization and inhibition of angiogenic and inflammatory disorders, *J Thromb Haemost* **1**: 1356–1370 (2003).

52. P. Carmeliet, L. Moons, A. Luttun, V. Vincenti, V. Compernolle, M. De Mol, Y. Wu, F. Bono, L. Devy, H. Beck, D. Scholz, T. Acker, T. DiPalma, M. Dewerchin, A. Noel, I. Stalmans, A. Barra, S. Blacher, T. Vanden Driessche, A. Ponten, U. Eriksson, K. H. Plate, J. M. Foidart, W. Schaper, D. S. Charnock-Jones, D. J. Hicklin, J. M. Herbert, D. Collen and M. G. Persico, Synergism between vascular endothelial growth factor and placental growth factor contributes to angiogenesis and plasma extravasation in pathological conditions, *Nat Med* **7**: 575–583 (2001).

53. K. Hattori, B. Heissig, Y. Wu, S. Dias, R. Tejada, B. Ferris, D. J. Hicklin, Z. Zhu, P. Bohlen, L. Witte, J. Hendrikx, N. R. Hackett, R. G. Crystal, M. A. Moore, Z. Werb, D. Lyden and S. Rafii, Placental growth factor reconstitutes hematopoiesis by recruiting VEGFR1(+) stem cells from bone-marrow microenvironment, *Nat Med* **8**: 841–849 (2002).

54. B. Li, E. E. Sharpe, A. B. Maupin, A. A. Teleron, A. L. Pyle, P. Carmeliet and P. P. Young, VEGF and PlGF promote adult vasculogenesis by enhancing EPC recruitment and vessel formation at the site of tumor neovascularization, *FASEB J* **20**: 1495–1497 (2006).

55. S. Kolakowski, Jr., M. F. Berry, P. Atluri, T. Grand, O. Fisher, M. A. Moise, J. Cohen, V. Hsu and Y. J. Woo, Placental growth factor provides a novel local angiogenic therapy for ischemic cardiomyopathy, *J Card Surg* **21**: 559–564 (2006).

56. X. S. Liu, P. Claus, M. Wu, G. Reyns, P. Verhamme, P. Pokreisz, S. Vandenwijngaert, C. Dubois, J. Vanhaecke, E. Verbeken, J. Bogaert and S. Janssens, Placental growth factor increases regional myocardial blood flow and contractile function in chronic myocardial ischemia, *Am J Physiol Heart Circ Physiol* **304**: H885–894 (2013).

57. C. Roncal, I. Buysschaert, E. Chorianopoulos, M. Georgiadou, O. Meilhac, M. Demol, J. B. Michel, S. Vinckier, L. Moons and P. Carmeliet, Beneficial effects of prolonged systemic administration of PlGF on late outcome of post-ischaemic myocardial performance, *J Pathol* **216**: 236–244 (2008).

58. T. Funatsu, Y. Sawa, S. Ohtake, T. Takahashi, G. Matsumiya, N. Matsuura, T. Nakamura and H. Matsuda, Therapeutic angiogenesis in the ischemic canine heart induced by myocardial injection of naked complementary DNA plasmid encoding hepatocyte growth factor, *J Thorac Cardiovasc Surg* **124**: 1099–1105 (2002).

59. T. Nakamura, T. Nishizawa, M. Hagiya, T. Seki, M. Shimonishi, A. Sugimura, K. Tashiro and S. Shimizu, Molecular cloning and expression of human hepatocyte growth factor, *Nature* **342**: 440–443 (1989).

60. T. Nakamura, S. Mizuno, K. Matsumoto, Y. Sawa, H. Matsuda and T. Nakamura, Myocardial protection from ischemia/reperfusion injury by endogenous and exogenous HGF, *J Clin Invest* **106**: 1511–1519 (2000).

61. T. Sato, Y. Tani, S. Murao, H. Fujieda, H. Sato, M. Matsumoto, T. Takeuchi and Y. Ohtsuki, Focal enhancement of expression of c-Met/hepatocyte growth factor receptor in the myocardium in human myocardial infarction, *Cardiovasc Pathol* **10**: 235–240 (2001).

62. V. Jayasankar, Y. J. Woo, L. T. Bish, T. J. Pirolli, S. Chatterjee, M. F. Berry, J. Burdick, T. J. Gardner and H. L. Sweeney, Gene transfer of hepatocyte

growth factor attenuates postinfarction heart failure, *Circulation* **108**: II230–236 (2003).

63. V. Jayasankar, Y. J. Woo, T. J. Pirolli, L. T. Bish, M. F. Berry, J. Burdick, T. J. Gardner and H. L. Sweeney, Induction of angiogenesis and inhibition of apoptosis by hepatocyte growth factor effectively treats postischemic heart failure, *J Card Surg* **20**: 93–101 (2005).

64. J. S. Kim, H. Y. Hwang, K. R. Cho, E. A. Park, W. Lee, J. C. Paeng, D. S. Lee, H. K. Kim, D. W. Sohn and K. B. Kim, Intramyocardial transfer of hepatocyte growth factor as an adjunct to CABG: phase I clinical study, *Gene Ther* **20**: 717–722 (2013).

65. B. Yuan, Z. Zhao, Y. R. Zhang, C. T. Wu, W. G. Jin, S. Zhao, W. Wang, Y. Y. Zhang, X. L. Zhu, L. S. Wang and J. Huang, Short-term safety and curative effect of recombinant adenovirus carrying hepatocyte growth factor gene on ischemic cardiac disease, *In Vivo* **22**: 629–632 (2008).

66. R. Mohle, F. Bautz, S. Rafii, M. A. Moore, W. Brugger and L. Kanz, The chemokine receptor CXCR-4 is expressed on CD34+ hematopoietic progenitors and leukemic cells and mediates transendothelial migration induced by stromal cell-derived factor-1, *Blood* **91**: 4523–4530 (1998).

67. J. Yamaguchi, K. F. Kusano, O. Masuo, A. Kawamoto, M. Silver, S. Murasawa, M. Bosch-Marce, H. Masuda, D. W. Losordo, J. M. Isner and T. Asahara, Stromal cell-derived factor-1 effects on *ex vivo* expanded endothelial progenitor cell recruitment for ischemic neovascularization, *Circulation* **107**: 1322–1328 (2003).

68. H. Zheng, G. Fu, T. Dai and H. Huang, Migration of endothelial progenitor cells mediated by stromal cell-derived factor-1alpha/CXCR4 via PI3K/Akt/eNOS signal transduction pathway, *J Cardiovasc Pharmacol* **50**: 274–280 (2007).

69. W. Hiesinger, M. J. Brukman, R. C. McCormick, J. R. Fitzpatrick, 3rd, J. R. Frederick, E. C. Yang, J. R. Muenzer, N. A. Marotta, M. F. Berry, P. Atluri and Y. J. Woo, Myocardial tissue elastic properties determined by atomic force microscopy after stromal cell-derived factor 1alpha angiogenic therapy for acute myocardial infarction in a murine model, *J Thorac Cardiovasc Surg* **143**: 962–966 (2012).

70. E. Chang, J. Paterno, D. Duscher, Z. N. Maan, J. S. Chen, M. Januszyk, M. Rodrigues, R. C. Rennert, S. Bishop, A. J. Whitmore, A. J. Whittam, M. T. Longaker and G. C. Gurtner, Exercise induces stromal cell-derived factor-1 alpha-mediated release of endothelial progenitor cells with increased vasculogenic function, *Plast Reconstr Surg* **135**: E340–350 (2015).

71. D. Y. Jo, S. Rafii, T. Hamada and M. A. Moore, Chemotaxis of primitive hematopoietic cells in response to stromal cell-derived factor-1, *J Clin Invest* **105**: 101–111 (2000).

72. K. Pillarisetti and S. K. Gupta, Cloning and relative expression analysis of rat stromal cell derived factor-1 (SDF-1)1: SDF-1 alpha mRNA is selectively induced in rat model of myocardial infarction, *Inflammation* **25**: 293–300 (2001).

73. P. Atluri, G. P. Liao, C. M. Panlilio, V. M. Hsu, M. J. Leskowitz, K. J. Morine, J. E. Cohen, M. F. Berry, E. E. Suarez, D. A. Murphy, W. M. Lee, T. J. Gardner, H. L. Sweeney and Y. J. Woo, Neovasculogenic therapy to augment perfusion and preserve viability in ischemic cardiomyopathy, *Ann Thorac Surg* **81**: 1728–1736 (2006).

74. P. Atluri, C. M. Panlilio, G. P. Liao, W. Hiesinger, D. A. Harris, R. C. McCormick, J. E. Cohen, T. Jin, W. Feng, R. D. Levit, N. Dong and Y. J. Woo, Acute myocardial rescue with endogenous endothelial progenitor cell therapy, *Heart Lung Circ* **19**: 644–654 (2010).

75. Y. J. Woo, J. C. Zhang, M. D. Taylor, J. E. Cohen, V. M. Hsu and H. L. Sweeney, One year transgene expression with adeno-associated virus cardiac gene transfer, *Int J Cardiol* **100**: 421–426 (2005).

76. J. D. Abbott, Y. Huang, D. Liu, R. Hickey, D. S. Krause and F. J. Giordano, Stromal cell-derived factor-1alpha plays a critical role in stem cell recruitment to the heart after myocardial infarction but is not sufficient to induce homing in the absence of injury, *Circulation* **110**: 3300–3305 (2004).

77. Y. L. Tang, W. Zhu, M. Cheng, L. Chen, J. Zhang, T. Sun, R. Kishore, M. I. Phillips, D. W. Losordo and G. Qin, Hypoxic preconditioning enhances the benefit of cardiac progenitor cell therapy for treatment of myocardial infarction by inducing CXCR4 expression, *Circ Res* **104**: 1209–1216 (2009).

78. A. T. Askari, S. Unzek, Z. B. Popovic, C. K. Goldman, F. Forudi, M. Kiedrowski, A. Rovner, S. G. Ellis, J. D. Thomas, P. E. DiCorleto, E. J. Topol and M. S. Penn, Effect of stromal-cell-derived factor 1 on stem-cell homing and tissue regeneration in ischaemic cardiomyopathy, *Lancet* **362**: 697–703 (2003).

79. J. W. MacArthur, Jr., B. P. Purcell, Y. Shudo, J. E. Cohen, A. Fairman, A. Trubelja, J. Patel, P. Hsiao, E. Yang, K. Lloyd, W. Hiesinger, P. Atluri, J. A. Burdick and Y. J. Woo, Sustained release of engineered stromal cell-derived factor 1-alpha from injectable hydrogels effectively recruits endothelial progenitor cells and preserves ventricular function after myocardial infarction, *Circulation* **128**: S79–86 (2013).

80. M. S. Penn, Importance of the SDF-1:CXCR4 axis in myocardial repair, *Circ Res* **104**: 1133–1135 (2009).

81. I. Elmadbouh, H. Haider, S. Jiang, N. M. Idris, G. Lu and M. Ashraf, *Ex vivo* delivered stromal cell-derived factor-1alpha promotes stem cell homing and induces angiomyogenesis in the infarcted myocardium, *J Mol Cell Cardiol* **42**: 792–803 (2007).

82. S. Sundararaman, T. J. Miller, J. M. Pastore, M. Kiedrowski, R. Aras and M. S. Penn, Plasmid-based transient human stromal cell-derived factor-1 gene transfer improves cardiac function in chronic heart failure, *Gene Ther* **18**: 867–873 (2011).

83. J. Tang, J. Wang, H. Song, Y. Huang, J. Yang, X. Kong, L. Guo, F. Zheng and L. Zhang, Adenovirus-mediated stromal cell-derived factor-1 alpha gene transfer improves cardiac structure and function after experimental myocardial infarction through angiogenic and antifibrotic actions, *Mol Biol Rep* **37**: 1957–1969 (2010).

84. M. M. Zaruba, H. D. Theiss, M. Vallaster, U. Mehl, S. Brunner, R. David, R. Fischer, L. Krieg, E. Hirsch, B. Huber, P. Nathan, L. Israel, A. Imhof, N. Herbach, G. Assmann, R. Wanke, J. Mueller-Hoecker, G. Steinbeck and W. M. Franz, Synergy between CD26/DPP-IV inhibition and G-CSF improves cardiac function after acute myocardial infarction, *Cell Stem Cell* **4**: 313–323 (2009).

85. T. Zhao, D. Zhang, R. W. Millard, M. Ashraf and Y. Wang, Stem cell homing and angiomyogenesis in transplanted hearts are enhanced by combined intramyocardial SDF-1alpha delivery and endogenous cytokine signaling, *Am J Physiol Heart Circ Physiol* **296**: H976–986 (2009).

86. M. S. Penn, F. O. Mendelsohn, G. L. Schaer, W. Sherman, M. Farr, J. Pastore, D. Rouy, R. Clemens, R. Aras and D. W. Losordo, An open-label dose escalation study to evaluate the safety of administration of nonviral stromal cell-derived factor-1 plasmid to treat symptomatic ischemic heart failure, *Circ Res* **112**: 816–825 (2013).

87. E. S. Chung, L. Miller, A. N. Patel, R. D. Anderson, F. O. Mendelsohn, J. Traverse, K. H. Silver, J. Shin, G. Ewald, M. J. Farr, S. Anwaruddin, F. Plat, S. J. Fisher, A. T. AuWerter, J. M. Pastore, R. Aras and M. S. Penn, Changes in ventricular remodelling and clinical status during the year following a single administration of stromal cell-derived factor-1 non-viral gene therapy in chronic ischaemic heart failure patients: the STOP-HF randomized Phase II trial, *Eur Heart J* **36**: 2228–2238 (2015).

88. G. A. McQuibban, G. S. Butler, J. H. Gong, L. Bendall, C. Power, I. Clark-Lewis and C. M. Overall, Matrix metalloproteinase activity inactivates the CXC chemokine stromal cell-derived factor-1, *J Biol Chem* **276**: 43503–43508 (2001).

89. V. F. Segers, T. Tokunou, L. J. Higgins, C. MacGillivray, J. Gannon and R. T. Lee, Local delivery of protease-resistant stromal cell derived factor-1 for stem cell recruitment after myocardial infarction, *Circulation* **116**: 1683–1692 (2007).

90. W. Hiesinger, J. R. Frederick, P. Atluri, R. C. McCormick, N. Marotta, J. R. Muenzer and Y. J. Woo, Spliced stromal cell-derived factor-1alpha analog stimulates endothelial progenitor cell migration and improves cardiac function in a dose-dependent manner after myocardial infarction, *J Thorac Cardiovasc Surg* **140**: 1174–1180 (2010).

91. W. Hiesinger, J. M. Perez-Aguilar, P. Atluri, N. A. Marotta, J. R. Frederick, J. R. Fitzpatrick, 3rd, R. C. McCormick, J. R. Muenzer, E. C. Yang, R. D. Levit, L. J. Yuan, J. W. Macarthur, J. G. Saven and Y. J. Woo, Computational protein design to reengineer stromal cell-derived factor-1alpha generates an effective and translatable angiogenic polypeptide analog, *Circulation* **124**: S18–26 (2011).

92. W. Hiesinger, A. B. Goldstone and Y. J. Woo, Re-engineered stromal cell-derived factor-1alpha and the future of translatable angiogenic polypeptide design, *Trends Cardiovasc Med* **22**: 139–144 (2012).

93. J. W. MacArthur, Jr., A. Trubelja, Y. Shudo, P. Hsiao, A. S. Fairman, E. Yang, W. Hiesinger, J. J. Sarver, P. Atluri and Y. J. Woo, Mathematically engineered stromal cell-derived factor-1alpha stem cell cytokine analog enhances mechanical properties of infarcted myocardium, *J Thorac Cardiovasc Surg* **145**: 278–284 (2013).

94. A. Trubelja, J. W. MacArthur, J. J. Sarver, J. E. Cohen, G. Hung, Y. Shudo, A. S. Fairman, J. Patel, B. B. Edwards, S. M. Damrauer, W. Hiesinger, P. Atluri and Y. J. Woo, Bioengineered stromal cell-derived factor-1alpha analogue delivered as an angiogenic therapy significantly restores viscoelastic material properties of infarcted cardiac muscle, *J Biomech Eng* **136** (2014).

95. J. W. Macarthur, Jr., J. E. Cohen, J. R. McGarvey, Y. Shudo, J. B. Patel, A. Trubelja, A. S. Fairman, B. B. Edwards, G. Hung, W. Hiesinger, A. B. Goldstone, P. Atluri, R. L. Wilensky, J. J. Pilla, J. H. Gorman, 3rd, R. C. Gorman and Y. J. Woo, Preclinical evaluation of the engineered stem cell chemokine stromal cell-derived factor 1alpha analog in a translational ovine myocardial infarction model, *Circ Res* **114**: 650–659 (2014).

96. C. J. Liu, D. S. Jones, 2nd, P. C. Tsai, A. Venkataramana and J. R. Cochran, An engineered dimeric fragment of hepatocyte growth factor is a potent c-MET agonist, *FEBS Lett* **588**: 4831–4837 (2014).

97. H. Wang, X. Zhang, Y. Li, Y. Ma, Y. Zhang, Z. Liu, J. Zhou, Q. Lin, Y. Wang, C. Duan and C. Wang, Improved myocardial performance in infarcted rat

heart by co-injection of basic fibroblast growth factor with temperature-responsive chitosan hydrogel, *J Heart Lung Transplant* **29**: 881–887 (2010).

98. G. Sakaguchi, K. Tambara, Y. Sakakibara, M. Ozeki, M. Yamamoto, G. Premaratne, X. Lin, K. Hasegawa, Y. Tabata, K. Nishimura and M. Komeda, Control-released hepatocyte growth factor prevents the progression of heart failure in stroke-prone spontaneously hypertensive rats, *Ann Thorac Surg* **79**: 1627–1634 (2005).

99. C. Bearzi, C. Gargioli, D. Baci, O. Fortunato, K. Shapira-Schweitzer, O. Kossover, M. V. Latronico, D. Seliktar, G. Condorelli and R. Rizzi, PlGF-MMP9-engineered iPS cells supported on a PEG-fibrinogen hydrogel scaffold possess an enhanced capacity to repair damaged myocardium, *Cell Death Dis* **5**: e1053 (2014).

100. E. Fathi, S. M. Nassiri, N. Atyabi, S. H. Ahmadi, M. Imani, R. Farahzadi, S. Rabbani, S. Akhlaghpour, M. Sahebjam and M. Taheri, Induction of angiogenesis via topical delivery of basic-fibroblast growth factor from polyvinyl alcohol-dextran blend hydrogel in an ovine model of acute myocardial infarction, *J Tissue Eng Regen Med* **7**: 697–707 (2013).

101. J. C. Garbern, E. Minami, P. S. Stayton and C. E. Murry, Delivery of basic fibroblast growth factor with a pH-responsive, injectable hydrogel to improve angiogenesis in infarcted myocardium, *Biomaterials* **32**: 2407–2416 (2011).

102. A. J. Rufaihah, S. R. Vaibavi, M. Plotkin, J. Shen, V. Nithya, J. Wang, D. Seliktar and T. Kofidis, Enhanced infarct stabilization and neovascularization mediated by VEGF-loaded PEGylated fibrinogen hydrogel in a rodent myocardial infarction model, *Biomaterials* **34**: 8195–8202 (2013).

103. S. B. Sonnenberg, A. A. Rane, C. J. Liu, N. Rao, G. Agmon, S. Suarez, R. Wang, A. Munoz, V. Bajaj, S. Zhang, R. Braden, P. J. Schup-Magoffin, O. L. Kwan, A. N. DeMaria, J. R. Cochran and K. L. Christman, Delivery of an engineered HGF fragment in an extracellular matrix-derived hydrogel prevents negative LV remodeling post-myocardial infarction, *Biomaterials* **45**: 56–63 (2015).

104. J. Wu, F. Zeng, X. P. Huang, J. C. Chung, F. Konecny, R. D. Weisel and R. K. Li, Infarct stabilization and cardiac repair with a VEGF-conjugated, injectable hydrogel, *Biomaterials* **32**: 579–586 (2011).

105. H. Zhu, X. Jiang, X. Li, M. Hu, W. Wan, Y. Wen, Y. He and X. Zheng, Intramyocardial delivery of VEGF via a novel biodegradable hydrogel induces angiogenesis and improves cardiac function after rat myocardial infarction, *Heart Vessels* **31**: 963–975 (2016).

106. C. B. Rodell, J. W. MacArthur, S. M. Dorsey, R. J. Wade, Y. J. Woo and J. A. Burdick, Shear-Thinning Supramolecular Hydrogels with Secondary Autonomous Covalent Crosslinking to Modulate Viscoelastic Properties, *Adv Funct Mater* **25**: 636–644 (2015).

107. Y. Izumiya, I. Shiojima, K. Sato, D. B. Sawyer, W. S. Colucci and K. Walsh, Vascular endothelial growth factor blockade promotes the transition from compensatory cardiac hypertrophy to failure in response to pressure overload, *Hypertension* **47**: 887–893 (2006).

108. I. Shiojima, K. Sato, Y. Izumiya, S. Schiekofer, M. Ito, R. Liao, W. S. Colucci and K. Walsh, Disruption of coordinated cardiac hypertrophy and angiogenesis contributes to the transition to heart failure, *J Clin Invest* **115**: 2108–2118 (2005).

109. M. C. Bajgelman, L. dos Santos, G. J. J. Silva, J. Nakamuta, R. A. Sirvente, M. Chaves, J. E. Krieger and B. E. Strauss, Preservation of cardiac function in left ventricle cardiac hypertrophy using an AAV vector which provides VEGF-A expression in response to p53, *Virology* **476**: 106–114 (2015).

110. X. Hao, E. A. Silva, A. Mansson-Broberg, K. H. Grinnemo, A. J. Siddiqui, G. Dellgren, E. Wardell, L. A. Brodin, D. J. Mooney and C. Sylven, Angiogenic effects of sequential release of VEGF-A165 and PDGF-BB with alginate hydrogels after myocardial infarction, *Cardiovasc Res* **75**: 178–185 (2007).

111. J. R. Fitzpatrick, 3rd, J. R. Frederick, R. C. McCormick, D. A. Harris, A. Y. Kim, J. R. Muenzer, A. J. Gambogi, J. P. Liu, E. C. Paulson and Y. J. Woo, Tissue-engineered pro-angiogenic fibroblast scaffold improves myocardial perfusion and function and limits ventricular remodeling after infarction, *J Thorac Cardiovasc Surg* **140**: 667–676 (2010).

112. T. M. Yau, K. Fung, R. D. Weisel, T. Fujii, D. A. Mickle and R. K. Li, Enhanced myocardial angiogenesis by gene transfer with transplanted cells., *Circulation* **104**: I218–222 (2001).

113. H. J. Chung, J. T. Kim, H. J. Kim, H. W. Kyung, P. Katila, J. H. Lee, T. H. Yang, Y. I. Yang and S. J. Lee, Epicardial delivery of VEGF and cardiac stem cells guided by 3-dimensional PLLA mat enhancing cardiac regeneration and angiogenesis in acute myocardial infarction, *J Control Release* **205**: 218–230 (2015).

114. J. R. Frederick, J. R. Fitzpatrick, 3rd, R. C. McCormick, D. A. Harris, A. Y. Kim, J. R. Muenzer, N. Marotta, M. J. Smith, J. E. Cohen, W. Hiesinger, P. Atluri and Y. J. Woo, Stromal cell-derived factor-1alpha activation of tissue-engineered endothelial progenitor cell matrix enhances ventricular function after myocardial infarction by inducing neovasculogenesis, *Circulation* **122**: S107–117 (2010).

8

ADIPOSE TISSUE ENGINEERING AND STEM CELLS

D. Adam Young, Brian Mailey, Jennifer Baker,
Anne M. Wallace and Karen L. Christman

1. Introduction

1.1 *Adipose tissue — composition and function*

Adipose tissue is a loose connective tissue that accounts for roughly 18% of the average human body mass.[1] It serves a variety of functions within the body, including energy metabolism, insulation, cushioning, and secretion of multiple cytokines.[1–3] There are two main types of adipose tissue, each serving different functions. White adipose tissue, or WAT, is the most predominant type, being localized within the subcutaneous space of the skin, around the organs of the trunk (known as visceral fat), or in the breast and intramuscular regions. WAT serves as an energy storage depot and endocrine organ in the subcutaneous and visceral locations. When localized around an organ, it serves as a specific nutrient store for highly metabolically active organs like the heart.[4] A second type of adipose tissue, known as brown adipose tissue, or BAT, functions differently from WAT. BAT generates thermal energy from oxidative phosphorylation of nutrients instead of converting it into the energy transport molecule, adenosine triphosphate (ATP).[5] Because of this characteristic, BAT is abundant in infants, which have not yet developed the mechanism for shivering to produce warmth. However, during development BAT is replaced with WAT, and has only recently been identified in small quantities in human adults, most commonly in the neck and surrounding the aorta and vertebrae.[5]

Adipose tissue is a vascularized tissue composed of several cell types surrounded by a basement membrane-like extracellular matrix (ECM). A majority of the tissue is composed of mature fat cells, known as adipocytes, which actively store triglycerides in their large lipid vacuoles (Fig. 1a). The majority of the cytoplasm is displaced by this lipid vacuole, pushing the nucleus to the edge of the cell and giving the tissue a characteristic honeycomb appearance in histology.[6] Other cell types include endothelial cells, fibroblasts, smooth muscle cells, and a population of adult stem cells, which will be discussed in more detail in the following section.[6] These cells release a variety of cytokines and adipokines, including but not limited to leptin, tumor necrosis factor alpha (TNFα), adiponectin, and adipsin.[3] The ECM of adipose tissue accounts for roughly 3% of the mass of the tissue and its composition is similar to that of basement membrane, constituted primarily of collagens, laminin, elastin, and various proteoglycans such as perlecan and heparan sulfate.[7] The tissue is organized into discrete lobules of adipocytes. Each adipocyte is surrounded by a reinforced basement membrane composed of a thin layer of collagen type IV and a thicker layer of collagens I, III, V, and VI, as well as laminin and several glycoproteins.[7] The tissue is further reinforced by long bundles of collagen I fibers, which are aligned with the major arteries and veins of the tissue.[8]

1.2 *Adipose-derived adult stem cells*

Within adipose tissue resides a small population of multipotent progenitor cells known as adipose-derived stem cells or ASCs (Fig. 1b). These cells are also referred to as adipose-derived regenerative cells (ADRCs), adipose-derived stromal cells (ADSCs), or adipose mesenchymal stem

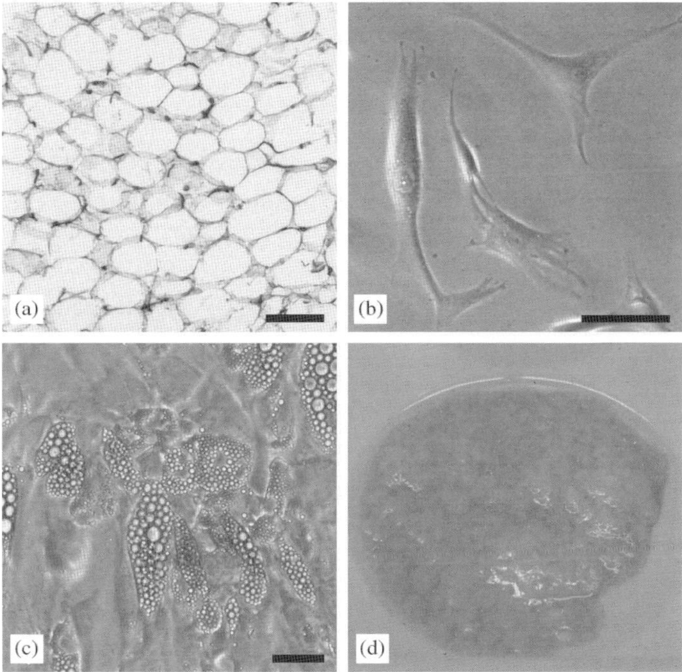

Fig. 1. Adipose tissue and adipose stem cells. Under the microscope, adipose tissue is easily identified by its signet ring, or honeycomb, appearance when stained with H&E (a). Adipocytes are filled with large lipid vacuoles, which push the nucleus to the side of the cell, thus creating this signet ring effect. Adipose tissue also contains a significant number of adipose-derived adult stem cells (ASCs), seen here under a brightfield scope (b). These are multipotent cells capable of differentiating into a variety of cell types, most notably new adipocytes, which are easily identifiable by the accumulating lipid droplets (c). ASCs can be readily obtained from human liposuctioned fat in clinically relevant numbers (d). Scale bars = 100 μm in A, 50 μm in B and C.

cells (AdMSCs). ASCs were first added to the stem cell repertoire in the early 2000s, when a team of researchers at UCLA established both the multipotent differentiation capacity and clonogenicity of a specific cell group within processed lipoaspirate; two of the required characteristics for a cell to earn the label as a "stem" cell.[9] ASCs behave similarly to the more commonly investigated bone marrow-derived mesenchymal stem cells (MSCs), but do have their own specific cell surface receptor profile and have different doubling times and differentiation efficiencies.[10] ASCs function in the body to repair many tissues following injury and are capable of regenerating the various components of the tissue that surround them. They exhibit multipotent plasticity, possessing ability to differentiate toward several different cell types including osteogenic, chondrogenic, adipogenic (Fig. 1c), myogenic, and vasculogenic lineages.[11] Some laboratory experiments have also suggested putative ability for ASCs to adopt characteristics of other complex tissues such as the heart, liver, and brain;[12–14] however, the exact mechanisms and whether differentiation into these cell types occurs remain intensely debated. The presence and activity of telomerase within ASCs have been disputed, though, and they do exhibit some degree of cell senescence, drawing questions regarding their progenitor status. However, their multipotency and upregulated gene expression of stem cell markers OCT4 and SOX2 have justified their classification as adult stem cells.[15,16]

The identification of an endogenous stem cell population within adipose tissue over the last decade has also brought significant attention to the field of adipose tissue engineering. ASCs are the predominant cell type used in cell-based adipose regeneration strategies, with only limited examples of embryonic stem cell use and lessening emphasis on bone marrow-derived adult stem cell use. ASCs are most commonly isolated from human liposuctioned fat (Fig. 1d) or other discarded sources of adipose tissue resulting from cosmetic or reconstructive surgeries, such as abdominoplasty or breast reduction. This generates a plentiful and readily available source from which to derive ASCs, which are present in higher density and, in the case of liposuctioned fat, are obtained in a much more minimally invasive manner than bone marrow-derived MSCs.[10] In fact, this ease of harvest has led several companies to begin developing point-of-care devices for the rapid isolation of ASCs at the patient's bedside. Furthermore, the availability of these cells within the clinic has led many surgeons to already consider utilizing them to treat their patients autologously, as will be discussed in Section 2.

2. Adipose Tissue Engineering in the Clinic

2.1 *Clinical options for replacing lost adipose tissue*

There are several choices in the clinician's toolbox for treating small-volume adipose deficits or augmentations, but few for large-volume repair. Most commonly utilized are the hyaluronic acid-based soft tissue fillers; although several other biopolymer and permanent fillers are also available. While these products have seen extensive usage and safety in the clinic, they often lose their effectiveness after six to nine months and require repetitive treatments for a consistent long-term effect. The high cost of these materials is also prohibitive for their use in larger volume applications. Accordingly, clinicians are seeking innovative solutions for adipose repair that harness the body's natural repair mechanisms instead of the temporary benefits offered by the non-regenerative filler materials currently on the market. Therefore, the restoration of soft tissue defects with autologous tissue remains the core challenge for plastic and reconstructive surgeons. Over the past century, advances in the field of tissue engineering and regenerative medicine have generated new options for treating challenging wounds and solving difficult reconstructive problems. These include dermal matrix systems and irradiated biological human and animal tissue scaffolds.[17,18] A promising advance came with the identification of multipotent cells in adult human fat, as described by Zuk *et al.* in 2001.[9,19] This finding was followed by a sharp rise in popularity of autologous fat grafting with a 10-fold increase in 2010 compared to 1997.[20] Fat grafting is a clinical procedure in which adipose tissue is removed from one location of the body, typically via liposuction procedures, and placed beneath the skin in another location of the same patient where more volume is desired. The initial success of these procedures has generally been attributed to the presence of endogenous ASCs in the grafted tissue. Fat-filling procedures now rank ninth among the most popular surgical cosmetic procedures.[20]

2.2 *Autologous fat transfer*

Autologous fat transplantation is a minimally invasive technique used for multiple applications in aesthetic and reconstructive surgery.[21–23] Adipose tissue is biocompatible, long lasting, and readily available, and it has a natural appearance. Short recovery times and the office-based nature of injections have further contributed to its popularity. The uses of fat grafting have evolved from that of small-volume utility, as in soft tissue fillers, to large-volume applications as a primary mode of reconstruction.[22]

This introduction of large-volume grafting has provided a new option for total facial rejuvenation or breast augmentation.[24] Today, fat grafting is part of every plastic surgeon's repertoire and considered an adjunct or primary technique for facial and breast aesthetic and reconstructive surgery.[22,25–28]

The efficacy of fat grafting is measured by long-term soft tissue retention. Percentage of graft survival varies among patients and fat grafting methods.[29] A wide range of retention rates have been reported in the literature from autologous fat grafts, with up to 70% of the initial fat-filling tissue volume being reabsorbed.[30,31] Ongoing investigations aim for determining optimal methods for harvest, injection, and understanding how individual patient factors might affect overall outcomes.[29] Currently, a meticulous technique with an exact attention to volume dispersion is felt to be a key factor for influencing engraftment.[22] Expert recommendations for optimizing engraftment include concentrating the fat per unit volume of graft delivered per area of native tissue.[22–24,32] This technique maximizes surface area of grafted fat into host tissue and provides maximum nutritional support to the autologous implant. However, the exact mechanisms mediating fat graft survival and resorption remain unclear. While the positive outcomes of these grafts have been partly attributed to the presence of ASCs, the variability in ASC density between patient grafts likely also contributes to the procedure's inconsistency.[33–37] This variability in graft retention, combined with the clinical time required to prepare the tissue for re-injection and availability of host donor tissue, argues for a new alternative that still harnesses the benefits of native tissue but offers off-the-shelf availability and more consistent performance.

2.3 *Expanding use of ASCs in the clinic*

As discussed in the previous section, ASCs have been shown *in vitro* to secrete growth factors, cytokines, chemokines, and extracellular matrix molecules that are important for neovascularization, regulation of apoptosis, modulation of immune response, and extracellular matrix remodeling.[38] The exact clinical promise *versus* the unsubstantiated hype of these cells is still unfolding, and good clinical and scientific studies are still needed and are currently underway. To date, there are 40 open and actively recruiting trials on www.clinical-trials.gov. Many of these trials are concentrating on the vasculogenic properties (e.g. venous stasis wounds, limb ischemia, cardiovascular disease) and regenerative capabilities (e.g. type-2 diabetes, Crohn's disease, breast reconstruction) of ASCs. In the field of plastic and reconstructive surgery, efforts are directed at identifying the influence of ASCs on clinical outcomes of transplanted fat; investigators are attempting to correlate fat graft survival with ASC count. The recent addition of 3D volumetrics has provided a practical method for quantifying percent fat retention (Fig. 2) in order to better evaluate these trials.[39,40]

2.4 *Point-of-care isolation of ASCs*

Isolating ASCs begins by obtaining fat via suction-assisted or direct lipectomy. The lipoaspirate or raw fat is enzymatically digested (typically via collagenase) to remove red blood cells and mature fat cells and produce a stromal vascular fraction (SVF). The SVF is a heterogeneous cell population including endothelial cells, pericytes, fibroblasts, and stem cells (ASCs). In 300 mL of lipoaspirate, there are estimated to be approximately 2–6×10^8 undifferentiated stem cells. These cells maintain the ability to self-renew and differentiate into various cell types.[41–44] The CD31–/CD34+ population of cells present in fat appears to hold the greatest potential for adipogenic differentiation and adipose tissue engineering.[45] A number of companies (e.g. Cytori Therapeutics, San Diego, CA) have produced commercially available enzyme systems to digest mature fat cells, releasing the stromal vascular fraction (including ASCs) to be used for re-implantation. These commercial enzymatic systems introduced a practical method for incorporating stem cells into clinical practice. However, current US Food and Drug Administration (FDA) regulations have limited the use of stem cell isolation equipment in the United States. To date, these devices are enrolled for study for eventual approval within the Center for Biologics Evaluation and Research (CBER). The crux of the argument revolves around whether the cells can be considered "minimally manipulated," and therefore acquitted of the stringent quality control regulations associated with drug production. Independent task forces created under the National Institutes of Health (NIH) and a separate

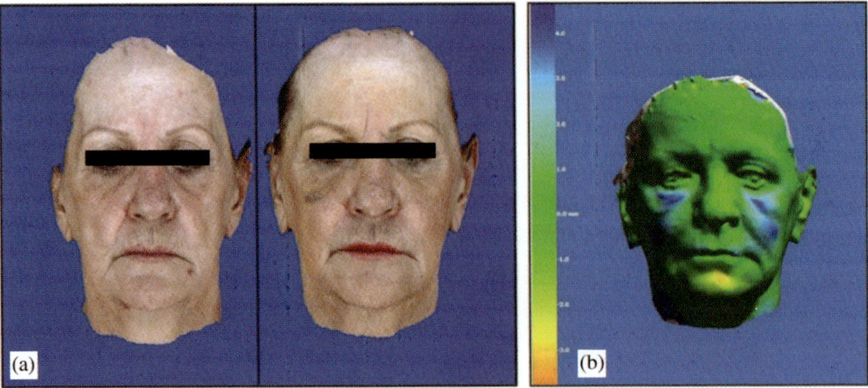

Fig. 2. Clinical analysis of fat grafting. Three-dimensional images of a 71-year-old female preoperative (a, left) and six-weeks postoperative (a, right) from autologous fat transfer to her temples, cheeks, tear-troughs, nasolabial folds, upper lip, chin, and marion-nette basins. Three-dimensional colorimetric analysis of the patient is seen in b. The postoperative picture is registered to the preoperative one using surface landmarks. Program software calculates areas of volume change and illustrates them with the color scheme shown at the left of the figure (e.g. areas of positive volume change are blue and negative volume change in yellow).

joint task force of the American Society of Plastic Surgeons and the American Society for Aesthetic Plastic Surgery were created to examine the scope of treatments and available scientific evidence to use stem cells in clinical practice.[46]

2.5 Stem cell-enriched fat

Since the identification of ASCs in fat tissue,[19] many scientists and surgeons have begun to use these cells to supplement or enrich fat grafts.[47] Pre-clinical and clinical studies have found these cells to be safe and effective.[48–50] A few anecdotal cases in the literature report negative consequences after stem cell therapy; however, the untoward effects appear to be due to contaminants of other additives, rather than the cells themselves.[51] Additionally, these cells may have altered behavior depending on their implanted environment. Published studies found stem cell-enriched fat (SCEF) grafts to have better retention and more normal histology as compared to traditional fat grafts.[48,52] This finding was also observed in direct comparisons of SCEF grafting for breast augmentation to conventional fat transfer techniques.[53,54] There is growing evidence to support positive outcomes with use of SCEF for facial grafting, but it is limited to few studies. A prospective study of 20 patients comparing SCEF grafts to conventional fat grafts found patients with SCEF grafts required less re-operations for additional fat filling.[55] Work from our group has demonstrated high patient satisfaction with an additional unexpected finding of improvement in skin pigmentation.[56] The ASC influence on skin pigmentation was mechanistically explained by *in vitro* and murine *in vivo* evidence demonstrating ASCs to have whitening effects by inhibiting the synthesis of melanin and the activity of tyrosinase.[57] These findings may implicate the use of ASCs in hostile tissue environments such as radiation wounds.[58–60] Future work is needed to fully understand the true clinical utility of ASCs.

3. Adipose Tissue Engineering in the Laboratory

3.1 *Elucidating adipogenesis in vitro*

Examining the mechanisms that drive the differentiation of ASCs (Fig. 3) in a laboratory setting is the first step for identifying important parameters to consider in the development of new adipose tissue engineering approaches. Adipogenic differentiation of ASCs is characterized by growth arrest and cytoskeletal remodeling, followed by an upregulated expression of the key adipose transcription factor, peroxisome proliferator-activated receptor gamma (PPARγ). Traditionally, ASCs have been coerced down an adipogenic pathway through the addition of chemical additives to the cell media, including isobutyl-methylxanthine (IBMX), dexamethasone, indomethacin, and insulin. IBMX, a phosphodiesterase inhibitor, and dexamethasone, a glucocorticoid, both encourage adipogenesis by inducing expression of another transcription factor, CCAAT enhancer binding protein (C/EBP), which in turn causes upregulation of PPARγ.[61] Indomethacin, a non-steroidal anti-inflammatory drug (NSAID), works in a different manner by precluding synthesis of prostaglandins, which have been shown to inhibit adipogenesis.[62] Insulin, arguably the most important component in this adipogenic cocktail, has a more complicated mode of action, having been implicated to be involved in both the

D. A. Young et al.

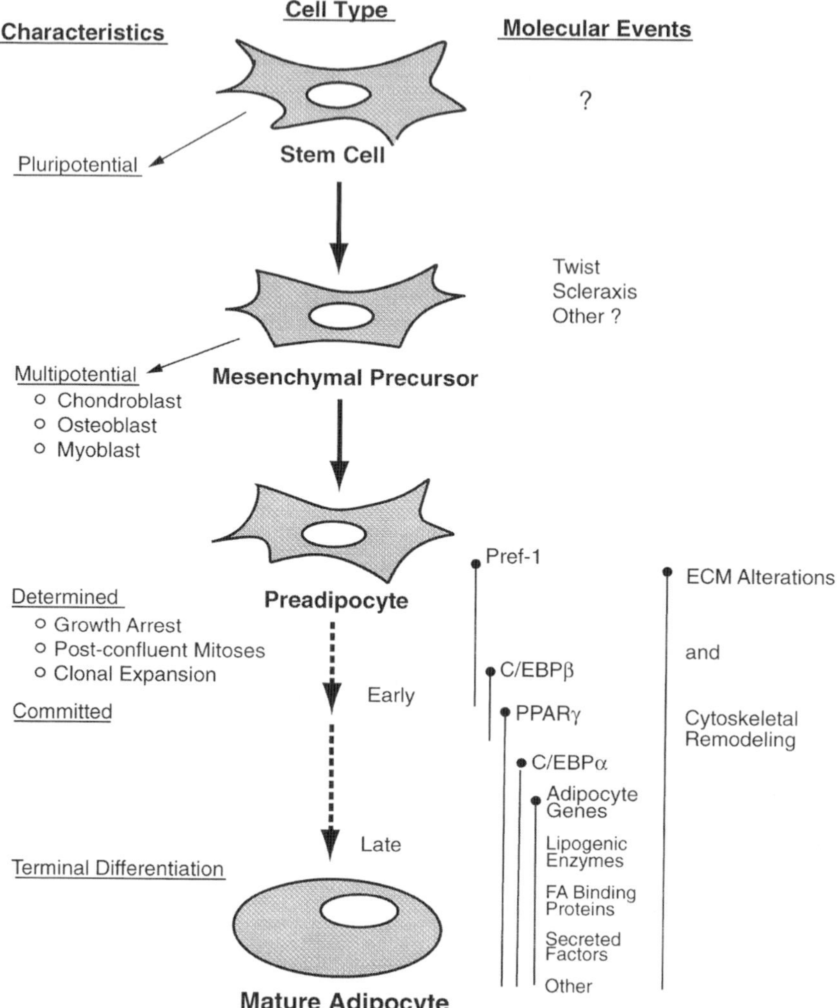

Fig. 3. Cellular changes during adipose differentiation. Stem cells undergo a variety of changes during adipogenic differentiation. Several molecular events occur as the cells begin to commit toward an adipogenic lineage, first expressing Pref-1 and C/EBPβ, but then switching these genes off and sequentially expressing PPARγ, C/EBPα, and other later markers of adipogenesis. The cell also undergoes cytoskeletal remodeling to adopt a more rounded morphology to accommodate the accumulating lipids. Figure adapted from Gregoire *et al.* with permission.[61]

activation of cyclic AMP and in the phosphatidylinositol 3-kinase and Akt signaling pathways.[63] This differentiation is most easily identified by the accumulation of lipid vacuoles within the intracellular space. These lipids can be visualized under a brightfield microscope by using Oil Red O staining, which will stain neutral lipids bright red. Further verification of differentiation can be obtained through polymerase chain reaction (PCR) analysis of several adipogenic genes, including PPARγ, C/EBP, and later markers such as leptin and fatty acid binding protein (aP2). The functional activity of the cells can be further evaluated by measuring glycerol-3 phosphate dehydrogenase (GPDH) activity, an indicator of lipid biosynthesis. For a more detailed explanation of the biological mechanisms involved in adipocyte differentiation, readers are directed to a thorough biological review published by Gregoire *et al.*[61]

While these chemicals have proven effective in differentiating ASCs *in vitro*, there has been significant interest in developing more translatable methods for differentiation by investigating how a cell's physical micro-environment can influence its behavior. Over the past decade, several groups have begun to establish the role that various extracellular matrix (ECM) proteins can exert on the differentiation of progenitor cells. In fact, every tissue has its own unique composition and structure of extracellular components. Researchers have therefore begun developing methods to extract ECM components from adipose tissue, hoping to use them to direct ASC adipogenesis. Using a variety of methods including detergent washes, enzymatic treatments, isopropanol rinses, and even simple homogenization, several groups have been able to remove the cellular content of adipose samples in order to isolate different forms of adipose ECM, a process known as decellularization.[64–67] This adipose ECM typically consists of various collagen isoforms, laminin, and a variety of proteoglycans such as perlecan, mimican, and heparin sulfate binding proteins.[64,68] Accordingly, by culturing ASCs on top of decellularized versions of adipose ECM, researchers have seen that the adipose ECM was indeed capable of stimulating the stem cells to mature toward an adipogenic lineage.[66,68,69] However, it is still unknown which ECM component or mixture of components is directly responsible for triggering this differentiation. A study by researchers at the O'Brien Institute in Australia was able to identify that a soluble protein fraction of the adipose ECM was responsible for this differentiation encouragement and not a factor of any of the collagenous components.[68] Further fractionation and detailed analysis of adipose ECM are still needed to elucidate which specific components are truly adipogenic.

In addition to the influence of ECM on differentiation, many tissue engineers have begun investigating the role of mechanics on ASC adipogenesis as well. In a seminal study by Engler *et al.*, bone marrow-derived mesenchymal stem cells (MSCs) were seen to differentiate toward a variety of lineages based solely on the stiffness of the underlying substrate.[70] In other words, MSCs cultured on substrates that mimicked the soft elasticity of brain tissue (1 kPa) were encouraged to differentiate toward a neural lineage whereas those cultured on stiffer substrates that represented bone tissue (34 kPa) were more likely to differentiate toward an osteogenic lineage. This same phenomenon has also been seen for ASCs, in which substrate stiffness was demonstrated to direct neurogenesis, myogenesis, or osteogenesis of the cells.[71] More recently, these concepts have been investigated specifically for adipogenesis. A study by Guvendiren and Burdick indicated that soft substrates (below 10 kPa) were capable of supporting MSC adipogenesis in either osteogenic media or a combined osteogenic/adipogenic media, but would lose this ability as the substrate stiffened, thus causing the cells to proceed toward an osteogenic lineage regardless of the media type.[72] Expanding upon these results, Young *et al.* demonstrated that ASCs cultured on 2 kPa substrates, which correspond to the specific stiffness of adipose tissue, showed increased adipogenesis compared to cells on a variety of stiffer substrates without the need for any additives in the media.[73] This study, along with another study by Yao *et al.*, demonstrated that the aspect ratio of the cell significantly influenced differentiation, with smaller aspect ratios favoring adipogenesis.[74] Furthermore, softer substrates facilitated the adoption of a more rounded cellular morphology and reduced intracellular tension, which drove adipogenic differentiation.[73] Collectively, these results highlight that materials, which more closely recapitulate the microenvironment of adipose tissue, both biochemically and biomechanically, are capable of stimulating stem cell adipogenesis. This suggests that incorporating these elements into new material designs may be important for promoting adipose regeneration, whether priming ASCs for direct implantation, or developing a material that encourages adipogenesis of host ASCs that are exogenously delivered or endogenously recruited to the treatment site.

3.2 *Exploring adipose tissue engineering in vivo*

The next step in the translational pathway for a tissue engineering strategy is to show *in vivo* biocompatibility and function. However, there is no currently established standard for *in vivo* models of adipose tissue engineering. This is highlighted in a thorough review by Patrick *et al.*, which examined all adipose tissue engineering animal models from 1998 to 2007 and reported a variety of species, strains, and implantation techniques that had been used.[75] Rodent models were the predominant choice, with mouse and rat species seeing roughly equivalent usage (Fig. 4). A majority of these studies involved injection or surgical implantation of the tissue-engineered material, with or without ASCs, into the dorsal subcutis of the animal. This subcutaneous placement of the material would mimic the injection site of the material in the clinic. Alternatively, several groups have begun to adopt a unique model in which a silicone chamber is wrapped around the femoral artery and vein of the animal and then filled with a given material.[76–78] While this model benefits from direct access to a mature vascular supply, the silicone chamber shields a majority of the material from the host, therefore confounding the conclusions about the host response to the material by initiating its own response and by restricting the surface area of the material available to interact with the host. A handful of pig models have also been utilized due to the similarity between porcine and human skin and subcutaneous tissue.[79] However, the paucity of these large animal studies and the disparate strategies for *in vivo* models highlight the less restrictive path through pre-clinical trials for adipose tissue engineering strategies compared to other tissue engineering fields.

Several materials have been shown to support or even encourage *de novo* adipogenesis in various animal models. A variety of synthetic materials have been used successfully as cell delivery vehicles. On their own, many synthetic and polymeric materials can initiate fibrous encapsulation or a foreign body response.[80–82] However, polymers such as poly(lactic-co-glycolic acid) (PLGA) and polyethylene glycol (PEG) have delivered ASCs that were pre-differentiated with adipogenic media and

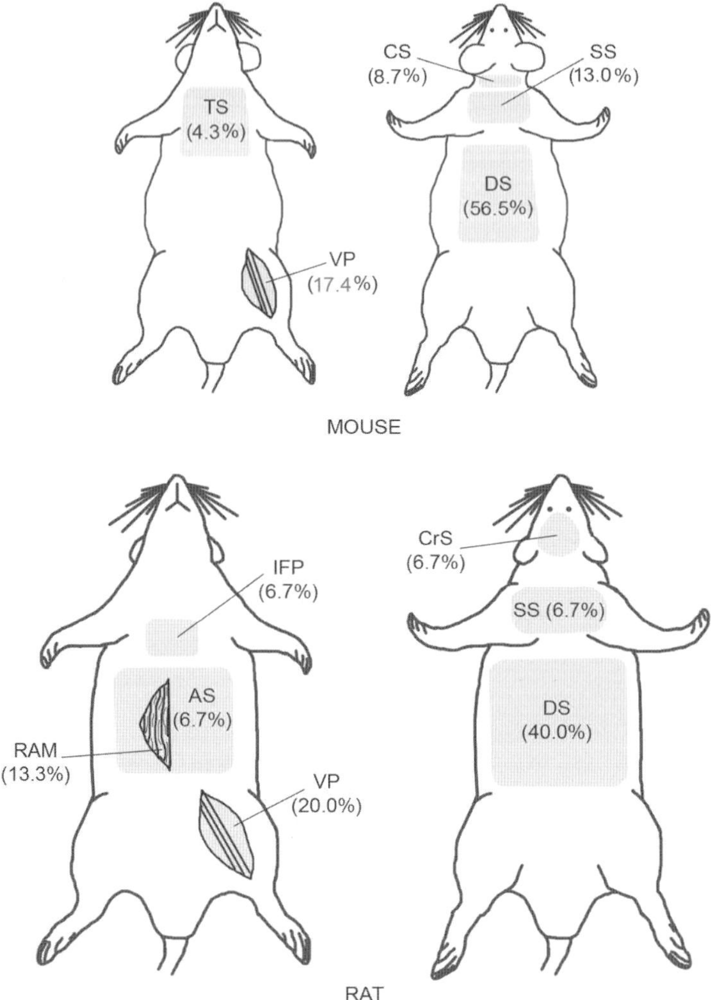

Fig. 4. Distribution of injection sites in adipose tissue engineering animal models. Several injection sites in mice and rats have been used in the literature for examining *in vivo* performance of adipose tissue engineering strategies. Most commonly utilized are injections into the dorsal subcutis or within tissue engineering chambers of the vascular pedicle. AS, abdominal subcutis; CS, cervical subcutis; CrS, cranial subcutis; DS, dorsal subcutis; IFP, inguinal fat pads; RAM, rectus abdomius muscle; SS, scapular subcutis; TS, thoracic subcutis; VP, vascular pedicle in groin. Figure adapted from Patrick *et al.* with permission.[75]

supported the continued maturation of these cells *in vivo*.[83–85] For example, Choi *et al.* designed 100 μm PLGA spheres to deliver differentiated MSCs subcutaneously to a mouse, and found substantial adipose differentiation after eight weeks *in vivo* compared to the cells or PLGA alone.[86] Several biopolymers, most notably collagen, fibrin, and hyaluronic acid (HA), have also been extensively studied in animal models for adipose tissue engineering. These materials have been used in a variety of forms as filler materials for adipose deficits, from injectable gels and powders to three-dimensional scaffolds. These biopolymers tend to be biocompatible but are rapidly broken down within four to nine months *in vivo*.[80] HA-based materials, after extensive crosslinking, appear to be best suited for tempo-rarily filling adipose voids with minimal immune response complications.[87,88] Collagen and fibrin-based materials, due to their shorter durability, have been better used as cell delivery vehicles, supporting the viability and maturation of pre-differentiated ASCs into adipose tissue *in vivo* (Fig 5).[89–92] However, none of these synthetic or bio-derived materials has been shown to encourage new adipogenesis on its own. Recently, as dis-cussed earlier, many groups have utilized various forms of adipose ECM for adipose tissue engineering strategies. These ECM-based materials have also been shown to be highly biocompatible and efficient delivery vehicles for ASCs into the subcutaneous region.[69,93] More importantly, however, adipose ECM materials have also been shown to stimulate new fat formation at the site of implantation without using ASCs.[65,68,94] This represents a critical step forward for the field of adipose tissue engineering, offering a material capable of natural adipose regeneration even without the aid of ASCs. However, these ECM materials do degrade over time *in vivo* and more research is needed to fully establish their degradation profile and potentially to extend their longevity if resorbed too quickly.

Overall, these animal studies have identified three main aspects that are crucial for *in vivo* adipose tissue engineering. Firstly, stimulating neo-vascularization appears to be a critical component for *in vivo* adipogenesis. Adipose tissue is highly vascularized and, like many regeneration strate-gies, increasing vascularization at the site of injury is a prerequisite for new adipose formation. This has been highlighted in several studies that utilize Matrigel, a gel composed of basement membrane proteins derived

                    D. A. Young et al.

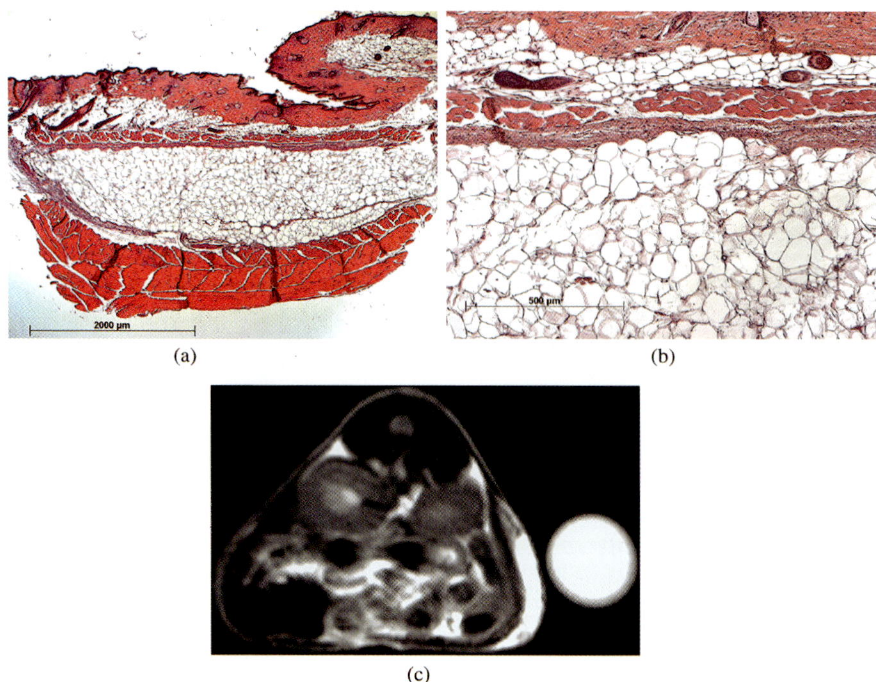

(a)

(b)

(c)

Fig. 5.   Adipose formation *in vivo*. Several studies have investigated the ability of various materials to support adipogenesis of ASCs *in vivo*. Seen here, a fibrin gel loaded with five million ASCs was injected subcutaneously into a mouse. After six months *in vivo*, the cells had developed into lipid-filled adipocytes (a, b). Magnetic resonance imaging (MRI) was also used to assess the volume of the implants in a minimally invasive manner (c). Volume maintenance is an important consideration for adipose tissue engineering strategies and MRI offers a non-invasive way to track this parameter. Figure adapted from Torio-Padron *et al.* with permission.[90]

from mouse sarcoma cells.[95] While Matrigel's tumorous origins preclude it from clinical usage, it has been shown in animal models to stimulate adipogenesis when delivered alongside basic fibroblast growth factor (bFGF). As a proangiogenic factor, bFGF stimulates new vascular ingrowth and, in many cases, has elicited some degree of adipogenesis in materials that otherwise would not stimulate any regeneration.[95,96] This correlation between neovascularization and adipogenesis has also fueled the popularity of the vascular chamber models discussed earlier, as they provide direct access to a mature blood supply.[77,78,97] Secondly, inclusion of a mixture of proteins such as laminin, fibronectin, and collagens I, III, and IV also seems to stimulate adipogenesis in animal models. These proteins are characteristic of adipose and basement membrane tissues and are a common link between all materials that have stimulated *in vivo* adipogenesis without the inclusion of ASCs. This has been seen in studies involving Matrigel and different forms of adipose ECM, and even in a study using extracted skeletal muscle proteins.[65,96,97] Lastly, studies have implicated that keeping mature adipose tissue in close proximity with an implanted material improves its adipogenic capability.[77,98] Whether this is more the result of migrating host progenitor cells or the release of angiogenic and adipogenic cytokines from host adipose tissue has not yet been clearly elucidated. However, it does argue that certain fatty anatomical sites should be targeted for animal models over ones that are relatively devoid of adipose tissue. Collectively, these three themes suggest that adipose tissue engineering strategies should mimic the native microenvironment of adipose tissue and stimulate neovascularization in order to adequately stimulate natural adipose regeneration.

4. Future Directions of the Field

The field of adipose tissue engineering has experienced substantial growth over the past decade. Part of this rapid progression can be attributed to the growing demand for soft tissue filler products and part of it is simply the extension of established tissue engineering principles into a field that was lacking adequate solutions for adipose deficits. The emergence of decellularized adipose tissue provides a promising, versatile biomaterial to be investigated for adipose regeneration (Fig. 6). It has already been shown to be a biocompatible scaffold capable of stimulating adipose differentiation of ASCs and *de novo* adipogenesis in animal models. It can also be fabricated into various forms, including self-assembling gels, powders, beads, and foams.[64,65,94,99] However, as with all biologically derived materials, it does break down over time in the body if it is not modified to some degree. Because a majority of the clinical applications for these materials will involve sustained fulfillment of an adipose deficit, it will be imperative to tightly control the degradation of these adipose ECM-derived materials. Future research will be needed to minimize the

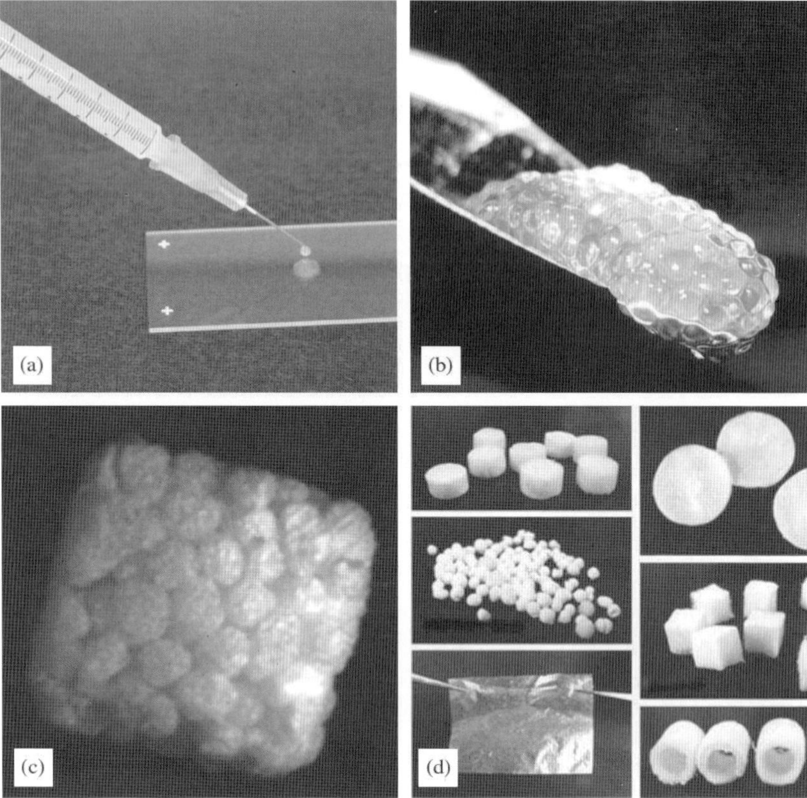

Fig. 6. Various forms of decellularized adipose tissue. Within just the past five years, the process of decellularization of human adipose tissues has become an important topic within the field of adipose tissue engineering, offering the ability to generate scaffolds from adipose ECM. In just the past few years, several different forms of this adipose ECM have been developed, ranging from injectable gels and powders (a), to beads (b), to microporous foams (c), and even solid three-dimensional structures in a variety of shapes and sizes (d). Figure adapted from Turner *et al.*, Yu *et al.*, and Choi *et al.* with permission.[64,67,94,99]

degradation of the scaffold or to ensure that the degradation is matched to the in-growth of regenerating tissue. Alternatively, decellularized adipose ECM could be combined with other biomaterials in order to match its biocompatibility and adipo-inductive properties with other desirable degradation or mechanical properties.

Alongside material development, there is also a need to design better animal models that better reflect the clinical setting. As seen previously, the diversity of species and strains currently being used in different laboratories precludes direct comparisons of the various materials and results being published. Furthermore, most laboratory rodents possess limited innate adipose tissue. Because several studies have demonstrated the influence of mature adipose tissue within the proximity of an experimental material, a standardized injection site should be established for mouse and rat models in order to better control confounding variables. Subcutaneously implanted materials will likely experience drastically different mechanical loading, cytokine stimulation, and degradation rates in caudal vs. cranial and ventral vs. dorsal locations. Additionally, the development of a breast tumor resection model would also be beneficial for assessing the performance and behavior of biomaterials within an implant site that has already experienced a traumatic injury. These improvements to adipose tissue engineering models will help narrow the gap between laboratory experiments and clinical translation.

Incorporation of ASCs into various adipose tissue engineering strategies has also become a popular trend recently. These cells, being derived from adipose tissue, are an ideal candidate for clinical adipose tissue engineering strategies. In fact, anecdotal clinical evidence has surfaced suggesting their ability to improve fat graft retention, stimulate subcutaneous vascularization, and even improve the appearance of skin. However, there are several ongoing regulatory debates that could dramatically affect the fate of ASC clinical usage. In terms of allogeneic usage, several issues must be addressed in terms of reproducible and sterile harvesting techniques, cell storage, and donor variability. In fact, the recent US District Court case of *United States v Regenerative Sciences* upheld the FDA's position that cultured ASCs, even if autologously derived, could be considered more than "minimally manipulated" and thus classified as a drug under FDA regulation.[100] The FDA has since issued a much debated Draft Guidance for Industry entitled "Human Cells, Tissues, and Cellular and Tissue-Based Products (HCT/Ps) from Adipose Tissue: Regulatory Considerations" that outlines their current position on cells and materials derived from adipose tissue. While these measures will help ensure the safety of stem cell procedures for patients, they will also significantly lengthen the timeline for establishing these procedures in the clinic and could encourage questionable "stem cell tourism" practices in locations outside of the United States. Therefore, while ASCs offer significant promise for adipose tissue engineering strategies, there is still much work to be done before they can become a clinically viable resource.

5. Conclusion

Adipose tissue engineering strives to develop novel materials and methods for repairing and regenerating fatty tissue deficits. There are a variety of materials, both synthetic and natural, available for filling adipose voids and for delivering ASCs to help regenerate the adipose tissue. Based on the current research, several design criteria have begun to emerge that can guide future adipose tissue engineering strategies. First of all, the material should be biocompatible, integrating well with the surrounding tissue without triggering a chronic inflammatory response. It should also be durable and provide a long-term solution to adipose deficits without the need for repeated treatment. Part of this solution could be to stimulate natural adipose regeneration at the site of implantation. In this case, it is also imperative for the strategy to facilitate neovascularization while promoting the formation of new adipose tissue. This last scenario would represent the pinnacle result for adipose tissue engineering, by providing the patient with a natural means to replace lost adipose tissue with healthy, mature, and functional fat tissue. With the development of decellularized adipose ECM materials and a better understanding of adipogenesis, this goal could become an achievable milestone for tissue engineering within the near future.

Acknowledgements

Funding was provided in part by the NIH Director's New Innovator Award Program, part of the NIH Roadmap for Medical Research, through grant number 1-DP2-OD004309. A. Y. would like to thank the National Science Foundation Graduate Research Fellowship Program and Engineering Innovation Fellowship Program. B. M., J. B., and A. W. would also like to recognize support from The Plastic Surgery Foundation.

References

1. T. M. Aktan, S. Duman and B. Cihantimur, Cellular and molecular aspects of adipose tissue, in *Adipose Stem Cells and Regenerative Medicine*, edited by Y. G. Illouz and A. Sterodimas (Springer, 2011).

2. J. H. Choi, J. M. Gimble, K. Lee, K. G. Marra, J. P. Rubin, J. J. Yoo, G. Vunjak-Novakovic and D. L. Kaplan, Adipose tissue engineering for soft tissue regeneration, *Tissue Eng Part B Rev* **16**: 413–426 (2010).

3. E. E. Kershaw and J. S. Flier, Adipose tissue as an endocrine organ, *J Clin Endocrinol Metab* **89**: 2548–2556 (2004).

4. S. Cinti, The adipose organ, *Prostaglandins Leukot Essent Fatty Acids* **73**: 9–15 (2005).

5. B. Cannon and J. Nedergaard, Yes, even human brown fat is on fire! *J Clin Invest* **122**: 486–489 (2012).

6. C. T. Gomillion and K. J. L. Burg, Stem cells and adipose tissue engineering, *Biomaterials* **27**: 6052–6063 (2006).

7. Y. Kubo, S. Kaidzu, I. Nakajima, K. Takenouchi and F. Nakamura, Organization of extracellular matrix components during differentiation of adipocytes in long-term culture, *In Vitro Cell Dev Biol Anim* **36**: 38–44 (2000).

8. K. Comley and N. A. Fleck, A micromechanical model for the Young's modulus of adipose tissue, *Int J Solids Struct* **47**: 2982–2990 (2010).

9. P. Zuk, M. Zhu, P. Ashjian, D. De Ugarte, J. Huang, H. Mizuno, Z. Alfonso, J. Fraser, P. Benhaim and M. Hedrick, Human adipose tissue is a source of multipotent stem cells. *Mol Biol Cell* **13**: 4279 (2002).

10. J. M. Gimble, A. J. Katz and B. A. Bunnell, Adipose-derived stem cells for regenerative medicine. *Circ Res* **100**: 1249–1260 (2007).

11. P. Zuk, M. Zhu, H. Mizuno, J. Huang, J. Futrell, A. Katz, P. Benhaim, H. Lorenz and M. Hedrick, Multilineage cells from human adipose tissue: implications for cell-based therapies, *Tissue Eng* **7**: 211–228 (2001).

12. V. Planat-Bénard, C. Menard, M. André, M. Puceat, A. Perez, J.-M. Garcia-Verdugo, L. Pénicaud and L. Casteilla, Spontaneous cardiomyocyte differentiation from adipose tissue stroma cells, *Circ Res* **94**: 223–229 (2004).

13. A. Banas, T. Teratani, Y. Yamamoto, M. Tokuhara, F. Takeshita, M. Osaki, T. Kato, H. Okochi and T. Ochiya, Rapid hepatic fate specification of adipose-derived stem cells and their therapeutic potential for liver failure, *J Gastroenterol Hepatol* **24**: 70–77 (2009).

14. D. X. Qian, H. T. Zhang, X. Ma, X. D. Jiang and R. X. Xu, Comparison of the efficiencies of three neural induction protocols in human adipose stromal cells, *Neurochem Res* **35**: 572–579 (2010).

15. P. C. Sachs, M. P. Francis, M. Zhao, J. Brumelle, R. R. Rao, L. W. Elmore and S. E. Holt, Defining essential stem cell characteristics in adipose-derived stromal cells extracted from distinct anatomical sites. *Cell Tissue Res* **349**: 505–515 (2012).

16. B. G. Jeon, B. M. Kumar, E. J. Kang, S. A. Ock, S. L. Lee, D. O. Kwack, J. H. Byun, B. W. Park and G. J. Rho, Characterization and comparison of telomere length, telomerase and reverse transcriptase activity and gene expression in human mesenchymal stem cells and cancer cells of various origins, *Cell Tissue Res* **345**: 149–161 (2011).

17. A. Cuadra, G. Correa, R. Roa, J. L. Pineros, H. Norambuena, S. Searle, R. Las Heras and W. Calderon, Functional results of burned hands treated with Integra(R), *J Plast Reconstr Aesthet Surg* **65**: 228–234 (2012).

18. I. Jones, L. Currie and R. Martin, A guide to biological skin substitutes, *Br J Plast Surg* **55**: 185–193 (2002).

19. P. A. Zuk, M. Zhu, H. Mizuno, J. Huang, J. W. Futrell, A. J. Katz, P. Benhaim, H. P. Lorenz and M. H. Hedrick, Multilineage cells from human adipose tissue: implications for cell-based therapies. *Tissue Eng* **7**: 211–228 (2001).

20. B. Mailey, A. Artinyan, J. Khalili, J. Denitz, N. Sanchez-Luege, C. L. Sun, S. Bhatia, N. Nissen, S. D. Colquhoun and J. Kim, Evaluation of absolute serum alpha-fetoprotein levels in liver transplant for hepatocellular cancer, *Arch Surg* **146**: 26–33 (2011).

21. ASPRS Ad-Hoc Committee on New Procedures, Report on autologous fat transplantation, *Plast Surg Nurs* **7**: 140–141 (1987).

22. S. R. Coleman, Structural fat grafting: more than a permanent filler, *Plast Reconstr Surg* **118**: 108S–120S (2006).

23. W. P. Coleman 3rd, Autologous fat transplantation, *Plast Reconstr Surg* **88**: 736 (1991).

24. S. R. Cohen and B. Mailey, Adipocyte-derived stem and regenerative cells in facial rejuvenation, *Clin Plast Surg* **39**: 453–464 (2012).

25. V. Cervelli, L. Palla, M. Pascali, B. De Angelis, B. C. Curcio and P. Gentile, Autologous platelet-rich plasma mixed with purified fat graft in aesthetic plastic surgery, *Aesthetic Plast Surg* **33**: 716–721 (2009).

26. T. G. Hardy, N. Joshi and M. H. Kelly, Orbital volume augmentation with autologous micro-fat grafts, *Ophthal Plast Reconstr Surg* **23**: 445–449 (2007).

27. Y. G. Illouz, The fat cell "graft": a new technique to fill depressions, *Plast Reconstr Surg* **78**: 122–123 (1986).

28. B. Phulpin, P. Gangloff, N. Tran, P. Bravetti, J. L. Merlin and G. Dolivet, Rehabilitation of irradiated head and neck tissues by autologous fat transplantation, *Plast Reconstr Surg* **123**: 1187–1197 (2009).

29. M. Zhu, Z. Zhou, Y. Chen, R. Schreiber, J. T. Ransom, J. K. Fraser, M. H. Hedrick, K. Pinkernell and H. C. Kuo, Supplementation of fat grafts with adipose-derived regenerative cells improves long-term graft retention. *Ann Plast Surg* **64**: 222–228 (2010).

30. I. Niechajev and O. Sevcuk, Long-term results of fat transplantation: clinical and histologic studies, *Plast Reconstr Surg* **94**: 496–506 (1994).

31. H. W. Horl, A. M. Feller and E. Biemer, Technique for liposuction fat reimplantation and long-term volume evaluation by magnetic resonance imaging, *Ann Plast Surg* **26**: 248–258 (1991).

32. S. R. Coleman, Hand rejuvenation with structural fat grafting, *Plast Reconstr Surg* **110**: 1731–1744; discussion 1745–1737 (2002).

33. L. S. Toledo and R. Mauad, Fat injection: a 20-year revision, *Clin Plast Surg* **33**: 47–53, vi (2006).

34. T. A. Moseley, M. Zhu and M. H. Hedrick, Adipose-derived stem and progenitor cells as fillers in plastic and reconstructive surgery, *Plast Reconstr Surg* **118**: 121S–128S (2006).

35. T. Tiryaki, N. Findikli and D. Tiryaki, Staged stem cell-enriched tissue (SET) injections for soft tissue augmentation in hostile recipient areas: a preliminary report, *Aesthetic Plast Surg* **35**: 965–971 (2011).

36. C. Tremolada, G. Palmieri and C. Ricordi, Adipocyte transplantation and stem cells: plastic surgery meets regenerative medicine, *Cell Transplant* **19**: 1217–1223 (2010).

37. P. Gir, G. Oni, S. A. Brown, A. Mojallal and R. J. Rohrich, Human adipose stem cells: current clinical applications, *Plast Reconstr Surg* **129**: 1277–1290 (2012).

38. B. M. Strem, K. C. Hicok, M. Zhu, I. Wulur, Z. Alfonso, R. E. Schreiber, J. K. Fraser and M. H. Hedrick, Multipotential differentiation of adipose tissue-derived stem cells, *Keio J Med* **54**: 132–141 (2005).

39. A. A. Davila, D. W. Buck 2nd, D. Chopp, C. M. Connor, S. Persing, V. Rawlani and J. Y. Kim, A novel prospective three-dimensional analysis of nasolabial fold augmentation, *Aesthet Surg J* **32**: 488–494 (2012).

40. B. Mailey, B. Buchberg, C. Prendergast, A. Artinyan, J. Khalili, N. Sanchez-Luege, S. D. Colquhoun and J. Kim, A disease-based comparison of liver transplantation outcomes, *Am Surg* **75**: 901–908 (2009).

41. D. Zipori, The nature of stem cells: state rather than entity, *Nat Rev Genet* **5**: 873–878 (2004).

42. J. A. Thomson, J. Itskovitz-Eldor, S. S. Shapiro, M. A. Waknitz, J. J. Swiergiel, V. S. Marshall and J. M. Jones, Embryonic stem cell lines derived from human blastocysts, *Science* **282**: 1145–1147 (1998).

43. I. L. Weissman, Stem cells: units of development, units of regeneration, and units in evolution, *Cell* **100**: 157–168 (2000).

44. L. Li and H. Clevers, Coexistence of quiescent and active adult stem cells in mammals, *Science* **327**: 542–545 (2010).

45. H. Li, L. Zimmerlin, K. G. Marra, V. S. Donnenberg, A. D. Donnenberg and J. P. Rubin, Adipogenic potential of adipose stem cell subpopulations, *Plast Reconstr Surg* **128**: 663–672 (2011).

46. B. Mailey, C. Truong, A. Artinyan, J. Khalili, N. Sanchez-Luege, J. Denitz, H. Marx, L. D. Wagman and J. Kim, Surgical resection of primary and metastatic hepatic malignancies following portal vein embolization, *J Surg Oncol* **100**: 184–190 (2009).

47. D. Matsumoto, K. Sato, K. Gonda, Y. Takaki, T. Shigeura, T. Sato, E. Aiba-Kojima, F. Iizuka, K. Inoue, H. Suga and K. Yoshimura, Cell-assisted lipotransfer: supportive use of human adipose-derived cells for soft tissue augmentation with lipoinjection, *Tissue Eng* **12**: 3375–3382 (2006).

48. C. Calabrese, L. Orzalesi, D. Casella and L. Cataliotti, Breast reconstruction after nipple/areola-sparing mastectomy using cell-enhanced fat grafting, *Ecancermedicalscience* **3**: 116 (2009).

49. R. Perez-Cano, J. J. Vranckx, J. M. Lasso, C. Calabrese, B. Merck, A. M. Milstein, E. Sassoon, E. Delay and E. M. Weiler-Mithoff, Prospective trial of adipose-derived regenerative cell (ADRC)-enriched fat grafting for partial mastectomy defects: the RESTORE-2 trial, *Eur J Surg Oncol* **38**: 382–389 (2012).

50. X. Shen, B. Mailey, J. D. Ellenhorn, P. G. Chu, A. M. Lowy and J. Kim, CC chemokine receptor 9 enhances proliferation in pancreatic intraepithelial neoplasia and pancreatic cancer cells, *J Gastrointest Surg* **13**: 1955–1962; discussion 1962 (2009).

51. M. K. Dobke, Are we witnessing the emergence of a superspecialty? *Clin Plast Surg* **39**: xiii–xiv (2012).

52. T. Kamakura and K. Ito, Autologous cell-enriched fat grafting for breast augmentation, *Aesthetic Plast Surg* **35**: 1022–1030 (2011).

53. K. Yoshimura, T. Shigeura, D. Matsumoto, T. Sato, Y. Takaki, E. Aiba-Kojima, K. Sato, K. Inoue, T. Nagase, I. Koshima and K. Gonda, Characterization of freshly isolated and cultured cells derived from the fatty and fluid portions of liposuction aspirates, *J Cell Physiol* **208**: 64–76 (2006).

54. K. Yoshimura, Y. Asano, N. Aoi, M. Kurita, Y. Oshima, K. Sato, K. Inoue, H. Suga, H. Eto, H. Kato and K. Harii, Progenitor-enriched adipose tissue transplantation as rescue for breast implant complications, *Breast J* **16**: 169–175 (2010).

55. A. Sterodimas, J. de Faria, B. Nicaretta and F. Boriani, Autologous fat transplantation versus adipose-derived stem cell-enriched lipografts: a study, *Aesthet Surg J* **31**: 682–693 (2011).

56. A. H. Hassanein, B. A. Mailey and M. K. Dobke, Robot-assisted plastic surgery, *Clin Plast Surg* **39**: 419–424 (2012).

57. W. S. Kim, S. H. Park, S. J. Ahn, H. K. Kim, J. S. Park, G. Y. Lee, K. J. Kim, K. K. Whang, S. H. Kang, B. S. Park and J. H. Sung, Whitening effect of adipose-derived stem cells: a critical role of TGF-beta 1, *Biol Pharm Bull* **31**: 606–610 (2008).

58. J. H. Jeong, Adipose stem cells and skin repair, *Curr Stem Cell Res Ther* **5**: 137–140 (2010).

59. H. Nakagawa, S. Akita, M. Fukui, T. Fujii and K. Akino, Human mesenchymal stem cells successfully improve skin-substitute wound healing, *Br J Dermatol* **153**: 29–36 (2005).

60. M. Cherubino, J. P. Rubin, N. Miljkovic, A. Kelmendi-Doko and K. G. Marra, Adipose-derived stem cells for wound healing applications, *Ann Plast Surg* **66**: 210–215 (2011).

61. F. M. Gregoire, C. M. Smas and H. S. Sul, Understanding adipocyte differentiation, *Physiol Rev* **78**: 783–809 (1998).

62. G. Shillabeer, V. Kumar, E. Tibbo and D. C. Lau, Arachidonic acid metabolites of the lipoxygenase as well as the cyclooxygenase pathway may be involved in regulating preadipocyte differentiation, *Metab Clin Exp* **47**: 461–466 (1998).

63. D. J. Klemm, J. W. Leitner, P. Watson, A. Nesterova, J. E. Reusch, M. L. Goalstone and B. Draznin, Insulin-induced adipocyte differentiation: activation of CREB rescues adipogenesis from the arrest caused by inhibition of prenylation, *J Biol Chem* **276**: 28430–28435 (2001).

64. D. A. Young, D. O. Ibrahim, D. Hu and K. L. Christman, Injectable hydrogel scaffold from decellularized human lipoaspirate, *Acta Biomater* **7**: 1040–1049 (2011).

65. I. Wu, Z. Nahas, K. A. Kimmerling, G. D. Rosson and J. H. Elisseeff, An injectable adipose matrix for soft-tissue reconstruction, *Plast Reconstr Surg* **129**: 1247–1257 (2012).

66. L. E. Flynn, The use of decellularized adipose tissue to provide an inductive microenvironment for the adipogenic differentiation of human adipose-derived stem cells, *Biomaterials* **31**: 4715–4724 (2010).

67. J. S. Choi, B. S. Kim, J. Y. Kim, J. D. Kim, Y. C. Choi, H.-J. Yang, K. Park, H. Y. Lee and Y. W. Cho, Decellularized extracellular matrix derived from human adipose tissue as a potential scaffold for allograft tissue engineering, *J Biomed Mater Res* **97**: 292–299 (2011).

68. C. J. Poon, M. V. Pereira E Cotta, S. Sinha, J. A. Palmer, A. A. Woods, W. A. Morrison and K. M. Abberton, Preparation of an adipogenic hydrogel from subcutaneous adipose tissue, *Acta biomater* **9**: 5609–5620 (2013).

69. A. E. B. Turner, C. Yu, J. Bianco, J. F. Watkins and L. E. Flynn, The performance of decellularized adipose tissue microcarriers as an inductive substrate for human adipose-derived stem cells, *Biomaterials* **33**: 4490–4499 (2012).

70. A. J. Engler, S. Sen, H. L. Sweeney and D. E. Discher, Matrix elasticity directs stem cell lineage specification, *Cell* **126**: 677–689 (2006).

71. Y. S. Choi, L. G. Vincent, A. R. Lee, M. K. Dobke and A. J. Engler, Mechanical derivation of functional myotubes from adipose-derived stem cells, *Biomaterials* **33**: 2482–2491 (2012).

72. M. Guvendiren and J. A. Burdick, Stiffening hydrogels to probe short- and long-term cellular responses to dynamic mechanics, *Nat Comms* **3**: 792 (2012).

73. D. A. Young, Y. S. Choi, A. J. Engler and K. L. Christman, Mimicking the stiffness of adipose tissue stimulates adipogenesis of adult adipose-derived stem cells, submitted (2013).

74. X. Yao, R. Peng and J. Ding, Effects of aspect ratios of stem cells on lineage commitments with and without induction media, *Biomaterials* **34**: 930–939 (2013).

75. C. W. Patrick, R. Uthamanthil, E. Beahm and C. Frye, Animal models for adipose tissue engineering, *Tissue Eng Part B Rev* **14**: 167–178 (2008).

76. K. J. Cronin, A. Messina, K. R. Knight, J. J. Cooper-White, G. W. Stevens, A. J. Penington and W. A. Morrison, New murine model of spontaneous autologous tissue engineering, combining an arteriovenous pedicle with matrix materials, *Plast Reconstr Surg* **113**: 260–269 (2004).

77. F. Stillaert, M. Findlay, J. Palmer, R. Idrizi, S. Cheang, A. Messina, K. Abberton, W. Morrison and E. W. Thompson, Host rather than graft origin of Matrigel-induced adipose tissue in the murine tissue-engineering chamber, *Tissue Eng* **13**: 2291–2300 (2007).

78. S. Uriel, J. J. Huang, M. L. Moya, M. E. Francis, R. Wang, S. Y. Chang, M. H. Cheng and E. M. Brey, The role of adipose protein derived hydrogels in adipogenesis, *Biomaterials* **29**: 3712–3719 (2008).

79. K. Hemmrich, K. Van de Sijpe, N. P. Rhodes, J. A. Hunt, C. Di Bartolo, N. Pallua, P. Blondeel and D. von Heimburg, Autologous *in vivo* adipose tissue engineering in hyaluronan-based gels: a pilot study, *J Surg Res* **144**: 82–88 (2008).

80. G. Lemperle, V. Morhenn and U. Charrier, Human histology and persistence of various injectable filler substances for soft tissue augmentation, *Aesthetic Plast Surg* **27**: 354–366; discussion 367 (2003).

81. L. Christensen, V. Breiting, M. Janssen, J. Vuust and E. Hogdall, Adverse reactions to injectable soft tissue permanent fillers, *Aesthetic Plast Surg* **29**: 34–48 (2005).

82. A. D. Lynn, T. R. Kyriakides and S. J. Bryant, Characterization of the *in vitro* macrophage response and *in vivo* host response to poly(ethylene glycol)-based hydrogels, *J Biomed Mater Res* **93**: 941–953 (2010).

83. Y. S. Choi, S. N. Park and H. Suh, The effect of PLGA sphere diameter on rabbit mesenchymal stem cells in adipose tissue engineering, *J Mater Sci Mater Med* **19**: 2165–2171 (2008).

84. S. W. Kang, S. W. Seo, C. Y. Choi and B. S. Kim, Porous poly(lactic-co-glycolic acid) microsphere as cell culture substrate and cell transplantation vehicle for adipose tissue engineering, *Tissue Eng Part C Methods* **14**: 25–34 (2008).

85. A. T. Hillel, S. Unterman, Z. Nahas, B. Reid, J. M. Coburn, J. Axelman, J. J. Chae, Q. Guo, R. Trow, A. Thomas, Z. Hou, S. Lichtsteiner, D. Sutton, C. Matheson, P. Walker, N. David, S. Mori, J. M. Taube and J. H. Elisseeff, Photoactivated composite biomaterial for soft tissue restoration in rodents and in humans, *Sci Transl Med* **3**: 93ra67 (2011).

86. Y. S. Choi, S. M. Cha, Y. Y. Lee, S. W. Kwon, C. J. Park and M. Kim, Adipogenic differentiation of adipose tissue derived adult stem cells in nude mouse, *Biochem Biophys Res Commun* **345**: 631–637 (2006).

87. A. Verpaele and A. Strand, Restylane SubQ, a non-animal stabilized hyaluronic acid gel for soft tissue augmentation of the mid- and lower face, *Aesthet Surg J* **26**: S10–S17 (2006).

88. I. B. Allemann and L. Baumann, Hyaluronic acid gel (Juvederm) preparations in the treatment of facial wrinkles and folds, *Clin Interv Aging* **3**: 629–634 (2008).

89. S. W. Cho, I. Kim, S. H. Kim, J. W. Rhie, C. Y. Choi and B. S. Kim, Enhancement of adipose tissue formation by implantation of adipogenic-differentiated preadipocytes, *Biochem Biophys Res Commun* **345**: 588–594 (2006).

90. N. Torio-Padron, N. Baerlecken, A. Momeni, G. B. Stark and J. Borges, Engineering of adipose tissue by injection of human preadipocytes in fibrin, *Aesthetic Plast Surg* **31**: 285–293 (2007).

91. Y. Kimura, M. Ozeki, T. Inamoto and Y. Tabata, Adipose tissue engineering based on human preadipocytes combined with gelatin microspheres containing basic fibroblast growth factor, *Biomaterials* **24**: 2513–2521 (2003).

92. J. P. Rubin, J. M. Bennett, J. S. Doctor, B. M. Tebbets and K. G. Marra, Collagenous microbeads as a scaffold for tissue engineering with adipose-derived stem cells, *Plast Reconstr Surg* **120**: 414–424 (2007).

93. J. S. Choi, H. J. Yang, B. S. Kim, J. D. Kim, J. Y. Kim, B. Yoo, K. Park, H. Y. Lee and Y. W. Cho, Human extracellular matrix (ECM) powders for injectable cell delivery and adipose tissue engineering, *J Control Release* **139**: 2–7 (2009).

94. C. Yu, J. Bianco, C. Brown, L. Fuetterer, J. F. Watkins, A. Samani and L. E. Flynn, Porous decellularized adipose tissue foams for soft tissue regeneration, *Biomaterials* **34**: 3290–3302 (2013).

95. K. Toriyama, N. Kawaguchi, J. Kitoh, R. Tajima, K. Inou, Y. Kitagawa and S. Torii, Endogenous adipocyte precursor cells for regenerative soft-tissue engineering, *Tissue Eng* **8**: 157–165 (2002).

96. N. Kawaguchi, K. Toriyama, E. Nicodemou-Lena, K. Inou, S. Torii and Y. Kitagawa, *De novo* adipogenesis in mice at the site of injection of basement membrane and basic fibroblast growth factor, *Proc Natl Acad Sci USA* **95**: 1062–1066 (1998).

97. K. M. Abberton, S. K. Bortolotto, A. A. Woods, M. Findlay, W. A. Morrison, E. W. Thompson and A. Messina, Myogel, a novel, basement membrane-rich, extracellular matrix derived from skeletal muscle, is highly adipogenic *in vivo* and *in vitro*, *Cells Tissues Organs* **188**: 347–358 (2008).

98. J. L. Kelly, M. W. Findlay, K. R. Knight, A. Penington, E. W. Thompson, A. Messina and W. A. Morrison, Contact with existing adipose tissue is inductive for adipogenesis in matrigel. *Tissue Eng* **12**: 2041–2047 (2006).

99. A. E. B. Turner and L. E. Flynn, Design and characterization of tissue-specific extracellular matrix-derived microcarriers, *Tissue Eng Part C Methods* **18**: 186–197 (2012).

100. D. Cyranoski, FDA challenges stem-cell clinic, *Nature* **466**: 909 (2010).

9

ENGINEERING CARTILAGE: FROM MATERIALS TO SMALL MOLECULES

Jeannine M. Coburn and Jennifer H. Elisseeff

1. Introduction

This chapter begins by reviewing the structural and compositional make-up of articular cartilage, followed by a brief discussion of osteoarthritis (OA). Defects of cartilage that may lead to OA are discussed along with associated surgical strategies for repair. The authors will also describe numerous tissue engineering strategies aimed to repair cartilage defects, including materials and small molecules that direct stem-cell differentiation toward a chondrogenic lineage. The goal of this chapter is to not be all-encompassing on each of these areas but to give an overview to the readers of the exciting research being performed in the area of cartilage tissue engineering and the current limitations.

2. Structure of Articular Cartilage of the Knee

Articular cartilage (also known as hyaline cartilage) lines the surface of all diarthrodial joints including the hips, knees, and shoulders. The diarthrodial joints are enclosed in a fibrous capsule. Lining the inner surface of the capsule is the synovium, which contains synoviocytes. The synoviocytes secrete the synovial fluid that provides the nutrients to the articular cartilage along with lubrication. Due to the smooth surface of the articular cartilage and the synovial fluid, minimal friction is exerted in the joint space during normal motion.[1]

Articular cartilage is predominantly composed of water and extracellular matrix (ECM). Additionally, a single-cell population resides within the articular cartilage, known as chondrocytes, which emerge from mesenchymal precursor cells during limb development.[2] The ECM is divided into two major categories, collagen and proteoglycans (PGs). The main type of collagen in articular cartilage is type II and to a lesser extent types IX and XI.[3] Collagen fibers are found throughout the cartilage and contribute to the tensile strength of the tissue.[1] Proteoglycans are composed of a core protein, typically aggrecan, with glycosaminoglycans (GAGs) bound to a serine residue via a trisaccharide linker. The predominant GAG molecules of articular cartilage are chondroitin sulfate and keratin sulfate. GAGs are polysaccharides, and each sugar moiety contains one or multiple negative charges resulting in elongation of the PGs due to charge repulsion. Aggregates of PGs are formed when multiple PGs bind to hyaluronic acid, a non-sulfated GAG molecule, via link proteins. Moreover, due to the highly negative charge, the molecules are extremely hydrophilic and trap in large amounts of water resulting in a highly elastic tissue.[1] The collagen network, with its high tensile strength, interacts with the PGs, resulting in a fiber reinforced composite with high compressive strength.

Similar to other connective tissues, articular cartilage functions *via* its ECM, offering its biological function and mechanical integrity. The resident chondrocytes play a key role in ECM production and turnover. They serve to produce the collagen and PGs of cartilage along with the molecules that remodel the matrix, specifically matrix metalloproteinases (MMPs) and disintegrin and metalloproteinase with thrombosondin motifs (ADAMTS) which predominantly degrade collagens and aggrecans, respectively. The anabolic and catabolic nature of chondrocytes results in natural cartilage turnover. An imbalance in the natural cartilage homoeostasis leads to arthritis.

There are four zones in articular cartilage: the superficial tangential zone, transition layer, deep zone, and calcified cartilage. The superficial tangential zone contains flattened chondrocytes. Collagen and cells of this layer are aligned tangential to the surface. This layer has been suggested to contain progenitor cells.[4] The transition layer contains randomly oriented chondrocytes and collagen followed by the deep zone. Within the deep zone, collagen and cells are aligned perpendicular to the surface. A smooth transition from articular cartilage to subchondral bone is marked by calcified cartilage starting at the tidemark. The tidemark is the basophilic line that separates the calcified and uncalcified cartilage. Due to the spatial orientation of the ECM and cells along with specialized mechanical properties, it is particularly difficult to engineer the full thickness of articular cartilage.

Articular cartilage is avascular, acquiring most of its nutrients from the synovial fluid. Because of the dense ECM that limits chondrocyte mobility, and the avascular nature, cartilage has a limited ability to self-repair. This is problematic when articular cartilage is damaged due to trauma or OA, making it an excellent candidate for tissue engineering strategies. The exact mechanism that triggers OA is not well understood; however, when patients have cartilage lesions caused by some form of trauma, they are at higher risk of developing OA. In the United States alone, 27 million people live with OA.[5]

3. Osteoarthritis of the Knee

Osteoarthritis is a disease of the whole organ system, including the synovium, subchondral bone, and articular cartilage.[6] It is unknown which pathological changes initiate OA. However, with OA there is an imbalance in ECM deposition and degradation resulting in a net loss of cartilage. Radiographic changes of the articular cartilage include thinning of the joint space and sclerosis of the subchondral bone. Histologically, a loss of proteoglycans, or fixed charges (via safranin-O or toluidine blue staining), is observed.[6] A loss in total collagen occurs at later stages of OA; however, it is preceded by increased swelling properties of the cartilage due to loosening of the collagen fibers[7] seen at the initial stages of OA. The first stages of OA are marked by fibrillation of the cartilage surface; proceeding to complete loss of articular cartilage such that the subchondral bone is exposed.[6] The tidemark is also seen to thicken which correlates to calcification of the deep zone, known as hypertrophy, and marked with an increase in type X collagen.[6] Additionally, the chondrocytes are seen to be clustered, as opposed to single cells, due to initiation of proliferation. Bony growths, known as osteophytes,[6] can also be found.

The two major classes of degradation molecules occurring naturally in articular cartilage are increased in OA. A disintegrin and metalloproteinase with thrombospondin motifs, specifically ADAMTS-4 and -5, play a crucial role in the degradation of aggrecan.[8] Due to degradation of aggrecan, the PGs disassemble and diffuse out of the articular cartilage, resulting in a decrease in compressive strength due to reduction of negative charge. Matrix metalloproteinases play a key role in the degradation of collagen and gelatin. The three main MMPs prevalent in OA are MMP-2, −9 and −13.[9] MMP-13 is a collagenase efficient at degrading type II collagen fibrils. Gelatinases, MMP-2 and MMP-9, follow up by degrading the newly formed gelatin. Both classes of enzymes play a pivotal role in matrix degradation that causes a reduction in the mechanical integrity of articular cartilage. They also pose a particular challenge when attempting to engineer the tissue as they may serve to inhibit the fill tissue formation — an aspect not typically modeled during *in vitro* testing of tissue engineering strategies.

Inflammation also plays an important role in cartilage degradation and the occurrence of OA.[6] Specifically, interleukin (IL)-1 and tumor necrosis factor (TNF) are inflammatory cytokines found to be increased in primary cultures of OA chondrocytes compared to healthy chondrocytes. *In vivo*, IL-1β and TNF-α have been found to be localized to the superficial zone of articular cartilage.[10] These cytokines induce the production of the proteases along with pro-inflammatory cytokines acting as a positive feedback loop.[11] Additionally, these cytokines are known to be associated with a decrease in chondrocyte-specific gene.[12] Both the increase in proteases and decrease in terminally differentiated chondrocytic gene expression contribute to the reduction in ECM.

These cytokines are also known to activate many intracellular signaling pathways including JNK and p38 MAPK, both of which activate NF-κB *via* degradation of the inhibitor of κB (IκB).[13,14] The abundant form of NF-κB is a heterodimer of p65 and p50. It is sequestered in the cytoplasm *via* interacting with IκB (either α or β).[13] The NF-κB heterodimer is released from IκB after IκB is phosphorylated and targeted for ubiquitin-mediated degradation. This allows for NF-κB translocation to the nucleus where it transcriptionally regulates many cytokines including pro-inflammatory cytokines IL-1 and TNF-α, and expression of IκB-α[13] as well as with further downregulation of chondrocyte-specific genes.

Patient symptoms associated with OA include pain and instability, which can result in immobility. Pain is felt in early OA coinciding with activity level. However, as OA progresses, chronic pain may become more persistent, even in the absence of activity.[15] The underlying cause of pain with OA is not fully understood, as articular cartilage is aneural. However, some have theorized the pain resides from swelling of the synovium, in turn affecting the peripheral nervous system.[11] Also, MRI imaging has shown lesions in the bone marrow which is better associated with patients experiencing pain than those who do not.[16] Non-surgical treatments for pain management include nonsteroidal anti-inflammatory drugs (NSAIDS) and intra-articular injections of hyaluronic acid or corticosteroids.

4. Surgical Strategies for Repairing Focal Cartilage Defects

Focal cartilage defects can occur due to trauma, repetitive impact, or progressive mechanical degeneration. There are three types of cartilage defects: (1) partial thickness, (2) full thickness, and (3) osteochondral which penetrate into the subchondral bone.[17] These defects, if left untreated, may result in further knee degeneration leading to OA. Occurrence of OA appears on average ten years sooner in patients with cartilage defects. Because cartilage has a limited ability to repair itself, surgical techniques are employed to repair the defect site by filling with biological material to reduce or eliminate the progression to OA.

Surgical treatments for OA are only employed when non-surgical treatments have failed. There are three primary techniques utilized to repair cartilage defects: (1) bone marrow stimulation, (2) mosaicplasty or osteochondral autograft transfer system (OATS), and (3) autologous chondrocyte implantation (ACI). All of these procedures result in inferior fibrocartilage formation and mechanical properties. This is due to the disorganization of the fill tissue, predominantly composed of type I collagen, whereas articular cartilage has organized cellularity and ECM.

4.1 *Bone marrow stimulation*

Bone marrow stimulation utilizes natural repair strategies to fill cartilage lesions. The microfracture technique is the most widely used.[18] This technique involves debridement of the defect area down to the subchondral bone followed by creating multiple punctures into the subchondral bone to allow bone marrow to fill the defect site.[18] A fibrin clot will form within the defect which contains mesenchymal stem cells (MSCs) and acts to facilitate migration of additional MSCs within days.[19] These cells will then go on to form new fibrocartilage tissue. There have been some positive clinical outcomes including reduction in pain. However, within 18 to 24 months, post-surgery deterioration is observed.[20,21]

4.2 *Mosaicplasty and osteochondral autograph transfer system*

Mosaicplasty and OATS procedures employ osteochondral plugs transplanted into the defect site. They are suitable for smaller-sized defects (less than 4 cm^2) due to surgical challenges and donor site morbidity.[22,23] One large graft, or in the case of mosaicplasty, multiple small grafts about 1 mm in diameter, is isolated from a non-load bearing area of the knee and transplanted to the defect site. The implanted graft maintains its articular cartilage structure. However, the bonding tissue between the implants is fibrocartilage in nature.

4.3 *Autologous chondrocyte implantation*

The final surgical procedure for cartilage repair is autologous chondrocyte implantation (ACI).[24] This is a two-stage process. In the first stage cartilage from a non-load bearing area of the joint is removed. The chondrocytes are then isolated from the ECM and expanded *in vitro*. Expanded cells are then re-implanted during the second surgery into the focal defect and held into place using a periosteal flap. The periosteal flap is believed to play a role in defect repair due to paracrine effects or serving as a source for autogenous cells.[25] This technique has been found to have comparable results to microfracture after two years.[26] In addition, histological evaluation of the fill tissue has shown some hyaline-like cartilage, with the rest being fibrocartilage.[24] However, many limitations to this procedure exist, including donor site morbidity, limited cell supply, and dedifferentiation of the isolated cells. Dedifferentiation occurs due to *in vitro* expansion in monolayer culture resulting in loss of cartilage-specific gene expression and ECM production. Redifferentiation of the expanded chondrocytes is necessary in order to obtain hyaline-like cartilage.

5. Scaffolds for Assisting Operative Techniques

Multiple scaffolds have been generated to enhance the regenerating cartilage. Materials have been utilized to anchor the cells in place and augment the operative procedures. Using tissue engineering strategies to augment operation procedures is of interest to the research community because it offers many opportunities to direct the fill tissue. Specifically, materials have been utilized to direct redifferentiation of chondrocytes, differentiation of stem cells, and fill tissue architecture to match that of native tissue. Cell-free and cell-laden scaffolds have been used in the clinical setting. Cell-laden scaffolds are homogenous in cell distribution, better recapitulate the native three-dimensional environment, and help to maintain or induce the chondrocytic phenotype.

5.1 *Collagen scaffolds for augmenting ACI*

Multiple collagen scaffolds have been developed and used clinically for augmenting the ACI procedure. A type I/III collagen flap has been designed to eliminate the need for a periosteal flap.[27] It had significant advantages compared to the original technique, including less invasive surgery, reduced surgery time, and decreased postoperative pain. This technique was limited in that suturing of the collagen membrane to the articular surface is required and tedious.

Due to enhanced histological findings on the collagen flap, a new technology was developed to eliminate the limitation of the collagen flap, a procedure known as matrix-induced autologous chondrocyte implantation (MACI®).[28,29] As opposed to injecting cells underneath the flap, cells are seeded directly onto a type I/III collagen membrane isolated from porcine peritoneal cavity. The cell-laden collagen membrane is then placed into the prepared defect using a thin layer of fibrin glue to secure it to the defect. The chondrocytes can penetrate into both the collagen membrane and the fibrin glue and retain their chondrocytic phenotype. The resulting tissue is hyaline-like and positive for type II collagen.[29] Similar results are seen with a porcine type I/III bilayer matrix, Chondro-Gide®.[30]

Another ACI augmenting collagen scaffold, NeoCart®, is a three-dimensional type I collagen scaffold seeded with expanded chondrocytes.[31] The seeded scaffold is then cultured in a bioreactor. The total time for implant development is 67 +/− 18 days. A proprietary tissue adhesive, CT3 (Histogenics), composed of collagen and polyethylene glycol, is used to secure the implant in place. Since this implant is fairly new, minimal clinical outcomes have been observed. However, the treated patient's pain score was lower than at baseline. In addition, range of motion and knee function were seen to be improved.

5.2 *Collagen scaffolds with autologous MSCs*

MSCs are a promising cell source for tissue engineering because they have the ability to differentiate into multiple lineages including bone, cartilage, fat, and astrocytes.[32] They can be isolated from individual patients in a minimally invasive manner and expanded *in vitro* without losing their ability to differentiate. Mesenchymal stem cells have been evaluated clinically for their ability to augment bone marrow stimulation. In this work, two millimeters of the subchondral bone were removed until bleeding was seen.[33,34] Perforation was then performed using 1.2-mm Kirshner wire to facilitate further bleeding. MSCs were previously isolated using standard procedures and expanded through one passage *in vitro*. The day before surgery, MSCs were lifted from the tissue culture plates (average 13 million) and embedding into 1.2 ml of 0.25% type I collagen from porcine tendon. The cell suspension was seeded onto collagen sheets (derived from bovine source) and allowed to gel. The collagen/MSC scaffold was cultured overnight in DMEM supplemented with autologous serum and antibiotics. The composite was placed into the defect with the collagen sheet covering the upper side. A perestium flap was used to secure the material in place. Control patients received the collagen scaffold with no cells and others did not receive the perforation technique. The resulting fill tissue was found to be mechanically weaker in all groups compared to the surrounding cartilage. However, histologically (using toluidine blue) the cell-laden group was superior to the cell-free group, exhibiting a metachromatic staining and some hyaline-like tissue.[33,34] Further work is necessary to understand the origin of the cell population residing within the scaffolds after implantation.

5.3 *Hyaluronic acid scaffolds*

Hyaluronic acid (HA) (or hyaluronan) is found in all soft tissues. HA is an unsulfated glycosaminoglycan of disaccharide repeat units, glucuronic acid, and N-acetylglucosamine. Its molecular weight ranges from 4,000 to 8×10^6 Da. As mentioned previously, it interacts with proteoglycans along with other proteins and molecules. In addition, cells expressing CD-44 can bind to and migrate on HA.[35] Degradation products of HA contribute to many biological activities including size-dependent effect on chondrogenic differentiation,[36] vascularization,[37] and angiogenesis.[38]

Modified HA has been used as a scaffold for articular cartilage defect repair, specifically Hyaff®. Hyaff® hyaluronic acid starts as a molecular weight of 180–200 kDa. Esterification of the glucoronic acid groups is performed. By varying the alcohol used in the esterification reaction and the degree of substitution bioresorbabilty, water solubility and residence time can be varied.

HYAFF® 11 is biocompatible and resorbable without the presence of an inflammatory response.[39] It has been used to augment the ACI procedure.[40] Isolated chondrocytes were expanded *in vitro* followed by seeding and culturing on the scaffold for an additional two weeks. Used in this way, it is marketed as Hyalograft® C. In animal models, these scaffolds have been shown to develop hyaline-like cartilage and integrate with the surround tissue. Clinical trials have shown that patients treated with Hyalograft® C had significant pain reduction and improvements in physical activity and knee function.[40]

5.4 *Fibrin scaffolds*

Fibrin glue is a material that synthetically mimics fibrin clots, part of natural tissue repair after injury. It is a two-component system composed of fibrinogen and thrombin. Once mixed together, the thrombin degrades fibrinogen to fibrin that acts as a tissue adhesive and three-dimensional scaffold. Fibrin glue is used as an adhesive to secure various scaffolds into cartilage defects due to its chondroinductive property.[41]

Scaffolds of fibrin can also be made and have been studied for cartilage tissue engineering.[42] Fibrin scaffolds have been used clinically, again, to augment the ACI procedure. In this treatment, the expanded chondrocytes are mixed with fibrinogen and thrombin to form a cell-seeded fibrin scaffold. The defect site is then debrided to the subchondral bone and the subchondral bone penetrated. A thin layer of fibrin glue is applied followed by molding and implantation of the cell-seeded fibrin scaffold. Another layer of fibrin glue is then applied. This technique is superior to ACI alone because it eliminates the need for a periosteal flap, reducing patient recovery time.[43]

Zimmer Inc. has also developed a product using fibrin glue.[44] The product, known as DeNovo® NT Graft, utilizes minced cartilage tissue from allogous juvenile cartilage. Viable cartilage pieces are mixed intraoperatively with fibrin and placed into the defect site. A thin layer of fibrin is then applied over the implant to maintain placement. Utilizing viable tissue results in a cell source that can migrate out of the minced pieces and fill the defect area. Advantages over the ACI procedure include only requiring a single operation and eliminating the need for costly *in vitro* expansion.

5.5 *Chitosan scaffold*

Chitosan is a polysaccharide derived from deacetylating chitin isolated from the exoskeleton of crustaceans. It is composed of D-glucosamine and N-acetyl-D-glucosamine linked *via* a $\beta[1-4]$ linkage. Due to its positive charge, it behaves as a bioadhesive and can adhere to negatively charged tissue.[45] BST-CarGel® is composed of a mixture of chitosan and uncoagulated whole blood that gels within ten minutes. BST-CarGel® has been evaluated for its ability to augment microfracture technique.[46] First, the microfracture technique is performed and a "dry field" is created. At least 5 ml of peripheral blood is obtained, 4.5 ml of which is mixed with the supplied BST-CarGel®. This mixture is the placed within the prepared cartilage defect and allowed to gel for 15 minutes. Multiple animal studies have been performed and have shown that using BST-CarGel® resulted in increased hyaline-like cartilage formation, GAG content, collagen content, and repair tissue volume when compared to microfracture alone.[46] Clinical studies have shown that BST-CarGel® decreased pain and stiffness along with increasing joint function.

5.6 *Polyester-based scaffolds*

Polyesters are synthetic polymers of repeating degradable ester groups with various amounts of carbons separating the esters and side chains. For example, poly(lactic acid) and poly(glyocolic acid) contain one carbon between the ester bonds with poly(lactic acid) containing a methyl group on the α-carbon. Another common polyester is poly(ϵ-caprolactone) which contains six carbons between the ester groups. Polydioxanone is a polyester-ether containing three carbons along with oxygen (making the ether) between each ester bond. The esters can be hydrolytically degraded to carboxylic acids and alcohols. Degradation rate of polyesters can be controlled by the specific polyester used, copolymerization, or mixing multiple types of polyesters together. These variations change the degradation rate due to altering the degree of crystallinity. The range of degradation that can be achieved is a couple of weeks up to years.

One example clinically evaluated is a polymer-based scaffold of polylactic/polyglycolic acid (polyglactin, vicryl) and polydioxanone collagen fleece coined Bioseed®-C.[47,48] Chondrocytes isolated and expanded from the patient are seeded within the scaffold and secured using fibrin. The scaffold is cut to the shape of the defect, armed with vicryl sutures at each corning and secured using k-wires passed through the femur. A press fitting technique is used to securely hold the Bioseed®-C scaffold in place. This technique is more stable than ACI and eliminates donor site morbidity associated with periosteal flaps. The procedure is also performed completely arthroscopically so it reduces adhesions associated with open surgeries. However, in a comparative clinical study between ACI and Bioseed®-C, no differences were seen in the clinical outcomes between the treatment groups.[47]

Another polyester system undergoing clinical evaluation utilizes minced cartilage pieces collected from the patient seeded onto a foam scaffold, coined cartilage autograft implantation system (CAIS).[44] The minced pieces are held in place with the use of fibrin adhesive. The scaffold is then stapled into place. Evaluation of a similar scaffold in goats after six months of implantation resulted in completely filled defects with hyaline-like cartilage that had more type II collagen and less type I collagen than scaffolds without minced cartilage.[49]

6. Mesenchymal Stem Cells for Cartilage Tissue Engineering

Limitations exist in using articular chondrocytes. First, cells must be isolated from the donor site during the initial surgery, necessitating two invasive procedures in order to repair the defected cartilage. Second, the number of chondrocytes that can be isolated is limited and therefore the cells must be expanded *in vitro*. This results in a costly procedure and dedifferentiation of the chondrocytes due to cell expansion.[50] Finally, donor site morbidity is of concern and limited research has been performed to investigate these effects.

Alternative cell sources exist for cartilage tissue engineering. These include embryonic stem cells (ESCs) and MSCs. ESCs are derived from the inner cell mass of an embryo. Due to the isolation technique, ethical issues surrounding ESCs have limited the research and extent of use of these cells. Alternatively, MSCs can be isolated from bone marrow, umbilical cord blood, adipose tissue, synovial cells, and peripheral blood.[51] The first three are the most common sources under investigation. Bone marrow-derived MSCs have been studied more extensively than any other source. Umbilical cord blood-derived MSCs are limited by availability because they can only be isolated at birth. Adipose-derived MSCs can easily be isolated from lipoaspirate after cosmetic liposuction.

This derivation technique has proven to result in higher cell yields than that of bone marrow and umbilical cord derived-MSCs. However, this cell source is limited to individuals with excess fat to isolate cells from. Human adipose-derived MSCs have also been shown to have an inferior ability to differentiation toward cartilage and bone when compared to human bone marrow derived-MSCs.

Differentiation of MSCs toward a chondrogenic lineage can be induced by a variety of physical and chemical factors. A three-dimensional culturing system is required in order for adequate chondrogenesis to occur. Examples of three-dimensional environments used are pellet culture, high-density culture, hydrogels and sponge-like scaffolds. Chemical factors are included into the medium to direct more specific differentiation. Dexamethasone is supplemented into the medium, though the exact mechanism of how it induces chondrogenesis is not well understood.[52] It is believed to increase Sox-9 gene expression, a transcription factor that activates type II procollagen gene expression.[53] Various growth factors and proteins are also supplemented into the medium to induce differentiation, including transforming growth factor (TGF)-β1 and -β3, IGF-1, bone morphogenic proteins (BMPs) and fibroblast growth factor (FGF)-2.[54] These factors are also known to be key regulators of chondrogenesis during skeletal development.[2]

7. Hydrogels for Directed Differentiation of Mesenchymal Stem Cells

Hydrogels have been used as a three-dimensional scaffold for studying cellular function and differentiation. As mentioned previously, bioactive materials have been used as scaffolds. Many bioinert materials have been studied as a means to encapsulate cells in a three-dimensional configuration. Examples of bioinert hydrogels include poly(ethylene glycol)-diacrylate (PEGDA), poly(vinyl alcohol) (PVA) and alginate. As a result, the soluble components contained in the medium are the primary contributors to directing cell fate. In the case of bioactive hydrogels, the cells can interact with the substrate resulting in materials that can be used to direct cellular activity. When encapsulating MSCs into hydrogels it is of particular interest to engineer materials that direct differentiation. In this way, scaffold-assisted differentiation occurs because of the material-cell interaction, resulting in a scaffold with potential to enhance *in vivo* differentiation, where factors contained in medium are not available.

7.1 *Functionalized poly(ethylene glycol) hydrogels*

Poly(ethylene glycol) (PEG) hydrogels have been investigated extensively as bioinert scaffolds for three-dimensional cell culture. PEG has been functionalized with acrylate groups to facilitate crosslinking via numerous methods, including ultraviolet light exposure and a free radical initiator, reduction-oxidation reaction, and Michael addition.[55] PEG hydrogels have been shown to support differentiation of MSCs toward a chondrogenic lineage via the factors present within the medium.

Numerous research groups have functionalized PEG hydrogels with arginine-glycine-aspartic acid (RGD), a cell-binding peptide found in many proteins. When anchorage-dependent cells are encapsulated into inert three-dimensional hydrogels, adhesion is not possible, resulting in a large percentage of apoptotic cells. Therefore, incorporating adhesion peptides, such as RGD, facilitates cell survival within the hydrogels. Nuttelman and colleagues[56] demonstrated that human MSC survival in PEG hydrogels increases from 15% to 75% when RGD was covalently incorporated into the network. Enhancing cellular survival is of particular interest because tissue formation will be facilitated within the scaffold due to an increase in total ECM deposition.

When evaluating human embryoid body-derived MSCs (MSCs derived from embryonic stem cells) Hwang and colleagues[57] found a significant increase in chondrogenesis in the presence of covalently linked RGD in PEG hydrogels. Specifically, increase in GAG and total collagen production was seen on a per cell basis when compared to PEG-only hydrogels. Increased safranin-O and types I and II collagen staining confirmed this. Though it is favorable to have an increase in type II collagen, type I collagen is less favorable. However, when gene expression was evaluated, type II collagen, aggrecan, and link protein increased in the RGD group; whereas type I collagen gene expression was similar to that of PEG hydrogels alone.

Interestingly, when alginate hydrogels were functionalized with RGD, chondrogenesis of bovine MSCs was inhibited in a concentration-dependent manner.[58] Similar inhibition of sulfated GAG production was observed in bovine chondrocytes encapsulated in alginate hydrogels with covalently linked RGD.[59] Silk scaffolds with covalently linked RGD also exhibited similar inhibitor effects on chondrogenesis of human MSCs.[60]

Still, RGD peptide does have a positive effect on cell viability. To combat the negative effect on chondrogenesis, Salinas and colleagues developed an RGD-containing peptide with an MMP-13 cleavage site.[61] Matrix metalloproteinase-13 activity was observed to peak at an intermediate culturing time when MSCs were undergoing chondrogenesis, which resulted in cleavage of the RGD sequence. In this way, the increase in viability can be harnessed earlier on, followed by removal of the RGD to eliminate the inhibitor effect seen with RGD and chondrogenesis. Though cell viability decreased at the time at which cleavage was believed to occur, GAG production relative to DNA increased in a time-dependent manner even after the RGD sequence was cleaved. In a control group, where the RGD peptide was not cleavable, GAG production decreased, supporting the hypothesis that RGD is important for cell viability but necessary to remove to enhance chondrogenesis.

Salinas and colleagues also evaluated a decorin-binding peptide, lysine-leucine-glutamic acid-arginine (KLER) in combination with RGD on the chondrogenesis of human MSCs.[62] Decorin is known to influence fibrillogenesis of collagen at two major sites, arginine-glutamic acid-leucine-histidine (RELH) and KLER. When both RGD and KLER were covalently linked to PEG hydrogels, collagen content increased over the time course of the experiment when compared to RGD and RGD plus scrambled peptide. Gene expression for aggrecan was highest in the RGD/KLER peptide group compared to the controls at early time points, whereas type II collagen expression was highest at later time points.

Another peptide has been evaluated for its effects on chondrogenesis. This peptide, known as collagen-mimetic peptide (CMP), immobilizes collagen via binding through a strand invasion route.[63] Using a synthetic peptide to bind endogenous collagen is favorable over collagen scaffolds because it eliminates possible immune response. The material in which the peptide is covalently linked can also be mechanically more robust than collagen gels, making them also favorable from a practical aspect. When compared to PEG hydrogels, CMP-containing hydrogels had an increase in cartilage matrix production (GAG and type II collagen) and cartilage-specific gene expression of differentiating goat MSCs. On the other hand, type I collagen production, marker for bone formation, and type X collagen gene expression were decreased.

In addition to peptides, GAG molecules can be covalently linked to PEG hydrogels to facilitate enhancement of chondrogenesis. One such GAG molecule, chondroitin sulfate, has been shown to enhance the chondrogenesis of goat MSCs compared to PEG hydrogels.[64] Visualization of the hydrogels after chondrogenesis showed a nodule-like appearance. When investigated further, the MSCs were found to be undergoing a mesenchymal condensation-like state based on gene expression for versican and cadherin 11 (markers found naturally in mesenchymal condensation). Chondroitin sulfate (CS)-based hydrogels also enhanced total collagen production and type II collagen deposition. In addition, hypertrophic markers were lower in the CS-based hydrogels. Overall, CS-based hydrogels have the ability to enhance chondrogenesis while reducing hypertrophy and may recapitulate the natural process of mesenchymal condensation.

7.2 *Naturally derived materials*

Natural derived biomaterials, such as HA, CS, and collagen, have been utilized as three-dimensional scaffolds for MSC differentiation.[64–67] These materials mimic the natural ECM of cartilage and better recapitulate the *in vivo* environment. Limitations to these types of materials include minimal control over spatial location and orientation of the biomolecule and mechanical weakness. In addition, because the materials are derived from natural sources (i.e. animal tissue), concerns of immune response exist.

HA hydrogels have been developed via functionalizing HA with methacrylate groups to facilitate free-radical crosslinking.[66] These HA hydrogels have been compared in parallel to PEG hydrogels (with similar mechanical properties). *In vitro* chondrogenesis of human MSCs was enhanced in the HA hydrogels based on an increase in type II collagen and proteoglycans deposition along with chondrocytic gene expression. *In vivo*, cartilage-specific gene expression was also seen to be higher in HA hydrogels. Three different groups underwent *in vivo* evaluation. The group not administered exogenous TGF-β3 or pre-differentiated for two weeks *in vitro* before implantation demonstrated the optimal gene expression profile. This shows that HA hydrogels may be capable of differentiating MSCs toward a chondrocytic lineage without external stimulation.

As mentioned in the previous section, CS-based hydrogels have been shown to enhance chondrogenesis of MSCs. Further developing on this concept of CS-guided cell fate determination, a modified electrospinning technique has been developed to engineer low-density, CS-based nanofiber scaffolds.[67] The low-density nanofibers enhanced chondrogenesis of MSCs *in vitro* as well as cartilage repair in a rat osteochondral defect model. These materials acted via physical and biochemical cues to direct MSC differentiation and cartilage repair.

As mentioned previously, collagen scaffolds have been used *in vivo*. They have been able to elicit a more hyaline-like cartilage tissue fill favorable for cartilage tissue engineering. A study comparing type I collagen, type II collagen, and a bioinert hydrogel, alginate, showed that type II collagen hydrogels enhanced chondrogenesis of bovine MSCs based on gene expression and ECM deposition.[68] Similarly, studies have shown enhancement of osteogenesis in type I collagen scaffolds (type I collagen being the predominant collagen type in bone).[69,70] These findings support the hypothesis that ECM component found in the native tissue of interest can facilitate enhancement of differentiation toward that tissue lineage.

8. Fiber-Hydrogel Composites

Hydrogels have been the predominant focus for researchers in developing tissue engineering strategies for articular cartilage. Mechanically, they better resemble the developing limb in comparison to the stiffer, mature cartilage. However, they are primitive mimics of articular cartilage as they lack the fibrous nature of native cartilage. Many research groups have investigated methods of producing fibrous scaffolds to act as a physical stimulus similar to that of the fibrous phase of articular cartilage.[71,72] However, these systems then lack the hydrogel phase. Fewer researchers have investigated fibrous-hydrogel composites for cartilage tissue engineering (Fig. 1).[73–76] These composites employ the

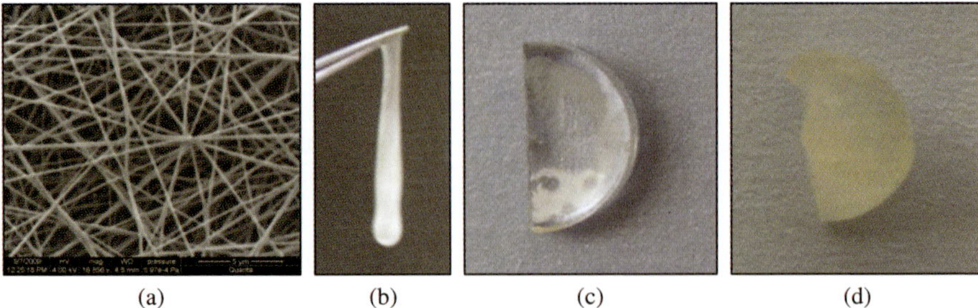

(a) (b) (c) (d)

Fig. 1. Electrospun fibers used to enhance hydrogel properties developed in our laboratory. (a) SEM of PVA-methacrylate fibers electrospun onto aluminum collector. (b) Electrospun PVA-methacrylate fibers that have been collected on a grounded ethanol bath. (c) PEGDA hydrogel having a transparent appearance. (d) PEGDA hydrogel with incorporated fibers from image b.

use of a hydrogel phase as a delivery vehicle for cell infiltration into the fibrous phase. The hydrogel phase also helps to maintain the round cellular morphology native to cartilage. Fiber-reinforced hydrogels have an added benefit of enhancing the mechanical integrity of the scaffold; this may help in addressing larger-size cartilage defects than currently feasible clinically.

Polyesters have been studied extensively as fibrous meshes for cartilage tissue engineering. Favorable results have been obtained for both chondrocytes and chondrogenic differentiation of MSCs seeded directly onto the meshes. Chondrocytes have been seeded onto electrospun poly(ε-caprolactone) (PCL) meshes and were found to maintain their chondrocytic phenotype.[77] Chondrogenic differentiation of MSCs in the presence of TGF-β1 resulted in rounded cell morphology similar to that of native chondrocytes.[78] In addition, favorable gene expression was found with chondrogenic MSCs, including increased type II collagen and decreased type X collagen. When compared to pellet culture, chondrogenic MSCs on nanofiber scaffolds were found to have a two-fold increase in GAG production. These findings imply that fibrous polyester scaffolds are good candidates for cartilage tissue engineering. However, limitations exist in high-throughput development of fibrous meshes large enough to fill cartilage defects. In addition, cell infiltration into electrospun mats is limited due to the small pore size.

A non-woven fleece of polyglactin, E210, has been implanted subcutaneously in athymic nude mice after infiltrating with alginate-containing bovine chondrocytes.[73] Crosslinking of the alginate was facilitated by immersing the composites in a calcium chloride solution prior to implantation. After explantation, E210 fleeces without cells and alginate were too weak to evaluate and broke when cut. E210 scaffolds seeded with cells, with and without aliginate, had similar levels of GAG and type II collagen production. However, the scaffolds with alginate had a more even distribution of cells and did not exhibit any shrinking.

A similar composite scaffold has been evaluated using a poly(glycolic acid) (PGA) mesh and either aliginate or type I collagen hydrogels.[74] Rabbit MSCs were seeded in a similar manner. At early time significant differences were seen with the alginate-fiber compared to collagen-fiber and fiber-only scaffolds. Cell morphology in the alginate scaffolds was maintained as rounded whereas in the collagen scaffolds a fan shape was seen. However, the fan-shaped cells became more polygonal over time, indicating chondrogenic differentiation. Cell proliferation and differentiation were seen to be delayed in the alginate scaffolds at three weeks. However, after six weeks of culture, the GAG content normalized to DNA was significantly higher in the alginate scaffolds compared to collagen scaffolds.

Finally, a woven scaffold composed of PGA yarn has been utilized as a reinforcement material for cartilage tissue engineering.[75] Fibers were oriented in the x, y, and z directions resulting in a highly interconnected fibrous network where the pore size could be controlled. The hydrogel phase was composed of either agarose or fibrin. The authors were able to show that fibers enhance the mechanical properties of the hydrogels independent of the type of hydrogel used. From a biological prospective, the authors were able to incorporate porcine articular chondrocytes homogenously throughout the fibrous scaffold with the use of vacuum-assisted infusion. In a later study, chondrogenesis of human adipose-derived MSCs in woven PCL scaffolds with fibrin and without exogenous growth factors was investigated.[76] Collagen-rich ECM was found to completely encapsulate the fiber/hydrogel scaffolds and fill the inner pores. On the other hand, the fiber-alone scaffolds were only encapsulated by the collagen-rich ECM after 28 days of culture.

9. Small Molecules for Directing the Chondrocyte Phenotype

Current systems for directing stem-cell differentiation toward a chondrogenic lineage have been lacking in their ability to produce tissue comparable to articular cartilage. Therefore, other methods of directing differentiation are necessary in order to obtain comparable tissue to the native environment. Small molecules are commonly supplemented into the culture medium for directing differentiation *in vitro*, including ascorbic acid and dexamethasone. However, other molecules have been explored, typically in the presence of other chondrogenic-inducing molecules. Small molecules are extremely useful for *in vivo* applications as they can be administered locally to the site of tissue repair. Controlled release systems can also be developed to allow the small molecule delivery over time, allowing for longer residence time.

9.1 *Glucosamine and its analogs*

Glucosamine has been studied extensively for its effects on chondrocytes.[79–82] Many studies have shown that glucosamine enhances cartilage-matrix deposition along with cartilage-specific gene expression of chondrocytes. A limited number of studies have been performed to evaluate the effects of glucosamine on chondrogenesis. Glucosamine and its analogs (N-acetyl glucosamine, glucosamine oligomers from chitosan, lactose, lactosamine and N-acetyl lactosamine) have been evaluated on a chondrogenic cell line, ATDC5 cells.[83] By evaluating alkaline phosphatase (ALP) activity, a marker for mineralization, glucosamine was found to be the only molecule capable of decreasing bone formation. Therefore, it was the only sugar evaluated in subsequent experiments. Glucosamine was shown to increase alcian blue staining for sulfate GAG and decrease gene expression for matrix gla protein (MGP), SMAD 2, and SMAD 4. MGP is a matrix protein that plays a role in mineralization of cartilage and bone. SMAD 2 is believed to regulate hypertrophy. SMAD 2 and SMAD 4 both play a role in chondrocyte differentiation. Therefore, glucosamine may have an effect on mineralization by affecting genes associated with the transition from cartilage to bone.

Hwang and colleagues[84] studied the effect of glucosamine (GlcN) on chondrogenesis of mouse embryoid bodies (derived from mouse embryonic stem cells). Glucosamine at concentrations of 2 mM and 10 mM (in the presence of TGF-β1) were shown to decrease metabolic activity and DNA content of embryoid bodies encapsulated in PEGDA hydrogels. However, at 2 mM the GAG production was significantly higher than control. Aggrecan gene expression was also found to be enhanced at this concentration. In addition, when TGF-β1 was removed from the culture medium, staining for sulfated GAG was only positive in the cells conditioned with 2 mM GlcN and the staining was comparable to groups treated with TGF-β1 alone. This is indicative that GlcN has chondroinductive properties at a concentration of 2 mM. Gibson and colleagues[85] found that local delivery of GlcN suppressed progression of experimental OA in a rat model based on a modified Mankin score which accounts for parameters of surface structure, proteoglycan content, and cellular changes of the articular cartilage and subchondral bone.

Coburn and colleagues[86,87] investigated the effects of short chain fatty acid (SCFA)-sugar analogs in an *in vitro* model of OA, namely continuous IL-1β stimulation of chondrocytes cultured in three-dimensional PEGDA hydrogels. The sugar, *N*-acetyl glucosamine, was modified on the 3, 4, and 6 hydroxyl positions with *n*-butyrate (3,4,6-*O*-Bu$_3$GlcNAc) or the 1, 3, and 4 hydroxyl positions with *n*-butyrate (1,3,4-*O*-Bu$_3$GlcNAc). 3,4,6-*O*-Bu$_3$GlcNAc was previously shown to decrease NFκB activity.[88] Twenty-four days of continued IL-1β stimulation reduced GAG and collagen secretion by the chondrocytes. 3,4,6-*O*-Bu$_3$GlcNAc treatment resulted in a concentration-dependent increase in GAG and collagen secretion after 21 days of exposure (three days pretreated with IL-1β followed by 21 days of co-treated with a sugar analog and IL-1β) whereas 1,3,4-*O*-Bu$_3$GlcNAc did not. Short-term, monolayer studies showed that 3,4,6-*O*-Bu$_3$GlcNAc decreased expression of the NFκB target genes, IL-1β, TNFα, IκBα, and NFκB1, as well as expression of the proteolytic enzymes MMP13, ADAMTS4, and ADAMTS5. Kim and colleagues[89] found that sustained-release delivery of 3,4,6-*O*-Bu$_3$GlcNAc into the articular space of a rat model of OA prevented disease progression and limited inflammation in diseased cells.

9.2 *NINDS library screening*

The National Institute of Neurological Disorders (NINDS) has put together a library of small molecules to aid in the discovery of new therapies for neurological disorders. In it is a collection of 1040 small molecules. Huang and colleagues[90] developed a system to perform high-throughput screening of chondrogenesis of bovine bone marrow-derived MSCs. Pellet culture was used as a three-dimensional model to minimize the number of cells required to analyze a single substance. Inducers of chondrogenesis were defined as those molecules capable of increasing GAG production by a factor of 1.5 or more compared to chondrogenic medium (in the absence of exogenous growth factors). Five small molecules were found to induce chondrogenesis at concentrations of 10 μM: doxylamine succinate, pergolide mesylate, perphazine, eszopiclone, and colforsin. Further investigation into each individual molecule is necessary in order to fully characterize the chondrogenic state of the cells and to optimize the concentration for induction.

10. Conclusion

Tissue engineering strategies have been developed to repair cartilage defects to ultimately reduce the prevalence of OA. These strategies include scaffolds for directing tissue formation and stem-cell differentiation in addition to small molecules for directing stem-cell differentiation and tissue repair. This chapter has been a brief summary of articular cartilage architecture, OA, and surgical techniques for repairing cartilage defects along with the current research areas designed to enhance the surgical techniques available.

Acknowledgements

This work was supported by NIH R01EB05517-01 and NIH F31AG033999.

References

1. V. C. Mow, A. Ratcliffe and A. R. Poole, Cartilage and diarthrodial joints as paradigms for hierarchical materials and structures, *Biomaterials* **13**: 67–97 (1992).
2. B. G. Mary, T. Kaneyuki and I. Kosei, The control of chondrogenesis, *J Cell Biochem* **97**: 33–44 (2006).
3. D. Eyre, Collagen of articular cartilage, *Arthritis Res* **4**: 30–35 (2002).

4. R. V. Patel and J. J. Mao, Microstructural and elastic properties of the extra-cellular matrices of the superficial zone of neonatal articular cartilage by atomic force microscopy, *Front Biosci* **8**: a18–25 (2003).

5. R. C. Lawrence, D. T. Felson, C. G. Helmick, L M. Arnold, H. Choi, R. A. Deyo, S. Gabriel, R. Hirsch, M. C. Hochberg, G. G. Hunder, J. M. Jordan, J. N. Katz, H. M. Kremers and F. Wolfe, Estimates of the prevalence of arthritis and other rheumatic conditions in the United States: part II, *Arthritis Rheum* **58**: 26–35 (2008).

6. H. I. Roach, T. Aigner, S. Soder, J. Haag and H. Welkerling, Pathobiology of osteoarthritis: pathomechanisms and potential therapeutic targets, *Curr Drug Targets* **8**: 271–282 (2007).

7. A. Maroudas, Balance between swelling pressure and collagen tension in normal and degenerate cartilage, *Nature* **260**: 808–809 (1976).

8. K. Huang and L. D. Wu, Aggrecanase and aggrecan degradation in osteoarthritis: a review, *J Int Med Res* **36**: 1149–1160 (2008).

9. H. Takaishi, T. Kimura, S. Dalal, Y. Okada and J. D'Armiento, Joint diseases and matrix metalloproteinases: a role for MMP-13, *Curr Pharm Biotechnol* 9: 47–54 (2008).

10. C. T. Lynne, J. A. Daman and E. W. David, Matrix metalloproteinase and proinflammatory cytokine production by chondrocytes of human osteoarthritic cartilage: associations with degenerative changes, *Arthritis Rheum* **44**: 585–594 (2001).

11. M. B. Goldring and S. R. Goldring, Osteoarthritis. *J Cell Physiol* **213**: 626–634 (2007).

12. M. B. Goldring, J. Birkhead, L. J. Sandell, T. Kimura and S. M. Krane, Interleukin 1 suppresses expression of cartilage-specific types II and IX collagens and increases types I and III collagens in human chondrocytes, *J Clin Invest* **82**: 2026–2037 (1988).

13. S. C. Sun, P. A. Ganchi, D. W. Ballard and W. C. Greene, NF-κB controls expression of IκBα: evidence for an inducible autoregulatory pathway, *Science* **259**: 1912–1915 (1993).

14. A. Liacini, J. Sylvester, L. W. Qing, W. Huang, F. Dehnade, M. Ahmad and M. Zafarullah, Induction of matrix metalloproteinase-13 gene expression by TNF-α is mediated by MAP kinases, AP-1, and NF-κB transcription factors in articular chondrocytes, *Exp Cell Res* **288**: 208–217 (2003).

15. G. A. Hawker, L. Stewart, M. R. French, J. Cibere, J. M. Jordan, L. March, M. Suarez-Almazor and R. Gooberman-Hill, Understanding the pain experience in hip and knee osteoarthritis: an OARSI/OMERACT initiative, *Osteoarthritis Cartilage* **16**: 415–422 (2008).

16. F. McCrae, J. Shouls, P. Dieppe and I. Watt, Scintigraphic assessment of osteoarthritis of the knee joint, *Ann Rheum Dis* **51**: 938–942 (1992).
17. T. M. Simon and D. W. Jackson, Articular cartilage: injury pathways and treatment options, *Sports Med Arthrosc* **14**: 146–154 (2006).
18. J. R. Steadman, W. G. Rodkey and J. J. Rodrigo, Microfracture: surgical technique and rehabilitation to treat chondral defects, *Clin Orthop* **391**: S362–369 (2001).
19. F. Shapiro, S. Koide and M. J. Glimcher, Cell origin and differentiation in the repair of full-thickness defects of articular cartilage, *J Bone Joint Surg Am* **75**: 532–553 (1993).
20. G. Knutsen, L. Engebretsen, T. C. Ludvigsen, J. O. Drogset, T, Grontvedt, E. Solheim, T. Strand, S. Roberts, V. Isaksen and O. Johansen, Autologous chondrocyte implantation compared with microfracture in the knee: a randomized trial, *J Bone Joint Surg Am* **86**: 455–464 (2004).
21. P. C. Kreuz, M. R. Steinwachs, C. Erggelet, S. J. Krause, G. Konrad, M. Uhl and N. Südkamp, Results after microfracture of full-thickness chondral defects in different compartments in the knee, *Osteoarthritis Cartilage* **14**: 1119–1125 (2006).
22. L. Hangody, G. Kish, Z. Kárpáti and R. Eberhart, Osteochondral plugs: autogenousosteochondral mosaicplasty for the treatment of focal chondral and osteochondral articular defects, *Oper Tech Orthop* **7**: 312–322 (1997).
23. L. Hangody, G. Kish, Z. Karpati, I. Udvarhelyi, I. Szigeti and M. Bely, Mosaicplasty for the treatment of articular cartilage defects: application in clinical practice, *Orthopedics* **21**: 751–756 (1998).
24. M. Brittberg, A. Lindahl, A, Nilsson, C. Ohlsson, O. Isaksson and L. Peterson, Treatment of deep cartilage defects in the knee with autologous chondrocyte transplantation, *N Engl J Med* **331**: 889–895 (1994).
25. S. W. O'Driscoll, Articular cartilage regeneration using periosteum, *Clin Orthop* **367**: S186–203 (1999).
26. G. Knutsen, L. Engebretsen, T. C. Ludvigsen, J. O. Drogset, T. Grontvedt, E. Solheim, T. Strand, S. Roberts, V. Isaksen and O. Johansen, Autologous chondrocyte implantation compared with microfracture in the knee. a randomized trial, *J Bone Joint Surg Am* **86**: 455–464 (2004).
27. T. W. Briggs, S. Mahroof, L. A. David, J. Flannelly, J. Pringle and M. Bayliss, Histological evaluation of chondral defects after autologous chondrocyte implantation of the knee, *J Bone Joint Surg Br* **85**: 1077–1083 (2003).
28. P. Cherubino, F. A. Grassi, P. Bulgheroni and M. Ronga, Autologous chondrocyte implantation using a bilayer collagen membrane: a preliminary report, *J Orthop Surg* **11**: 10–15 (2003).

29. M. H. Zheng, C. Willers, L. Kirilak, P. Yates, J. Xu, D. Wood and A. Shimmin, Matrix-induced autologous chondrocyte implantation (MACI): biological and histological assessment, *Tissue Eng* **13**: 737–746 (2007).

30. P. Behrens, T. Bitter, B. Kurz and M. Russlies, Matrix-associated autologous chondrocyte transplantation/implantation (MACT/MACI): 5-year follow-up, *Knee* **13**: 194–202 (2006).

31. D. C. Crawford, C. M. Heveran, W. D. Cannon, Jr., L. F. Foo and H. G. Potter, An autologous cartilage tissue implant NeoCart for treatment of grade III chondral injury to the distal femur: prospective clinical safety trial at 2 years, *Am J Sports Med* **37**: 1334–1343 (2009).

32. A. I. Caplan, Mesenchymal stem cells, in *Handbook of Stem Cells*, edited by R. Lanza (Elsevier Academic Press, San Diego, 2004), pp. 299–308.

33. S. Wakitani, T. Mitsuoka, N. Nakamura, Y. Toritsuka, Y. Nakamura and S. Horibe, Autologous bone marrow stromal cell transplantation for repair of full-thickness articular cartilage defects in human patellae: two case reports, *Cell Transplant* **13**: 595–600 (2004).

34. R. Kuroda, K. Ishida, T. Matsumoto, T. Akisue, H. Fujioka, K. Mizuno, H. Ohgushi, S. Wakitani and M. Kurosaka, Treatment of a full-thickness articular cartilage defect in the femoral condyle of an athlete with autologous bone-marrow stromal cells, *Osteoarthritis Cartilage* **15**: 226–231 (2007).

35. A. Aruffo, I. Stamenkovic, M. Melnick, C. B. Underhill and B. Seed, CD44 is the principal cell surface receptor for hyaluronate, *Cell* **61**: 1303–1313 (1990).

36. M. J. Kujawa, D. A. Carrino and A. I. Caplan, Substrate-bonded hyaluronic acid exhibits a size-dependent stimulation of chondrogenic differentiation of stage 24 limb mesenchymal cells in culture, *Dev Biol* **114**: 519–528 (1986).

37. P. Rooney, S. Kumar, J. Ponting and M. Wang, The role of hyaluronan in tumour neovascularization (review), *Int J Cancer* **60**: 632–636 (1995).

38. D. C. West, I N. Hampson, F. Arnold and S. Kumar, Angiogenesis induced by degradation products of hyaluronic acid, *Science* **228**: 1324–1326 (1985).

39. E. Tognana, A. Borrione, C. De Luca and A. Pavesio, Hyalograft C: hyaluronan-based scaffolds in tissue-engineered cartilage, *Cells Tissues Organs* **186**: 97–103 (2007).

40. S. Nehrer, S. Domayer, R. Dorotka, K. Schatz, U. Bindreiter and R. Kotz, Three-year clinical outcome after chondrocyte transplantation using a hyaluronan matrix for cartilage repair, *Eur J Radiol* **57**: 3–8 (2006).

41. A. Sage, A. A. Chang, B. L. Schumacher, R. L. Sah and D. Watson, Cartilage outgrowth in fibrin scaffolds, *Am J Rhinol Allergy* **23**: 486–491 (2009).

42. T. A. E. Ahmed, E. V. Dare and M. Hincke, Fibrin: a versatile scaffold for tissue engineering applications, *Tissue Eng Part B Rev* **14**: 199–215 (2008).

43. M. K. Kim, S. W. Choi, S. R. Kim, I. S. Oh and M. H. Won, Autologous chondrocyte implantation in the knee using fibrin, *Knee Surg Sports Traumatol Arthrosc* **18**: 528–534 (2010).

44. F. McCormick, A. Yanke, M. Provencher and B. Cole, Minced articular cartilage: basic science, surgical technique, and clinical application, *Sports Med Arthrosc* **16**: 217–220 (2008).

45. C. D. Hoemann, M. Hurtig, E. Rossomacha, J. Sun, A. Chevrier, M. S. Shive and M. D. Buschmann, Chitosan-glycerol phosphate/blood implants improve hyaline cartilage repair in ovine microfracture defects, *J Bone Joint Surg Am* **87**: 2671–2686 (2005).

46. M. S. Shive, C. D. Hoemann, A. Restrepo, M. B. Hurtig, N. Duval, P. Ranger, W. Stanish and M. D. Buschmann, BST-CarGel: in situ chondroinduction for cartilage repair, *Oper Tech Orthop* **16**: 271–278 (2006).

47. C. Erggelet, P. Kreuz, E. Mrosek, J. Schagemann, A. Lahm, P. Ducommun and C. Ossendorf, Autologous chondrocyte implantation versus ACI using 3D-bioresorbable graft for the treatment of large full-thickness cartilage lesions of the knee, *Arch Orthop Trauma Surg* **130**: 957–964 (2010).

48. P. C. Kreuz, S. Muller, C. Ossendorf, C. Kaps and C. Erggelet, Treatment of focal degenerative cartilage defects with polymer-based autologous chondrocyte grafts: four-year clinical results, *Arthritis Res Ther* **11**: R33 (2009).

49. L. Yiling, D. Sridevi, W. Ziwei, M. B. Dino, M. B. Steven, J. C. Brian and B. Francois, Minced cartilage without cell culture serves as an effective intra-operative cell source for cartilage repair, *J Orthop Res* **24**: 1261–1270 (2006).

50. V. Lefebvre, C. Peeters-Joris and G. Vaes, Production of collagens, collagenase and collagenase inhibitor during the dedifferentiation of articular chondrocytes by serial subcultures, *Biochim Biophys Acta* **1051**: 266–275 (1990).

51. C. Csaki, P. R. A. Schneider and M. Shakibaei, Mesenchymal stem cells as a potential pool for cartilage tissue engineering, *Ann Anat* **190**: 395–412 (2008).

52. D. Assia, L. P. Geraldine, J. H. David and S. T. Rocky, Glucocorticoids promote chondrogenic differentiation of adult human mesenchymal stem cells by enhancing expression of cartilage extracellular matrix genes, *Stem Cells* **24**: 1487–1495 (2006).

53. I. Sekiya, P. Koopman, K. Tsuji, S. Mertin, V. Harley, Y. Yamada, K. Shinomiya, A. Nifuji and M. Noda, Dexamethasone enhances SOX9 expression in chondrocytes, *J Endocrinol* **169**: 573–579 (2001).

54. C. Gaissmaier, J. L. Koh and K. Weise, Growth and differentiation factors for cartilage healing and repair, *Injury* **39**: 88–96 (2008).

55. W. E. Hennink and C. F. van Nostrum, Novel crosslinking methods to design hydrogels, *Adv Drug Deliv Rev*, **54**: 13–36 (2002).

56. C. R. Nuttelman, M. C. Tripodi and K. S. Anseth, Synthetic hydrogel niches that promote hMSC viability, *Matrix Biol* **24**: 208–218 (2005).

57. N. S. Hwang, S. Varghese, Z. Zhang and J. Elisseeff, Chondrogenic differentiation of human embryonic stem cell-derived cells in arginine-glycine-aspartate-modified hydrogels, *Tissue Eng* **12**: 2695–2706 (2006).

58. J. T. Connelly, A. J. Garcia and M. E. Levenston, Inhibition of *in vitro* chondrogenesis in RGD-modified three-dimensional alginate gels, *Biomaterials* **28**: 1071–1083 (2007).

59. N. Genes, J. Rowley, D. Mooney and L. Bonassar, editors. *Culture Of Chondrocytes In RGD-Alginate: Effects On Mechanical And Biosynthetic Properties* (Orthopedic Research Society, Orlando, Florida, 2000).

60. L. Meinel, S. Hofmann, V. Karageorgiou, L. Zichner, R. Langer, D. Kaplan and G. Vunjak-Novakovic, Engineering cartilage-like tissue using human mesenchymal stem cells and silk protein scaffolds, *Biotechnol Bioeng* **88**: 379–391 (2004).

61. C. N. Salinas and K. S. Anseth, The enhancement of chondrogenic differentiation of human mesenchymal stem cells by enzymatically regulated RGD functionalities, *Biomaterials* **29**: 2370–2377 (2008).

62. C. N. Salinas and K. S. Anseth, Decorin moieties tethered into PEG networks induce chondrogenesis of human mesenchymal stem cells, *J Biomed Mater Res A* **90**: 456–464 (2009).

63. H. J. Lee, C. Yu, T. Chansakul, N. S. Hwang, S. Varghese, S. M. Yu and J. H. Elisseeff, Enhanced chondrogenesis of mesenchymal stem cells in collagen mimetic peptide-mediated microenvironment, *Tissue Eng Part A* **14**: 1843–1851 (2008).

64. S. Varghese, N. S. Hwang, A. C. Canver, P. Theprungsirikul, D. W. Lin and J. Elisseeff, Chondroitin sulfate based niches for chondrogenic differentiation of mesenchymal stem cells, *Matrix Biol* **27**: 12–21 (2008).

65. U. Noth, L. Rackwitz, A. Heymer, M. Weber, B. Baumann, A. Steinert, N. Schutze, F. Jakob and J. Eulert, Chondrogenic differentiation of human mesenchymal stem cells in collagen type I hydrogels, *J Biomed Mater Res A* **83**: 626–635 (2007).

66. C. Chung and J. A. Burdick, Influence of three-dimensional hyaluronic acid microenvironments on mesenchymal stem cell chondrogenesis, *Tissue Eng Part A* **15**: 243–254 (2009).

67. J. M. Coburn, M. Gibson, S. Monagle, Z. Patterson and J. H. Elisseeff, Bioinspired nanofibers support chondrogenesis for articular cartilage repair, *Proc Natl Acad Sci USA* **109**: 10012–10017 (2012).

68. D. Bosnakovski, M. Mizuno, G. Kim, S. Takagi, M. Okumura and T. Fujinaga, Chondrogenic differentiation of bovine bone marrow mesenchymal stem cells (MSCs) in different hydrogels: influence of collagen type II extracellular matrix on MSC chondrogenesis, *Biotechnol Bioeng* **93**: 1152–1163 (2006).

69. M. P. Lynch, J. L. Stein, G. S. Stein and J. B. Lian, The influence of type I collagen on the development and maintenance of the osteoblast phenotype in primary and passaged rat calvarial osteoblasts: modification of expression of genes supporting cell growth, adhesion, and extracellular matrix mineralization, *Exp Cell Res* **216**: 35–45 (1995).

70. M. Mizuno, M. Shindo, D. Kobayashi, E. Tsuruga, A. Amemiya and Y. Kuboki, Osteogenesis by bone marrow stromal cells maintained on type I collagen matrix gels *in vivo*, *Bone* **20**(2): 101–107 (1997).

71. N. Ashammakhi, A. Ndreu, L. Nikkola, I. Wimpenny and Y. Yang, Advancing tissue engineering by using electrospun nanofibers, *Regen Med* **3**: 547–574 (2008).

72. K. Tuzlakoglu and R. L. Reis, Biodegradable polymeric fiber structures in tissue engineering, *Tissue Eng Part B Rev*, **15**: 17–27 (2008).

73. W. J. Marijnissen, G. J. van Osch, J. Aigner, S. W. van der Veen, A. P. Hollander, H. L. Verwoerd-Verhoef and J. A. N. Verhaar, Alginate as a chondrocyte-delivery substance in combination with a non-woven scaffold for cartilage tissue engineering, *Biomaterials* **23**: 1511–1517 (2002).

74. D. Hannouche, H. Terai, J. R. Fuchs, S. Terada, S. Zand, B. A. Nasseri, H. Petite, L. Sedel and J. P. Vacanti, Engineering of implantable cartilaginous structures from bone marrow-derived mesenchymal stem cells, *Tissue Eng* **13**: 87–99 (2007).

75. F. T. Moutos, L. E. Freed and F. Guilak, A biomimetic three-dimensional woven composite scaffold for functional tissue engineering of cartilage, *Nat Mater* **6**: 162–167 (2007).

76. F. T. Moutos and F. Guilak, Functional properties of cell-seeded three-dimensionally woven poly(ε-caprolactone) scaffolds for cartilage tissue engineering, *Tissue Eng Part A* **16**: 1291–1301 (2010).

77. W. J. Li, K. G. Danielson, P. G. Alexander and R. S. Tuan, Biological response of chondrocytes cultured in three-dimensional nanofibrous poly(ε-caprolactone) scaffolds, *J Biomed Mater Res A* **67A**: 1105–1114 (2003).

78. W. J. Li, R. Tuli, C. Okafor, A. Derfoul, K. G. Danielson, D. J. Hall and R. S. Tuan, A three-dimensional nanofibrous scaffold for cartilage tissue engineering using human mesenchymal stem cells, *Biomaterials* **26**: 599–609 (2005).

79. S. Varghese, P. Theprungsirikul, S. Sahani, N. Hwang, K. J. Yarema and J. H. Elisseeff, Glucosamine modulates chondrocyte proliferation, matrix synthesis, and gene expression, *Osteoarthritis Cartilage* **15**: 59 68 (2007).

80. Y. C. Lin, Y. C. Liang, M. T. Sheu, Y. C. Lin, M. T. Hsieh, T. F. Chen and C. H. Chen, Chondroprotective effects of glucosamine involving the p38 MAPK and Akt signaling pathways, *Rheumatol Int* **28**: 1009–1016 (2008).

81. A. R. Shikhman, D. C. Brinson, J. Valbracht and M. K. Lotz, Differential metabolic effects of glucosamine and N-acetylglucosamine in human articular chondrocytes, *Osteoarthritis Cartilage* **17**: 1022–1028 (2009).

82. A. S. d'Abusco, V. Calamia, C. Cicione, B. Grigolo, L. Politi and R. Scandurra, Glucosamine affects intracellular signalling through inhibition of mitogen-activated protein kinase phosphorylation in human chondrocytes, *Arthritis Res Ther* **9**: R104 (2007).

83. S. Nakatani, H. Mano, R. Im, J. Shimizu and M. Wada, Glucosamine regulates differentiation of a chondrogenic cell line, ATDC5, *Biol Pharm Bull* **30**: 433–438 (2007).

84. N. S. Hwang, S. Varghese, P. Theprungsirikul, A. Canver and J. Elisseeff, Enhanced chondrogenic differentiation of murine embryonic stem cells in hydrogels with glucosamine, *Biomaterials* **27**: 6015–6023 (2006).

85. M. Gibson, H. Li, J. Coburn, L. Moroni, Z. Nahas, C. Bingham, 3rd, K. Yarema and J. Elisseeff, Intra-articular delivery of glucosamine for treatment of experimental osteoarthritis created by a medial meniscectomy in a rat model, *J Orthop Res* **32**: 302–309 (2014).

86. J. M. Coburn, N. Bernstein, R. Bhattacharya, U. Aich, K. J. Yarema and J. H. Elisseeff, Differential response of chondrocytes and chondrogenic-induced mesenchymal stem cells to C1-OH tributanoylated N-acetylhexosamines, *PLoS One* **8**: e58899 (2013).

87. J. M. Coburn, L. Wo, N. Bernstein, R. Bhattacharya, U. Aich, C. O. Bingham, 3rd, K. J. Yarema and J. H. Elisseeff, Short-chain fatty acid-modified hexosamine for tissue-engineering osteoarthritic cartilage. *Tissue Eng Part A* **19**: 2035–2044 (2013).

88. C. T. Campbell, U. Aich, C. A. Weier, J. J. Wang, S. S. Choi, M. M. Wen, K. Maisel, S. G. Sampathkumar and K. J. Yarema, Targeting pro-invasive oncogenes with short chain fatty acid-hexosamine analogues inhibits the mobility of metastatic MDA-MB-231 breast cancer cells, *J Med Chem* **51**: 8135–8147 (2008).

89. C. Kim, O. H. Jeon, D. H. Kim, J. J. Chae, L. Shores, N. Bernstein, R. Bhattacharya, J. M. Coburn, K. J. Yarema and J. H. Elisseeff, Local delivery of a carbohydrate analog for reducing arthritic inflammation and rebuilding cartilage, *Biomaterials* **83**: 93–101 (2016).
90. A. Huang, N. Motlekar, A. Stein, S. Diamond, E. Shore and R. Mauck, High-throughput screening for modulators of mesenchymal stem cell chondrogenesis, *Ann Biomed Eng* **36**: 1909–1921 (2008).

10

ADULT STEM CELLS FOR ARTICULAR CARTILAGE TISSUE ENGINEERING

Sushmita Saha, Jennifer Kirkham, David Wood,
Stephen Curran and Xuebin B. Yang

1. Introduction

Healthy articular cartilage is vital for proper functioning of our joints. However, the absence of neural tissue and blood vessels coupled with a sparse cell population and slow metabolic activity greatly limits the reparative process of this connective tissue.[1] Articular cartilage can be damaged in numerous ways, including congenital pathologies such as pediatric growth plate disorders, trauma-induced injuries, and age-related degenerative joint disorders such as osteoarthritis. Current treatment for cartilage damage has been primarily symptomatic, ranging from the use of analgesics and anti-inflammatory drugs to surgical interventions that include debridement, cell therapy/grafting, biological resurfacing, joint replacement, and joint fusion.[2,3] These methods have had varying success with either partial healing or the formation of fibrocartilage which is mechanically inferior to native hyaline cartilage.

Functional tissue engineering offers hope to the field of cartilage regeneration/repair by combining three basic elements[4,5]: (1) a suitable source of cells that have a high chondrogenic potential, are easily expandable, and can be maintained for long periods of time in culture; (2) a biocompatible and biodegradable porous scaffold to act as a carrier to deliver and/or support cell growth; and (3) bioactive regulators which could be either biomolecules such as transforming growth factor β3 (TGF-β3) or bone morphogenic proteins (BMPs) and/or mechanical factors such as hydrostatic pressure, stress, or strain which can stimulate cell proliferation and differentiation.

Successful repair of cartilage defects by autologous chondrocyte transplantation was first reported in 1994 by Brittberg *et al.*[6] A year later Genzyme employed this technique and produced Carticel®, an autologous chondrocyte implantation (ACI) treatment approved by the US Food and Drug Administration (FDA) for patients with knee cartilage injuries.[7] Autologous grafts or chondrocyte transplantation have been shown to be effective; however, they involve an invasive procedure with limited cell availability, loss of differentiation capability *in vitro* and a risk of donor site morbidity. Owing to their intrinsic properties, stem cells are now emerging as a favorable cell source for use in cartilage tissue regeneration. However, the application of embryonic stem cells (ESCs) is not ideal, due to ethical issues and problems in the regulation of cell differentiation *in vivo*. For these reasons, adult mesenchymal stem cells (MSCs) are preferred for cell-based therapies.[8]

A stem cell is defined as a cell which possesses the ability to self-renew and differentiate into multiple mature cell lineages.[9] Adult stem cells/multipotent cells possess limited self-renewal capabilities, unlike embryonic stem cells which are considered to be pluripotent. Many researchers are investigating the transdifferentiation capacity/plasticity of adult stem cells which remains a highly controversial subject.[10,11] In spite of the numerous questions regarding their transdifferentiation capabilities, MSCs have emerged as a popular cell source in cartilage tissue engineering. MSCs cultured *in vitro* have been documented to lack major histocompatibility complex (MHC) class II cell surface markers. These cells possess only class I MHC surface markers without any co-stimulator molecules, making them an ideal source in auto/allo/xenogenic therapeutic applications.[12] It is also well understood that MSCs have a trophic effect, i.e. depending on their local environment and activity status, they secrete large quantities of bioactive molecules which help bring about a therapeutic response.[13] The presence of MSC immunomodulatory and trophic functions *in vitro* has opened up a new era of cell-mediated therapies.[14] In the last five to ten years, a plethora of sources of stem cells with chondrogenic potential has been reported. However, there is no uniformly accepted definitive phentotype and/or cell surface markers to assist with MSC isolation. Thus, in order to be classified as an MSC by the International Society for Cellular Therapy, the cell population must: (1) adhere to plastic under standard culture conditions; (2) express cluster of differentiation (CD) markers — CD105, CD7, and CD90; and (3) be able to differentiate down chondro-/osteo- and adipo-genic lineages.[15] This chapter will present a concise synopsis of adult stem cell sources, functions, and identities (e.g. bone marrow, fat pad, synovial fluid, dental pulp, periosteum) with potential for use in cartilage tissue engineering along with a discussion on the more recent discoveries of stem cell sources.

2. Human Bone Marrow Mesenchymal Stem Cells

An estimated one in every 10,000 nucleated cells in the bone marrow of newborns is a stem cell.[16] Human bone marrow mesenchymal stem cells (hBMMSCs) form the non-hematopoietic component of the stem cells found in the bone marrow. Friedenstein *et al.* (1966) provided the earliest evidence of the existence of stem cells in the bone marrow when they generated a hematopoietic ossicle by transplanting a whole bone marrow under the kidney capsule in mice.[17] Since then, the field of MSC research has gained immense popularity, with scientists regularly reporting new sources for MSCs. Bone marrow mesenchymal stem cells (BMMSCs) still remain the most studied and best understood stem cell source used in cartilage tissue engineering. Bone marrow aspirates (containing the BMMSCs) are easily accessed by introducing a needle directly into the bone marrow, mostly at the iliac crest.[18,19] The ability of BMMSCs to undergo successful chondrogenesis has been studied in a variety of models. The most popular are high-density cell pellets/micromasses or cell three-dimensional scaffold constructs *in vitro* and *in vivo* under various mechanical conditions with/without the addition of bioactive regulators. A typical chondroinductive cocktail contains dexamethasone (Dex) and ascorbic acid with/without a growth factor in a serum-free culture medium. Positive identification of an appropriate cartilaginous phenotype is based mainly upon observed expression of the transcription factor Sox9 and the extracellular matrix proteins collagen type II, aggrecan, and cartilage oligomeric matrix protein, amongst others.[20]

Numerous studies have looked into the factors that contribute to the chondrogenic differentiation of hBMMSCs. For example, chondrogenic preconditioning of hBMMSCs in aggregate cultures with fibroblast growth factor 2 (FGF-2) was observed to enhance chondrogenesis in tissue-engineered constructs while withdrawal of TGF-β3 was observed to differentiate the hBMMSCs to a hypertrophic state.[21,22] Enhanced glycosaminoglycans (GAGs) synthesis was observed when hBMMSCs were cultured in a hyaluronan-alginate layer culture system.[23]

Evaluation of the efficacy of hBMMSCs to undergo chondrogenesis *in vivo* has yielded mixed results in animal models. BMMSCs seeded on various scaffolds and implanted into rabbit cartilage defects resulted in hypertrophy of the tissue with large areas of bone replacement visible.[24,25] BMMSCs seeded on a poly(lactic acid)(PLA)-alginate amalgam in a canine model led to the formation of fibrous tissue.[26] Chen and colleagues (2005) showed the formation of cartilage-like tissue in a sheep defect model when they seeded autologous BMMSCs on a poly(lactic-co-glycolic) acid (PLGA) scaffold. Cartilage-like tissue formation was observed from four to eight weeks without any visible signs of bone formation.[27] Autologous hBMMSCs seeded on a collagen gel and implanted in the patello-femoral joint covered with autologous periosteum or synovium led to an improvement of clinical symptoms, with the defect being fully repaired after 12 months with formation of fibrocartilage tissue in one of the patients.[28] Yang *et al.* have shown the potential of using hBMMSCs in combination with biomimetic biomaterial scaffolds to form cartilage-like tissues in different *in vivo* models (Fig. 1).[29–31]

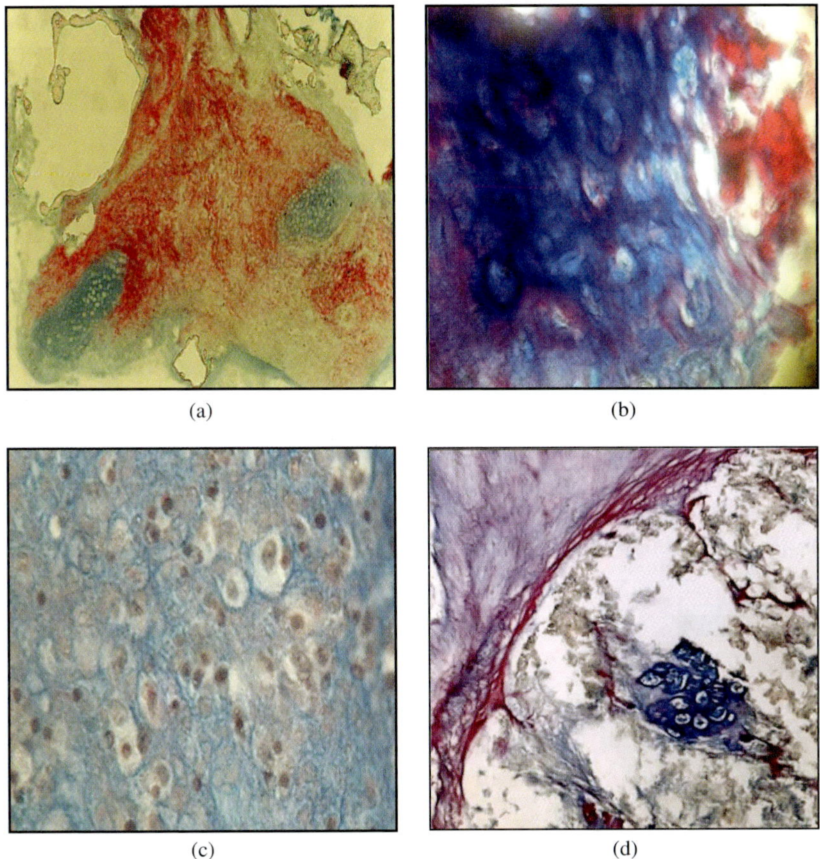

Fig. 1. Cartilage tissue formation by hBMMSCs *in vivo*: (a) after seven days in a Chorioallantoic membrane assay model; (b) after ten weeks in a diffusion chamber model intraperitoneal implantation in Nu/Nu mice (X. B. Yang *et al.*, 2004)[31]; (c) in a diffusion chamber model by hBMMSCs after adenoviral BMP-2 gene transfer (from Partridge *et al.*, *Biochem Biophys Res Commun* **292**: 144–152 (2002), reprinted with permission from Elsevier)[30]; (d) in a diffusion chamber model enhanced by pleiotrophin (from Yang *et al.*, *J Bone Miner Res* **18**: 47–57 (2003), reprinted with permission from the American Society for Bone and Mineral Research).[29]

One of the major drawbacks observed when working with hBMMSCs is the difficulty in maintaining their plasticity *in vitro*. The 'Hayflick limit' in hBMMSCs has been observed where successive passaging *in vitro* leads to a reduced proliferation and differentiation capacity.[32] Another drawback is the lack of a definitive *in vivo* model using hBMMSCs to generate high quality hyaline cartilage, which requires further research for the understanding of the basic biology of hBMMSCs and their behavior in a physiological condition.

3. Adipose-Derived Stem Cells

The use of adipose tissue (mesodermal origin) has been a recent hot bed for the discovery of stem cells.[33] Adipose-derived stem cells (ASCs) were first formally documented by Zuk and colleagues in 2001 in human lipoaspirates.[34] The white adipose tissue (WAT) present mainly in the intra-abdominal viscera and subcutaneous tissues contains higher numbers of ASCs with enhanced proliferation and plasticity properties in comparison to brown adipose tissue (BAT) which is generally present in abundance in newborns.[35]

Similar to hBMMSCs, ASCs have been found to be immunosuppressive due to the absence of MHC class II antigens on their surface. In addition, ASCs have been observed to exert an inhibitory effect over allogenic lymphocytes *in vitro*. However, freshly isolated ASCs are capable of eliciting a T-cell proliferative response at passages 0 and 1 which disappears in latter passages.[36,37]

Two-dimensional culture of ASCs is unfavourable for maintaining chondrogenic phenotype, but culturing these cells on an elastin-like polypeptide was observed to significantly enhance chondrogenesis.[38] Using a chondroinductive cocktail including TGF-β, expression of collagen type II, aggrecan, and sulphated proteoglycans was observed. Enhanced chondrogenesis of ASCs was observed when the cells were cultured with BMP-6. In contrast, hBMMSCs underwent osetogenesis upon BMP-6 exposure.[39] Hennig *et al.* (2007) further showed that a combination of TGF-β with BMP-6 was the ideal cocktail for chondroinduction of ASCs.[40,41]

Fibrin glue scaffolds seeded with ASCs and implanted *in vivo* for eight weeks showed aggrecan and type II collagen expression.[42] Rada *et al.* (2013) showed the potential of using two specific ASC subpopulations, encapsulated in a novel gellan gum hydrogel, to enhance chondrogenic differentiation and new cartilage tissue formation in nude mice after six weeks of subcutaneous implantation.[43]

Much controversy lies in determining whether ASCs have higher chondroinductive abilities in comparison to hBMMSCs. hBMMSCs were reported to possess a higher chondrogenic potential than that of adipose tissue-derived MSCs in the pellet culture system in the presence of TGF-β1/β2.[44] However, Lee *et al.* (2004) reported that ASCs were superior to bone marrow stromal cells (BMSCs) with respect to maintenance of proliferation ability. Microarray analysis of gene expression revealed differentially expressed genes between ASCs and BMSCs but their phenotypes and the overall gene expression profiles were similar.[45]

4. Periosteum-Derived Progenitor Cells/Periosteum-Derived Stem Cells

The outer surface of bones at the non-articulating junctions are covered by a thin layer of connective tissue known as the periosteum. The outer fibrous layer of the periosteum is made up of fibroblasts, while the inner cambial layer contains progenitor cells responsible for increasing bone width.[46] The presence of progenitor cells in the periosteum that are capable of differentiating into three mesodermal lineages (osteo-/adipo-/chondrogenic) has been demonstrated since the early 1990s.[47] De Bari *et al.* (2006) isolated and characterized periosteum-derived progenitor cells (PDPCs) with MSC surface markers. These PDPCs were observed to possess MSC-like multipotentiality when analysed *via* single cell lineage analysis, and they underwent chondrogenesis when cultured as micromasses.[48]

A major plus point to using PDPCs is their ability to proliferate at a higher rate compared to BMMSCs.[49] PDPCs are routinely used in the clinic to regenerate bone; however, their clinical use in repairing cartilage is undocumented so far.[50,51] The ability of PDPCs to undergo chondrogenesis *in vivo* was demonstrated in a rabbit model where the formation of ectopic cartilaginous tissue was observed 20 days following dissection of the periosteum (7×15 mm^2 periosteum defect). However, the newly formed cartilaginous tissue turned to bone by day 40. The advantage of such a technique is the elimination of the need for cell culture.[52] In another report, a similar approach was used to create an artificial space ("bioreactor") between the tibia and periosteum in which angiogenesis was inhibited (via local administration of the anti-angiogenic factor suramin in a hyaluronic acid (HA)-based gel matrix) to promote a more hypoxic environment within the "*in vivo* bioreactor" space. Cartilage formation was observed within the *in vivo* bioreactor after ten days.[53]

While these cells possess MSC-like multipotentiality and higher proliferation rates than hBMMSCs, a major disadvantage of using periosteum-derived stem cells (PDSCs) is the need to carry out a surgical procedure for their isolation, limiting their application as a potential autograft transplant. The probability of cell loss in the cambial layer when harvesting the periosteal cells coupled with the variation in MSC numbers in the cambial layer depending on donor age yields inconsistent results.

5. Synovium-Derived Mesenchymal Stem Cells

The non-articular surfaces of diarthrodial joints are lined by a thin membranous tissue known as the synovium. The main function of this tissue is to maintain the synovial fluid cavity which nourishes the articulating cartilage. De Bari and colleagues (2001) successfully extracted MSCs from the synovial membrane for the first time.[54] The immunogenicity of these cells has been found to be similar to that of hBMMSCs. Synovium-derived mesenchymal stem cells (SMSCs) are negative for the expression of class II MHC molecules and suppress T-cell proliferation in a mixed lymphocyte population.[55]

Numerous evidences point to the chondrogenic potential of these multipotent cells. Crawford *et al.* (2006) reported the ability of cell cultures isolated from synovial nodules from a patient suffering from primary synovial chromatosis (PSC) to undergo chondrogenic differentiation. The benign synovial metaplasia in PSC is thought to arise due to the proliferation of mesenchymal progenitor cells (MPCs).[56] The expression of chondrogenic markers such Sox9, aggrecan, and cartilage oligomeric matrix protein (COMP) has been documented to be 2–25% higher in synovial calf tissue when compared to committed articular cartilage.[57]

SMSCs have the highest proliferation ability in the presence of TGF-β1, IGF-I, and FGF-2, while synergistic action of TGF-β1 and IGF-I enhanced SMSC chondrogenesis.[58] Shirasawa *et al.* (2006) documented enhanced human SMSC chondrogenesis in pellet culture as a result of the synergistic effects of TGF-β3, Dex, and BMP-2.[59] However, further studies reported an inhibitory effect of Dex on BMP-2 chondroinduced synovial cells.[1,60]

In vivo, like PDSCs, SMSCs migrated from the synovium to partial thickness articular cartilage defects in adult rabbits and yucatan pigs, but the cells differentiated into fibroblasts and filled the cavities with a fibrous connective tissue.[1] Pei *et al.* (2009) seeded allogenic SMSCs on a polyglycolic acid (PGA) mesh, cultured the resulting constructs *in vitro* for one month, and then implanted the neo-tissue into full thickness femoral condyle cartilage defects in rabbits. They observed the formation of hyaline-like cartilage tissue with no detectable levels of collagen type I.[61]

Although the main advantage of using SMSCs is their high proliferative capacity and enhanced chondrogenic differentiation capabilities compared to other MSCs,[59,62] the invasive surgery needed to access these stem cells coupled with an unclear understanding of their differentiation mechanism may hinder their popular use.

6. Human Dental Pulp Stem Cells

Human dental pulp stem cells (HDPSCs) have been isolated mainly from the pulp tissues of third permanent molar teeth and can be maintained for up to 25 passages. A number of studies have reported that HDPSCs could be a suitable cell source for bone tissue engineering, but there are few references reporting their potential in cartilage regeneration.[63] A STRO-1(+) DPSC population has been thought to possess a higher multilineage potential compared to non-sorted cells, probably because of their homogeneous nature. These STRO-1(+) cells can be differentiated down neurogenic, osteogenic/odontogenic, adipogenic, and myogenic lineages. However, chondrogenic differentiation of these primary cells was found to be inferior to myogenic differentiation by the same cells.[64]

Wei and colleagues (2008) demonstrated multilineage differentiation in all HDPSC samples with collagen type II expression levels being significantly upregulated after chondroinduction.[65] The main advantage of using HDPSCs lies in the ease of obtaining these cells without any invasive surgery by banking milk teeth. However, their superiority in terms of chondrogenic differentiation over the other stem cell sources available still needs to be determined by a set of robust comparative studies.

7. Umbilical Cord/Cord Blood-Derived Stem Cells

Wharton's Jelly in the human umbilical cord has recently been reported to possess MSC-like cells (human umbilical cord perivascular cells, HUCPVCs),[66] and depending upon how these cells are harvested, they can differentiate down the neuronal or cardiac pathways. These HUCPVCs are located close to the vasculature of the cord and have a colony-forming fibroblast frequency of about 1:300, expressing an immunoprivileged and immunomodulatory phenotype.[67] A comparison of the chondrogenic potential between these cells and hBMMSCs revealed that when cultured as pellets for 21 days along with TGF-β3, HUCPVCs formed larger-sized pellets than HBMMSC pellets. The sulphanated glycosaminoglycan (sGAG) content and alcian blue-picrosirius staining intensities were comparable between the two cell types, indicating similar matrix formation. Unlike hBMMSCs, these cells demonstrate the potential to grow in multi-layers, overlying cellular aggregates. It is, however, important that independent studies investigate the possibility of these cells transforming to a cancerous phenotype and identify the molecular mechanism involved in such multi-layered cellular growth.

Unrestricted somatic stem cells (USSCs) derived from cord blood are believed to be a promising source of therapeutic stem cells, and can be differentiated down all three germ layers (e.g. ectoderm, mesoderm, and endoderm).[68] However, when compared to hBMMSCs cultured as pellets for 21 days, there was a significant decrease in sGAG content of the USSC compared to hBMMSCs pellets, suggesting reduced matrix formation.[69] Perhaps as a consequence of this, there have been few publications reporting on the use of USSCs for cartilage tissue engineering in recent years.

8. Other Potential Cell Sources with a Chondrogenic Potential

As an alternative to MSCs, MPC populations were recently isolated from traumatised muscle tissue that had been surgically debrided due to an orthopaedic wound. The MPCs had similar morphological, proliferation, and differentiation capacities as hBMMSCs.[70] However, before research using these cells can progress, it is important to demonstrate that they can undergo robust *in vitro* expansion and maintain their properties and differentiation capabilities in an *in vivo* environment.

The anterior cruciate ligament (ACL) has been identified as another source for MSCs with chondrogenic potential. Stem/stromal cells from ACL can be enzymatically released and cultured as monolayers on plastic dishes.[71,72] These cells were observed to have similar phenotypic characteristics as hBMMSCs. FGF-2 and TGF-β1 can enhance the proliferation of these cells along with extracellular matrix protein production.[73] However, in other studies, only one cell line out of six donors was found to be capable of tri-potent differentiation. ACL-derived cells were observed to be more beneficial in the formation of ligament fibroblasts than cartilage tissue.[72]

The dogma of articular cartilage being a non-regenerative tissue is now increasingly being challenged with the hypothesis of the presence of a progenitor/stem cell population in the superficial layer of articular cartilage itself.[74] Morphological variations are observed in the different layers of articular cartilage: cells of the flattened superficial zone secrete lubricin; the rounded and columnar arranged middle zone cells produce cartilage intermediate layer protein (CILP); while the considerably larger deep zone cells express type X collagen and alkaline phosphatase.[75,76] Hayes *et al.* (2001) demonstrated the presence of slow-cycling cells, akin to progenitor cells, in the superficial zone of articular cartilage through the use of bromo-deoxyuridine (BrdU) injections. These cells expressed notch homolog 1 (NOTCH-1) and possessed a high colony-forming efficiency.[77]

Grogan and colleagues (2009) demonstrated the localization of progenitor cells in healthy cartilage using NOTCH-1, STRO-1, and vascular cell adhesion molecule-1 (VCAM-1) as stem cell markers with highest frequencies of labeled cells being observed in the superficial zone. Progenitor cells were also isolated from osetoarthritic cartilage. The frequency of NOTCH-1, STRO-1, and VCAM-1 positive cells was found to be higher in the middle and not the superficial zone for osetoarthritic tissue. However, a similar frequency (0.14 ± 0.05%) of progenitor cells was observed in healthy and osetoarthritic cartilage. These progenitor cells possessed chondrogenic and osteogenic capabilities but not adipogenic differentiation potential. However, the authors suggested that NOTCH-1, STRO-1, or VCAM-1 may not be useful in identifying progenitors in cartilage and that their increased expression in osetoarthritic cartilage perhaps indicates their involvement in osteoarthritis.[78] Similarly, Ozbey and colleagues (2014) reported the presence of resident stem-like cells in adult human articular cartilage[79] and Li *et al.* (2016) suggested that the superficial cells within the articular cartilage of juvenile mice represent self-renewing chondrocyte progenitors.[80]

While the presence of these cell populations highlights the possibility of cartilage possessing an intrinsic regenerative capacity, further research is required to elucidate their role before they can be used in transplant studies.

9. Conclusion and Future Directions

A myriad of sources of MSCs with chondrogenic potential have been identified. Under appropriate conditions, these cells can be maintained and differentiated down the chondrogenic lineage. Some of these MSC populations are more easily isolated and differentiated than others, yet they all offer promising solutions to the field of cartilage tissue engineering. Before selecting an appropriate MSC source for cell therapy, it is important to consider whether the cell of choice can be expanded/maintained/ differentiated *in vitro*, cultured on a scaffold, and easily assessed *via* conventional methods of histology. At present there is a need for more robust tests to characterise cells classified as stem cells by using clonogenic-lineage-specific gene marking with multipotentiality visible under *in vivo* conditions. Upon fully understanding the biology of the stem cells isolated, we can go on to design methodologies to stimulate these cells to repair the damaged articular cartilage tissue in the body. However, the translation of research from lab to bedside could be another challenge for functional articular cartilage tissue engineering.

Acknowledgements

Sushmita Saha was supported by a PhD studentship from Smith & Nephew Ltd. Xuebin B. Yang was supported by the European Union ([FP7/2007–2013] [FP7/2007–2011]) under agreement no. [318553] — "SkelGEN".

References

1. E. B. Hunziker and L. C. Rosenberg, Repair of partial-thickness defects in articular cartilage: cell recruitment from the synovial membrane, *J Bone Joint Surg Am* **78**: 721–733 (1996).

2. T. Andriacchi, Requirements for biological replacement of the articular cartilage at the knee joint, in *Functional Tissue Engineering*, edited by F. Guilak, D. L. Butler, S. A. Goldstein and D. J. Mooney (Springer, New York, 2003), pp. 106–116.

3. C. Pascual-Garrido, E. Daley, N. N. Verma, B. J. Cole, A comparison of the outcomes for cartilage defects of the knee treated with biologic resurfacing versus focal metallic implants, *Arthroscopy* **33**: 364–373 (2017).

4. X. B. Yang and R. O. Oreffo, Bone tissue engineering, in *Current Topics in Bone Biology*, edited by H. W. Deng, Y. Z. Liu and C. Y. Guo (World Scientific Publishing Company, Singapore, 2005), pp. 435–460.

5. Z. Man, L. Yin, Z. Shao, X. Zhang, X. Hu, J. Zhu, L. Dai, H. Huang, L. Yuan, C. Zhou, H. Chen and Y. Ao, The effects of co-delivery of BMSC-affinity peptide and rhTGF-beta1 from coaxial electrospun scaffolds on chondrogenic differentiation, *Biomaterials* **35**: 5250–5260 (2014).

6. M. Brittberg, A. Lindahl, A. Nilsson, C. Ohlsson, O. Isaksson and L. Peterson, Treatment of deep cartilage defects in the knee with autologous chondrocyte transplantation, *N Engl J Med* **331**: 889–895 (1994).

7. Vericel Corporation. Carticel®. Use your patients' own chondrocytes to repair their own knee cartilage with Carticel. (Last access date: 7 February 2017. Available from: http://www.carticel.com/.)

8. Y. Lin, E. Luo, X. Chen, L. Liu, J. Qiao, Z. Yan, Z. Li, W. Tang, X. Zheng and W. Tian, Molecular and cellular characterization during chondrogenic differentiation of adipose tissue-derived stromal cells *in vitro* and cartilage formation *in vivo*, *J Cell Mol Med* **9**: 929–939 (2005).

9. M. F. Pittenger, A. M. Mackay, S. C. Beck, R. K. Jaiswal, R. Douglas, J. D. Mosca, M. A. Moorman, D. W. Simonetti, S. Craig and D. R. Marshak, Multilineage potential of adult human mesenchymal stem cells, *Science* **284**: 143–147 (1999).

10. A. Prasad, D. B. Teh, F. R. Shah Jahan, J. Manivannan, S. M. Chua and A. H. All, Direct conversion through trans-differentiation: efficacy and safety, *Stem Cells Dev* **26**: 154–165 (2017).

11. S. Filip, D. English and J. Mokry, Issues in stem cell plasticity, *J Cell Mol Med* **8**: 572–577 (2004).

12. A. I. Caplan, Why are MSCs therapeutic? New data: new insight, *J Pathol* **217**: 318–324 (2009).

13. A. I. Caplan and J. E. Dennis, Mesenchymal stem cells as trophic mediators, *J Cell Biochem* **98**: 1076–1084 (2006).

14. M. B. Murphy, K. Moncivais and A. I. Caplan, Mesenchymal stem cells: environmentally responsive therapeutics for regenerative medicine, *Exp Mol Med* **45**: e54 (2013).

15. M. Dominici, K. Le Blanc, I. Mueller, I. Slaper-Cortenbach, F. Marini, D. Krause, R. Deans, A. Keating, D. Prockop and E. Horwitz, Minimal criteria for defining multipotent mesenchymal stromal cells: the International Society for Cellular Therapy position statement. *Cytotherapy* **8**: 315–317 (2006).

16. S. E. Haynesworth, V. M. Goldberg and A. L. Caplan, Diminution of the number of mesenchymal stem cells as a cause for skeletal aging, in *Musculoskeletal Soft-Tissue Aging: Impact on Mobility*, edited by J. A. Buckwalter, V. M. Goldberg and Y. Woo (American Academy of Orthopaedic Surgeons, Rosemont, Illinois, 1994), pp. 79–87.

17. A. J. Friedenstein, I. I. Piatetzky-Shapiro and K. V. Petrakova, Osteogenesis in transplants of bone marrow cells, *J Embryol Exp Morphol* **16**: 381–390 (1966).

18. P. Bianco and P. Gehron Robey, Marrow stromal stem cells, *J Clin Invest* **105**: 1663–1668 (2000).

19. H. Y. Nam, P. Karunanithi, W. C. Loo, S. Naveen, H. Chen, P. Hussin, L. Chan and T. Kamarul, The effects of staged intra-articular injection of cultured autologous mesenchymal stromal cells on the repair of damaged cartilage: a pilot study in caprine model, *Arthritis Res Ther* **15**: R129 (2013).

20. S. Saha, J. Kirkham, D. Wood, S. Curran and X. B. Yang, Informing future cartilage repair strategies: a comparative study of three different human cell types for cartilage tissue engineering, *Cell Tissue Res* **352**: 495–507 (2013).

21. K. J. Penick, L. A. Solchaga and J. F. Welter, High-throughput aggregate culture system to assess the chondrogenic potential of mesenchymal stem cells, *Biotechniques* **39**: 687–691 (2005).

22. A. M. Mackay, S. C. Beck, J. M. Murphy, F. P. Barry, C. O. Chichester and M. F. Pittenger, Chondrogenic differentiation of cultured human mesenchymal stem cells from marrow, *Tissue Eng* **4**: 415–428 (1998).

23. K. W. Kavalkovich, R. E. Boynton, J. M. Murphy and F. Barry, Chondrogenic differentiation of human mesenchymal stem cells within an alginate layer culture system, *In Vitro Cell Dev Biol Anim* **38**: 457–466 (2002).

24. R. Abdul Rahman, N. Mohamad Sukri, N. Md Nazir, M. A. Ahmad Radzi, A. H. Zulkifly, A. Che Ahmad, A. A. Hashi, S. Abdul Rahman and M. Sha'ban, The potential of 3-dimensional construct engineered from

poly(lactic-co-glycolic acid)/fibrin hybrid scaffold seeded with bone marrow mesenchymal stem cells for *in vitro* cartilage tissue engineering, *Tissue Cell* **47**: 420–430 (2015).

25. T. Yanai, T. Ishii, F. Chang and N. Ochiai, Repair of large full-thickness articular cartilage defects in the rabbit: the effects of joint distraction and autologous bone marrow-derived mesenchymal cell transplantation. *J Bone Joint Surg Br* **87**: 721–729 (2005).

26. J. S. Wayne, C. L. McDowell, K. J. Shields and R. S. Tuan, *In vivo* response of polylactic acid-alginate scaffolds and bone marrow-derived cells for cartilage tissue engineering, *Tissue Eng* **11**: 953–963 (2005).

27. J. Chen, C. Wang, S. Lu, J. Wu, X. Guo, C. Duan, L. Dong, Y. Song, J. Zhang, D. Jing, L. Wu, J. Ding and D. Li, *In vivo* chondrogenesis of adult bone marrow-derived autologous mesenchymal stem cells, *Cell Tissue Res* **319**: 429–438 (2005).

28. S. Wakitani, M. Nawata, K. Tensho, T. Okabe, H. Machida and H. Ohgushi, Repair of articular cartilage defects in the patello-femoral joint with autologous bone marrow mesenchymal cell transplantation: three case reports involving nine defects in five knees, *J Tissue Eng Regen Med* **1**: 74–79 (2007).

29. X. Yang, R. S. Tare, K. A. Partridge, H. I. Roach, N. M. Clarke, S. M. Howdle, K. M. Shakesheff and R. O. Oreffo, Induction of human osteoprogenitor chemotaxis, proliferation, differentiation, and bone formation by osteoblast stimulating factor-1/pleiotrophin: osteoconductive biomimetic scaffolds for tissue engineering, *J Bone Miner Res* **18**: 47–57 (2003).

30. K. Partridge, X. Yang, N. M. Clarke, Y. Okubo, K. Bessho, W. Sebald, S. M. Howdle, K. M. Shakesheff and R. O. Oreffo, Adenoviral BMP-2 gene transfer in mesenchymal stem cells: *in vitro* and *in vivo* bone formation on biodegradable polymer scaffolds, *Biochem Biophys Res Commun* **292**: 144–152 (2002).

31. X. B. Yang, M. J. Whitaker, W. Sebald, N. Clarke, S. M. Howdle, K.M. Shakesheff and R. O. Oreffo, Human osteoprogenitor bone formation using encapsulated bone morphogenetic protein 2 in porous polymer scaffolds, *Tissue Eng* **10**: 1037–1045 (2004).

32. A. I. Caplan, Mesenchymal stem cells, *J Orthop Res* **9**: 641–650 (1991).

33. W. Lu, K. Ji, J. Kirkham, Y. Yan, A. R. Boccaccini, M. Kellett, Y. Jin and X. B. Yang, Bone tissue engineering by using a combination of polymer/ bioglass composites with human adipose-derived stem cells, *Cell Tissue Res* **356**: 97–107 (2014).

34. P. A. Zuk, M. Zhu, H. Mizuno, J. Huang, J. W. Futrell, A. J. Katz, P. Benhaim, H. P. Lorenz and M. H. Hedrick, Multilineage cells from human adipose tissue: implications for cell-based therapies, *Tissue Eng* **7**: 211–228 (2001).

35. B. Prunet-Marcassus, B. Cousin, D. Caton, M. Andre, L. Penicaud and L. Casteilla, From heterogeneity to plasticity in adipose tissues: site-specific differences, *Exp Cell Res* **312**: 727–736 (2006).

36. B. Puissant, C. Barreau, P. Bourin, C. Clavel, J. Corre, C. Bousquet, C. Taureau, B. Cousin, M. Abbal, P. Laharrague, L. Penicaud, L. Casteilla and A. Blancher, Immunomodulatory effect of human adipose tissue-derived adult stem cells: comparison with bone marrow mesenchymal stem cells, *Br J Haematol* **129**: 118–129 (2005).

37. K. McIntosh, S. Zvonic, S. Garrett, J. B. Mitchell, Z. E. Floyd, L. Hammill, A. Kloster, Y. Di Halvorsen, J. P. Ting, R. W. Storms, B. Goh, G. Kilroy, X. Wu and J. M. Gimble, The immunogenicity of human adipose-derived cells: temporal changes *in vitro*, *Stem Cells* **24**: 1246–1253 (2006).

38. H. Betre, S. R. Ong, F. Guilak, A. Chilkoti, B. Fermor and L. A. Setton, Chondrocytic differentiation of human adipose-derived adult stem cells in elastin-like polypeptide, *Biomaterials* **27**: 91–99 (2006).

39. B. T. Estes, A. W. Wu and F. Guilak, Potent induction of chondrocytic differentiation of human adipose-derived adult stem cells by bone morphogenetic protein 6, *Arthritis Rheum* **54**: 1222–1232 (2006).

40. T. Hennig, H. Lorenz, A. Thiel, K. Goetzke, A. Dickhut, F. Geiger and W. Richter, Reduced chondrogenic potential of adipose tissue derived stromal cells correlates with an altered TGFbeta receptor and BMP profile and is overcome by BMP-6, *J Cell Physiol* **211**: 682–6891 (2007).

41. R. A. Somoza, J. F. Welter, D. Correa and A. I. Caplan, Chondrogenic differentiation of mesenchymal stem cells: challenges and unfulfilled expectations, *Tissue Eng Part B Rev* **20**: 596–608 (2014).

42. J. L. Dragoo, G. Carlson, F. McCormick, H. Khan-Farooqi, M. Zhu, P. A. Zuk and P. Benhaim, Healing full-thickness cartilage defects using adipose-derived stem cells, *Tissue Eng* **13**: 1615–1621 (2007).

43. T. Rada, P. P. Carvalho, T. C. Santos, A. G. Castro, R. L. Reis and M. E. Gomes, Chondrogenic potential of two hASCs subpopulations loaded onto gellan gum hydrogel evaluated in a nude mice model, *Curr Stem Cell Res Ther* **8**: 357–364 (2013).

44. G. I. Im, Y. W. Shin and K. B. Lee, Do adipose tissue-derived mesenchymal stem cells have the same osteogenic and chondrogenic potential as bone marrow-derived cells? *Osteoarthritis Cartilage* **13**: 845–853 (2005).

45. R. H. Lee, B. Kim, I. Choi, H. Kim, H. S. Choi, K. Suh, Y. C. Bae and J. S. Jung, Characterization and expression analysis of mesenchymal stem cells from human bone marrow and adipose tissue, *Cell Physiol Biochem* **14**: 311–324 (2004).

46. G. Augustin, A. Antabak and S. Davila, The periosteum. Part 1: Anatomy, histology and molecular biology, *Injury* **38**: 1115–1130 (2007).

47. T. Nakase, H. Nakahara, M. Iwasaki, T. Kimura, K. Kimata, K. Watanabe, A. I. Caplan and K. Ono, Clonal analysis for developmental potential of chick periosteum-derived cells: agar gel culture system, *Biochem Biophys Res Commun* **195**: 1422–1428 (1993).

48. C. De Bari, F. Dell'Accio, J. Vanlauwe, J. Eyckmans, I. M. Khan, C. W. Archer, E. A. Jones, D. McGonagle, T. A. Mitsiadis, C. Pitzalis and F. P. Luyten, Mesenchymal multipotency of adult human periosteal cells demonstrated by single-cell lineage analysis, *Arthritis Rheum* **54**: 1209–1221 (2006).

49. A. M. Ng, A. B. Saim, K. K. Tan, G. H. Tan, S. A. Mokhtar, I. M. Rose, F. Othman and R. B. Idrus, Comparison of bioengineered human bone construct from four sources of osteogenic cells, *J Orthop Sci* **10**: 192–199 (2005).

50. BioTissue Technologies. Chondrotissue®, The implant for cartilage injuries in the knee, ankle, and hip. (Last access date: 7 February 2017. Available from: www.biotissue.de.)

51. T. A. Ahmed and M. T. Hincke, Mesenchymal stem cell-based tissue engineering strategies for repair of articular cartilage, *Histol Histopathol* **29**: 669–689 (2014).

52. P. J. Emans, D. A. Surtel, E. J. Frings, S. K. Bulstra and R. Kuijer, *In vivo* generation of cartilage from periosteum, *Tissue Eng* **11**: 369–377 (2005).

53. M. M. Stevens, R. P. Marini, D. Schaefer, J. Aronson, R. Langer and V. P. Shastri, *In vivo* engineering of organs: the bone bioreactor, *Proc Natl Acad Sci USA* **102**: 11450–11455 (2005).

54. C. De Bari, F. Dell'Accio, P. Tylzanowski and F. P. Luyten, Multipotent mesenchymal stem cells from adult human synovial membrane, *Arthritis Rheum* **44**: 1928–1942 (2001).

55. K. Le Blanc, L. Tammik, B. Sundberg, S. E. Haynesworth and O. Ringden, Mesenchymal stem cells inhibit and stimulate mixed lymphocyte cultures and mitogenic responses independently of the major histocompatibility complex, *Scand J Immunol* **57**: 11–20 (2003).

56. A. Crawford, A. Frazer, J. M. Lippitt, D. J. Buttle and T. Smith, A case of chondromatosis indicates a synovial stem cell aetiology, *Rheumatology (Oxford)* **45**: 1529–1533 (2006).

57. N. Shintani, T. Kurth and E. B. Hunziker, Expression of cartilage-related genes in bovine synovial tissue, *J Orthop Res* **25**: 813–819 (2007).

58. M. Pei, F. He and G. Vunjak-Novakovic, Synovium-derived stem cell-based chondrogenesis, *Differentiation* **76**: 1044–1056 (2008).

59. S. Shirasawa, I. Sekiya, Y. Sakaguchi, K. Yagishita, S. Ichinose and T. Muneta, *In vitro* chondrogenesis of human synovium-derived mesenchymal stem cells: optimal condition and comparison with bone marrow-derived cells, *J Cell Biochem* **97**: 84–97 (2006).

60. T. Kurth, E. Hedbom, N. Shintani, M. Sugimoto, F. H. Chen, M. Haspl, S. Martinovic and E. B. Hunziker, Chondrogenic potential of human synovial mesenchymal stem cells in alginate, *Osteoarthritis Cartilage* **15**: 1178–1189 (2007).

61. M. Pei, F. He, B. M. Boyce and V. L. Kish, Repair of full-thickness femoral condyle cartilage defects using allogeneic synovial cell-engineered tissue constructs. *Osteoarthritis Cartilage* **17**: 714–722 (2009).

62. B. A. Jones and M. Pei, Synovium-derived stem cells: a tissue-specific stem cell for cartilage engineering and regeneration, *Tissue Eng Part B Rev* **18**: 301–311 (2012).

63. R. El-Gendy, J. Kirkham, P. J. Newby, Y. Mohanram, A. R. Boccaccini and X. B. Yang. Investigating the vascularization of tissue-engineered bone constructs using dental pulp cells and 45S5 Bioglass® scaffolds. *Tissue Eng Part A* **21**: 2034–2043 (2015).

64. W. Zhang, X. F. Walboomers, S. Shi, M. Fan and J. A. Jansen, Multilineage differentiation potential of stem cells derived from human dental pulp after cryopreservation, *Tissue Eng* **12**: 2813–2823 (2006).

65. X. Wei, L. P. Wu, J. Q. Ling and L. Liu, [Multilineage differentiation of human dental pulp cells and periodontal ligament cells *in vitro*], *Zhonghua Kou Qiang Yi Xue Za Zhi* **43**: 495–499 (2008).

66. E. Chen, M. K. Tang, Y. Yao, W. W. Yau, L. M. Lo, X. Yang, Y. L. Chui, J. Chan and K. K. Lee, Silencing BRE expression in human umbilical cord perivascular (HUCPV) progenitor cells accelerates osteogenic and chondrogenic differentiation, *PLoS One* **8**: e67896 (2013).

67. R. Sarugaser, D. Lickorish, D. Baksh, M. M. Hosseini and J. E. Davies, Human umbilical cord perivascular (HUCPV) cells: a source of mesenchymal progenitors, *Stem Cells* **23**: 220–229 (2005).

68. G. Kogler, S. Sensken, J. A. Airey, T. Trapp, M. Muschen, N. Feldhahn, S. Liedtke, R. V. Sorg, J. Fischer, C. Rosenbaum, S. Greschat, A. Knipper, J. Bender, O. Degistirici, J. Gao, A. I. Caplan, E. J. Colletti, G. Almeida-Porada, H. W. Muller, E. Zanjani and P. Wernet, A new human somatic stem cell from placental cord blood with intrinsic pluripotent differentiation potential, *J Exp Med* **200**: 123–135 (2004).

69. O. Degistirici, M. Jager and A. Knipper, Applicability of cord blood-derived unrestricted somatic stem cells in tissue engineering concepts. *Cell Prolif* **41**: 421–440 (2008).

70. W. M. Jackson, A. B. Aragon, F. Djouad, Y. Song, S. M. Koehler, L. J. Nesti and R. S. Tuan, Mesenchymal progenitor cells derived from traumatized human muscle, *J Tissue Eng Regen Med* **3**: 129–38 (2009).

71. A. F. Steinert, M. Kunz, P. Prager, T. Barthel,F. Jakob, U. Noth, M. M. Murray, C. H. Evans and R. M. Porter, Mesenchymal stem cell characteristics of human anterior cruciate ligament outgrowth cells, *Tissue Eng Part A* **17**: 1375–1388 (2011).

72. T. F. Huang, Y. T. Chen, T. H. Yang, L. L. Chen, S. H. Chiou, T. H. Tsai, C. C. Tsai, M. H. Chen, H. L. Ma and S. C. Hung, Isolation and characterization of mesenchymal stromal cells from human anterior cruciate ligament, *Cytotherapy* **10**: 806–814 (2008).

73. M. T. Cheng, H. W. Yang, T. H. Chen and O. K. Lee, Modulation of proliferation and differentiation of human anterior cruciate ligament-derived stem cells by different growth factors, *Tissue Eng Part A* **15**: 3979–3989 (2009).

74. C. Karlsson and A. Lindahl, Articular cartilage stem cell signalling, *Arthritis Res Ther* **11**: 121 (2009).

75. P. Lorenzo, M. T. Bayliss and D. Heinegard, A novel cartilage protein (CILP) present in the mid-zone of human articular cartilage increases with age, *J Biol Chem* **273**: 23463–23468 (1998).

76. B. L. Schumacher, J. A. Block, T. M. Schmid, M. B. Aydelotte and K. E. Kuettner, A novel proteoglycan synthesized and secreted by chondrocytes of the superficial zone of articular cartilage, *Arch Biochem Biophys* **311**: 144–152 (1994).

77. A. J. Hayes, S. MacPherson, H. Morrison, G. Dowthwaite and C. W. Archer, The development of articular cartilage: evidence for an appositional growth mechanism, *Anat Embryol (Berl)* **203**: 469–479 (2001).

78. S. P. Grogan, S. Miyaki, H. Asahara, D. D. D'Lima and M. K. Lotz, Mesenchymal progenitor cell markers in human articular cartilage: normal distribution and changes in osteoarthritis, *Arthritis Res Ther* **11**: R85 (2009).

79. O. Ozbey, Z. Sahin, N. Acar, F. T. Ozcelik, A.M. Ozenci, S. Koksoy and I. Ustunel, Characterization of colony-forming cells in adult human articular cartilage, *Acta Histochem* **116**: 763–770 (2014).

80. L. Li, P. T. Newton, T, Bouderlique, M. Sejnohova, T. Zikmund, E. Kozhemyakina, M. Xie, J. Krivanek, J. Kaiser, H. Qian, V. Dyachuk, A B Lassar, M. L. Warman, B. Barenius, I. Adameyko and A. S. Chagin, Superficial cells are self-renewing chondrocyte progenitors, which form the articular cartilage in juvenile mice, *FASEB J* (2016) [Epub ahead of print].

FASEB J (2015) [Epub ahead of print].

11

STEM CELLS FOR DISC REPAIR

Ann Ouyang, Aliza A. Allon, Zorica Buser,
Sigurd Berven and Jeffrey C. Lotz

1. Introduction

The intervertebral disc forms an avascular, fibrocartilaginous joint between adjacent vertebral bodies and provides flexibility while routinely supporting forces up to several multiples of body weight.[1] The disc is composed of three major sub-tissues: the gelatinous nucleus pulposus (NP), the fibrous annulus fibrous, and cartilaginous endplates (Fig. 1). The NP is centrally located and composed primarily of sulfated glycosamino-glycan (GAG), type II collagen, and water. The NP serves as the osmotic mechanism that generates volume and hydrostatic pressure because the high GAG content makes the tissue very hydrophilic. The annulus fibro-sus is firmly attached to the vertebral edges and the endplate to serve both as a ligament to guide intervertebral movement, and as a barrier to contain

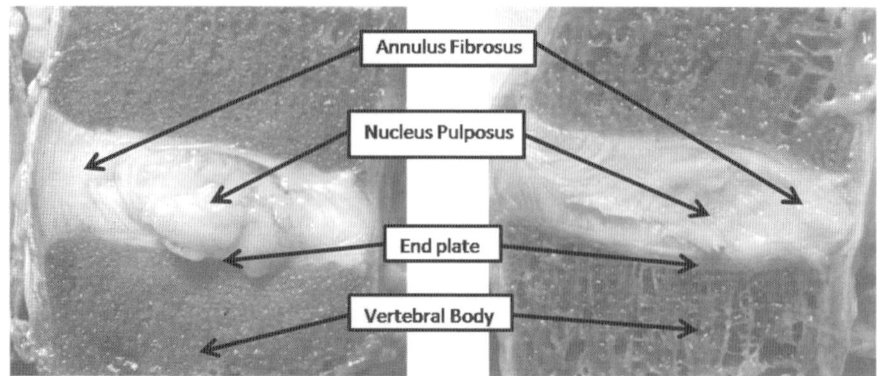

Fig. 1. Mid-sagittal sections through a healthy (*left*) and moderately degenerate (*right*) human disc. In the healthy state, the nucleus possesses a significant capacity to swell and support spinal forces. The early stages of degeneration are characterized by nuclear fibrosis and annular fissuring.

nuclear swelling and thereby allow disc pressurization.[2] The endplate is a thin (0.1–1.6 mm) hyaline cartilage layer that separates the NP from the adjacent vertebra. The endplate functions as a semi-permeable membrane to allow diffusive communication between disc nuclear cells and vertebral vasculature, as well as to prevent large molecular weight GAG from leaving the nuclear space.

Pain of spinal origin afflicts most adults at some point in their lives: the annual US incidence of acute and/or chronic back pain is approximately 100 million.[3] Intervertebral disc degeneration underlies several painful low back disorders including intervertebral disc herniation (IVDH), degenerative spondylolisthesis (DS), spinal stenosis (SS), and degenerative disc disease (DDD). For the first three (IVDH, DS, SS), recent randomized prospective clinical studies have demonstrated advantages of surgical care compared with non-operative care.[3,4] However, DDD management remains the most difficult challenge because the underlying source of pain is unclear, causing uncertainty when developing guidelines for operative and non-operative care and therapies with improved efficacy.[5] As a result, estimates suggest there are between 1.5 and 4 million adults in the United States with DDD-related chronic low back pain (CLBP) that have failed conservative management and await therapeutic intervention, of which there are few options beyond spinal fusion.[4–6]

CLBP is a common indication for spine fusion surgery, and more recently total disc replacement. Although success rates from spinal fusion are in the range of 70%, it has several disadvantages including a decrease in spinal range of motion and acceleration of degeneration at adjacent levels.[7,8] Disc replacement (where a prosthetic disc is implanted between vertebrae) is an attractive alternative to fusion since it maintains near-physiologic movement, but there are currently barriers to widespread use, such as limited insurance reimbursement, potential surgical and implant-related complications, and the risk of adjacent-level degeneration.[9–11] Spine fusion will likely remain the standard of care for advanced disease, where arthritic changes prevent return of normal function even with newer motion-sparing technologies. The concept of minimally invasive biologic disc repair for less advanced cases of degeneration has grown in recent years. These include gene therapy, injection of various growth factors, and cell implantation (with or without scaffold).

2. The Arduous Intervertebral Disc Environment

The process of age-related disc degeneration can be considered chronic dysfunctional matrix remodeling in response to physical inputs such as impaired transport and/or abnormal mechanical loading, with the response to both likely modulated by yet undefined familial risk factors.[12] At an early age (before ten years) there is a marked decrease in endplate vascularity and beginnings of structural disorganization. After age 20, the disc becomes sealed off from the vertebral blood supply by the cartilage endplates and subchondral bone.[13,14] Thereafter, disc cell survival is dependent on diffusion from capillaries in the adjacent vertebra (for nuclear cells) and surrounding vascularized tissues (for annular cells).[15] The capillaries within the vertebra terminate just above the hyaline cartilage endplate, providing a continuous capillary bed across the bone disc interface.[9] Once nutrients reach the endplate, movement of small solutes (e.g. glucose and oxygen) pass through disc matrix primarily by diffusion.[15–19] Larger solutes may also be influenced by convective fluid flow created by mechanical disc compression and recovery. Cells compete for nutrition, making it difficult to sustain high cell densities at the long distances from the nutrition source typical of human lumbar discs (approximately 8 mm).[16–19]

This disc transport limitation affects the disc environment in several key ways. Tissue oxygen concentrations are low, in the range of 0.5–5%.[20] These hypoxic conditions inhibit matrix synthesis,[15,20] but nucleus pulposus cells (NPCs) may have adaptations for low-oxygen conditions that improve cell proliferation and survival.[21] Stem cells also may have some beneficial responses to hypoxia, such as increased chondrogenic differentiation and GAG synthesis.[22–25] Because of limited oxygen, the NPCs produce energy through anaerobic glycolysis, which utilizes glucose and generates lactic acid as a byproduct.[15,18] Accumulation of lactic acid decreases disc pH (to near pH 6.3) and is detrimental to the matrix as it decreases glycosaminoglycan production, tissue inhibitor of metalloproteinases (TIMP) production, and cell viability.[22,26] The dependence on anaerobic glycolysis for cell production of ATP makes glucose a critical nutrient. Disc glucose concentrations are typically considered to be in the range of 0.5–5 mM, where disc cells die within 24 hours at concentrations below 0.2 mM.[18] Other factors in the serum are also important as serum deprivation results in decreased cell proliferation and increased cell senescence.[21] Moreover, the influence of these factors is not necessarily independent: hypoxia supports nucleus cell survival during serum withdrawal and mediates cellular responses to changes in pH and growth factor concentrations.[21,22]

Another main feature of the disc nucleus is osmolality. The disc functions biomechanically by using a high osmotic pressure (generated by proteoglycan fixed charge) to attract water and produce physical pressure to support spinal compression. This osmotic stress (in the range of 250–500 mOsm) causes changes in cell volume and stimulates cell behavior via cytoskeleton rearrangement.[18] Some studies suggest that hyper-osmotic conditions downregulate matrix synthesis in NPCs,[19] but others have found that high osmolality may actually improve GAG content and mechanical properties of matrix synthesized by mature NPCs.[27] NPCs may also be adapted to high osmolarity because it is part of the normal disc environment, but stem cells may have more negative responses, such as decreased proliferation and proteoglycan synthesis.[28,29]

During disc degeneration, hypoxia, low glucose, and acidic pH of the disc are exacerbated, while osmolarity decreases due to GAG loss.[22,28] The degenerative disc environment also has a high concentration of pro-inflammatory cytokines and matrix fragments, which further impair regenerative potential.[6] The interaction of all these environmental factors and the response of both NPCs and stem cells affect the outcome of cell therapies in a degenerative microenvironment.

3. Evaluating Stem Cell-Based Therapy

3.1 *In vitro outcome measures and assessments*

Since disc degeneration is considered to initiate in the nucleus, regenerative strategies target nucleus regeneration. As a result, stem cell-based therapies focus their efforts on replicating the characteristics of the native NPCs. Stem cell performance is typically evaluated by characterizing gene expression and the matrix synthesis. In order to assess differentiation stage and the cell fate, the gene expression levels are measured for the positive chondrogenic marker Sox9, as well as several negative markers including the fibroblastic marker collagen I, the hypertrophic marker collagen X, and the osteogenic marker Runx2. Major matrix proteins aggrecan and collagen II are often evaluated to assess the cells' level of protein synthesis. In addition, catabolic genes often upregulated in degeneration, such as matrix metalloproteinases (MMPs) and disintegrins and metalloproteinases with thrombospondin motifs (ADAMTSs), are generally measured, in particular MMP-2, -9, and -13 and ADAMTS-1, −4, and −15.[30,31] Tissue inhibitors of metalloproteinases (TIMPs) also play an important part in balancing catabolic activity in the disc. Measuring the gene expression of TIMPs, particularly TIMP-1 and -3, can provide a more complete picture of cell activity.[31] At the protein level, proteoglycans are typically measured using a dimethylmethylene blue assay.[32,33] Extracellular matrix (ECM) deposition can be also be assessed qualitatively using histologic staining with Safranin-O or immunohistochemistry techniques.[32]

Biological efficacy should be demonstrated in models of increasing complexity. Two- and three-dimensional cell culture systems can be useful for initially demonstrating cellular effects, dosing, and toxicity. Three-dimensional systems are preferable to maintain cell phenotype; and augmenting stimuli including other disc mimetic conditions such as pressure, hypoxia, and inflammation are important as these factors can significantly influence cell function.[34] The degenerative disc conditions will vary with the degree of degeneration; however, the cells implanted in a human degenerative disc will generally experience hypoxia (about 4% O_2), high pressure (350 KPa at rest), inflammatory cytokines (particularly TNF-α and Il-1β), and a drop in pH (as low as pH 6.7). To simulate those conditions, a number of different methods can be used. The oxygen content can be controlled with a hypoxic incubator and a pressurized environment can be stimulated with the aid of a bioreactor. In addition, the inflammatory cytokines and pH levels can be replicated in the culture media. Therefore, the *in vivo* degenerative disc environment can be mimicked *in vitro* and should be considered during testing and optimization. A more comprehensive biochemical and mechanical environment can also be created through loaded organ culture of whole discs.[35] Recent studies have greatly improved cell survival in long-term organ culture models.[36]

3.2 *Models for efficacy and safety: In vivo pre-clinical models*

Ultimately, small animal studies are a critical next step because of the important *in situ* interactions between disc cells and spatially varying host features unique to the healing disc environment: pressure, hypoxia, degraded matrix, cytokines, other stromal and inflammatory cells, plus systemic factors. An increasingly complex model brings challenges in response interpretation. Outcome measures should be coupled to the designed treatment mechanisms, but should also include overall indices of disc quality such as histology and biomechanics. Time dependence of the outcomes is critical to establish whether the therapeutic response is persistent above the background degenerative response typically triggered by the therapy delivery. Design of experiments (DOE) techniques for study design and statistical analyses can help establish sample size and efficiently optimize treatment parameters.[37] Furthermore, it is also important to consider species and age differences, as well as the methodology for inducing degeneration, when interpreting the results of animal studies.[38]

Safety also needs to be established in pre-clinical models. The avascular nature of the disc environment can lead to persistence of active therapeutic agents, secreted cytokines, as well as carrier degradation products. Consequently, even though a growth factor or carrier has an established use track record in other tissues, they need to be evaluated in the unique NP environment. Adverse reactions can manifest through interdiscal toxicity, inflammatory cell recruitment, and matrix erosion. For example, cytokines and scaffold degradation products can diffuse from the disc and incite a sclerotic reaction in the adjacent vertebral endplates, along the delivery wound site, or outside the annulus fibrosus.[39] Equally important is to establish the reaction to extradiscal placement of the therapeutic materials and delivery vehicles. It is likely that these can escape from the disc during surgery or early in the post-surgical period. Inflammation and the mass effect induced by these materials can adversely affect adjacent nerve roots and other paraspinal tissues.[40] The safety of injected materials or cells should also be confirmed under the worst-case scenario to avoid a catastrophic event, such as the one seen for chemonucleolysis.[41]

After efficacy and safety are established *in vitro* and in small animals, large animal studies are required to motivate clinical use, principally because of size effects on disc transport[17] and biomechanics. Typical animals used for this purpose include goats, sheep, and mini-pigs. As with other pre-clinical models, efficacy may be difficult to establish due to a lack of relevant starting points and clinical metrics that match the intended patient population. Yet, biologic plausibility should be supported as well as safety through histological, biochemical, and biomechanical assays. Comparisons to negative controls (surgical procedure without treatment delivery) and untreated levels can help judge effect size and potential clinical relevance.

4. Non-Stem Cell-Based Regeneration Strategies

4.1 *Gene therapy and growth factors*

The most straightforward acellular biologic strategy is to inject growth factors into the disc. Several studies have reported *in vivo* effects of osteogenic protein-1 (OP-1), transforming growth factor β (TGF-β), growth differentiation factor 5 (GDF-5), and recombinant human bone morphogenetic protein-2 (rhBMP-2) that have led to an increase in disc height and GAG content and reduced degeneration.[42-44] Although promising results have been reported using these techniques, the relative acellularity of human degenerated discs raises the concern that the patient's own disc cells may be insufficient to mount a therapeutic response. Other concerns regarding growth factors include the short half-life of bioactive molecules in the NP, and poor growth factor retention under normal mechanical loading, especially in cases of annular tears.[42,45]

Both *in vivo* and *in vitro* gene therapies have shown promising results to treat disc degeneration.[46,47] Growth factors such as TGF-β, BMP2; transcription factor Sox9; inhibitors interleukin-1 receptor anatagonist (IL1Ra), and TIMPs have been successfully delivered to the NPCs. However, one risk is the activation of the host immune response due to the presence of viral vectors during gene therapy. Another approach that modulates cells at the gene level is RNA interference which is designed to silence genes that are the potential cause of disc degeneration without needing a viral delivery vehicle.[48]

Recently, several new types of gene editing technology have revolutionized the field of gene therapy, including zinc-finger nucleases (ZFN), transcription activator-like effector nucleases (TALENs), and clustered regularly interspaced short palindromic repeats (CRISPR-Cas).[49] In particular, CRISPR-Cas is a breakthrough technology that uses bacterial mechanisms to provide a simpler and more efficient method of gene editing in mammalian cells, including human stem cells.[50] CRISPR has been successfully used to edit genomes in human stem cells, and could therefore be a useful tool to optimize tissue engineering and repair.[51,52] In musculoskeletal tissues, a CRISPR-mediated knockdown of N-cadherin was used to study the protein's role in NPC phenotype and morphology.[53] CRISPR has also been used to target cytokine-mediated tissue degradation, which is a major issue in both osteoarthritis and intervertebral disc degeneration. Knocking down or repressing inflammatory cytokine receptors in induced pluripotent stem cells (iPSCS) and human adipose-derived stem cells (hADSCs) improved chondrogenesis under inflammatory conditions, suggesting that gene therapy could confer resistance to inflammation in engineered tissues.[54,55] However, these technologies are still under development, so a more nuanced understanding of their effects is still required before widespread clinical use.

4.2 *Autologous nucleus pulposus cells*

The introduction of cells capable of surviving within the intervertebral disc and producing appropriate amounts of matrix is an important component of disc tissue engineering. One type of cell considered is an autologous (derived from the patient) NPC cell-line. Even with some promising data in a canine model and clinical study (EuroDISC),[56,57] there is legitimate clinical concern over donor-site morbidity, since harvesting the patient's own cells requires damage to an adjacent disc, which will likely induce degeneration in that level. Also, disc acellularity will require a slow *in vitro* cell-culture expansion step to obtain sufficient cell numbers. Furthermore, autologous cells will be similarly aged to the diseased level and potentially limited in their ability to mount therapeutic repair response. One way to "activate" and improve the viability of autologous NPCs may be to co-culture them with stem cells prior to transplantation, a technique that is also currently undergoing clinical trials.[58]

4.3 *Biomaterials*

Biomaterials-based regeneration strategies fulfill multiple roles. When inserted into the disc space, they can preserve nucleus volume and defend against scar tissue encroachment by adjacent annular tissue during early healing. In particular, scaffolds and incompressible hydrogels can enhance acute biomechanical stability by providing resistance to compression. In addition to these biomechanical functions, biomaterials need to provide an environment that supports regenerative activity of either native disc cells or additional injected cells. Unfortunately, biomechanical and biological roles may create conflicting design constraints, with stiffer materials being more suitable for biomechanical retention and stability, and porous, pliant materials being more appropriate for nutrient transport and a three-dimensional milieu conducive to a disc cell phenotype. Importantly, these materials, which may or may not degrade over time, have to be synergistic with cell function over the long term. If non-degrading, they need to be biocompatible and non-migratory under complex loading/pressures. If degrading, the degradation kinetics should ideally be timed with cell matrix synthesis. Also, because of disc size and avascularity, degradation products may have a longer persistence and achieve higher concentrations than observed in other applications.[59]

Both hydrogels and solid scaffolds can be used to stimulate intervertebral disc (IVD) regeneration, and within these categories, both synthetic and natural materials have been explored. Examples include synthetic polymers such as poly(lactic-co-glycolic) acid (PLGA),[60] polyglycolic acid (PGA),[61] polylactide (PLA),[62] poly-N-isopropylacrylamide (pNIPAAM),[63] polyethylene glycol (PEG),[64] and polyurethane (PU).[65] Many natural hydrogel and scaffold materials are also promising, such as hyaluronan,[66] collagen/atelocollagen,[67] chitosan,[68] alginate,[69] agarose,[70] calcium polyphosphate,[71] demineralized bone particles,[72] fibrin sealant,[60,73] and small intestine submucosa (SIS)[74] — a natural ECM. Many of these materials can also be used in combination with each other, or as delivery systems for cells and bioactive agents such as growth factors, cytokine inhibitors, or antibiotics.[59] Using a biomaterial as a cell carrier or delivery vehicle can improve cell and protein retention in the high-pressure disc nucleus.

5. Stem Cells for Disc Repair

5.1 *Autologous versus allogenic*

Adult mesenchymal stem cells (MSCs) are attractive for disc tissue engineering since they can differentiate into a variety of cell types, including NPC-like cells. Depending on the therapy, the MSCs can be autologous or allogenic (derived from a donor). In the case of autologous transplantation, the patient would have a preliminary outpatient procedure where MSCs would be harvested from bone marrow or adipose tissue. Since MSCs represent only a small percentage of the cells in either of these donor tissues, the MSCs would need to be separated and expanded *in vitro* to have sufficient numbers desired for the therapy, with the second implantation procedure several weeks later. Clinical-scale expansion of MSCs may cause unexpected phenotypic changes and is subject to additional Food and Drug Administration (FDA) restrictions.[75,76] In the case of an allogenic transplant, the patient would be treated with MSCs from an organ donor in a one-procedure, cost-effective approach. While host rejection of allogenic cells is a concern, MSCs generally do not elicit a rejection response in transplantation studies. They were therefore thought to be immunoprivileged, but may actually be "immune evasive" and immunosuppressive.[77] In addition, factors such as age and gender of the donor can affect MSC gene expression and functional properties.[78]

5.2 *Differentiation of stem cells before implantation*

A primary concern regarding the use of MSCs for disc repair is whether they can survive the harsh *in vivo* conditions and appropriately differentiate *in situ*. Although the NPC lineage is not fully characterized, it is generally agreed to closely match that of chondrocytes.[79,80] While MSCs are known to readily differentiate into this cell type under controlled conditions *in vitro*, it must still be established whether they will spontaneously differentiate and thrive *in situ*, or alternatively require augmentation with supplemental differentiation factors.

Environmental cues can provide MSCs with important differentiation signals. Several studies have traced labeled MSCs implanted within discs and observed that they persist, integrate with host tissue, and differentiate over time.[81,82] Similarly, *in vitro* studies have shown that MSC cultured in three-dimensional scaffolds also exhibit some levels of differentiation.[83,84] Yet, several studies have shown greater proliferation and matrix deposition with MSCs that are first pre-differentiated or implanted along with stimulatory factors.[85,86] Consequently, the introduction of key bioactive molecules is commonly used to enhance desired MSC differentiation.

The use of adenoviral vectors encoding chondrogenic growth factors or anti-apoptotic genes in MSCs is one possible strategy to ensure the survival, differentiation, and sustained performance of implanted MSCs. Gene transfer therapy enables the sustained synthesis of the encoded bioactive transgene products that may be effective since these factors would not likely be present in a degenerative *in vivo* environment. The transfer of key genes, including TGF-β1 and BMP-2, to MSCs *in vitro* have led to the sustained upregulation of key matrix production and differentiation genes.[87,88] Transfected cells can be implanted alone or in combination with

274   *A. Ouyang et al.*

untransfected cells.[89] However, the use of viral vectors and genetically modified cells represent important safety hurdles when considered for clinical application.[86,90] These safety concerns will likely slow clinical adoption.

Exposing MSCs to key growth factors either before or during implantation is a more popular strategy. The most common approach is to culture MSCs in three-dimensional alginate bead culture with TGF-β1 supplemented media, where MSCs have robust differentiation and matrix synthesis.[91] This method is also convenient because the cells can easily be released from alginate using sodium citrate washes and re-implanted without compromising cell viability. Controlled release of growth factors such as TGF-β1 and other such molecules can also be incorporated into scaffolds that are seeded with cells and implanted.[92,93] The main concern regarding TGF-β-induced differentiation of MSCs is the progression of MSCs toward hypertrophy, whereby the MSCs begin secreting collagen X and MMP-13.[94] This is a concern since the mechanical properties of matrix secreted by hypertrophic cells do not match those desired for disc regeneration.[95]

5.3 *Co-culture*

Co-culture was originally used to explore interactions between implanted stem cells and host cells, but synergistic effects were observed, suggesting potential therapeutic applications. Currently, co-culture of MSCs with mature instructive cells has been shown to improve regeneration of many tissue types, including neural, cardiovascular, hepatic, renal, and musculoskeletal, and has been used in some clinical studies.[58,96]

Many studies have explored co-culture of MSCs with NPCs as well as with a similar cell type, articular chondrocytes. Experimental conditions such as cell density, cell type ratios, media conditions, culture time, and culture configurations vary widely among different studies, but many find common benefits of co-culture. These benefits include improved chondrogenesis and ECM production,[32,97–104] improved mechanical properties,[101] increased cell proliferation,[99,100,103,105] reduced hypertrophy,[97,106,107] and increased resilience to inflammation.[33,108,109] Furthermore, *in vivo* studies have verified co-culture effects in a more complex environment.[110–112]

However, observations from these studies have raised two key questions: first, whether benefits of co-culture are mediated by direct cell-cell contact or by released soluble factors, and second, whether MSCs play a synthetic or trophic role in tissue regeneration.

To determine if co-culture benefits are mediated by direct contact or soluble factors, various groups have experimented with direct and indirect co-culture systems. Watanabe *et al.* found that MSCs and NPCs in direct contact had increased proliferation and proteoglycan synthesis compared to cells cultured in the same well without direct contact,[103] which was similar to results observed with chondrocytes.[99,100,102] Direct co-cultures also had a reduced response to inflammation compared with indirect co-cultures.[109] Possible mechanisms suggested for cell-cell communication in direct co-cultures include spontaneous cell fusion, exchange of membrane components, and gap junctions.[102,113,114]

On the other hand, indirect co-culture systems and conditioned media also may increase cell proliferation and chondrogenesis.[105,112,115,116] Paracrine signaling in indirect co-cultures may be mediated by soluble growth factors: TGF-β1,-3; FGF-1,-4,-6; IGF-1; BMP-2; EGF; and PDGF have all been shown to increase in co-culture.[100,112,117,118]

Direct contact and soluble factors may also not be mutually exclusive mechanisms of interaction. Yamamoto *et al.* found that growth factor concentrations increased with both indirect and direct co-culture, but were higher in groups with direct contact.[117] Other studies have also found that closer proximity of co-cultured cell types increases matrix production.[101,119] Optimal tissue engineering outcomes may rely on a combination of contact-mediated mechanisms and improved paracrine signaling.

Three-dimensional pellets have been identified as a particular direct co-culture configuration that may enhance chondrogenesis. Pellet culture is commonly used to promote chondrogenic differentiation *in vitro*,[23] and even in single cell-type cultures, direct cell-cell contact between MSCs increases chondrogenesis.[24,120] While most co-culture pellets are formed from randomly mixed cells, a structured co-culture pellet may provide additional advantages and more closely mimic developmental processes. The bilaminar cell pellet (BCP) is a co-culture pellet composed of an inner sphere of MSC enclosed in an outer shell of NPC, allowing for homotypic interactions between cells of the same type within the layer and for

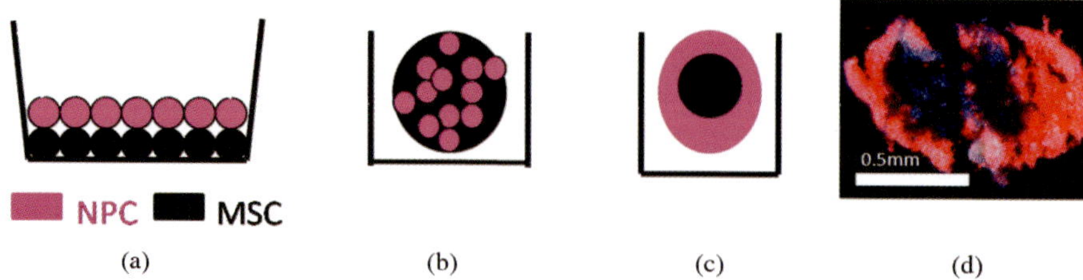

Fig. 2. (a) Co-culture in monolayer by Yamamoto[117] and Richardson[122] (b) Random three-dimensional co-culture pellet.[32,73,124] (c) Bilaminar co-culture pellet (BCP).[32,33,107,122] (d) Frozen section histology of BCP with MSC dyed with DiI (red).[122]

heterotypic interactions between different cell types across a defined interface (Fig. 2). This organization mimics the developmental processes of condensation, where cell aggregates form, and induction, where a mature layer of tissue directs the differentiation of a naïve one. The two cell types provide one another with stimulatory signaling, which allows for enhanced chondrogenesis and increased cell proliferation.[32,121] A similar bilaminar pellet with co-cultured MSCs and articular chondrocytes also exhibited less hypertrophy than MSC-only pellets.[107]

One interesting phenomenon observed in co-cultured pellets is self-organization. Both randomly organized and bilaminar pellets experienced budding of satellite pellets, and all satellite pellets had a bilayered structure with an MSC core and NPC outer layer, regardless of origin.[121] Self-organization with an MSC core was also observed in co-cultured pellets containing randomly mixed MSCs and primary chondrocytes.[99] We observed the same MSC core and NPC shell self-organization in large batches of randomly seeded co-cultured micropellets, and hypothesize that the self-organization may be mediated by differences in intracellular cohesivity among the two cell types (Fig. 3).[109] Cell type ratio may also influence the orientation of the organization.[104]

Another question that arises from recent co-culture studies is whether MSCs play a synthetic or trophic role in tissue regeneration. The traditional theory is that NPCs provide biochemical and/or physical cues to direct MSC differentiation into a chondrogenic phenotype. This is supported by upregulation of collagen II, aggrecan, and Sox9 gene expression without hypertrophy in MSCs after co-culture,[32,122,123] and chondrogenic

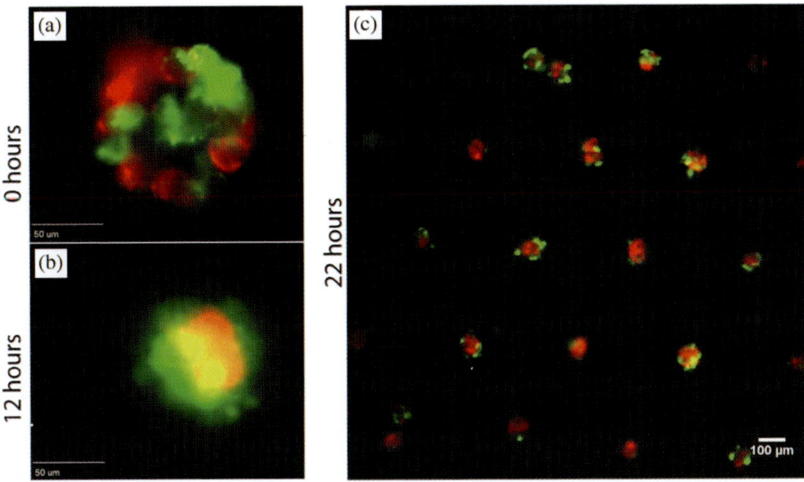

Fig. 3. Bilaminar self-organization in co-culture micropellets. MSCs (red) and NPCs (green) were labeled with DiI and DiO membrane dyes. Micropellets were seeded with a random 50:50 cell mixture (a), but condensed after 12 hours to form a bilaminar structure with an MSC core and NPC outer layer (b, c). Adapted from: A. Ouyang *et al.*, Effects of cell type and configuration on anabolic and catabolic activity in 3D co-culture of mesenchymal stem cells and nucleus pulposus cells, *J Orthop Res* **35**: 61–73 (2017).

phenotypes of MSCs injected into an *in vivo* disc environment.[124] Immunostaining in co-culture BCPs also suggests that while NPCs initially produce the majority of collagen II and aggrecan in the pellet, over time MSCs in the center of the pellets also synthesize ECM.[32,107] In co-culture with chondrocytes, MSCs also express chondrogenic markers and proteins, synthesize ECM, and exhibit chondrocyte-like morphology and surface markers.[101,110,112,115,125,126]

However, researchers are increasingly questioning the theory of MSC differentiation because MSCs often have a beneficial effect without detectable engraftment or differentiation.[127,128] In animal studies targeting IVD and cartilage regeneration, MSCs are retained well in subcutaneous studies,[110,112] but disappear or migrate to other regions when injected *in situ*.[104,129] Many *in vitro* studies have also observed a decrease in MSC-chondrocyte ratio over time with relatively constant overall cell numbers, suggesting that MSCs undergo cell death and are replaced by chondrocytes.[97,99,100,102] We observed a similar pattern in MSC and NPC co-cultures (unpublished data). In addition, several studies were unable to detect chondrogenic genes from MSCs, suggesting that they did not play a synthetic role in the co-culture.[97,99,130]

Although MSCs do not appear to persist and differentiate in the above examples, their initial presence still leads to increased matrix production and other benefits, suggesting that they have a trophic effect and provide stimulatory signals to co-cultured cells. In recent years, MSCs have been proposed to influence regeneration with bioactive factors rather than differentiation in many different tissues and organ systems.[131,132] Yang *et al.* theorized that MSCs injected into a mouse DDD model stimulated endogenous notochordal cells to prevent cell death and increase matrix production.[124] Other studies also found that MSCs induced cell proliferation of NPCs and chondrocytes in co-culture, possibly by secreting the growth factor FGF-1.[99,117,118]

Another trophic role of MSCs, downmodulating inflammatory responses, has particular significance in therapeutic settings.[133] MSC suppression of lymphocytes is well documented and thought to be mediated by cell-cell contact and secreted anti-inflammatory molecules such as prostaglandin E2 (PGE-2), TGF-β, and IL-10.[134,135] Although the nucleus pulposus is normally immune-privileged, disc degeneration is characterized by infiltration of immune cells and high concentrations of pro-inflammatory molecules such as TNF-α and IL-1β.[136,137] As part of the degenerative cascade, NPCs both secrete pro-inflammatory cytokines and respond to these cytokines by upregulation of catabolic factors, primarily MMPs and ADAMTS proteases. ECM degradation by these proteases compromises mechanical stability of the disc and continues the "vicious circle" of degeneration.[138]

MSCs have been shown to modulate a similar degenerative cascade in osteoarthritic chondrocytes by downregulating gene expression of inflammatory and catabolic factors.[108,116] When co-cultured with IVD fragments and lymphocytes, MSCs also reduced lymphocyte activation and proliferation in a donor-dependent manner.[139] When Allon *et al.* tested co-cultured BCPs in a simulated degenerative environment with hypoxia, pressure, and inflammatory cytokines, they also observed that MSCs played an immunomodulatory role. When cultured alone, NPCs were highly sensitive to the inflammatory environment and had reduced proteoglycan synthesis.

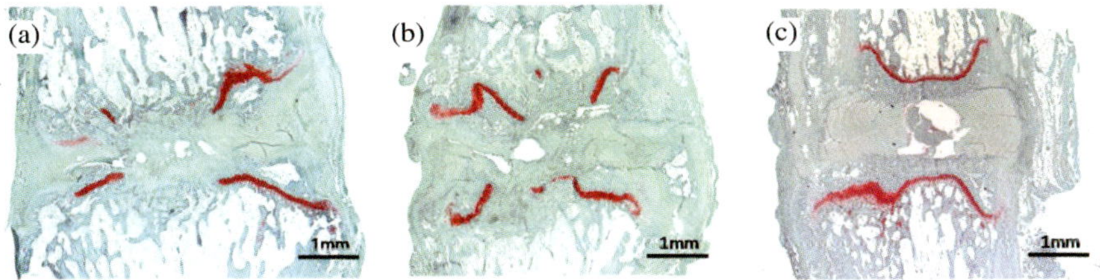

Fig. 4. Rat disc paraffin section histology with Safranin-O staining five weeks after surgery. (a) This disc was a control disc with no treatment. (b) This disc was injected with the fibrin carrier alone. In images a and b, the disc has collapsed and there is no proteoglycan in the disc space. The end plate and growth plate are severely disrupted. (c) This disc was treated with a BCP and fibrin carrier. The disc height is maintained with some proteoglycan staining in the disc space. The end plate and growth plate are both continuous. Adapted from: A. A. Allon *et al.*, Structured coculture of stem cells and disc cells prevent disc degeneration in a rat model, *Spine J* **10**: 1089–1097 (2010).

Co-cultured BCPs produced more proteoglycan than both NPC-only and MSC-only pellets, suggesting that they are a promising cell therapy for degenerative conditions.[33] BCPs also demonstrated resilience in an inflammatory environment in a rat degeneration model (Fig. 4) and in a similar study with chondrocytes.[107,111] We also observed that co-culture with MSCs downmodulates catabolic gene induction in an inflammatory and hypoxic microenvironment, and that the downmodulation is enhanced by direct contact in a micropellet configuration.[109]

As with direct contact and soluble factor signaling, MSC differentiation and trophic effects may not be mutually exclusive, and studies may find evidence of both.[32,100,107,124] Additionally, environmental factors such as growth factors, preconditioning, and mechanical stimuli can influence cell behavior and are difficult to standardize among studies. A clearer understanding of these complex variables is crucial toward developing an optimal strategy for disc repair.

6. Conclusion

The successful design of cell-based treatments for low back pain is confounded by ambiguities of disease and pain mechanisms in patients, and lack of consensus regarding ideal pre-clinical models. In particular, the primary clinical endpoint — pain relief — is currently not directly testable in animals. Yet, these therapies can be advanced by establishing biologic plausibility of efficacy and safety using models of increasing complexity, starting with cell culture, small animals (rats and rabbits), then large animals (goat and mini-pig) that more closely mimic nutritional, biomechanical, and surgical realities of human application. Ultimately, success will hinge on carefully designed clinical trials with well-defined patient selection criteria and objective outcome metrics that demonstrate significant benefits relative to gold-standard control treatments, such as spinal fusion.

References

1. J. C. Lotz, A. J. Fields and E. C. Liebenberg, The role of the vertebral end plate in low back pain, *Global Spine J* **3**: 153–164 (2013).
2. Y. S. Nosikova, J. P. Santerre, M. Grynpas, G. Gibson and R. A. Kandel, Characterization of the annulus fibrosus–vertebral body interface: identification of new structural features, *J Anat* **221**: 577–589 (2012).
3. J. N. Weinstein, J. D. Lurie, T. D. Tosteson, B. Hanscom, A. N. Tosteson, E. A. Blood, N. J. Birkmeyer, A. S. Hilibrand, H. Herkowitz, F. P. Cammisa, T. J. Albert, S. E. Emery, L. G. Lenke, W. A. Abdu, M. Longley, T. J. Errico and S. S. Hu, Surgical versus nonsurgical treatment for lumbar degenerative spondylolisthesis, *N Engl J Med* **356**: 2257–2270 (2007).
4. J. N. Weinstein, J. D. Lurie, T. D. Tosteson, A. N. Tosteson, E. A. Blood, W. A. Abdu, H. Herkowitz, A. Hilibrand, T. Albert and J. Fischgrund, Surgical versus nonoperative treatment for lumbar disc herniation: four-year results for the Spine Patient Outcomes Research Trial (SPORT). *Spine (Phila Pa 1976)* **33**: 2789–2800 (2008).
5. J. Schafer, D. O'Connor, S. Feinglass and M. Salive, Medicare Evidence Development and Coverage Advisory Committee Meeting on lumbar fusion surgery for treatment of chronic back pain from degenerative disc disease. *Spine (Phila Pa 1976)* **32**: 2403–2404 (2007).
6. L. J. Smith, N. L. Nerurkar, K. S. Choi, B. D. Harfe and D. M. Elliott, Degeneration and regeneration of the intervertebral disc: lessons from development, *Dis Model Mech* **4**: 31–41 (2011).
7. A. C. Disch, W. Schmoelz, G. Matziolis, Higher risk of adjacent segment degeneration after floating fusions: long-term outcome after low lumbar spine fusions, *J Spinal Disord Tech* **21**: 79–85 (2008).

8. J. H. Min, J. S. Jang, B. J. Jung, H. Y. Lee, W. C. Choi, C. S. Shim, G. Choiand and S. H. Lee, The clinical characteristics and risk factors for the adjacent segment degeneration in instrumented lumbar fusion, *J Spinal Disord Tech* **21**: 305–309 (2008).

9. R. D. Guyer and D.D. Ohnmeiss, Intervertebral disc prostheses, *Spine (Phila Pa 1976)* **28**: S15–23 (2003).

10. R. C. Huang and H. S. Sandhu, The current status of lumbar total disc replacement, *Orthop Clin North Am* **35**: 33–42 (2004).

11. J. J. Costi, B. J. C. Freeman and D. M. Elliott, Intervertebral disc properties: challenges for biodevices, *Expert Rev Med Devices* **8**: 357–376 (2011).

12. J. C. Lotz, A. Staples, A. Walsh and A. H. Hsieh, Mechanobiology in intervertebral disc degeneration and regeneration, *Conf Proc IEEE Eng Med Biol Soc* **7**: 5459 (2004).

13. A. L. Nachemson, J. C. A. Bjure, L. G. Grimby, Physical fitness in young women with idiopathic scoliosis before and after an exercise program, *Arch Phys Med Rehabil* **51**: 95–98 (1970).

14. S. Bernick and R. Cailliet, Vertebral end-plate changes with aging of human vertebrae, *Spine (Phila Pa 1976)* **7**: 97–102 (1982).

15. E. M. Bartels, J. C. Fairbank, C. P. Winlove and J. P. Urban, Oxygen and lactate concentrations measured *in vivo* in the intervertebral discs of patients with scoliosis and back pain, *Spine (Phila Pa 1976)* **23**: 1–7; discussion 8 (1998).

16. S. Holm, A. Maroudas, J. P. Urban, G. Selstam and A. Nachemson, Nutrition of the intervertebral disc: solute transport and metabolism, *Connect Tissue Res* **8**: 101–119 (1981).

17. J. C. Lotz, Animal models of intervertebral disc degeneration: lessons learned, *Spine (Phila Pa 1976)* **29**: 2742–2750 (2004).

18. T. Grunhagen, G. Wilde, D. M. Soukane, S. A. Shirazi-Adl and J. P. Urban, Nutrient supply and intervertebral disc metabolism, *J Bone Joint Surg Am* **88**: 30–35 (2006).

19. J. P. Urban, S. Smith and J. C. Fairbank, Nutrition of the intervertebral disc, *Spine (Phila Pa 1976)* **29**: 2700–2709 (2004).

20. J. W. Stairmand, S. Holm and J.P. Urban, Factors influencing oxygen concentration gradients in the intervertebral disc: a theoretical analysis, *Spine (Phila Pa 1976)* **16**: 444–449 (1991).

21. W. E. Johnson, S. Stephan and S. Roberts, The influence of serum, glucose and oxygen on intervertebral disc cell growth *in vitro*: implications for degenerative disc disease, *Arthritis Res Ther* **10**: R46 (2008).

22. S. M. Naqvi and C. T. Buckley, Bone marrow stem cells in response to intervertebral disc-like matrix acidity and oxygen concentration:

implications for cell-based regenerative therapy, *Spine* (*Phila Pa 1976*) **41**: 743–750 (2016).

23. X. Tang, L. Fan, M. Pei, L. Zeng and Z. Ge, Evolving concepts of chondrogenic differentiation: history, state-of-the-art and future perspectives, *Eur Cell Mater* **30**: 12–27 (2015).

24. B. Cao, Z. H. Li, R. Peng and J. D. Ding, Effects of cell–cell contact and oxygen tension on chondrogenic differentiation of stem cells, *Biomaterials* **64**: 21–32 (2015).

25. J. R. Choi, B. Pingguan-Murphy, W. A. Wan Abas, Impact of low oxygen tension on stemness, proliferation and differentiation potential of human adipose-derived stem cells, *Biochem Biophys Res Commun* **448**: 218–224 (2014).

26. H. A. Horner and J.P. Urban, 2001 Volvo Award Winner in Basic Science Studies: effect of nutrient supply on the viability of cells from the nucleus pulposus of the intervertebral disc, *Spine* (*Phila Pa 1976*) **26**: 2543–2549 (2001).

27. G. D. O'Connell, I. B. Newman and M. A. Carapezza, Effect of long-term osmotic loading culture on matrix synthesis from intervertebral disc cells, *Biores Open Access* **3**: 242–249 (2014).

28. K. Wuertz, K. Godburn, C. Neidlinger-Wilke, J. Urban and J. C. Iatridis, Behavior of mesenchymal stem cells in the chemical microenvironment of the intervertebral disc, *Spine* (*Phila Pa 1976*) **33**: 1843–1849 (2008).

29. U. Potocar, S. Hudoklin, M. E. Kreft, J. Završnik, K. Božikov and M. Fröhlich, Adipose-derived stem cells respond to increased osmolarities, *PLoS One* **11**: e0163870 (2016).

30. B. E. Bachmeier, A. Nerlich, N. Mittermaier, C. Weiler, C. Lumenta, K. Wuertz and N. Boos, Matrix metalloproteinase expression levels suggest distinct enzyme roles during lumbar disc herniation and degeneration, *Eur Spine J* **18**:1573–1586 (2009).

31. N. V. Vo, R. A. Hartman, T. Yurube, L. J. Jacobs, G. A. Sowa and J. D. Kang, Expression and regulation of metalloproteinases and their inhibitors in intervertebral disc aging and degeneration, *Spine J* **13**: 331–341 (2013).

32. A. A. Allon, K. Butcher, R. Schneider and J. C. Lotz, Structured coculture of mesenchymal stem cells and disc cells enhances differentiation and proliferation, *Cells Tissues Organs* **196**: 99–106 (2012).

33. A. A. Allon, K. Butcher, R. Schneider and J.C. Lotz, Structured bilaminar coculture outperforms stem cells and disc cells in a simulated degenerate disc environment, *Spine* (*Phila Pa 1976*) **37**: 813–818 (2012).

34. H. S. An and K. Masuda, Relevance of *in vitro* and *in vivo* models for intervertebral disc degeneration, *J Bone Joint Surg Am* **88**: 88–94 (2006).

35. B. Gantenbein, S. Illien-Jünger, S. C. W. Chan, J. Walser, L. Haglund, S. J. Ferguson, J. C. Iatridis and S. Grad, Organ culture bioreactors: platforms to study human intervertebral disc degeneration and regenerative therapy, *Curr Stem Cell Res Ther* **10**: 339–352 (2015).

36. M. Grant, L. M. Epure, O. Salem, N. AlGarni, O. Ciobanu, M. Alaqeel, J. Antoniou and F. Mwale, Development of a large animal long-term intervertebral disc organ culture model that includes the bony vertebrae for *ex vivo* studies, *Tissue Eng Part C Methods* **22**: 636–643 (2016).

37. H. Tye, Application of statistical "design of experiments" methods in drug discovery, *Drug Discov Today* **9**: 485–491 (2004).

38. C. Daly, P. Ghosh, G. Jenkin, D. Oehme and T. Goldschlager, A review of animal models of intervertebral disc degeneration: pathophysiology, regeneration, and translation to the clinic, *Biomed Res Int* (2016) [Epub ahead of print].

39. J. A. Ulrich, E. Liebenberg, D. Thuillier, ISSLS prize winner: repeated disc injury causes persistent inflammation, *Spine (Phila Pa 1976)* **32**: 2812–2819 (2007).

40. K. Olmarker and K. Larsson, Tumor necrosis factor alpha and nucleus-pulposus-induced nerve root injury, *Spine (Phila Pa 1976)* **23**: 2538–2544 (1998).

41. M. D. Brown, Update on chemonucleolysis, *Spine (Phila Pa 1976)* **21**: 62–68S (1996).

42. K. Masuda, Biological repair of the degenerated intervertebral disc by the injection of growth factors, *Eur Spine J* **17**: 441–451 (2008).

43. C. Feng, H. Liu, Y. Yang, B. Huang and Y. Zhou, Growth and differentiation factor-5 contributes to the structural and functional maintenance of the intervertebral disc, *Cell Physiol Biochem* **35**: 1–16 (2015).

44. H. Inoue, S. R. Montgomery, B. Aghdasi, The effect of bone morphogenetic protein-2 injection at different time points on intervertebral disk degeneration in a rat tail model, *J Spinal Disord Tech* **28**: E35–44 (2015).

45. K. Masuda, T. R. J. Oegema and H. S. An, Growth factors and treatment of intervertebral disc degeneration, *Spine (Phila Pa 1976)* **29**: 2757–2760 (2004).

46. K. Nishida, T. Suzuki, K. Kakutani, T. Yurube, K. Maeno, M. Kurosaka and M. Doita, Gene therapy approach for disc degeneration and associated spinal disorders, *Eur Spine J* **17**: 459–466 (2008).

47. B. Yue, Y. Lin, X. Ma, H. Xiang, C. Qiu, J. Zhang, L. Li and B. Chen, Survivin-TGFB3-TIMP1 gene therapy *via* lentivirus vector slows the course of intervertebral disc degeneration in an *in vivo* rabbit model, *Spine (Phila Pa 1976)* **41**: 926–934 (2016).

48. T. Suzuki, K. Nishida, K. Kakutani, K. Maeno, T. Yurube, T. Takada, M. Kurosaka and M. Doita, Sustained long-term RNA interference in nucleus pulposus cells *in vivo* mediated by unmodified small interfering RNA, *Eur Spine J* **18**: 263–270 (2009).

49. T. Gaj, C.A. Gersbach and C. F. R. Barbas, ZFN, TALEN, and CRISPR/Cas-based methods for genome engineering, *Trends Biotechnol* **31**: 397–405 (2013).

50. S. N. Waddington, R. Privolizzi, R. Karda and H. C. O'Neill, A Broad Overview and Review of CRISPR-Cas Technology and Stem Cells, *Curr Stem Cell Rep* **2**: 9–20 (2016).

51. F. González, CRISPR/Cas9 genome editing in human pluripotent stem cells: harnessing human genetics in a dish, *Dev Dyn* **245**: 788–806 (2016).

52. G. Schwank, B. K. Koo, V. Sasselli, J. F. Dekkers, I. Heo, T. Demircan, N. Sasaki, S. Boymans, E. Cuppen, C. K. van der Ent, E. E. Nieuwenhuis, J. M. Beekman and H. Clevers, Functional repair of CFTR by CRISPR/Cas9 in intestinal stem cell organoids of cystic fibrosis patients, *Cell Stem Cell* **13**: 653–658 (2013).

53. P. Y. Hwang, L. F. Jing, J. Chen, F. L. Lim, R. H. Tang, H. W. Choi, K. M. Cheung, M. V. Risbud, C. A. Gersbach, F. Guilak, V. Y. Leung and L. A. Setton, N-cadherin is key to expression of the nucleus pulposus cell phenotype under selective substrate culture conditions, *Sci Rep* **6** (2016).

54. J. M. Brunger, A. Zutshi, V. P. Willard, C. A. Gersbach and F. Guilak, CRISPR/Cas9 editing of induced pluripotent stem cells for engineering inflammation-resistant tissues, *Arthritis Rheumatol* (2016) [Epub ahead of print].

55. N. Farhang, J. M. Brunger, J. D. Stover, P. I. Thakore, B. Lawrence, F. Guilak, C. A. Gersbach, L. A. Setton and R. D. Bowles, CRISPR-based epigenome editing of cytokine receptors for the promotion of cell survival and tissue deposition in inflammatory environments, *Tissue Eng Part A* (2017) [Epub ahead of print].

56. H. J. Meisel, V. Siodla, T. Ganey, Y. Minkus, W. C. Hutton and O. J. Alasevic, Clinical experience in cell-based therapeutics: disc chondrocyte transplantation, *Biomol Eng* **24**: 5–21 (2007).

57. C. Hohaus, T. M. Ganey, Y. Minkus and H. J. Meisel, Cell transplantation in lumbar spine disc degeneration disease, *Eur Spine J* **17**: 492–503 (2008).

58. J. Mochida, D. Sakai, Y. Nakamura, T. Watanabe, Y. Yamamoto and S. Kato, Intervertebral disc repair with activated nucleus pulposus cell transplantation: a three-year, prospective clinical study of its safety, *Eur Cell Mater* **29**: 202–212 (2015).

59. E. M. Schutgens, M.A. Tryfonidou, T.H. Smit, F. Cumhur Öner, A. Krouwels, K. Ito and L.B. Creemers, Biomaterials for intervertebral disc regeneration: past performance and possible future strategies, *Eur Cell Mater* **30**: 210–231 (2015).

60. M. Sha'ban, S. J. Yoon, Y. K. Ko, H. J. Ha, S. H. Kim, J. W. So, R. B. Idrus and G. Khang, Fibrin promotes proliferation and matrix production of intervertebral disc cells cultured in three-dimensional poly(lactic-co-glycolic acid) scaffold, *J Biomater Sci Polym Ed* **19**: 1219–1237 (2008).

61. A. Abbushi, M. Endres, M. Cabraja, S. N. Kroppenstedt, U. W. Thomale, M. Sittinger, A. A. Hegewald, L. Morawietz, A. J. Lemke, V. G. Bansemer, C. Kaps and C. Woiciechowsky, Regeneration of intervertebral disc tissue by resorbable cell-free polyglycolic acid-based implants in a rabbit model of disc degeneration, *Spine (Phila Pa 1976)* **33**: 1527–1532 (2008).

62. H. Mizuno, A. K. Roy, C. A. Vacanti, K. Kojima, M. Ueda and L. J. Bonassar, Tissue-engineered composites of anulus fibrosus and nucleus pulposus for intervertebral disc replacement, *Spine (Phila Pa 1976)* **29**: 1290–1297; discussion 1297–1298 (2004).

63. N. Willems, H. Y. Yang, M. L. P. Langelaan, A. R. Tellegen, G. C. M. Grinwis, H. J. C. Kranenburg, F. M. Riemers, S. G. M. Plomp, E. G. M. Craenmehr, W. J. A. Dhert, N. E. Papen-Botterhuis, B. P. Meij, L. B. Creemers and M. A. Tryfonidou, Biocompatibility and intradiscal application of a thermoreversible celecoxib-loaded poly-N-isopropylacrylamide MgFe-layered double hydroxide hydrogel in a canine model, *Arthritis Res Ther* **17**: 214 (2015).

64. K. Benz, C. Stippich, C. Osswald, C. Gaissmaier, N. Lembert, A. Badke, E. Steck, W. K. Aicher and J. A. Mollenhauer, Rheological and biological properties of a hydrogel support for cells intended for intervertebral disc repair, *BMC Musculoskelet Disord* **13**: 54 (2012).

65. J. Hu, B. Chen, F. Guo, J. Du, P. Gu, X. Lin, W. Yang, H. Zhang, M. Lu, Y. Huang and G. Xu, Injectable silk fibroin/polyurethane composite hydrogel for nucleus pulposus replacement, *J Mater Sci Mater Med* **23**: 711–722 (2012).

66. M. Alini, W. Li, P. Markovic, M. Aebi, R. C. Spiro and P. J. Roughley, The potential and limitations of a cell-seeded collagen/hyaluronan scaffold to engineer an intervertebral disc-like matrix, *Spine (Phila Pa 1976)* **28**: 446–454; discussion 453 (2003).

67. D. Sakai, J. Mochida, T. Iwashina, T. Watanabe, K. Suyama, K. Ando and T. Hotta, Atelocollagen for culture of human nucleus pulposus cells forming nucleus pulposus-like tissue *in vitro*: influence on the proliferation and proteoglycan production of HNPSV-1 cells, *Biomaterials* **27**: 346–353 (2006).

68. X. Shao and C. J. Hunter, Developing an alginate/chitosan hybrid fiber scaffold for annulus fibrosus cells, *J Biomed Mater Res A* **82**: 701–710 (2007).

69. A. I. Chou and S. B. Nicoll, Characterization of photocrosslinked alginate hydrogels for nucleus pulposus cell encapsulation, *J Biomed Mater Res A* **91**: 187–194 (2009).

70. H. E. Gruber, G. L. Hoelscher, K. Leslie and E. N. Hanley, Three-dimensional culture of human disc cells within agarose or a collagen sponge: assessment of proteoglycan production, *Biomaterials* **27**: 371–376 (2006).

71. C. A. Seguin, M. D. Grynpas, R. M. Pilliar, S. D. Waldman, and R. A. Kandel, Tissue engineered nucleus pulposus tissue formed on a porous calcium polyphosphate substrate, *Spine (Phila Pa 1976)* **29**: 1299–1306; discussion 1306–1307 (2004).

72. S. H. Kim, S. J. Yoon, B. Choi, H. J. Ha, J. M. Rhee, M. S. Kim, Y. S. Yang, H. B. Lee and G. Khang, Evaluation of various types of scaffold for tissue engineered intervertebral disc, *Adv Exp Med Biol* **585**: 167–181 (2006).

73. H. E. Gruber, K. Leslie, J. Ingram, H. J. Norton and E. N. Hanley, Cell-based tissue engineering for the intervertebral disc: *in vitro* studies of human disc cell gene expression and matrix production within selected cell carriers, *Spine J* **4**: 44–55 (2004).

74. C. Le Visage, S. H. Yang, L. Kadakia, A. N. Sieber, J. P. Kostuik and K. W. Leong, Small intestinal submucosa as a potential bioscaffold for intervertebral disc regeneration, *Spine (Phila Pa 1976)* **31**: 2423–2430; discussion 2431 (2006).

75. G. R. Fajardo-Orduña, H. Mayani, M, E. Castro-Manrreza, E. Flores-Figueroa, P. Flores-Guzmán, L. Arriaga-Pizano, P. Piña-Sánchez, E. Hernández-Estévez, A. E. Castell-Rodríguez, A. K. Chávez-Rueda, M. V. Legorreta-Haquet, E. Santiago-Osorio and J. J. Montesinos, Bone marrow mesenchymal stromal cells from clinical scale culture: *in vitro* evaluation of their differentiation, hematopoietic support, and immunosuppressive capacities, *Stem Cells Dev* **25**: 1299–1310 (2016).

76. J. Zeckser, M. Wolff, J. Tucker and J. Goodwin, Multipotent mesenchymal stem cell treatment for discogenic low back pain and disc degeneration, *Stem Cells Int* **2016**: 1–13 (2016).

77. J. A. Ankrum, J. F. Ong and J. M. Karp, Mesenchymal stem cells: immune evasive, not immune privileged, *Nat Biotechnol* **32**: 252–260 (2014.)

78. G. Siegel, T. Kluba, U. Hermanutz-Klein, K. Bieback, H. Northoff and R. Schafer, Phenotype, donor age and gender affect function of human bone marrow-derived mesenchymal stromal cells, *BMC Med* **11**: 146 (2013).

79. C. R. Lee, D. Sakai, T. Nakai, K. Toyama, J. Mochida, M. Alini and S. Grad, A phenotypic comparison of intervertebral disc and articular cartilage cells in the rat, *Eur Spine J* **16**: 2174–2185 (2007).

80. J. Rutges, L. B. Creemers, W. Dhert, S. Milz, D. Sakai, J. Mochida, M. Alini and S. Grad, Variations in gene and protein expression in human nucleus pulposus in comparison with annulus fibrosus and cartilage cells: potential associations with aging and degeneration, *Osteoarthritis Cartilage* **18**: 416–423 (2010).

81. D. Sakai, J. Mochida, T. Iwashina, T. Watanabe, T. Nakai, K. Ando and T. Hotta, Differentiation of mesenchymal stem cells transplanted to a rabbit degenerative disc model: potential and limitations for stem cell therapy in disc regeneration, *Spine (Phila Pa 1976)* **30**: 2379–2387 (2005).

82. G. Crevensten, A. J. Walsh, D. Ananthakrishnan, P. Page, G. M. Wahba, J. C. Lotz and S. Berven, Intervertebral disc cell therapy for regeneration: mesenchymal stem cell implantation in rat intervertebral discs, *Ann Biomed Eng* **32**: 430–434 (2004).

83. H. J. Lee, C. Yu, T. Chansakul, N. S. Hwang, S. Varghese, S. M. Yu and J. H. Elisseeff, Enhanced chondrogenesis of mesenchymal stem cells in collagen mimetic peptide-mediated microenvironment, *Tissue Eng Part A* **14**: 1843–1851 (2008).

84. N. S. Hwang, S. Varghese and J. Elisseeff, Controlled differentiation of stem cells, *Adv Drug Deliv Rev* **60**: 199–214 (2008).

85. J. Wang, Y. Tao, X. Zhou, H. Li, C. Liang, F. Li and Q. Chen, The potential of chondrogenic pre-differentiation of adipose-derived mesenchymal stem cells for regeneration in harsh nucleus pulposus microenvironment, *Exp Biol Med (Maywood)* **241**: 2104–2111 (2016).

86. A. Wei, B. Shen, L. Williams and A. Diwan, Mesenchymal stem cells: potential application in intervertebral disc regeneration, *Transl Pediatr* **3**: 71–90 (2014).

87. M. R. Pagnotto, Z. Wang, J. C. Karpie, M. Ferretti, X. Xiao and C. R. Chu, Adeno-associated viral gene transfer of transforming growth factor-beta1 to human mesenchymal stem cells improves cartilage repair, *Gene Ther* **14**: 804–813 (2007).

88. A. J. Nixon, L. R. Goodrich, M. S. Scimeca, T. H. Witte, L. V. Schnabel, A. E. Watts and P. D. Robbins, Gene therapy in musculoskeletal repair, *Ann NY Acad Sci* **1117**: 310–327 (2007).

89. C. J. Xian and B.K. Foster, Repair of injured articular and growth plate cartilage using mesenchymal stem cells and chondrogenic gene therapy, *Curr Stem Cell Res Ther* **1**: 213–229 (2006).

90. C. J. Wallach, J. S. Kim, S. Sobajima, C. Lattermann, W. M. Oxner, K. McFadden, P. D. Robbins, L. G. Gilbertson and J. D. Kang, Safety assessment of intradiscal gene transfer: a pilot study, *Spine J* **6**: 107–112 (2006).

91. A. T. Mehlhorn, H. Schmal, S. Kaiser, G. Lepski, G. Finkenzeller, G. B. Stark and N. P. Südkamp, Mesenchymal stem cells maintain TGF-beta-mediated chondrogenic phenotype in alginate bead culture, *Tissue Eng* **12**: 1393–1403 (2006).

92. J. Sohier, D. Hamann, M. Koenders, M. Cucchiarini, H. Madry, C. van Blitterswijk, K. de Groot and J. M. Bezemer, Tailored release of TGF-beta1 from porous scaffolds for cartilage tissue engineering, *Int J Pharm* **332**: 80–89 (2007).

93. A. J. DeFail, C. R. Chu, N. Izzo and K. G. Marra, Controlled release of bioactive TGF-beta 1 from microspheres embedded within biodegradable hydrogels, *Biomaterials* **27**: 1579–1585 (2006).

94. M. B. Mueller, and R. S. Tuan, Functional characterization of hypertrophy in chondrogenesis of human mesenchymal stem cells, *Arthritis Rheum* **58**: 1377–1388 (2008).

95. K. Pelttari, A. Winter, E. Steck, K. Goetzke, T. Hennig, B. G. Ochs, T. Aigner and W. Richter, Premature induction of hypertrophy during *in vitro* chondrogenesis of human mesenchymal stem cells correlates with calcification and vascular invasion after ectopic transplantation in SCID mice, *Arthritis Rheum* **54**: 3254–3266 (2006).

96. N. K. Paschos, W. E. Brown, R. Eswaramoorthy, J. C. Hu and K. A. Athanasiou, Advances in tissue engineering through stem cell-based co-culture, *J Tissue Eng Regen Med* **9**: 488–503 (2015).

97. V. V. Meretoja, R. L. Dahlin, F. K. Kasper and A. G. Mikos, Enhanced chondrogenesis in co-cultures with articular chondrocytes and mesenchymal stem cells, *Biomaterials* **33**: 6362–6369 (2012).

98. K. Tsuchiya, G, Chen, T. Ushida, T. Matsuno and T. Tateshi, The effect of coculture of chondrocytes with mesenchymal stem cells on their cartilaginous phenotype *in vitro*, *Mater Sci Eng C* **24**: 391–396 (2004).

99. L. Wu, J. C. Leijten, N. Georgi, J. N. Post, C. A. van Blitterswijk and M. Karperien, Trophic effects of mesenchymal stem cells increase chondrocyte proliferation and matrix formation, *Tissue Eng Part A* **17**: 1425–1436 (2011).

100. C. Acharya, A. Adesida, P. Zajac, M. Mumme, J. Riesle, I. Martin and A. Barbero, Enhanced chondrocyte proliferation and mesenchymal stromal cells chondrogenesis in coculture pellets mediate improved cartilage formation, *J Cell Physiol* **227**: 88–97 (2012).

101. L. Bian, D. Y. Zhai, R. L. Mauck and J. A. Burdick, Coculture of human mesenchymal stem cells and articular chondrocytes reduces hypertrophy and enhances functional properties of engineered cartilage, *Tissue Eng Part A* **17**: 1137–1145 (2011).

102. T. S. de Windt, B. F. Saris Daniel, I. C. M. Slaper-Cortenbach, M. H. P. van Rijen, D. Gawlitta, L. B. Creemers, R. A. de Weger, J. A. Dhert Wouter and L. A. Vonk, Direct cell–cell contact with chondrocytes is a key mechanism in multipotent mesenchymal stromal cell-mediated chondrogenesis, *Tissue Eng Part A* **21**: 2536–2547 (2015).

103. T. Watanabe, D. Sakai, Y. Yamamoto, T. Iwashina, K. Serigano, F. Tamura and J. Mochida, Human nucleus pulposus cells significantly enhanced biological properties in a coculture system with direct cell-to-cell contact with autologous mesenchymal stem cells, *J Orthop Res* **28**: 623–630 (2010).

104. S. Sobajima, G. Vadala, A. Shimer, J. S. Kim, L. G. Gilbertson and J. D. Kang, Feasibility of a stem cell therapy for intervertebral disc degeneration, *Spine J* **8**: 888–896 (2008).

105. S. H. Yang, C. C. Wu, T. T. F. Shih, Y. H. Sun and F. H. Lin, *In vitro* study on interaction between human nucleus pulposus cells and mesenchymal stem cells through paracrine stimulation, *Spine* **33**: 1951–1957 (2008).

106. J. Fischer, A. Dickhut, M. Rickert and W. Richter, Human articular chondrocytes secrete parathyroid hormone-related protein and inhibit hypertrophy of mesenchymal stem cells in coculture during chondrogenesis, *Arthritis Rheum* **62**: 2696–2706 (2010).

107. M. E. Cooke, A. A. Allon, T. Cheng, A. C. Kuo, H. T. Kim, T. P. Vail, R. S. Marcucio, R. A. Schneider, J. C. Lotz and T. Alliston, Structured three-dimensional co-culture of mesenchymal stem cells with chondrocytes promotes chondrogenic differentiation without hypertrophy, *Osteoarthritis Cartilage* **19**: 1210–1218 (2011).

108. C. Manferdini, M. Maumus, E. Gabusi, A. Piacentini, G. Filardo, J. A. Peyrafitte, C. Jorgensen, P. Bourin, S. Fleury-Cappellesso, A. Facchini, D. Noël and G. Lisignoli, Adipose-derived mesenchymal stem cells exert antiinflammatory effects on chondrocytes and synoviocytes from osteoarthritis patients through prostaglandin E2, *Arthritis Rheum* **65**: 1271–1281 (2013).

109. A. Ouyang, A. E. Cerchiari, X. Tang, E. Liebenberg, T. Alliston, Z. J. Gartner and J. C. Lotz, Effects of cell type and configuration on anabolic and catabolic activity in 3D co-culture of mesenchymal stem cells and nucleus pulposus cells, *J Orthop Res* **35**: 61–73 (2017).

110. H. N. Yang, J. S. Park, K. Na, D. G. Woo, Y. D. Kwon and K. H. Park, The use of green fluorescence gene (GFP)-modified rabbit mesenchymal stem

cells (rMSCs) co-cultured with chondrocytes in hydrogel constructs to reveal the chondrogenesis of MSCs, *Biomaterials* **30**: 6374–6385 (2009).

111. A. A. Allon, N. Aurouer, B. B.Yoo, E. C. Liebenberg, Z. Buser and J. C. Lotz, Structured coculture of stem cells and disc cells prevent disc degeneration in a rat model, *Spine J* **10**: 1089–1097 (2010).

112. X. Liu, H. Sun, D. Yan, L. Zhang, X. Lv, T. Liu, W. Zhang, W. Liu, Y. Cao and G. Zhou, *In vivo* ectopic chondrogenesis of BMSCs directed by mature chondrocytes, *Biomaterials* **31**: 9406–9414 (2010).

113. C. C. Niu, L. J. Yuan, S. S. Lin, L. H. Chen and W. J. Chen, Mesenchymal stem cell and nucleus pulposus cell coculture modulates cell profile, *Clin Orthop Relat Res* **467**: 3263–3272 (2009).

114. S. Strassburg, N. W. Hodson, P. I. Hill, S. M. Richardson and J. A. Hoyland, Bi-directional exchange of membrane components occurs during co-culture of mesenchymal stem cells and nucleus pulposus cells, *PLoS One* **7**: e33739 (2012).

115. A. Aung, G. Gupta, G. Majid and S. Varghese, Osteoarthritic chondrocyte-secreted morphogens induce chondrogenic differentiation of human mesenchymal stem cells, *Arthritis Rheum* **63**: 148–158 (2011).

116. J. Platas, M. I. Guillén, M. D. Pérez del Caz, F. Gomar, V. Mirabet and M. J. Alcaraz, Conditioned media from adipose-tissue-derived mesenchymal stem cells downregulate degradative mediators induced by interleukin-1β in osteoarthritic chondrocytes, *Mediators Inflamm* **2013**: 357014 (2013).

117. Y. Yamamoto, J. Mochida, D. Sakai, T. Nakai, K. Nishimura, H. Kawada and T. Hotta, Upregulation of the viability of nucleus pulposus cells by bone marrow-derived stromal cells: significance of direct cell-to-cell contact in coculture system, *Spine (Phila Pa 1976)* **29**: 1508–1514 (2004).

118. L. Wu, J. Leijten, C. A. van Blitterswijk and M. Karperien, Fibroblast growth factor-1 is a mesenchymal stromal cell-secreted factor stimulating proliferation of osteoarthritic chondrocytes in co-culture, *Stem Cells Dev* **22**: 2356–2367 (2013).

119. J. H. Lai, G. Kajiyama, R. L. Smith, W. Maloney and F. Yang, Stem cells catalyze cartilage formation by neonatal articular chondrocytes in 3D bio-mimetic hydrogels, *Sci Rep* **3**: 3553 (2013).

120. L. Xie, M. Mao, L. Zhou and B. Jiang, Spheroid mesenchymal stem cells and mesenchymal stem cell-derived microvesicles: two potential therapeutic strategies, *Stem Cells Dev* **25**: 203–213 (2016).

121. A. A. Allon, R. A. Schneider and J. C. Lotz, Co-culture of adult mesenchymal stem cells and nucleus pulposus cells in bilaminar pellets for intervertebral disc regeneration, *SAS J* **3**: 41–49 (2009).

122. S. M. Richardson, R. V. Walker, S. Parker, N. P. Rhodes, J. A. Hunt, A. J. Freemont and J. A. Hoyland, Intervertebral disc cell-mediated mesenchymal stem cell differentiation, *Stem Cells* **24**: 707–716 (2006).

123. G. Vadala, R. K. Studer, G. Sowa, F. Spiezia, C. Iucu, V. Denaro. L. G. Gilbertson and J. D. Kang, Coculture of bone marrow mesenchymal stem cells and nucleus pulposus cells modulate gene expression profile without cell fusion, *Spine (Phila Pa 1976)* **33**: 870–876 (2008).

124. F. Yang, V. Y. Leung, K. D. Luk, D. Chan and K. M. Cheung, Mesenchymal stem cells arrest intervertebral disc degeneration through chondrocytic differentiation and stimulation of endogenous cells, *Mol Ther* **17**: 1959–1966 (2009).

125. W. H. Chen, M. T. Lai, A. T. Wu, C. C. Wu, J. G. Gelovani, C. T. Lin, S. C. Hung, W. T. Chiu and W. P. Deng, *In vitro* stage-specific chondrogenesis of mesenchymal stem cells committed to chondrocytes, *Arthritis Rheum* **60**(2): 450–459 (2009).

126. Y. H. Yang, A. J. Lee and G.A. Barabino, Coculture-driven mesenchymal stem cell-differentiated articular chondrocyte-like cells support neocartilage development, *Stem Cells Transl Med* **1**: 843–854 (2012).

127. D. J. Prockop, "Stemness" does not explain the repair of many tissues by mesenchymal stem/multipotent stromal cells (MSCs), *Clin Pharmacol Ther* **82**: 241–243 (2007).

128. E. M. Horwitz and W.R. Prather, Cytokines as the major mechanism of mesenchymal stem cell clinical activity: expanding the spectrum of cell therapy, *Isr Med Assoc J* **11**: 209–211 (2009).

129. N. Ozeki, T. Muneta, H. Koga, Y. Nakagawa, M. Mizuno, K. Tsuji, Y. Mabuchi, C. Akazawa, E. Kobayashi, K. Matsumoto, K. Futamura, T. Saito and I. Sekiya, Not single but periodic injections of synovial mesenchymal stem cells maintain viable cells in knees and inhibit osteoarthritis progression in rats, *Osteoarthritis and Cartilage* **24**: 1061–1070 (2016).

130. X. T. Mo, S. C. Guo, H. Q. Xie, L. Deng, W. Zhi, Z. Xiang, X. Q. Li and Z. M. Yang, Variations in the ratios of co-cultured mesenchymal stem cells and chondrocytes regulate the expression of cartilaginous and osseous phenotype in alginate constructs, *Bone* **45**: 42–51 (2009).

131. A. I. Caplan and J. E. Dennis, Mesenchymal stem cells as trophic mediators, *J Cell Biochem* **98**: 1076–1084 (2006).

132. P. R. Baraniak and T.C. McDevitt, Stem cell paracrine actions and tissue regeneration, *Regen Med* **5**: 121–143 (2010).

133. N. G. Singer and A.I. Caplan, Mesenchymal stem cells: mechanisms of inflammation, *Annu Rev Pathol* **6**: 457–478 (2011).

134. L. S. Meirelles, A. M. Fontes, D. T. Covas and A. I. Caplan, Mechanisms involved in the therapeutic properties of mesenchymal stem cells, *Cytokine Growth Factor Rev* **20**: 419–427 (2009).
135. A. I. Caplan and D. Correa, The MSC: an injury drugstore, *Cell Stem Cell* **9**: 11–15 (2011).
136. Z. Sun, M. Zhang, X. H. Zhao, Z. H. Liu, Y. Gao, D. Samartzis, H. Q. Wang and Z. J. Luo, Immune cascades in human intervertebral disc: the pros and cons, *Int J Clin Exp Pathol* **6**: 1009–1014 (2013).
137. M. V. Risbud and I.M. Shapiro, Role of cytokines in intervertebral disc degeneration: pain and disc content, *Nat Rev Rheumatol* **10**: 44–56 (2014).
138. P. P. Vergroesen, I. Kingma, K. S. Emanuel, R. J. Hoogendoorn, T. J. Welting, B. J. van Royen, J. H. van Dieën and T. H. Smit, Mechanics and biology in intervertebral disc degeneration: a vicious circle, *Osteoarthritis Cartilage* **23**: 1057–1070 (2015).
139. A. Bertolo, T. Thiede, N. Aebli, M. Baur, S. J. Ferguson and J. V. Stoyanov, Human mesenchymal stem cell co-culture modulates the immunological properties of human intervertebral disc tissue fragments *in vitro*, *Eur Spine J* **20**: 592–603 (2011).

12

CLINICAL APPLICATIONS OF A STEM CELL-BASED THERAPY FOR ORAL BONE RECONSTRUCTION

V. Thomas Eshraghi and Bradley McAllister

1. Introduction

We have observed a rapid evolution in both surgical techniques and materials utilized in the promotion of regenerative therapy for oral reconstructive procedures. In particular, bone augmentation has been promoted through different methods which include the use of growth and differentiation factors, particulate and block grafting materials,[1–4] distraction osteogenesis,[5–7] as well as membrane-assisted guided bone regeneration (GBR).[8] A central theme to all of these oral reconstructive technologies is the need for rapid and complete revascularization.

Early research that introduced us to the concept of using barrier membranes in what is referred to as GBR outlined the need for exclusion of undesirable soft tissue cellular contents and provision of a

293

secluded space into which osteogenic cells from various sources can migrate for successful bone healing.[9–11] The pattern of bone regeneration involves angiogenesis and ingress of osteogenic cells from the defect periphery toward the center to create a well-vascularized granulation tissue. This provides a scaffold for woven bone proliferation and bone apposition within the defect.[12] The size of the defect influences the bone healing capacity. In circumstances where the defect size is too large to generate a biomechanically stable central scaffold, bone formation will become limited to the marginal stable zone with a central zone of disorganized loose connective tissue. Critical to the outcome of GBR is maintenance of primary wound closure throughout the healing period.[13,14]

Perforation of the cortical bone layer has been advocated in GBR as it has been postulated that this will increase the vascularity of the wound and release growth factors and cells with angiogenic and osteogenic potential. Aggressive recipient bed preparation with decortication and intra-marrow penetration have also been supported due to increases in the rate of revascularization, the availability of osteoprogenitor cells, and the increased rate of remodeling.[12,15] In addition to the surrounding bone, the periosteum is also widely considered as an important source of cells with osteogenic capability. Despite the desirable graft containment and cellular exclusion characteristics of most barrier membranes, limiting the access of periosteal-derived osteogenic cells during the early phases of wound healing may not be of benefit to the healing of larger-sized defects.

The autograft, allograft, alloplast, and xenograft materials all have reported success either alone, or in combination, for particulate bone augmentation.[16] The particulate autograft is currently recognized as the gold standard for most bone grafting, including the treatment of dental implant-related defects. Several studies have demonstrated the effectiveness of particulate autograft due the availability of cells with osteogenic potential, as well as osteoinductive and osteoconductive properties.[17,18] However, autografts have recognized limitations, such as donor site morbidity, increased cost, potential resorption, size mismatch, an inadequate volume of graft material, as well as the unpredictability in the quantities of osteogenic precursor cells.

Allografts have the advantage of being available in higher quantities, and eliminate the morbidity associated with a second surgical site. Biochemical extraction techniques have shown that growth and differentiation factors are present in demineralized freeze-dried bone allograft (DFDBA) preparations; however, quantities have been shown to be variable from lot to lot indicating a potential variation in performance.[19–21] Thus, allografts primarily act as a scaffold for the in-growth of capillaries, perivascular tissues, and osteoprogenitor cells from the adjacent recipient bed.

The absence of differentiating precursor cells or osteoblasts in adequate quantities will ensue in limited bone formation. Osteoblasts contain the cellular machinery for production of bone matrix, but they are unable to undergo further division and have limited migratory capacity. This limits the expected benefits of cells contained in oral-derived autografts, for example, which exhibit high variability in the numbers of cells with osteogenic potential.

Based on our current understanding of graft healing and the prerequisites for optimal bone regeneration, tissue engineering research has been focused on providing the necessary cellular machinery, namely the mesenchymal stem cells (MSCs) and osteoprogenitor cells, directly in sites that require bone regeneration. It is this concept that has been utilized in the processing of the commercially available graft material that will be discussed here. Historically the majority of efforts for bone grafting with MSCs and osteoprogenitor cells have focused on the concept of harvesting these cells followed by *in vitro* culture expansion for later implantation.[22,23] The following methodology section describes a novel approach that leaves the MSCs and osteoprogenitor cells found within allogeneic bone and substantially depletes unwanted cells. To date, this product has been used successfully in oral regeneration, foot, ankle, and spinal surgery.[24–35]

2. Procurement Methodology for Stem Cell-Containing Allograft

One of the cellular allografts in this chapter (Osteocel®) is commercially prepared for NuVasive, Inc.™ from cadavers recovered by licensed tissue procurement agencies (AlloSource) and distributed into the dental market by ACE Surgical Supply. Cadaver tissues are rushed to the processing facility on wet ice and processing is begun within 24 hours of the donors' death. In parallel rigorous safety testing, donor screening and evaluation for bacterial, fungal, and spore contamination begin. Screening measures consist of physical examination and evaluation of both medical and social history, including a next-of-kin interview. Comprehensive serological and microbial testing are also performed which include nucleic acid testing (NAT) for hepatitis-C and HIV. Donor assessment culminates with a complete medical record review by a licensed physician. Cortical bone is separated and processed into demineralized bone particles for adding back to the cellular graft component. A process of selective immunodepletion, that involves several extensive wash steps, is initiated to remove undesirable cells, such as red blood cells and lymphocytes that can provoke an immune response. These unwanted cells are substantially depleted, leaving the remaining cell-rich cancellous bone matrix. The cellular cancellous bone component then undergoes a broad-spectrum anti-microbial treatment (vancomycin, gentamicin sulfate, and amphotericin) designed to eliminate potential contamination while preserving the viability of the cells. These remaining viable MSCs and osteoprogenitor cells remain attached to the cancellous bone matrix. Approximately 20% demineralized cortical bone particulate from the same donor is then combined to the cell-containing cancellous bone matrix component. A standard cryo-preservation solution containing 10% DMSO with human serum albumen (HSA) is added and the product is stored at −80 ±5 degrees Celsius (°C), permitting a five-year shelf life.

During the product processing validation, fluorescence-activated cell sorting (FACS) testing was performed to confirm the retention of MSCs that are positive for cluster of differentiation 105 (CD105) and CD166 while being negative for CD45.36 Fig. 1 depicts a representative FACS Scatter Plot of the cellular allograft. While there is not a single identifiable surface marker for MSCs, this marker combination profile is indicative of MSCs and osteoprogenitor cells. Quality testing is performed on every lot of Osteocel® to validate a minimum cell count of 50,000 cells/cc, and a minimum cellular viability of 70% of the enzymatically released cells. Another iteration of the cellular allograft product (Osteocel®Plus), which

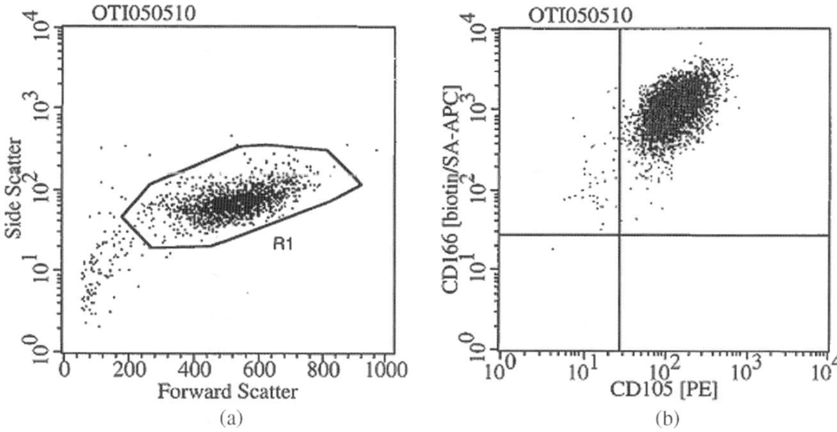

Fig. 1. Representative FACS Scatter Plot of Osteocel®. (a) Forward Scatter and Side Scatter dot plot from the donor 2 population of Osteocel®Plus cells with [R1] 94.44%, [R2] 99.93%, and [R1 + R2] 94.42%. (b) Dot plot showing positive expression of CD105 and CD166 markers, after gating for CD45- from the donor 2 cells. Quadrant gating UL 0.82%, UR 99.14%, LL 0.04%, and LR 0.00%.

was not utilized in the clinical and histological aspect of this chapter, has quality testing for a minimum cell count of 250,000 cells/cc. The cell count and viability are determined on released cells by a Trypan Blue dye exclusion test with a hemocytometer. Cellular osteogenic activity of each lot is also validated by performing *in vitro* cell differentiation and alkaline phosphatase assays (Fig. 2).

The cellular bone graft material is stored at −80°C and shipped to the clinic on dry ice where it is prepared as per the manufacturer's recommendation. Since the graft contains vital cells, the maximum temperature of the water bath used during the thawing process should not exceed 37°C. After the cryopreserved cells are thawed, the liquid is decanted and the cells are rinsed with sterile saline. The cell-containing graft is then ready for implantation, with a working window of four hours (Fig. 2). Depending on the defect treated, the particle size (1–3 mm) is often found to be too large for oral reconstructive procedures. In such instances, rongeurs can be used to carefully reduce the particle size.

Scanning electron microscopy (SEM) images from this particulate graft have consistently revealed the cellular component of the allograft

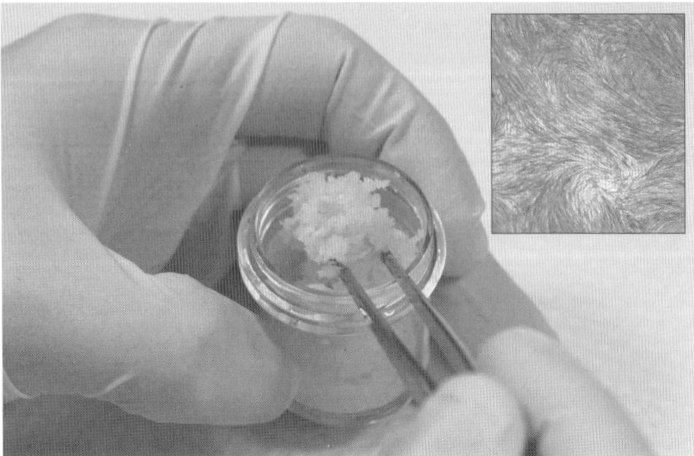

Fig. 2.   Macroscopic aspect of the Osteocel®product. Following the appropriate thawing of the cellular allograft, there is a recommended four-hour window for its use. (*Inset*) Cellular image from the Osteocel®derived cells demonstrating positive osteogenic activity (staining positive for alkaline phosphatase activity).

together with the extracellular matrix surrounding them. Figure 3 contains example SEMs at different magnifications. The SEM preparation starts with a 0.1M cacodylate buffer containing 5% sucrose rinse. The cells are fixed with 2.5% gluteraldehyde for one hour. Following three rinses with cacodylate buffer, the samples are soaked in 1% osmium tetraoxide solution in water for one hour at 4°C. The samples are subsequently dehydrated step-wise in increasing concentrations of ethanol starting at 50% and continuing to 100% for approximately two minutes at each step. A critical-point dryer is used to dry the samples. The bone particles are attached to the SEM plates with adhesive and silver paint followed by a gold/platinum sputter coating prior to imaging. Images were captured utilizing a Quanta Model 600 (FEI, Hillsboro, OR) with a Tungsten filament at high vacuum mode, and Soft Image Solutions (Olympus, Inc., Germany) software was employed for image collection.

3. Product Evolution

Similar to the product growth and configuration variety that the general allograft industry experienced years ago, the cellular allograft industry has evolved with at least six different companies making products and

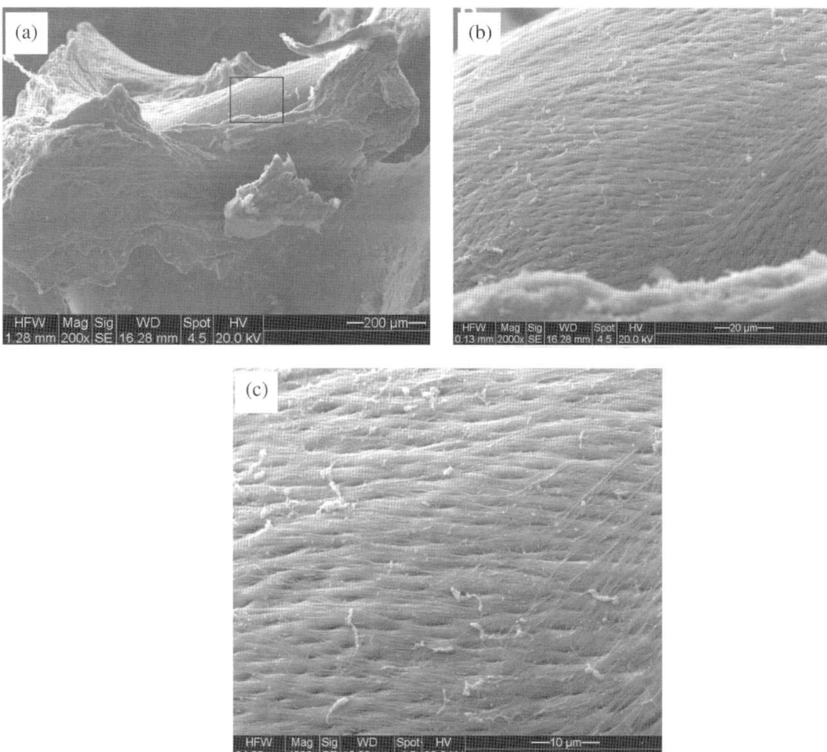

Fig. 3. (a) SEM image showing the cancellous bone coated with native cells. The cells are covered by extracellular matrix. (b) Higher magnification (2000×) of the box in (a). (c) Higher magnification (4000×).

approximately eight different configurations. The first cellular allograft which became available over 10 years ago, Osteocel® (Nuvasive, San Diego), came as a particulate graft material with a guaranteed cell count of 50,000 cells/cc. The follow-up product was Osteocel®Plus (Nuvasive, San Diego) which was the same particulate configuration with a guaranteed cell count of 250,000 cells/cc. The characterization paper by Jones *et al.*[36] demonstrated that native cell counts of a million cells/cc is possible, without any amplification process or cell addition. Osteocel®Plus has been characterized in more detail with immunofluorescent staining as described in the Neman publication.[37]

As a follow-up product, the putty form of Osteocel®Plus is now available under the label Osteocel®Pro (Nuvasive, San Diego).

This product contains an integrated cryopreservative and comes with 2.5 times the amount of demineralized bone which yields a putty characteristic which can improve surgical handling in certain applications.

Additional cellular allograft products which have come on the market in recent years include LifeNet's product Vivigen, Biomet's Cellentra, Orthofix products Trinity Evolution and Trinity Elite, the Osiris product Bio4 distributed by Stryker, and lastly, the Allosource Allostem product line. Trinity Elite is a more fibrillar version of the particulate product Trinity Evolution which allows for clumping of the graft material which can aid in the product handling, but has no improvement in the space maintenance.

Bio4 has the addition of periosteum which improves handling over particulate products, and having the periosteum adds to the cellular component. The addition of the periosteum also reportedly increases a variety of factors known to have a positive influence on angiogenesis. With the exception of Allostem, all of the described cellular allograft products use a cell depletion strategy where the unwanted cells are washed out of the cancellous cadaver bone marrow, leaving behind a bone matrix rich in mesenchymal stem cells, pre-osteoblasts, osteoblasts, and osteocytes. In contrast, Allostem uses a different minimal manipulation strategy where adipose-derived stem cells from the same cadaver are harvested and then added back into multiple configurations of bone products including particulate and demineralized bone sheets.

4. Ridge Augmentation

With the greater acceptance and awareness of dental implant therapy as the strongest method of tooth replacement, amongst practitioners and patients alike, it is not uncommon to encounter reconstructive scenarios that require bone augmentation in the overall treatment plan. This is particularly seen in cases with long-standing edentulism, trauma, and infection. Augmentation of the alveolar ridge for the ideal placement of an implant thus becomes necessary for an optimal esthetic outcome.[38] The various techniques and materials employed in ridge augmentation procedures have been discussed elsewhere and is beyond the focus of this chapter.[4] Of relevance, Osteocel®, as a cellular-based grafting material, provides an attractive option for ridge augmentation, particularly in larger defects.

It is hypothesized that the cellular contents within Osteocel® would benefit from a rapid revascularization. This revascularization process takes place at a much faster rate from the periosteal source than bone. To exploit this notion, use of a classic barrier membrane should ideally be avoided, when possible, to facilitate this process during the initial stages of healing. Other graft materials containing molecular enhancement products have likewise been shown in certain studies to undergo a slower healing rate when used in conjunction with a barrier membrane.[39,40] Figure 4 shows the

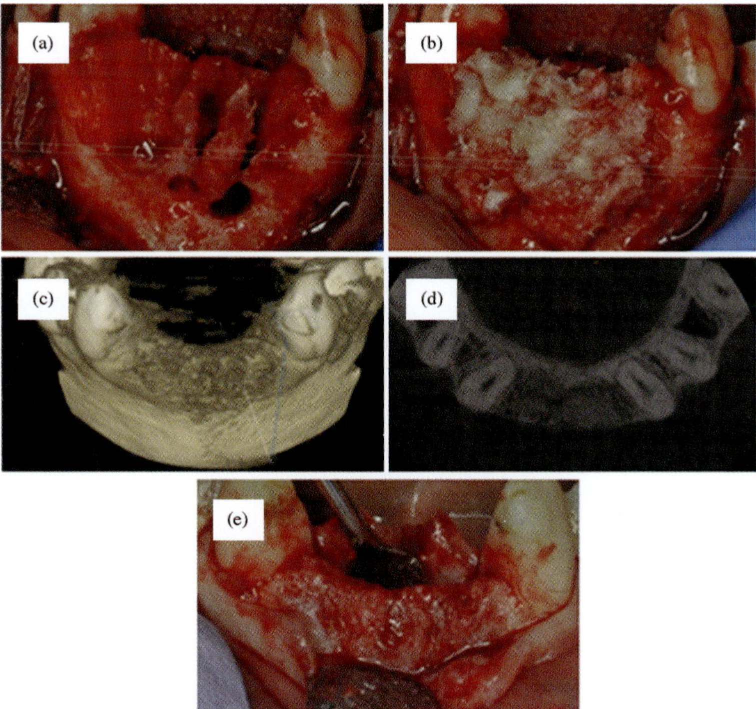

Fig. 4. Horizontal bone augmentation with Bio4 cellular allograft. (a) Clinical view showing the severe horizontal bone loss following removal of mandibular incisors. (b) Clinical view with cellular allograft in place. No membrane was utilized and the flap was sutured to tension-free primary closure with Vicryl™ sutures. (c–d) The four-month radiograph and CBCT rendering showing significant horizontal fill of the defect to the height of the residual lingual wall. (e) Clinical view showing the regeneration of 5–6 mm-thick horizontal augmentation.

successful use of Osteocel® alone for a small defect. The use of barrier membranes, however, may be deemed necessary when treating larger non-space-maintaining defects for graft containment. In larger-sized defects, as well as in cases where control of the location of the graft is critical, such as grafting against dental implants, a space-maintaining device in the form of a titanium mesh has been successfully employed.[4,41–43]

Titanium mesh offers resistance to bone graft collapse without the compromise in revascularization found with many products. However, when faced with a thinner tissue biotype, caution is necessary when using any non-resorbable device as ensuing soft tissue dehiscences can lead to a compromised outcome.[14] Recently Pieri and colleagues demonstrated titanium mesh use in combination with a mixture of intraoral autogenous bone and xenograft on 16 partially edentulous patients.[44] They reported a mean horizontal augmentation gain of 4.2 mm. Only one of the cases showed early exposure of the mesh device. Furthermore, titanium mesh and a cellular allograft have been described in the literature by this chapter's authors[43] and by Levine and McAllister.[45]

In cases where it was desired to regenerate more than 3 mm of horizontal bone, we have predictably used the graft in conjunction with a titanium mesh for both space maintenance and graft containment, preventing the cellular allograft from collapsing (Fig. 5).

While titanium mesh offers ideal space maintenance properties, the challenges associated with thin soft tissue biotypes have led to additional use of titanium-reinforced d-PTFE membranes. The soft membranes embed the titanium substructure, promoting easier handling and reducing soft tissue dehiscences; due to submicron (0.2 μm) porosity size, if exposed due to wound retraction, the density protects the underlying graft material and/or implant from bacterial contamination.[46] In cases with thin soft tissue biotype and the need for graft space maintenance and containment, we have successfully used d-PTFE membranes with cellular allografts (Fig. 6).

5. Sinus Augmentation

The posterior maxilla represents an area that has historically posed a challenge for treatment with dental implants. These range from comprising

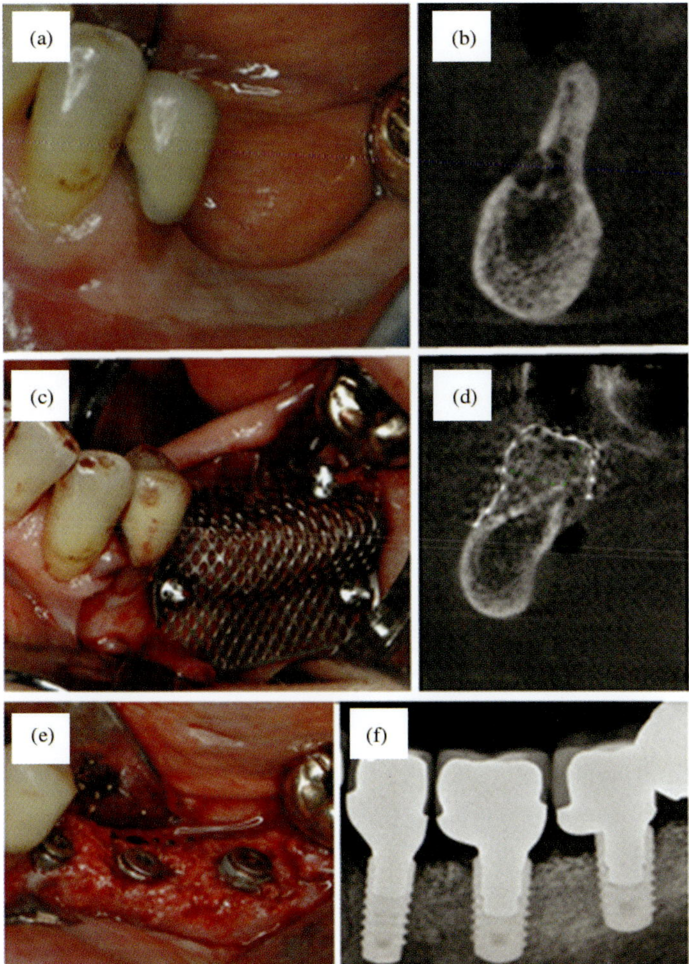

Fig. 5. Horizontal and vertical bone augmentation with titanium mesh and cellular allograft. (a) Preoperative clinical view illustrating severe vertical and horizontal defect above the mental bundle. (b) CBCT cross-section with deficient bone volume above the mental bundle. (c) Clinical view of shaped and fixated titanium mesh with cellular allograft. (d) UR-CBCT cross-section with outline of mesh and interior cellular allograft. (e) Clinical view of Straumann implants placed and significant horizontal gains of 4–5 mm. (f) One-year follow-up of restored implants with stable crestal bone levels.

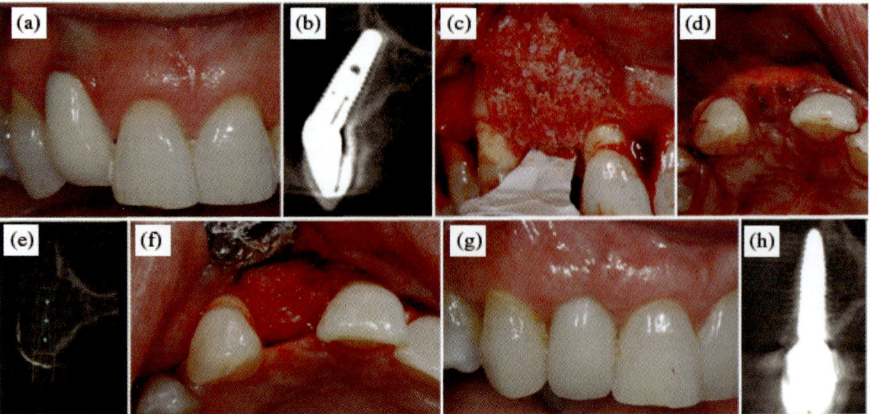

Fig. 6. A bone regeneration case utilizing a d-PTFE membrane and cellular allograft. Clinical (a) and radiographic (b) presentation prior to implant removal and bone augmentation. Note the lack of facial bone and severe angulation of implant. (c) Cellular allograft in position before final positioning of d-PTFE barrier over graft. (d) Note the thin tissue biotype and membrane show-through. (e) Four-month post-grafting CBCT and implant positioning analysis. (f) Four-month surgical re-entry post-membrane removal. (g) Final restoration three months post-implant placement. (h) CBCT following implant restoration.

sites with poor bone quality to unfavorable bucco-lingual resorption patterns and inadequacy in the vertical dimension of available bone following extraction of the teeth.[47,48] In addition, bone regeneration within the graft material is dependent upon it being populated by osteogenic cells that primarily originate from the osseous floors and walls, and to a smaller degree from the Schneiderian membrane.[49,50] Thus, cellular infiltration, vascularization, *de novo* bone formation, and graft replacement often require long healing times to produce bone of adequate quantity and quality for implant placement in the posterior maxilla.

Ongoing maxillary sinus pneumatization and normal post-extraction bone atrophy have been managed successfully by the sinus augmentation procedure either before or simultaneously with implant placement. The literature is inundated with reports describing this procedure using a variety of grafting materials. The use of a variety of materials has shown a varied bone quality and quantity with reported percentage bone areas that range from as low as 5% to over 40%.[51,52] In addition, studies have demonstrated that it can take in excess of nine months to achieve optimal bone formation for implant stability.[53,54]

Dental implant technology has endeavored for a faster osseointegration period to allow for more rapid restoration of the lost dentition. The concept of molecular enhancement of graft materials with either growth or differentiation factors for a more rapid regenerative outcome becomes desirable.[55,56] Cellular enhanced bone graft materials potentially offer this timing advantage as well. Our experiences with Osteocel® as a sinus augmentation bone graft material have consistently provided promising outcomes.[57] This initial report based on histomorphometric analysis of grafted sinuses with Osteocel® showed an average vital bone content of 33% (range 22–40%) and an average residual graft content of 6% (range 3–7%) for cases that had an average healing period of 4.1 months (range 3–4.75 months). These results were confirmed in a recent larger multi-center study.[58] A faster graft healing time with respect to new bone formation in adequate quantities has encouraged an earlier initiation of implant placement and restoration. A sinus augmentation case with Osteocel® is shown with the radiographic follow-up (Fig. 7) and histological evaluation after four months (Fig. 8).

6. Periodontal Bone Defects

Bone loss around teeth as a manifestation of periodontal disease continues to pose challenges for regenerative therapy. Three unique tissue types (cementum, periodontal ligament, and bone) require regeneration to satisfy replacement of the full natural tooth attachment apparatus.

Historically, guided tissue regeneration (GTR) utilizing non-resorbable and resorbable barrier membranes and bone grafting with or without additional growth factors such as bone morphogenic proteins (BMP), enamel matrix derivative (EMD), and platelet-derived growth factor (PDGF) have been used with varying levels of success.[59–64] Although these approaches have been shown to be successful in carefully selected cases, such as in two- or three-wall angular defects where greater vascularity exists, they remain a less predictable treatment option when faced with more compromised patterns of disease progression around natural teeth, such as class II or class III furcations, significant horizontal bone loss, and one-wall

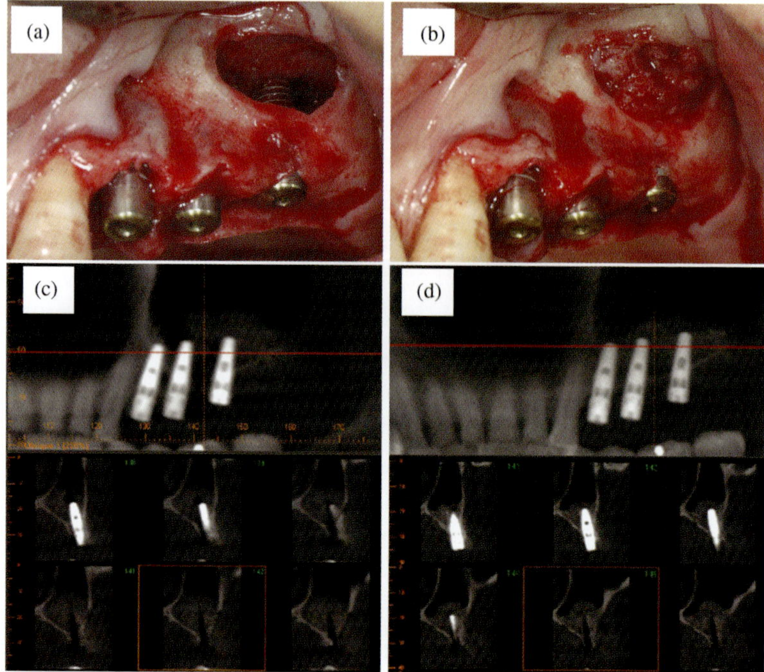

Fig. 7. Sinus augmentation with Osteocel®. (a) Sinus access after a classic lateral window approach with simultaneous implant placement. (b) Sinus after grafting with the cellular allograft. No membrane was used to cover the bone graft and lateral wall access window. (c) A CT scan of the grafted sinus immediately after graft and implant placement. (d) A CT scan of the same grafted sinus after four months of healing. Note the radiographic evidence of increased bone density in the bone graft region surrounding the dental implant. The radiographic findings of significant bone formation are consistent with the histologic data showing average percent bone areas in excess of 30%.

vertical bone defects.[64] These challenging defects have reduced vascularity because of the presence of the avascular tooth, and cellular recruitment is limited by the larger distance from the bony crest, which has been shown to reduce clinical outcomes.[65] It is not uncommon today to find that clinicians opt for removal of a large number of teeth exhibiting significant periodontal disease.[66]

Recently, we have demonstrated that allogeneic bone marrow-derived MSCs can enhance the regeneration of complex bone defects.[43,67] This has been further supported in current studies.[64,65] Given our understanding

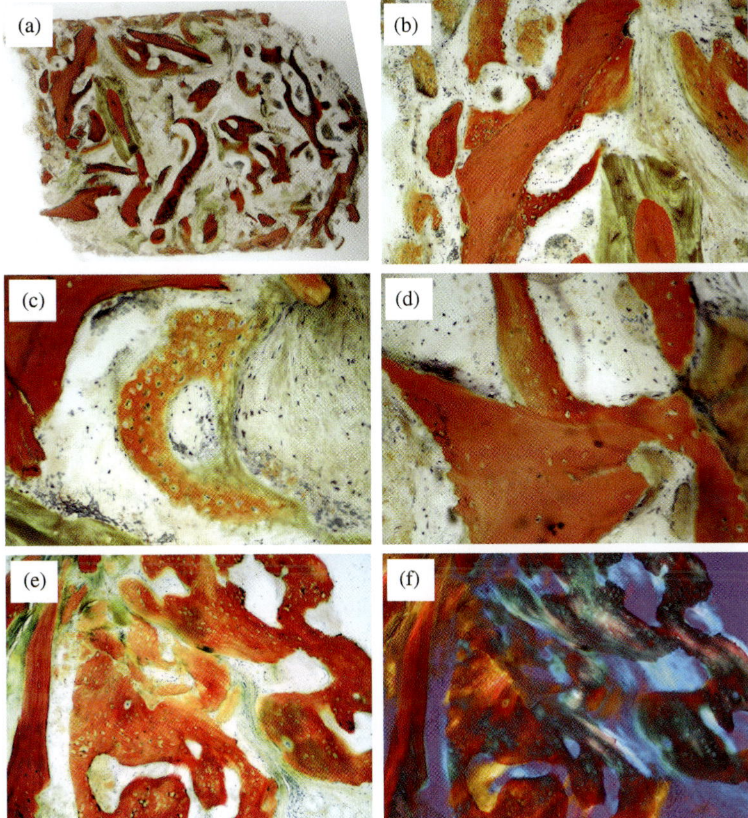

Fig. 8. Histological evaluation of the healed bone after sinus augmentation with cellular allograft residual. (a) Representative mineralized histologic core of a cellular allograft residual grafted sinus. The portion of the core shown goes from the most superior aspect (*left side*) to the original sinus floor (*right side*). It was harvested four months following the cellular allograft sinus grafting procedure. The red stained tissue is either the residual mineralized cellular allograft material (lighter red, osteocyte nuclei not always visible) or newly formed bone (darker red, osteocyte nuclei visible). The green stained tissue is the residual demineralized allograft material (no cells visible, non-vital bone). New bone formation can be appreciated throughout the core. Bone has formed directly on the residual mineralized and demineralized allograft particulate as well as in areas without residual graft material (original magnification of 20×). (b) New bone of varying levels of maturation can be appreciated in this higher magnification view (original magnification of 100×). (c) Osteoblasts can be seen lining the newly formed bone (original magnification of 200×). (d) Several areas of bone graft resorption can be appreciated in this view. This osteoclastic activity along with the low original packing density is likely responsible for

of stem cell-based therapy, the use of cellular allograft materials in potentially challenging periodontal defects may provide the armamentarium required for more predictable outcomes.[65,66,68–71]

While these are the first published reports employing allogeneic MSCs for the treatment of periodontal defects, different techniques of presenting autogenous-derived cellular bone grafts containing MSCs to periodontal defects have been reported previously.[72–75] Yamada *et al.*[72] described a technique where autogenous MSCs from iliac crest marrow aspirates were purified and expanded in the laboratory prior to surgical implantation into periodontal defects. A concern with this technology is the changes to the MSCs that have been reported during culture expansion.[76] Schallhorn *et al.*[73] and Dragoo and Irwin[74] harvested autogenous iliac crest marrow using a Turkel trephine and directly implanted the whole marrow containing MSCs and other cells. This technique showed a substantial correction in the periodontal defects treated.[73] Root resorption can be a negative outcome of this technique using whole marrow, especially in the presence of inflammation resulting from poor plaque control.[77,78]

The approach of using autogenous hip marrow bone grafts derived from aspirate techniques has four potentially significant disadvantages when compared to the cellular allograft material described previously.[79] First, the actual autogenous harvest procedures add significant discomfort, morbidity, and patient logistic difficulties. Second, the number of MSCs per unit volume is only around 1,000 cells/mL with the hip aspirate technique.[80–82] Third, the unprocessed autogenous hip marrow grafts contain cells of hematopoietic origin, which can result in osteoclastic activity.[80,81] Fourth, these unwanted cells, which comprise the majority of the cells present, compete for nutrients in the surgical wound healing environment.

Fig. 8. (*Continued*) the small percentage area that is the original cellular graft material. Multiple multinucleated cells can be seen (original magnification of 200×). (e) The difference in the level of bone maturity with areas of immature woven bone (*right side*) and areas of mature lamellar bone (*left side*) can be appreciated (original magnification of 100×). (f) Previous field (e) seen under polarized light. The demineralized bone areas (green) did not refract the polarized light to the extent the mineralized bone areas did. The difference in the level of bone maturity with areas of immature woven bone (*right side*) and areas of mature lamellar bone (*left side*) can also be appreciated with respect to the level of polarized light refraction (original magnification of 100×).

While positive results have been shown using autogenous hip marrow bone grafts for treatment of periodontal defects, these disadvantages have resulted in little use of this approach recently.

Cellular allografts, in contrast, eliminate the harvest requirement and undergo a selective immunodepletion process that results in a graft rich in MSCs and osteoprogenitor cells without cells of hematopoietic lineage. Due to procurement differences, root resorption should not be a problem.

In the recent studies utilizing cellular allografts, no ankylosis or root resorption was observed along with clinical efficacy of probing resistance and reduced signs of inflammation. This is suggestive of all three tissue types (cementum, periodontal ligament, and bone) of regeneration. The addition of stem cells to the periodontal defect is ideal because, in theory,

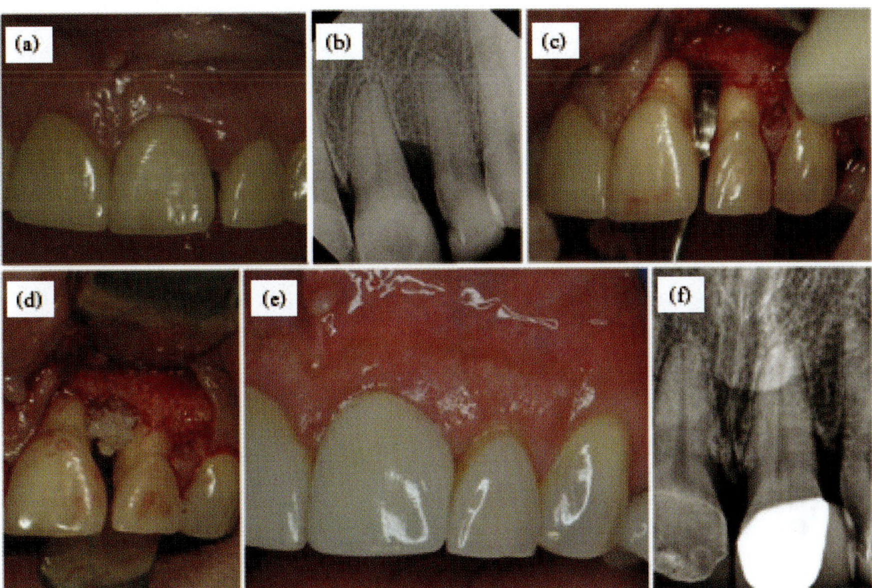

Fig. 9. (a) Clinical and (b) radiographic presentation prior to initiation of surgical therapy. (c) Following flap reflection, the horizontal bone loss between the maxillary left incisors was confirmed. (d) Placement of the cellular allograft after particle size reduction. (e) Clinical presentation at 36 months showing preservation of a significant amount of the interproximal soft tissue in spite of the original loss of supporting bone. (f) Radiographic presentation at 36 months showing approximately 4 mm of vertical bone growth in the interproximal area between the maxillary left incisors.

these cells can differentiate into all three critical tissue types to regenerate the lost tooth attachment apparatus without recruitment from the adjacent tissues.

A bone regeneration case of a significant horizontal bone defect is presented in Fig. 9.

7. Discussion

Stem cells can be derived from a variety of sources, including bone marrow, and are used in a variety of medical therapies. In the future, medical researchers anticipate being able to use technologies derived from stem cell research to treat a wider variety of systemic diseases, in addition to site-specific repair as was described in this chapter. To optimize these exciting applications for stem cells, a thorough characterization and understanding of stem cell biology will be required. Although stem cells can be isolated based on a distinctive set of cell surface markers, *in vitro* culture conditions can alter the behavior of cells, making it unclear whether the cells will behave in a similar manner *in vivo*. In fact, debate exists whether some proposed adult stem cell populations are truly stem cells. Because of their combined abilities of unlimited expansion and pluripotency, embryonic stem cells remain a theoretically viable source for regenerative medicine and tissue replacement after injury or disease. Differentiating embryonic stem cells into usable cells and ultimately organs is a challenge that tissue engineering researchers will face for years to come. Additionally, the use of embryonic stem cells is more controversial than adult stem cells.

Stem cells and progenitor cells act as a repair system for the body, not only replenishing specialized cells, but also maintaining the normal turnover of regenerative organs and tissues. In spite of this important function, pluripotent adult stem cells are rare and generally small in number within the body, with most being lineage-restricted (multipotent). They have also been shown to decrease in number with age. Bone marrow contains numerous cell types from both the hematopoietic stem cell lineage (for example platelets, osteoclasts) and the non-hematopoietic stem cell lineage (for example MSCs and osteoblasts).[81] During the Osteocel® procurement procedures, immunogenic cells and tissues are substantially depleted. The potential for immune response is why typical fresh-frozen bone allografts are not as attractive of a grafting option, even though in a few reports success for oral reconstructive procedures has been demonstrated. One recent study on 21 patients reported high success with dental implant placement into fresh-frozen bone allograft regenerated bone.[83] Since only a disinfection process is performed with typical fresh-frozen bone allografts and not a complete immunodepletion process, there are likely large numbers of undesirable cells remaining in the graft. This could potentially impact the graft performance and the host immune response.

For multiple reasons, MSCs enjoy a hypoimmunogenic host response. MSCs lack MHC-II and co-stimulatory molecule expression. MSCs are also immunomodulatory in that they prevent T-cell responses indirectly through modulation of dendritic cells and directly by disrupting NK as well as CD8+ and CD4+ cell function.[84] MSCs also induce a suppressive local microenvironment through cytokine production (prostaglandins and interleukin-10) and expression of indoleamine 2,3-dioxygenase which depletes the local milieu of tryptophan. One of the major proteins produced by MSCs is transforming growth factor-beta (TGF-beta) that regulates the host T-cells by promoting T-regulatory cells.

The cellular content of autogenous bone grafts varies based on the individual patient's medical profile, the harvest technique (aspiration or open harvest), anatomic location of the harvest (intraoral or extraoral), type of bone harvested (cortical or cancellous), age and gender. The cellular content has also been shown to have an effect on the bone graft performance.[85] Therefore, the identification of MSCs and osteoprogenitor cells and determination of their concentrations in different anatomical tissues have been an area of recent investigation.[79,80,86] Evaluation of bone marrow aspirates from the anterior iliac crest revealed a fairly small count of MSCs, although a higher percentage of cells that tested positive for CD105 was found in the iliac crest aspirates when compared to peripheral blood.[80] McLain and colleagues compared the osteoprogenitor cell concentrations between iliac crest and vertebral body aspirates.[86] Their findings show that vertebral aspirates (465 cells/cc marrow) have a slightly higher mean concentration than the iliac crest aspirates (356 cells/cc marrow). The process used to prepare the cellular allograft bone matrix described in this chapter involves the selective removal of immunogenic cells in hematopoietic lineage from cell-rich cancellous bone, while retaining the osteopotent cells in the mesenchymal lineage. The minimum number of MSCs and osteoprogenitor cells found in the commercially available Osteocel® product is 50,000 cells/cc and for Osteocel®Plus product is 250,000 cells/cc. Clearly these cell counts are a dramatic improvement over the cell counts for aspiration harvests and may result in an enhanced clinical result. Recently Cuomo and colleagues have conducted a pre-clinical trial evaluating MSCs from human bone marrow aspirates in combination with demineralized bone matrix in athymic rat femur critical size defects.[82] Unprocessed MSC concentrations were found to vary from 64 to 2933 cells/ml with an average of 1010 +/−960 cells/ml as determined from fibroblast colony forming unit (CFU-F) culture assays. MSC-enriched bone marrow aspirate (centrifugation concentration) improved the yield to an average of 6150 cells/ml. Interestingly, the bone-forming capabilities were found to be only comparable to the demineralized carrier alone. The authors mentioned that the cell number may be insufficient to stimulate a robust bone formation. The clinical ramification of cell number has also been discussed by Hernigou and colleagues when treating tibia non-unions with marrow aspirates. They have identified 30,000 cells/ml as the minimum number of progenitor cells necessary to induce healing in this indication.[87] A recent comparative study demonstrated that 1 gm of cellular allograft had the same cell content as 45 ml of bone marrow aspirate.[88] Another possibility is that the MSCs have not had influence on the healing. All of these considerations could have clinical implications when performing oral reconstructive surgery with bone marrow aspirates as has been presented by some clinicians.[80]

This exciting new cellular allograft technology will further assist us in the management of challenging oral regenerative procedures. Ongoing research is aimed at optimization of the clinical techniques and determining the long-term success of their application.

Acknowledgments

The Osteocel® sinus augmentation research project was supported by Ace Surgical, Brockton, MA. The authors gratefully acknowledge the assistance of Hari Prasad, Senior Research Scientist, Hard Tissue Research

Laboratory, University of Minnesota School of Dentistry, and Dr. Michael Rohrer, Professor, Director of the Hard Tissue Research Laboratory and Oral Pathology Laboratories, University of Minnesota, School of Dentistry, for the preparation of the specimens and the histological data.

The authors would also like to thank Dr. Tim Moseley, Chief Scientist NuVasive Inc.™, for the technical assistance in preparing this chapter.

References

1. C. M. Misch, Comparison of intraoral donor sites for onlay grafting prior to implant placement, *Int J Oral Maxillofac Implants* **12**: 767–776 (1997).

2. M. A. Pikos, Block autografts for localized ridge augmentation: Part I. The posterior maxilla, *Implant Dent* **8**: 279–285 (1999).

3. M. A. Pikos, Block autografts for localized ridge augmentation: Part II. The posterior mandible, *Implant Dent* **9**: 67–75 (2000).

4. B. S. McAllister and K. Haghighat, Bone augmentation techniques, *J Periodontol* **78**: 377–396 (2007).

5. M. S. Block, A. Chang and C. Crawford, Mandibular alveolar ridge augmentation in the dog using distraction osteogenesis, *J Oral Maxillofac Surg* **54**: 309–314 (1996).

6. G. A. Ilizarov, The tension-stress effect on the genesis and growth of tissues: Part I. The influence of stability of fixation and soft-tissue preservation, *Clin Orthop Relat Res* **238**: 249–281 (1989).

7. B. S. McAllister and T. E. Gaffaney, Distraction osteogenesis for vertical bone augmentation prior to oral implant reconstruction, *Periodontology 2000* **33**: 54–66 (2003).

8. J. T. Mellonig and M. Nevins, Guided bone regeneration of bone defects associated with implants: an evidence-based outcome assessment, *Int J Periodontics Restorative Dent* **15**: 168–185 (1995).

9. D. Buser, K. Dula, U. Belser, H. P. Hirt and H. Berthold, Localized ridge augmentation using guided bone regeneration. 1. Surgical procedure in the maxilla, *Int J Periodontics Restorative Dent* **13**: 29–45 (1993).

10. D. Buser, K. Dula, U. C. Belser, H. P. Hirt and H. Berthold, Localized ridge augmentation using guided bone regeneration. II. Surgical procedure in the mandible, *Int J Periodontics Restorative Dent* **15**: 10–29 (1995).

11. C. Dahlin, J. Gottlow, A. Linde and S. Nyman, Healing of maxillary and mandibular bone defects using a membrane technique: an experimental study in monkeys, *Scand J Plast Reconstr Surg Hand Surg* **24**: 13–19 (1990).

12. R. K. Schenk, D. Buser, W. R. Hardwick and C. Dahlin, Healing pattern of bone regeneration in membrane-protected defects: a histologic study in the canine mandible, *Int J Oral Maxillofac Implants* **9**: 13–29 (1994).

13. E. E. Machtei, The effect of membrane exposure on the outcome of regenerative procedures in humans: a meta-analysis, *J Periodontol* **72**: 512–516 (2001).

14. M. Simion, M. Baldoni, P. Rossi and D. Zaffe, A comparative study of the effectiveness of e-PTFE membranes with and without early exposure during the healing period, *Int J Periodontics Restorative Dent* **14**: 166–180 (1994).

15. P. S. de Carvalho, L. W. Vasconcellos and J. Pi, Influence of bed preparation on the incorporation of autogenous bone grafts: a study in dogs, *Int J Oral Maxillofac Implants* **15**: 565–570 (2000).

16. J. O. Hollinger, J. Brekke, E. Gruskin and D. Lee, Role of bone substitutes, *Clin Orthop Relat Res* **324**: 55–65 (1996).

17. R. Mowlem, Cancellous chip bone grafts: report on 75 cases, *Lancet* **2**: 746–748 (1944).

18. J. B. Mulliken and J. Glowacki, Induced osteogenesis for repair and construction in the craniofacial region, *Plast Reconstr Surg* **65**: 553–560 (1980).

19. P. V. Hauschka, A. E. Mavrakos, M. D. Iafrati, S. E. Doleman and M. Klagsbrun, Growth factors in bone matrix: isolation of multiple types by affinity chromatography on heparin-Sepharose, *J Biol Chem* **261**: 12665–12674 (1986).

20. T. K. Sampath, N. Muthukumaran and A. H. Reddi, Isolation of osteogenin, an extracellular matrix-associated, bone-inductive protein, by heparin affinity chromatography, *Proc Natl Acad Sci USA* **84**: 7109–7113 (1987).

21. Y. Shigeyama, J. A. D'Errico, R. Stone and M. J. Somerman, Commercially prepared allograft material has biological activity *in vitro*, *J Periodontol* **66**: 478–487 (1995).

22. R. Malekzadeh, J. O. Hollinger, D. Buck, D. F. Adams and B. S. McAllister, Isolation of human osteoblast-like cells and *in vitro* amplification for tissue engineering, *J Periodontol* **69**: 1256–1262 (1998).

23. S. P. Bruder, K. H. Kraus, V. M. Goldberg and S. Kadiyala, The effect of implants loaded with autologous mesenchymal stem cells on the healing of canine segmental bone defects, *J Bone Joint Surg Am* **80**: 985–996 (1998).

24. R. T. Scott and C. F. Hyer, Role of cellular allograft containing mesenchymal stem cells in high-risk foot and ankle reconstructions, *J Foot Ankle Surg* **1**: 32–35 (2013).

25. J. M. Ammerman, J. Libricz and M. D. Ammerman, The role of osteocel plus as a fusion substrate in minimally invasive instrumented transforaminal lumbar interbody fusion, *Clin Neurol Neurosurg* **7**: 991–994 (2013).

26. J. R. Bryant, R. Eid, J. C. Spann and A. S. Miller 3rd, Chest wall reconstruction with creation of neoribs using mesenchymal cell bone allograft and porcine small intestinal submucosa, *Plast Reconstr Surg* **3**: 148–149e (2010).

27. B. S. McAllister and T. E. Gaffaney, Distraction osteogenesis for vertical bone augmentation prior to oral implant reconstruction, *Periodontology 2000* **33**: 54–66 (2003).

28. S. M. Hollawell, Allograft cellular bone matrix as an alternative to autograft in hindfoot and ankle fusion procedures, *J Foot Ankle Surg* **51**: 222–225 (2012).

29. S. M. Rush, G. A. Hamilton and L. M. Ackerson, Mesenchymal stem cell allograft in revision foot and ankle surgery: a clinical and radiographic analysis, *J Foot Ankle Surg* **48**: 163–169 (2009).

30. T. A. Brosky, C. R. Menke and D. Xenos, Reconstruction of the first metatarsophalangeal joint following post-cheilectomy avascular necrosis of the first metatarsal head: a case report, *J Foot Ankle Surg* **48**: 61–69 (2009).

31. J. R. Clements, Use of allograft cellular bone matrix in multistage talectomy with tibiocalcaneal arthrodesis: a case report, *J Foot Ankle Surg* **51**: 83–86 (2012).

32. S. J. McAnany, J. Ahn, I. M. Elboghdady, A. Marquez-Lara, N. Ashraf, B. Svovrlj, S. C. Overley, K. Singh and S. A. Quereshi, Mesenchymal stem cell allograft as a fusion adjunct in one- and two-level anterior cervical discectomy and fusion: a matched cohort analysis, *Spine J* **16**: 163–167 (2016).

33. A. J. Sindler, S. Behmanesh and M. A. Reynolds, Evaluation of allogenic cellular bone graft for ridge augmentation: a case report, *Clinic Adv Periodontics* **3**: 159–165 (2013).

34. P. S. Rosen, S. J. Froum and D. W. Cohen, Consecutive case series using a composite allograft containing mesenchymal cells with an amnion-chorion barrier to treat mandibular class III/IV furcations, *Int J Periodontics Restorative Dent* **35**: 453–460 (2015).

35. B. Skovrlj, J. Z. Guzman, M. Al Maaieh, S. K. Cho, J. C. Iatridis and S. A. Quereshi, Cellular bone matrices: viable stem cell-containing bone graft substitutes, *Spine J* **14**: 2763–2772 (2014).

36. E. Jones, A. English, S. M. Churchman, D. Kouroupis, S. A. Boxall, S. Kinsey, P. G. Giannoudis, P. Emery and D. McGonagle, Large-scale extraction and characterization of CD271+ multipotential stromal cells from trabecular bone in health and osteoarthritis: implications for bone regeneration strategies based on uncultured or minimally cultured multipotential stromal cells, *Arthritis Rheum* **62**: 1944–1954 (2010).

37. J. Neman, V. Duenas, C. M. Kowolik, A. C. Hambrecht, M. Y. Chen and R. Jandial, Lineage mapping and characterization of the native progenitor population in cellular allograft, *J Spine* **13**: 162–174 (2013).

38. D. P. Tarnow, R. N. Eskow and J. Zamzok, Aesthetics and implant dentistry, *Periodontology 2000* **11**: 85–94 (1996).

39. M. Simion, I. Rocchietta, D. Kim, M. Nevins and J. Fiorellini, Vertical ridge augmentation by means of deproteinized bovine bone block and rhPDGF-BB: a histological study in a dog model, *Int J Periodontics Restorative Dent* **26**: 415–423 (2006).

40. A. A. Jones, D. Buser, R. Schenk, J. Wozney and D. L. Cochran, The effect of rhBMP-2 around endosseous implants with and without membranes in the canine model, *J Periodontol* **77**: 1184–1193 (2006).

41. C. Maiorana, F. Santoro, M. Rabagliati and S. Salina, Evaluation of the use of iliac cancellous bone and anorganic bovine bone in the reconstruction of the atrophic maxilla with titanium mesh: a clinical and histologic investigation, *Int J Oral Maxillofac Implants* **16**: 427–432 (2001).

42. L. Malchiodi, A. Scarano, M. Quaranta and A. Piattelli, Rigid fixation by means of titanium mesh in edentulous ridge expansion for horizontal ridge augmentation in the maxilla, *Int J Oral Maxillofac Implants* **13**: 701–705 (1998).

43. B. S. McAllister and V. T. Eshraghi, Alveolar ridge augmentation with allograft stem cell-based matrix and titanium mesh, *Clinic Adv Periodontics* **2**: 1–7 (2013).

44. F. Pieri, G. Corinaldesi, M. Fini, N. N. Aldini, R. Giardino and C. Marchetti, Alveolar ridge augmentation with titanium mesh and a combination of autogenous bone and anorganic bovine bone: a 2-year prospective study, *J Periodontol* **79**: 2093–2103 (2008).

45. R. A. Levine, B. S. McAllister, Implant site development using ti-mesh and cellular allograft in the esthetic zone for restorative-driven implant placement: a case report, *Int J Periodontics Restorative Dent* **3**: 373–381 (2016).

46. H. D. Barber, J. Lignelli, B. M. Smith and B. K. Bartee, Using a dense PTFE membrane without primary closure to achieve bone and tissue regeneration, *J Oral Maxillofac Surg* **65**: 748–752 (2007).

47. S. Lundgren, P. Moy, C. Johansson and H. Nilsson, Augmentation of the maxillary sinus floor with particulated mandible: a histologic and histomorphometric study, *Int J Oral Maxillofac Implants* **11**: 760–766 (1996).

48. R. S. Truhlar, I. H. Orenstein, H. F. Morris and S. Ochi, Distribution of bone quality in patients receiving endosseous dental implants, *J Oral Maxillofac Surg* **55**: 38–45 (1997).

49. S. W. Kim, I. K. Lee, K. I. Yun, C. H. Kim and J. U. Park, Adult stem cells derived from human maxillary sinus membrane and their osteogenic differentiation, *Int J Oral Maxillofac Implants* **24**: 991–998 (2009).

50. B. S. McAllister, M. D. Margolin, A. G. Cogan, M. Taylor and J. Wollins, Residual lateral wall defects following sinus grafting with recombinant human osteogenic protein-1 or Bio-Oss in the chimpanzee, *Int J Periodontics Restorative Dent* **18**: 227–239 (1998).

51. P. K. Moy, S. Lundgren and R. E. Holmes, Maxillary sinus augmentation: histomorphometric analysis of graft materials for maxillary sinus floor augmentation, *J Oral Maxillofac Surg* **51**: 857–862 (1993).

52. S. S. Wallace, Lateral window sinus augmentation using bone replacement grafts: a biologically sound surgical technique, *Alpha Omegan* **98**: 36–46 (2005).

53. S. S. Wallace, S. J. Froum and D. P. Tarnow, Histologic evaluation of a sinus elevation procedure: a clinical report, *Int J Periodontics Restorative Dent* **16**: 46–51 (1996).

54. Y. M. Lee, S. Y. Shin, J. Y. Kim, S. B. Kye, Y. Ku and I. C. Rhyu, Bone reaction to bovine hydroxyapatite for maxillary sinus floor augmentation: histologic results in humans, *Int J Periodontics Restorative Dent* **26**: 471–481 (2006).

55. P. J. Boyne, L. C. Lilly, R. E. Marx, P. K. Moy, M. Nevins, D. B. Spagnoli and R. G. Triplett, *De novo* bone induction by recombinant human bone morphogenetic protein-2 (rhBMP-2) in maxillary sinus floor augmentation, *J Oral Maxillofac Surg* **63**: 1693–1707 (2005).

56. M. Nevins, D. Garber, J. J. Hanratty, B. S. McAllister, M. L. Nevins, M. Salama, P. Schupbach, S. Wallace, S. M. Bernstein and D. M. Kim, Human histologic evaluation of anorganic bovine bone mineral combined with recombinant human platelet-derived growth factor BB in maxillary sinus augmentation: case series study, *Int J Periodontics Restorative Dent* **29**: 583–591 (2009).

57. B. S. McAllister, K. Haghighat and A. Gonshor, Histologic evaluation of a stem cell-based sinus-augmentation procedure, *J Periodontol* **80**: 679–686 (2009).

58. A. Gonshor, B. S. McAllister, S. S. Wallace and H. Prasad, Histologic and histomorphometric evaluation of an allograft stem cell-based matrix sinus augmentation procedure, *Int J Oral Maxillofac Implants* (in press).

59. S. Nyman, J. Lindhe, T. Karring and H. Rylander, New attachment following surgical treatment of human periodontal disease, *J Clin Periodontol* **9**: 290–296 (1982).

60. M. A. Reynolds, M. E. Aichelmann-Reidy and G. L. Branch-Mays, Regeneration of periodontal tissue: bone replacement grafts, *Dent Clin North Am* **54**: 55–71 (2010).

61. C. C. Villar and D. L. Cochran, Regeneration of periodontal tissues: guided tissue regeneration, *Dent Clin North Am* **54**: 73–92 (2010).

62. R. A. Yukna and J. T. Mellonig, Histologic evaluation of periodontal healing in humans following regenerative therapy with enamel matrix derivative: a 10-case series, *J Periodontol* **71**: 752–759 (2000).

63. G. Bowers, F. Felton, C. Middleton, D. Glynn, S. Sharp, J. Mellonig, R. Corio, J. Emerson, S. Park, J. Suzuki, S. Ma, E. Romberg and A. H. Reddi, Histologic comparison of regeneration in human intrabony defects when osteogenin is combined with demineralized freeze-dried bone allograft and with purified bovine collagen, *J Periodontol* **62**: 690–702 (1991).

64. J. T. Mellonig, P. Valderrama Mdel and D. L. Cochran, Histological and clinical evaluation of recombinant human platelet-derived growth factor combined with beta tri-calcium phosphate for the treatment of human class III furcation defects, *Int J Periodontics Restorative Dent* **29**: 169–177 (2009).

65. G. M. Bowers, R. G. Schallhorn, P. K. McClain, G. M. Morrison, R. Morgan and M. A. Reynolds, Factors influencing the outcome of regenerative therapy in mandibular class II furcations: Part I. *J Periodontol* **80**: 476–491 (2009).

66. G. Avila, P. Gailindo-Moreno, S. Soehren, C. E. Misch, T. Morelli and H. L. Wang, A novel decision-making process for tooth retention or extraction, *J Periodontol* **80**: 476–491 (2009).

67. B. S. McAllister, Stem cell-containing allograft matrix enhances periodontal regeneration: case presentations. *Int J Periodontics Restorative Dent* **21**: 149–155 (2011).

68. S. Koo, A. Alshihri, N. Karimbux and M. Maksoud, Cellular allograft in the treatment of a severe periodontal intrabony defect: a case report, *Clinic Adv Periodontics* **2**: 35–39 (2012).

69. P. Rosen, A case report on combination therapy using a composite allograft containing mesenchymal cells with an amnion-chorion barrier to treat a mandibular class III furcation, *Clinic Adv Periodontics* **3**: 64–69 (2013).

70. G. Intinini, Future approaches in periodontal regeneration: gene therapy, stem cells, and RNA interference, *Dent Clin North Am* **54**: 141–155 (2010).

71. M. Tobita and H. Mizuno, Periodontal disease and periodontal tissue regeneration, *Curr Stem Cell Res Ther* **55**: 168–174 (2010).

72. Y. Yamada, M. Ueda, H. Hibi and S. Baba, A novel approach to periodontal tissue regeneration with mesenchymal stem cells and platelet-rich plasma using tissue engineering technology: a clinical case report, *Int J Periodontics Restorative Dent* **26**: 363–369 (2006).

73. R. G. Schallhorn, W. H. Haitt and W. Boyce, Iliac transplants in periodontal therapy, *J Periodontol* **41**: 566–580 (1970).

74. M. R. Dragoo and R. K. Irwin, A method of procuring cancellous iliac bone utilizing a trephine needle, *J Periodontol* **44**: 599–613 (1973).

75. M. R. Dragoo and H. C. Sullivan, A clinical and histological evaluation of autogenous iliac bone grafts in humans: I. Wound healing 2 to 8 months, *J Periodontol* **44**: 599–613 (1973).

76. A. Banfi, A. Muraglia, B. Dozin, M. Mastrogiacomo, R. Cancedda and R. Quarto, Proliferation kinetics and differentiation potential of *ex vivo* expanded human bone marrow stromal cells: implications for their use in cell therapy, *Exp Hematol* **28**: 707–715 (2000).

77. M. R. Dragoo and H. C. Sullivan, A clinical and histological evaluation of autogenous iliac bone grafts in humans: II. External root resorption, *J Periodontol* **44**: 614–625 (1973).

78. R. G. Schallhorn, Postoperative problems associated with iliac transplants, *J Periodontol* **43**: 3–9 (1972).

79. G. F. Muschler, H. Nitto, C. A. Boehm and K. A. Easley, Age- and gender-related changes in the cellularity of human bone marrow and the prevelance of osteoblastic progenitors, *J Orthop Res* **19**: 117–125 (2001).

80. D. Smiler, M. Soltan and M. Albitar, Toward the identification of mesenchymal stem cells in bone marrow and peripheral blood for bone regeneration, *Implant Dent* **17**: 236–247 (2008).

81. M. Soltan, D. Smiler and J. H. Choi, Bone marrow: orchestrated cells, cytokines, and growth factors for bone regeneration, *Implant Dent* **18**: 132–141 (2009).

82. A. V. Cuomo, M. Virk, F. Petrigliano, E. F. Morgan and J. R. Lieberman, Mesenchymal stem cell concentration and bone repair: potential pitfalls from bench to bedside, *J Bone Joint Surg Am* **91**: 1073–1083 (2009).

83. F. Carinci, G. Brunelli, I. Zollino, M. Franco, A. Viscioni, L. Rigo, R. Guidi and L. Strohmenger, Mandibles grafted with fresh-frozen bone: an evaluation of implant outcome, *Implant Dent* **18**: 86–95 (2009).

84. J. M. Ryan, F. P. Barry, J. M. Murphy and B. P. Mahon, Mesenchymal stem cells avoid allogeneic rejection, *J Inflamm (Lond)* **2**: 8 (2005).

85. E. J. Caterson, L. J. Nesti, T. Albert, K. Danielson and R. Tuan, Application of mesenchymal stem cells in the regeneration of musculoskeletal tissues, *MedGenMed* **5**: E1 (2001).

86. R. F. McLain, J. E. Fleming, C. A. Boehm and G. F. Muschler, Aspiration of osteoprogenitor cells for augmenting spinal fusion: comparison of progenitor cell concentrations from the vertebral body and iliac crest, *J Bone Joint Surg Am* **87**: 2655–2661 (2005).

87. P. Hernigou, A. Poignard, F. Beaujean and H. Rouard, Percutaneous autologous bone-marrow grafting for nonunions. Influence of the number and concentration of progenitor cells, *J Bone Joint Surg Am* **87**: 1430–1437 (2005).

88. T. G. Baboolal, S. A. Boxall, Y. M. El-Sherbiny, T. A. Moseley, R. J. Cuthbert, P. V. Giannoudis, D. McGonagle and E. Jones, Multipotential stromal cell abundance in cellular bone allograft: comparison with fresh age-matched iliac crest bone and bone marrow aspirate, *Regen Med* **9**: 593–607 (2014).

13

POTENTIAL OF TISSUE ENGINEERING AND NEURAL STEM CELLS IN THE UNDERSTANDING AND TREATMENT OF NEURODEGENERATIVE DISEASES

*Caroline Auclair-Daigle and François Berthod**

1. Introduction

Neurodegenerative diseases consist of an heterogeneous assembly of pathological conditions in which specific regions of the central nervous system (CNS) relatively slowly and progressively deteriorate, resulting in movement and/or cognitive impairments. The specific etiology and causes of those disorders mainly affect elderly people. On the economic level, these types of neurodegenerative disorders are generating economical cost that sum up, each year, to hundreds of billions of dollars in developed countries. Given that the cause of most of these diseases remains unknown, the *in vitro* reconstruction of tissue-engineered models that mimic the diseases could be very helpful in understanding the subtle biological alterations that occur in comparison to the controls. In addition, since neurons lack the ability to regenerate, stem cells therapy may offer an opportunity to replace cells lost through damage or degeneration.

2. Neurodegenerative Diseases and their Current Treatments

2.1 *Parkinson's disease (PD)*

The second most common neurodegenerative disease, inflicting debilitating troubles to close to 1% of the population aged 60 and over. The suffering individuals are mainly affected by motor symptoms such as rigidity (muscle stiffness), bradykinesia (severe uncontrollable movement disorder), and shivers while resting.[1] PD is a progressive disease characterized by a degeneration of dopaminergic neurons in the substantia nigra, and a subsequent deficit of dopamine release in the striatum. The affected individuals are becoming severely disturbed and depressed, in addition to the development of dementia over the span of the disorder (greater than 70% in advanced PD patients).[2] There is currently no cure and no effective long-term treatment for PD, as the precise etiology of the neuronal loss is still unknown and the pathogenesis not fully understood.

The main treatment has been consisting, ever since the late 1960s, in the pharmacological lessening of the striatal dopamine deficit by administration of the dopamine precursor L-Dopa, which crosses the blood-brain barrier and enters neurons that convert it into dopamine. Alternative drugs were developped over the years, but failed to match the levodopa's efficacy, which still remains the gold standard for tratment of PD. However, its chronic administration is associated with motor complications reflecting fluctuations of the drug concentration in the plasma, and its efficacy lessens with time, with re-emerging of parkinsonian symptoms.[3,4] Neurosurgical procedures have also been used as alternative approaches, consisting of ablative procedures of specific regions of the brain, deep brain stimulation using electrical currents,[5] GDNF delivery in the striatum (a neuroprotective factor) through osmotic pump, slow-release beads or gene therapy, and cell replacement of the diseased neurons using fetal mesencephalic neurons.[6,7]

2.2 *Alzheimer's disease (AD)*

A progressive neurodegenerative disorder of the cortical regions of the brain, first affecting memory functions and then gradually affecting all cognitive functions with behavioral impairments, leading to the irreversible loss of neurons and dementia. AD is the most common cause of dementia worldwide, accounting for 50–60% of all cases, some of its risk factors including older age, family history, lower education level, and female gender. In AD, some cholinergic neurons lose their ability to function, reducing acetylcholine level. Current treatments include the use of drugs such as acetylcholinesterase inhibitors, in conjunction or not with NMDA-antagonists, but they have not been shown to delay institutionalization or functional decline.[8] Life expectancy following diagnostic usually range between 3–15 years, but may be more limited.

2.3 *Huntington's disease (HD)*

An autosomal dominant hereditary neurodegeneration initially described by psychiatrist G. Huntington in 1872. It consists of a disorder primarily affecting selective neuronal subtypes, and particularly GABAergic medium spiny neurons, the main neuronal subtype in the striatum. The mutant gene encodes the huntingtin protein, and the disease is believed to be due to a gain of toxic function of the mutant protein.[9] The prevalence of this condition varies in the range of 2–10 cases/100 000, and the onset of the disease usually occurs in people aged between 30–50, but has also been observed in the elderly. At this point, there is a current lack of effective treatment for HD. Since the abnormal huntingtin protein and its cellular functions have been identified in 1993, drugs have been elaborated to reduce the ailment's magnitude. Strains of mice have been created with the identical gene responsible for HD in humans, leading to the development of promising drugs, such as drugs that block the glutamine chains from clustering.[10] The drug tetrabenazine proved to lower dopamine release, limiting writhing movements.

2.4 *Amyotrophic Lateral Sclerosis (ALS)*

A progressive neurodegenerative disorder characterized by selective loss of lower spinal and brainstem motor neurons and upper motor neurons. Of the major physical consequences observed are paresis of skeletal and bulbar muscles, amyotrophies, fasciculations, spasticity and ultimately, paralysis. There appear to have no identifiable underlying cause other than genetic basis in familial cases, and there is currently no effective treatment available. The disease normally progresses rapidly, and survival rate rarely exceed 3–5 years after the onset. Death occurs as a result of diaphragm weakness, or from pulmonary infections. The only drug treatment currently approved for ALS is riluzole, a glutamate-release inhibitor. The drug possibly will expand life from 4 to 18 months and postpone the need for tracheostomy.[11,12]

2.5 *Multiple sclerosis (MS)*

An idiopathic primary demyelinating disease of the CNS, whose main condition generates destruction of normally developed myelin sheaths. It is a chronic progressive disorder characterized by disseminated neurological symptoms and, usually, several relapses during the course of its debut stage. MS was initially described by J. Charcot in 1866, defining the disease as a grouping of intention tremor, spastic paraplegia, speech impairment, visual loss and nystagmus. It currently stands as the most studied demyelinating disease. The most common theory for its presupposed etiology is multifactorial, stipulating that MS develops due to the combined presence of exogenous factors in genetically predisposed people, autoimmune response, and demyelinating lacerations in the white matter of CNS.[9] This episodic neurological disease normally strikes people aged 20–40, although 10% are recollected in people aged over 50. Current treatment for relapsing-remitting MS are not curative, but include interferon-β-1a or interferon β-1b, and glatiramer acetate (namely copolymer 1), a combination of random polymers that mimic the amino acid composition of myelin basin protein, which could both be postponing the onset of major disability. Other pharmaceutical agents such as α-4 integrin antagonists, intravenous immunoglobulin infusions and corticosteroids may limit the relapse rate.[13]

3. Tissue Engineering as a Tool to Better Understand Neurodegenerative Diseases

As previously discussed, the causes of most neurodegenerative diseases do remain unknown. The usual method to analyse such disorders is to perform histological and immunohistochemical analysis on post-mortem brain biopsies obtained from the patients, compared to normal tissues. These observations revealed the formation of β-amyloid plaques in AD, Lewy bodies in PD or superoxyde dismutase-1 (SOD-1) aggregates in ALS, but they did not specifically point out the role of these structures as being either a cause or a consequence of the disease. Genetic studies can also be performed with these post-mortem tissues to identify a potential mutation as the cause of the disease, such as in HD.

In ALS, 20% of the patients have at least another family member affected by the disease, even though only approximately 10% of these individuals possess a SOD-1 mutation, and 40%, a *C9orf72* hexanucleotide repeat expansion.[14–16] As for the sporadic cases of ALS, a combination of environmental factors with potential genetic susceptibility, in addition to other causes such as microtraumas, are currently being suspected.

The use of animal models that mimic the diseases can be of great help, but a minimal understanding of the etiology of the diseases need to be reached before those models can be generated. PD can be induced in rats following destruction of their substantia nigra, and ALS transgenic mouse model can be generated by overexpression of the human mutated SOD-1 gene. However, studying the effects of the disease on an entire animal remains complex and does not always bring a clear answer to the questions raised. In ALS for example, it is still not well understood how the SOD-1 or *C9orf72* mutations induce motor neuron degeneration.

Thus, *in vitro* culture systems could be very helpful to perform dynamic studies at the cellular level. However, culturing neurons faces major limitations, because neurons are nearly impossible to extract from the adult brain or spinal cord, and these cells do not proliferate *in vitro*. This is why neurons are usually extracted from mouse or rat embryos at an early stage of development (E12–E14), before they have had the time to establish too many connections. In addition, since neurons do not proliferate, cell extraction from embryos needs to be performed again for each experiment.

Finally, a broad limitation for these cultures, in common with most of the other cell types, consists of the vast difference in environment that occurs between a cell cultured on a plastic dish in two-dimensions, compared to the three-dimensional environment *in situ*. Both the cell-cell and cell-matrix contacts are totally different, as well as the global cell behavior, especially for the nervous system in which multiple connections between neurons are constantly established.

3.1 *Two-dimensional in vitro models of neural cell culture*

In vitro cultures of various types of neurons (hippocampal, motor, sympathetic, etc.) have been performed for several years now, by extracting cells from the nervous system of fetuses. These culture systems allow for the isolation of purified neurons from their surrounding environment, which greatly facilitates studies such as mRNA or protein analysis. However, it has been noted that neurons usually only survive over a short period of time in the absence of glial cells.

Mixed culture of spinal cord cells can be maintained for an extended period of time, and neurons can be microinjected with a plasmid to over-express a protein of interest (mutant or not) to study its impact on cell physiology or survival.[17]

Double compartment cultures have been developed by using neurons cultured on a plastic dish, and glial cells attached to a glass slide and then flipped over and put onto the neurons using spacers so that cells would be as close as possible from each other for studying the paracrine effects occurring without direct cell-cell contact.[18]

Some *in vitro* models have also been developed to mimic a traumatic injury, through the stretching of cortical neurons cultured on elastic substrates and subsequent analysis of the intracellular calcium concentration.[19]

3.2 *Three-dimensional tissue-engineered models of the nervous system*

Spinal cord or brain tissue slices can be maintained as organotypic cultures *in vitro*, preserving the three-dimensional organization of the nervous tissue with all of its cell-cell contacts. These models are very useful as they are highly physiological, allowing electrophysiological studies to be performed only a few hours following the tissue harvesting.[20]

However, they do not provide any sort of control on the cell types present in the tissue of interest, and the specific cells' proportion and condition.

What could be really helpful to study neurodegenerative diseases would be to reconstruct the nervous system by independently combining each cell type. This will be particularly useful with the multiple transgenic mouse models developed to mimic various disorders, such as the overexpression of the human mutant SOD-1 gene in the G93A ALS mouse model.[21] It would be even more interesting if we could be using the patient's own stem cells differentiated into specific types of neurons and glial cells, providing that these cells would keep their diseased phenotype over the full course of differentiation.

Some three-dimensional neural constructs were developed by coculturing neurons with astrocytes in a 500–800 µm thick 3D Matrigel™. Cells within these constructs displayed extensive 3D process outgrowths and a high viability over multiple weeks. In addition, neurons in this model can be tested through patch-clamp techniques, and were shown to be able to display electrophysiological action potentials and functional synapse formation.[22]

However, a drawback with the use of Matrigel™ is that it consists of a material containing multiple active molecules, such as laminin, collagen IV, entactin, heparan sulphate and cytokines, in undetermined concentrations.

Other biological polymers characterized by better controlled compositions, such as collagen, fibrin, methylcellulose or agarose, have also been used as culture scaffolds to reconstruct 3D neural tissues.[22]

We developed a tissue-engineered model of motor neuron culture to study the axonal migration and myelination processes in a three-dimensional environment. Since most of the motor neuron's cell surface is located outside of the spinal cord, coupled with an axon that can be measured up to one meter long, the process of axonal migration represents a major issue in the study of motor neurons. Moreover, this axon is also being myelinated by Schwann cells, the main glial cell type of the peripheral nervous system.

To closely mimic the 3D environment surrounding motor neurons, this peripheral nerve migration aspect should be taken into account. To recapitulate the tissues through which nerves make their way to the muscle, we developed a 3D connective tissue made of fibroblasts cultured in a collagen sponge and maturated for 2 weeks in order to promote extracellular matrix deposition. Mouse motor neurons extracted from E12 mouse embryos were purified through density gradient centrifugation,[23] and then seeded on top of the reconstructed connective tissue (Fig. 1). The motor neurons formed a thick and dense cell layer over it. To promote axonal migration, the sponge was lifted to the air-liquid interface, and a cocktail of neurotrophic factors was added to the culture medium underneath. As shown by immunohistochemistry, a large number of neurofilament M-positive neurites was observed migrating down from the neurons layer to the bottom of the connective tissue for over up to 1mm long in distance. In addition, when mouse Schwann cells were co-cultured with the fibroblasts in the sponge, they migrated alongside with the neurites, as shown by the Myelin Basic Protein double-staining with Neurofilament-M.[24] An analysis of these neurites, performed by transmission electron microscopy, showed that a thick myelin sheath was wrapped around some of the neurites after 28 days of *in vitro* maturation (Fig. 2). Thus, this tissue-engineered model was shown to promote axonal migration, axon myelination by Schwann cells, and spontaneous myelin sheath formation for the first time *in vitro*.[24]

This model should greatly facilitate studies on diseases related to motor neuron axon demyelination, such as MS or Charcot-Marie-Tooth disease of the peripheral nerves.

In addition, it could be very valuable to study ALS with such a model, by using different combinations of spinal cord cells obtained from SOD-1 mutant mice versus wild-type mice, all the while measuring motor neuron degeneration.

Finally, this model could be even more interesting if it could be reconstructed using the patient's own cells, to mimic the human disease *in vitro*.

Since living motor neurons cannot be extracted from an adult spinal cord, i.e. from post-mortem tissues, other alternatives should be explored in order to generate these cells from the patients. One of the best solution appears to proceed with the differentiation of motor neurons from adult stem cells.

In addition, the differentiation of neurons from autologous stem cells could be applied for cell replacement therapy applications, as a novel approach to treat neurodegenerative disorders by implantation of new and functional neurons.

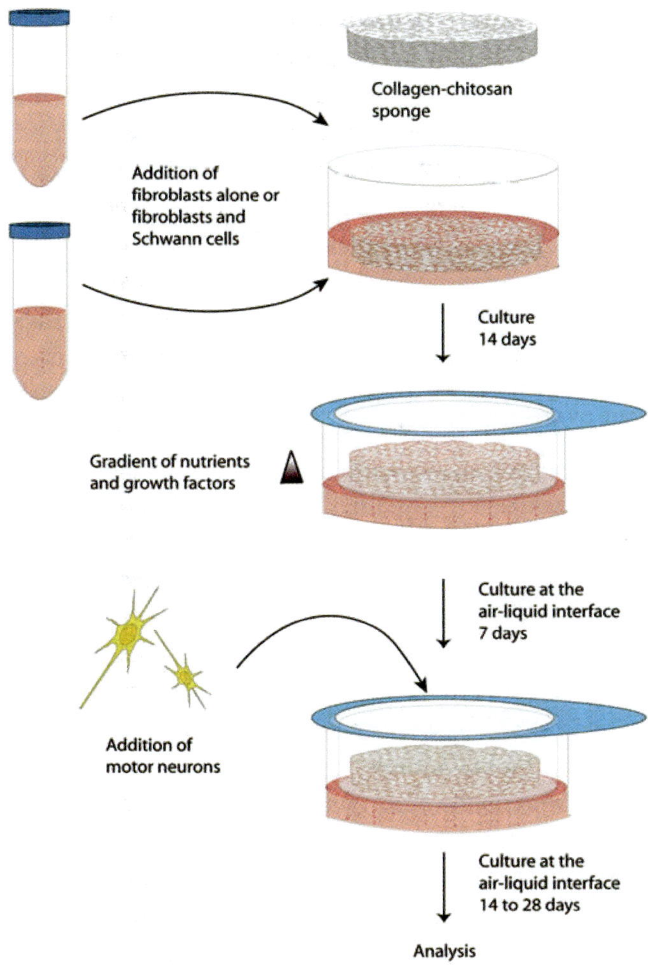

Fig. 1. Preparation of a tissue-engineered model to study axonal migration and myelination of motor neurons through a 3D connective tissue (from Gingras *et al.*, *Glia* **56**: 354–364, 2008).

4. Neural Stem Cells to Treat Neurodegenerative Diseases

After molecular biology, gene therapy, nanotechnology and tissue engineering, stem cells are now considered as representing the new emerging field that could revolutionize the future of medicine. This is particularly true in neurosciences, field in which the work with human neural cells is

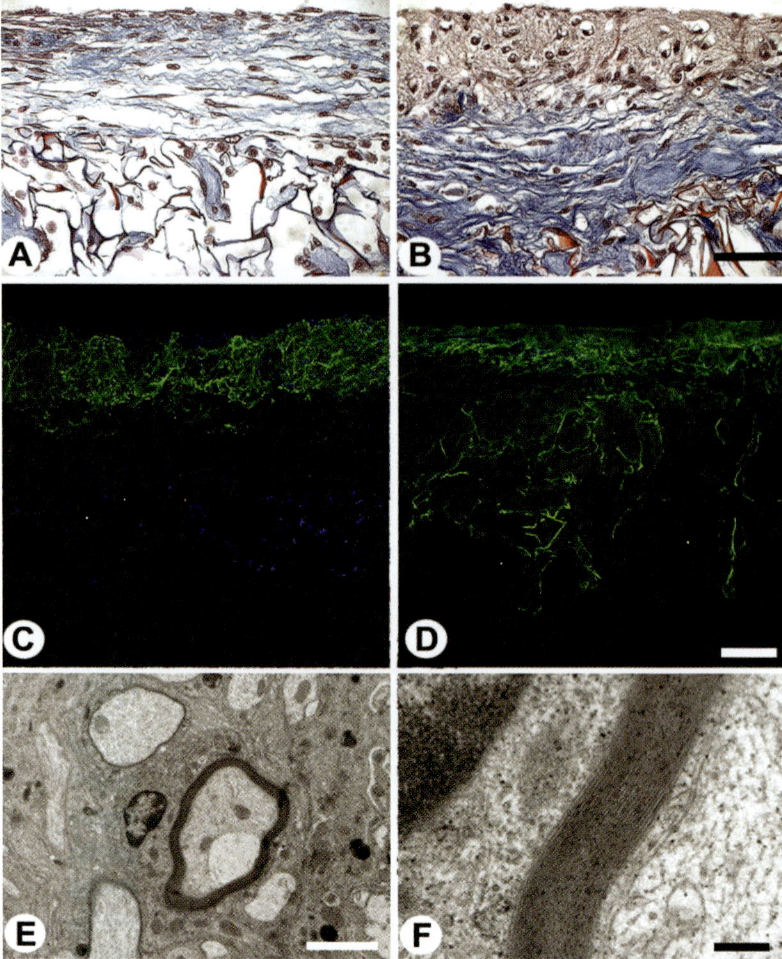

Fig. 2.   In (a), fibroblasts and Schwann cells were co-cultured for 21 days in the collagen sponge, and were observed by histology staining with Masson's trichrome. In (b), mouse motor neurons were seeded on top of the sponge and cultured for an additional 14 days (the dash line shows the limit between the upper neuron layer and the connective tissue underneath). In (c), the neurons layer (stained with an antibody against Neurofilament-M) was cultured at the air-liquid interface, but without the addition of neurotrophic factors; neurites did not migrate through the sponge. In (d), the culture medium underneath the sponge was supplemented with neurotrophic factors, which promoted neurite migration through the connective tissue. In (e), a transmission electron microscopic picture of the sponge showed that some neurites were wrapped with thick myelin sheaths, observed with a higher magnification in (f). Bar in B, 60 µm for (a–b); bar in (d), 100 µm for (c–d); bar in e, 2 µm; bar in f, 0.2 µm (Modified from Gingras *et al.*, *Glia* **56**: 354–364, 2008).

still impracticable, and for which the use of neural cells differentiated from human stem cells may be the new standard in a near future. To fully understand the current known potential and limitations of stem cells, as well as their diversity and specific related issues, a rapid overview of fetal neurons, as well as a comparative standpoint of embryonic versus adult stem cells, will be presented.

4.1 *Human fetal neural transplantations, a proof of concept for cell replacement*

Even though cell transplantation procedures have already been in clinical practice for numerous organs, attempts at replacing cells within the CNS do remain experimental. PD readily represents a candidate disease of choice for cell replacement therapy since the vast majority of its lost dopaminergic neurons come from a circumscribed area, the substantia nigra compacta.[25] This specificity made it the first disease to be considered and treated by cellular therapy. The first attempt was made using heterologous fetal neurons obtained from aborted fetuses, mainly because only fetal neurons can survive to tissue extraction. This initial series of transplantation studies provided us with insights suggesting that the grafted cells could, at least partially, survive, integrate the brain, and improve the disease's symptoms for an extended period of time.[26]

4.1.1 *Parkinson's Disease*

It was during the late 1980s that the very first methodical clinical transplantation trials using fetal dopaminergic neurons in patients with PD were elaborated.[27–29] Mesencephalic tissue (an area in the developing CNS rich in immature dopaminergic neurons) of 6 to 9 weeks old aborted fetuses was transplanted into the striatum.[7] Various immunosuppression policies were adopted, ranging from none,[30] to cyclosporin A for 6 months[31] and to a cocktail mix of cyclosporin A, steroids, and azathioprine.[32]

Of the open-labeled trials, without the presence of a control placebo group or blinding procedure, the initial results following fetal nigral grafts in PD patients showed graft survival and clinical improvement in these PD patients.[27–29] From that timeframe, over 400 PD patients throughout the world have been grafted with nigral tissue,[33] and improvements were reported from none to long-lasting dramatic benefits.[34]

Of the double-blind placebo controlled trials, in which there was presence of a sham group consisting of the PD patients being anesthetized, their skull perforated but without the actual cell transplantation step occurring, both the patients and observers remained blinded to this condition. In one study, 19/40 patients received a tissue strand cell transplantation, after which a subjective global rating scale test was conducted one year post-transplantation.[30] Slight improvement was noted in the transplanted group in comparison to the sham one, but 15% of the patients ended up with severe off-dyskinesias. In another study, solid pieces of embryonic ventral mesencephalon have been transplanted into the putamen of both sides of the brain. Eleven patients were sham controls. After 2 years, no net improvement was noted, and 56% of the patients resulted with off -dyskinesias.[31]

No real positive effect using cell transplantation was detected in those studies. Possible reasons for this poor outcome may be that the clinical assessments were not standardized accordingly, or the tissue preparation, long-term *in vitro* storage, immunosuppression use, and surgical approaches had design flaws.[30,34]

Interestingly, analysis of post-mortem tissues more than 10 years after graft of fetal neurons showed that, whereas some of these cells undergo pathological changes similar to Parkinson disease (expression of Lewy bodies), the majority of them did not display evidence of functionnal impairment or degeneration.[35,36]

4.1.2 *Huntington's Disease*

Fetal neural transplants have also been used to treat patients suffering from HD.[7] A recent study about fetal neural transplants reports cases of 3 HD patients autopsies, performed 10 years post-transplantation.[37] In 2 out of those 3 patients, there seems to have been a differentiation into adequate cell types and proper glutamatergic and dopaminergic innervating projections from the recipients' brain. Yet, degeneration was observed in the grafted cells. The grafts also showed astrogliosis, inflammatory infiltrates, and microglial activation. Given that only temporary and minor benefits, along with considerable graft degeneration were observed, this would limit the use of fetal cells for eventual studies conducted for HD.[37]

Human fetal neural transplantations applied to hundreds of PD patients over the last two decades have set new standards for cell transplantation into the brain. However, the risks linked with human fetal transplantation trials may very well be higher than initially expected. In fact, there are obvious logistic problems linked with the amount of donors needed for each patient, on top of all of the ethical concerns that have been raised concerning the use of cells obtained from aborted fetuses. Put together, these factors are widely restricting the application of fetal tissue for neural transplantation. Nevertheless, even if this cell replacement therapy could be proven successful in only a few cases, by promoting cell survival and integration into the neuronal circuitry and, ultimately, for long-term benefits for the disease's symptoms, it still shows that cell therapy applications could be successful, once a better control of the graft parameters will have been established.

4.2 *Embryonic stem cells (hESC)-derived neural precursor cells*

Fetal neural transplantations are limited, not only by the availability of fetuses but also largely by ethical concerns. One alternative could be the development of pluripotent stem cell banks generated from embryos at an earlier developmental stage (supernumeraries not used during *in vitro* fertilization), which may be not as ethically controversial as fetuses, and which could generate large amounts of various cell types due to their high proliferative and differentiation potentials (Table 1).

The use of hESC-derived NPCs led to some promising results in the treatment of PD and HD. However, concerns similar to those related to fetal neural transplants have been raised for those cells: the hESC-derived NPCs are heterologous and thus could be targeted by the host immune system, and they also face major limiting ethical concerns. Meanwhile, hESCs undoubtedly hold the highest proliferative potential and the largest and most efficient differentiation capacity in comparison to adult NPCs.

Table 1.   Comparison of human ESCs vs adult NPCs, iPS-derived NPCs and iN cells

	ESCs	Adult NPCs	iPS-derived NPCs	iN cells
Pros	Pluripotency (extended potential)	Multipotency	Pluripotency	Best preserved aging and disease characteristics
	Multilineage differentiation	Limited risk of tumor formation	Extensive self-renewal	Limited risk of tumor formation
	Extensive self-renewal	Autologous cells	Autologous cells	Autologous cells
	Access to early neural development stage	Availability (within tissues)	Availability (from somatic cells)	Limited ethical concerns
	Ease for inducing stable genetic changes	Limited ethical concerns	Limited ethical concerns	Easy and fast production
Cons	Major ethical concerns	Fate-limited potentiality	Complex and expensive production	Needs specific genetic reprograming for each neuronal type
	Risk of tumor formation	Uncommitted	Instability in culture	Genetic alteration
	Risk of immune rejection by the host	Limited proliferative capacity	Genetic alteration	No cell expansion after induction (for neurons)
	Limited availability of embryonic tissue	Limited amount of readily available cells	Risk of tumor formation	
		Variability in results reproducibility		

4.3 *Adult tissue-derived neural precursor cells (NPCs)*

The idea of growing cells in culture is particularly attractive if the cells consist of adult tissue-derived NPCs, capable of differentiation into various neural cell types. Studies are currently being done to develop methods aiming at modifying adult stem cells into precursor cells, opening up the possibility to harvest a patient's own cells and rendering them suitable for transplantation into the nervous system (Table 1).

4.3.1 *Brain-derived NPCs*

NPCs were initially defined as the self-renewing, multipotent cells that generate the main phenotypes of the nervous system, of both neuronal and glial subpopulations. Ever since the first subpopulation of mitotic NPCs was identified in the adult mice brain tissue,[38] NPCs have been isolated from various species including human.[39]

The maintenance of NPCs is ensured by the NPC niche, in which microenvironmental cues, but also interactions implicating the extracellular matrix and the cellular membranes, cell–cell interactions and the proximity to blood vessels, aid in maintaining cell proliferation, fate specification and differentiation.[40]

In the past 20 years, research groups have put a lot of efforts onto NPCs derived from various regions of the developing brain in order to stimulate the promotion of the functional recuperation process by dopaminergic differentiation in PD rat animal models.[41] Even if characteristics such as graft survival, neuronal and astrocytic differentiation, cell migration, and axonal extensions have been demonstrated, these types of studies had not yet shown evidence that a significant dopaminergic differentiation had set place.[42] However, given the actual location of these NPCs within the brain, there are some limitations that prevent their use for applications in cellular therapy.

4.3.2 *Bone Marrow-derived NPCs*

Other cell types generated from non-brain derived tissues, such as bone marrow, have been approached, having been shown to express, for example, certain dopaminergic markers. Mesenchymal stem cells

(MSC) obtained from adult rodent have been shown to differentiate, both *in vitro* and *in vivo,* into cells with mesenchymal, visceral mesoderm, neuroectoderm, and endoderm characteristics.[43] A subtype of these MSC was isolated and termed multipotent adult progenitor cells (MAPC). Surprisingly, it has been reported that as many as 30% of mouse MAPCs differentiated *in vitro* into tyrosine hydroxylase (TH, the rate-limiting enzyme in the production of dopamine)-expressing neurons. Those cells were presupposedly pluripotent, opening the possibility for MSC to be a cell source of choice for neurotransplantation in PD. Human bone marrow stromal cells were also shown to be able to convert into a NPC-like population that allowed for the expression of neuronal markers once differentiated, of which about 11% of the cells were TH-positive and were shown to release dopamine upon membrane depolarization.[44] However, the functionality of those cells remains to be shown, and so additional studies in PD animal models are still necessary.

4.3.3 *Skin-derived and adipose-derived NPCs*

It has previously been shown that differentiation of NPCs isolated from human adult skin could generate neural (and mesodermal) derivatives, while potentially providing a readily accessible source of human adult NPCs for transplantation: the skin-derived progenitor or precursor cells (SKP).[45–47] These cells possess distinct surface markers in common with MSCs, while preferentially differentiating into neural cell types. It appears that these SKPs can be passaged for up to 1 year without showing any significant senescence and seem to behave similarly to NPCs extracted from the brain in the way that they have the ability to form floating spherical colonies called neurospheres (Fig. 1a).[48] The achievability of expansion and possibility for long-term culturing of SKPs, coupled with their versatility, give them a certain appeal for use in therapy for disorders of the nervous system.[49–51] SKPs have also been shown to generate Schwann cells. The differentiated Schwann cells were transplanted into rat spinal cords following an induced traumatic injury, and appeared to survive within the injured environment and to myelinate host axons, all the while improving locomotor recovery.[52]

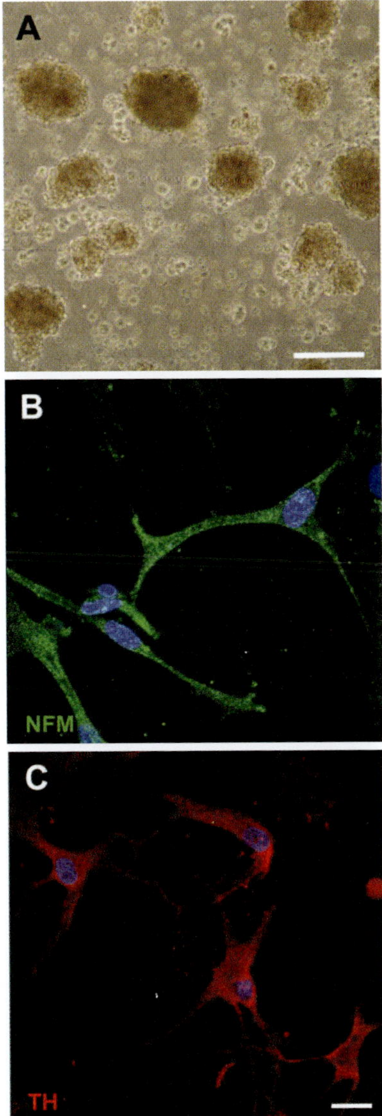

Fig. 3.   (a): Neurospheres-forming human skin-derived NPCs (Phase contrast micros-copy). (b): Human skin-derived NPCs differentiated in Neurofilament-M-expressing neurons (Immunohistochemistry). (c): Human skin-derived NPCs differentiated in TH-expressing neurons. Nuclei were stained with Hoechst. Bar in a, 200 µm; Bar in c, 25 µm for b–c.

Adipose tissue has been identified in 2001 as an alternative source of multipotent stromal MSCs, which can be obtained by a less invasive method and in larger quantities compared with skin or bone marrow NPCs (through liposuction).[53] These cells can also be a source of NPCs. The ability of Adipose-Derived Adult Stromal (ADAS) to neuroprotect or restore function in an injured dopaminergic pathway was investigated after transplantation of naive or neurally-induced ADAS into the striatum of parkinsonian rats. ADAS failed to generate stable dopaminergic neurons *in situ*, but gene expression analyses showed that both naive and differentiated ADAS cells express neuroprotective and trophic factors at the lesion site.[54]

4.3.4 *Induced Pluripotent Stem (iPS) cells derived-NPCs*

In recent years, several approaches have been developed in order to reprogram differentiated adult cells into pluripotent stem cells, named induced pluripotent stem (iPS) cells. Somatic cells have first been reprogrammed into iPS by viral transduction of four transcription factors, c-Myc, Oct4, Sox-2 and Klf4[55,56] and then by the expression of only Oct-4 and either Klf-4 or c-Myc.[57] More recently, it has been shown that the generation of adult iPS cells was achievable by no means of genetic modifications.[58]

A study has shown that iPS cells dedifferentiated from skin fibroblasts could undergo a differentiation into NPCs, then into dopaminergic neurons. After transplantation into the brains of parkinsonian rats, the grafted cells showed neuronal extensions throughout the neighboring brain regions, all the while lessening the PD-like symptoms.[59]

Autologous transplants using a patient's own cells are now at hands reach for the potential development of human adult NPC-based therapies for neurodegenerative diseases. It would now be possible to generate pure and highly proliferative colonies of iPS cells that, in theory, would have an equivalent potential as embryonic NPCs, additionally to being autologous, non-ethically restricted cells. But these avenues also come with a major setback: they also hoist the likelihood to induce tumor formation. Taking that iPS cells may be proven to be just as or even more performing than human adult NPCs, and all along being easier to extract, culture and control, they may very well become a cell type of choice for the treatment of neurodegenerative disorders.

4.3.5 *Induced Neural (iN) cells cells derived-NPCs*

Instead of generating in a first step iPS cells to be able to differentiate them into NPCs in a second step, a much easier and more efficient strategy would be to directly derive somatic cells (usually fibroblasts) into neurons. This approach was first developed in 2010 by the expression of three factors (*Ascl1*, *Brn2* and *Myt1l*) into fibroblasts to induce their differentiation into functional neurons.[60] This strategy was also applied to generate different types of neurons. Dopaminergic neurons were obtained through the combined expression into fibroblasts of the three factors previously used with two additional ones, Lmx1 and FoxA2, involved in dopamine neuron generation.[61] These neurons were shown to be functional *in vitro* and expressed action potentials.

In addition, fibroblasts engineered to express these inducible factors were able to directly convert into neurons after transplantation in the adult rodent brain.[62] Motor neurons can also be obtained from fibroblasts expressing seven factors (*Ascl1*, *Brn2*, *Myt1l*, *Lhx3*, *Hb9*, *Isl1*, and *Ngn2*).[63]

The major advantage of this approach is to bypass the generation of stem cells, which is a very demanding step in terms of efficiency, and a major concern for a clinical application. Indeed, using stem cells to obtain neurons through differentiation imply a risk to maintain few contaminant stem cells into the neuron population. These cells having high proliferative and differentiation potentials may induce teratoma or tumor formation. Starting from normal patient's derived fibroblasts to generate non-dividing cells such as neurons would markedly reduce that risk. Using iN cells instead of iPS would also be more appropriate to investigate the mechanism of diseases since the direct induction of fibroblasts into neurons would be susceptible to better maintain the disease and aging characteristics. Finally, the production of iN cells is much easier, faster and cheaper than for iPS cells. However, one drawback is that a specific differentiation cocktail has to be developed for each type of neuron. In addition, it will be necessary to produce iN cells from heterologous cells to perform a cell therapy to treat neurodegenerative diseases induced by genetic mutations. Otherwise, autologous cells would reproduce the mutated phenotype. Similar limitations are shared by iN and iPS cells, such as the concern of manipulating the genetic background of these cells.

4.4 *Advantages and limitations of NPC culture*

The main advantage with the use of human tissue-derived NPCs consists of the possibility to isolate cells from an autologous tissue, whether it is skin, fat or any other somatic tissues for iPS or iN cells. Patients can give their informed consent, hence circumventing the ethical concern. Harvesting skin, fat tissue or even bone marrow can be done quickly, through a minimally painful procedure with local anesthesia. The next steps are more challenging. First, NPC purification remains a major issue since no cell surface antigen allowing for immunoprecipitation enrichment has been identified so far, but some interesting purification steps can readily be achieved through floating neurospheres selection (Fig. 1a).[51]

Another limiting step that still remains is the NPC proliferation. Since the amount of autologous tissue that can be harvested is limited, the proliferation step is crucial to generate enough NPCs for transplantation purposes. The steps for deriving and expanding human adult NPCs are difficult to standardize, extremely complex and expensive to perform, and have not been really efficient.

Also, once NPCs have been grown, they need to be differentiated into the adequate subtype of neurons, depending on the neurodegenerative disorder that needs to be addressed. In that regard, the differentiation potential of human adult NPCs seems to still be more restricted than that of hESCs.[64]

Overall, human adult NPCs appear to be more complicated to culture and consequently it is more challenging to control their fate while trying to generate specific types of neurons.[65] One of the next major challenges would consists of generating enough human adult NPCs to use in therapeutic replacement of damaged tissues in the nervous system.

Most of these limitations will be overcome with the use of iPS and iN cells which seem to have a large potential in terms of proliferation, differentiation and purification, but raise additional concerns related to their long-term stability and potential tumorigenicity.

However, issues such as cell survival after transplantation, appropriate synaptic integration into the host brain and long-term behavioral recovery need to be addressed. Nevertheless, these limitations may be compensated by the autologous nature of these cells, and the absence of ethical concern.

5. Conclusion

Human adult NPCs now represent a true possibility and a open door to several other studies to come on not only cell transplantation as therapeutic approach for various disorders of the nervous system, but also on the creation by tissue engineering of 3D models to study the diseases.[24] The combination of these models with the use of human adult neuronal and glial cells generated from the differentiation of NPCs isolated directly from the patients and from readily accessible tissues sources (skin, fat, bone marrow) or derived from iPS or iN cells, will enable the development of powerful models to better understand human neurodegenerative diseases. Lastly, with the use of autologous human adult NPCs, therapeutic transplants may become one of the next best long-term hopes for reversing neurodegenerative diseases.

Acknowledgments

This work was supported by the Muscular Dystrophy Association (www. mda.org) and the Canadian Institutes of Health Research.

References

1. A. Samii, J. G. Nutt and B. R. Ransom, Parkinson's disease, *Lancet* **363**: 1783–1793 (2004).
2. D. Aarsland, K. Andersen, J. P. Larsen, A. Lolk and P. Kragh-Sorensen, Prevalence and characteristics of dementia in Parkinson disease: an 8-year prospective study, *Arch. Neurol.* **60**: 387–392 (2003).
3. W. Poewe, Treatments for Parkinson disease — past achievements and current clinical needs, *Neurology* **72**: S65–73 (2009).
4. O. Rascol, C. Goetz, W. Koller, W. Poewe and C. Sampaio, Treatment interventions for Parkinson's disease: an evidence based assessment, *Lancet* **359**: 1589–1598 (2002).
5. F. M. Weaver, K. Follett, M. Stern, K. Hur, C. Harris, W. J. Marks, Jr., J. Rothlind, O. Sagher, D. Reda, C. S. Moy, R. Pahwa, K. Burchiel, P. Hogarth, E. C. Lai, J. E. Duda, K. Holloway, A. Samii, S. Horn, J. Bronstein, G. Stoner, J. Heemskerk and G. D. Huang, Bilateral deep brain stimulation vs best medical therapy for patients with advanced Parkinson disease: a randomized controlled trial, *JAMA* **301**: 63–73 (2009).

6. T. Deierborg, D. Soulet, L. Roybon, V. Hall and P. Brundin, Emerging restorative treatments for Parkinson's disease, *Prog. Neurobiol.* **85**: 407–432 (2008).

7. A. Bjorklund and O. Lindvall, Cell replacement therapies for central nervous system disorders, *Nat Neurosci* **3**: 537–544 (2000).

8. C. Johnston, G. Harper and C. Landerfeld, Geriatric Disorders. In: *Current Medical Diagnostic and Treatment*, eds. S. McPhee and M. Papadakis (McGrawHill, New York, 2009), pp. 56–71.

9. V. N. Kornienko and I. N. Pronin, *Diagnostic Neuroradiology*, (Springer, New York, 2009).

10. X. Zhang, D. L. Smith, A. B. Meriin, S. Engemann, D. E. Russel, M. Roark, S. L. Washington, M. M. Maxwell, J. L. Marsh, L. M. Thompson, E. E. Wanker, A. B. Young, D. E. Housman, G. P. Bates, M. Y. Sherman and A. G. Kazantsev, A potent small molecule inhibits polyglutamine aggregation in Huntington's disease neurons and suppresses neurodegeneration *in vivo*, *Proc. Natl. Acad. Sci. USA* **102**: 892–897 (2005).

11. P. Van Damme and W. Robberecht, Recent advances in motor neuron disease, *Curr. Opin. Neurol.* **22**: 486–492 (2009).

12. F. Gros-Louis, C. Gaspar and G. A. Rouleau, Genetics of familial and sporadic amyotrophic lateral sclerosis, *Biochim. Biophys. Acta* **1762**: 956–972 (2006).

13. R. Simon, D. Greenberg and M. Aminoff, *Clinical Neurology*, (McGraw-Hill Professional, New York, 2009).

14. V. V. Belzil, P. N. Valdmanis, P. A. Dion, H. Daoud, E. Kabashi, A. Noreau, J. Gauthier, P. Hince, A. Desjarlais, J. P. Bouchard, L. Lacomblez, F. Salachas, P. F. Pradat, W. Camu, V. Meininger, N. Dupre and G. A. Rouleau, Mutations in FUS cause FALS and SALS in French and French Canadian populations, *Neurology* **73**: 1176–1179 (2009).

15. A. D. Gitler and H. Tsuiji, There has been an awakening: Emerging mechanisms of C9orf72 mutations in FTD/ALS, *Brain Res.* **1647**: 19–29 (2016).

16. X. Wen, T. Westergard, P. Pasinelli and D. Trotti, Pathogenic determinants and mechanisms of ALS/FTD linked to hexanucleotide repeat expansions in the C9orf72 gene, *Neurosci. Lett.* **636**: 16–26 (2017).

17. M. L. Tradewell, H. D. Durham, W. E. Mushynski and B. J. Gentil, Mitochondrial and axonal abnormalities precede disruption of the neurofilament network in a model of charcot-marie-tooth disease type 2E and are prevented by heat shock proteins in a mutant-specific fashion, *J. Neuropathol. Exp. Neurol.* **68**: 642–652 (2009).

18. B. Viviani, E. Corsini, M. Binaglia, C. L. Galli and M. Marinovich, Reactive oxygen species generated by glia are responsible for neuron death induced by human immunodeficiency virus-glycoprotein 120 *in vitro*, *Neuroscience* **107**: 51–58 (2001).

19. D. M. Geddes-Klein, K. B. Schiffman and D. F. Meaney, Mechanisms and consequences of neuronal stretch injury *in vitro* differ with the model of trauma, *J. Neurotrauma* **23**: 193–204 (2006).

20. S. Cho, A. Wood and M. R. Bowlby, Brain slices as models for neurodegenerative disease and screening platforms to identify novel therapeutics, *Curr. Neuropharmacol.* **5**: 19–33 (2007).

21. F. Berthod and F. Gros-Louis, *In Vivo* and *In Vitro* Models to Study Amyotrophic Lateral Sclerosis. In: *Amyotrophic Lateral Sclerosis*, ed. M. Maurer (InTech, Rijeka, 2012), pp. 81–124.

22. H. R. Irons, D. K. Cullen, N. P. Shapiro, N. A. Lambert, R. H. Lee and M. C. Laplaca, Three-dimensional neural constructs: a novel platform for neurophysiological investigation, *J. Neural Eng.* **5**: 333–341 (2008).

23. M. Gingras, V. Gagnon, S. Minotti, H. D. Durham and F. Berthod, Optimized protocols for isolation of primary motor neurons, astrocytes and microglia from embryonic mouse spinal cord, *J. Neurosci. Methods* **163**: 111–118 (2007).

24. M. Gingras, M. M. Beaulieu, V. Gagnon, H. D. Durham and F. Berthod, *In vitro* study of axonal migration and myelination of motor neurons in a three-dimensional tissue-engineered model, *Glia* **56**: 354–364 (2008).

25. H. Braak, E. Ghebremedhin, U. Rub, H. Bratzke and K. Del Tredici, Stages in the development of Parkinson's disease-related pathology, *Cell Tissue Res.* **318**: 121–134 (2004).

26. R. A. Barker, J. Drouin-Ouellet and M. Parmar, Cell-based therapies for Parkinson disease-past insights and future potential, *Nat Rev Neurol* **11**: 492–503 (2015).

27. O. Lindvall, S. Rehncrona, B. Gustavii, P. Brundin, B. Astedt, H. Widner, T. Lindholm, A. Bjorklund, K. L. Leenders, J. C. Rothwell and et al., Fetal dopamine-rich mesencephalic grafts in Parkinson's disease, *Lancet* **2**: 1483–1484 (1988).

28. I. Madrazo, V. Leon, C. Torres, M. C. Aguilera, G. Varela, F. Alvarez, A. Fraga, R. Drucker-Colin, F. Ostrosky, M. Skurovich and et al., Transplantation of fetal substantia nigra and adrenal medulla to the caudate nucleus in two patients with Parkinson's disease, *N. Engl. J. Med.* **318**: 51 (1988).

29. O. Lindvall, P. Brundin, H. Widner, S. Rehncrona, B. Gustavii, R. Frackowiak, K. L. Leenders, G. Sawle, J. C. Rothwell, C. D. Marsden and *et al.*, Grafts of

     C. Auclair-Daigle and F. Berthod

fetal dopamine neurons survive and improve motor function in Parkinson's disease, *Science* **247**: 574–577 (1990).

30. C. R. Freed, P. E. Greene, R. E. Breeze, W. Y. Tsai, W. DuMouchel, R. Kao, S. Dillon, H. Winfield, S. Culver, J. Q. Trojanowski, D. Eidelberg and S. Fahn, Transplantation of embryonic dopamine neurons for severe Parkinson's disease, *N. Engl. J. Med.* **344**: 710–719 (2001).

31. C. W. Olanow, C. G. Goetz, J. H. Kordower, A. J. Stoessl, V. Sossi, M. F. Brin, K. M. Shannon, G. M. Nauert, D. P. Perl, J. Godbold and T. B. Freeman, A double-blind controlled trial of bilateral fetal nigral transplantation in Parkinson's disease, *Ann. Neurol.* **54**: 403–414 (2003).

32. P. Piccini, O. Lindvall, A. Bjorklund, P. Brundin, P. Hagell, R. Ceravolo, W. Oertel, N. Quinn, M. Samuel, S. Rehncrona, H. Widner and D. J. Brooks, Delayed recovery of movement-related cortical function in Parkinson's disease after striatal dopaminergic grafts, *Ann. Neurol.* **48**: 689–695 (2000).

33. P. Brundin, S. Dunnett, A. Bjorklund and G. Nikkhah, Transplanted dopaminergic neurons: more or less?, *Nat. Med.* **7**: 512–513 (2001).

34. D. E. Redmond, Jr., Cellular replacement therapy for Parkinson's disease — where we are today?, *Neuroscientist* **8**: 457–488 (2002).

35. J. Y. Li, E. Englund, J. L. Holton, D. Soulet, P. Hagell, A. J. Lees, T. Lashley, N. P. Quinn, S. Rehncrona, A. Bjorklund, H. Widner, T. Revesz, O. Lindvall and P. Brundin, Lewy bodies in grafted neurons in subjects with Parkinson's disease suggest host-to-graft disease propagation, *Nat. Med.* **14**: 501–503 (2008).

36. I. Mendez, A. Vinuela, A. Astradsson, K. Mukhida, P. Hallett, H. Robertson, T. Tierney, R. Holness, A. Dagher, J. Q. Trojanowski and O. Isacson, Dopamine neurons implanted into people with Parkinson's disease survive without pathology for 14 years, *Nat. Med.* **14**: 507–509 (2008).

37. F. Cicchetti, S. Saporta, R. A. Hauser, M. Parent, M. Saint-Pierre, P. R. Sanberg, X. J. Li, J. R. Parker, Y. Chu, E. J. Mufson, J. H. Kordower and T. B. Freeman, Neural transplants in patients with Huntington's disease undergo disease-like neuronal degeneration, *Proc. Natl. Acad. Sci. USA* **106**: 12483–12488 (2009).

38. J. Altman and G. D. Das, Post-natal origin of microneurons in the rat brain, *Nature* **207**: 953–956 (1965).

39. P. Taupin and F. H. Gage, Adult neurogenesis and neural stem cells of the central nervous system in mammals, *J. Neurosci. Res.* **69**: 745–749 (2002).

40. J. C. Conover and R. Q. Notti, The neural stem cell niche, *Cell Tissue Res.* **331**: 211–224 (2008).

41. J. W. Shim, C. H. Park, Y. C. Bae, J. Y. Bae, S. Chung, M. Y. Chang, H. C. Koh, H. S. Lee, S. J. Hwang, K. H. Lee, Y. S. Lee, C. Y. Choi and S. H. Lee, Generation of functional dopamine neurons from neural precursor cells isolated from the subventricular zone and white matter of the adult rat brain using Nurr1 overexpression, *Stem Cells* **25**: 1252–1262 (2007).

42. R. J. Armstrong, C. B. Hurelbrink, P. Tyers, E. L. Ratcliffe, A. Richards, S. B. Dunnett, A. E. Rosser and R. A. Barker, The potential for circuit reconstruction by expanded neural precursor cells explored through porcine xenografts in a rat model of Parkinson's disease, *Exp. Neurol.* **175**: 98–111 (2002).

43. Y. Jiang, B. N. Jahagirdar, R. L. Reinhardt, R. E. Schwartz, C. D. Keene, X. R. Ortiz-Gonzalez, M. Reyes, T. Lenvik, T. Lund, M. Blackstad, J. Du, S. Aldrich, A. Lisberg, W. C. Low, D. A. Largaespada and C. M. Verfaillie, Pluripotency of mesenchymal stem cells derived from adult marrow, *Nature* **418**: 41–49 (2002).

44. A. Hermann, R. Gastl, S. Liebau, M. O. Popa, J. Fiedler, B. O. Boehm, M. Maisel, H. Lerche, J. Schwarz, R. Brenner and A. Storch, Efficient generation of neural stem cell-like cells from adult human bone marrow stromal cells, *J. Cell Sci.* **117**: 4411–4422 (2004).

45. J. G. Toma, M. Akhavan, K. J. Fernandes, F. Barnabe-Heider, A. Sadikot, D. R. Kaplan and F. D. Miller, Isolation of multipotent adult stem cells from the dermis of mammalian skin, *Nat. Cell Biol.* **3**: 778–784 (2001).

46. K. J. Fernandes, I. A. McKenzie, P. Mill, K. M. Smith, M. Akhavan, F. Barnabe-Heider, J. Biernaskie, A. Junek, N. R. Kobayashi, J. G. Toma, D. R. Kaplan, P. A. Labosky, V. Rafuse, C. C. Hui and F. D. Miller, A dermal niche for multipotent adult skin-derived precursor cells, *Nat. Cell Biol.* **6**: 1082–1093 (2004).

47. J. G. Toma, I. A. McKenzie, D. Bagli and F. D. Miller, Isolation and characterization of multipotent skin-derived precursors from human skin, *Stem Cells* **23**: 727–737 (2005).

48. B. A. Reynolds and S. Weiss, Generation of neurons and astrocytes from isolated cells of the adult mammalian central nervous system, *Science* **255**: 1707–1710 (1992).

49. M. Belicchi, F. Pisati, R. Lopa, L. Porretti, F. Fortunato, M. Sironi, M. Scalamogna, E. A. Parati, N. Bresolin and Y. Torrente, Human skin-derived stem cells migrate throughout forebrain and differentiate into astrocytes after injection into adult mouse brain, *J. Neurosci. Res.* **77**: 475–486 (2004).

50. A. Joannides, P. Gaughwin, C. Schwiening, H. Majed, J. Sterling, A. Compston and S. Chandran, Efficient generation of neural precursors

from adult human skin: astrocytes promote neurogenesis from skin-derived stem cells, *Lancet* **364**: 172–178. (2004).

51. M. Gingras, M. F. Champigny and F. Berthod, Differentiation of human adult skin-derived neuronal precursors into mature neurons, *J. Cell. Physiol.* **210**: 498–506 (2007).

52. J. Biernaskie, J. S. Sparling, J. Liu, C. P. Shannon, J. R. Plemel, Y. Xie, F. D. Miller and W. Tetzlaff, Skin-derived precursors generate myelinating Schwann cells that promote remyelination and functional recovery after contusion spinal cord injury, *J. Neurosci.* **27**: 9545–9559 (2007).

53. A. P. Franco Lambert, A. Fraga Zandonai, D. Bonatto, D. Cantarelli Machado and J. A. Pegas Henriques, Differentiation of human adipose-derived adult stem cells into neuronal tissue: does it work?, *Differentiation* **77**: 221–228 (2009).

54. M. K. McCoy, T. N. Martinez, K. A. Ruhn, P. C. Wrage, E. W. Keefer, B. R. Botterman, K. E. Tansey and M. G. Tansey, Autologous transplants of Adipose-Derived Adult Stromal (ADAS) cells afford dopaminergic neuro-protection in a model of Parkinson's disease, *Exp. Neurol.* **210**: 14–29 (2008).

55. K. Takahashi and S. Yamanaka, Induction of pluripotent stem cells from mouse embryonic and adult fibroblast cultures by defined factors, *Cell* **126**: 663–676 (2006).

56. M. Nakagawa, M. Koyanagi, K. Tanabe, K. Takahashi, T. Ichisaka, T. Aoi, K. Okita, Y. Mochiduki, N. Takizawa and S. Yamanaka, Generation of induced pluripotent stem cells without Myc from mouse and human fibroblasts, *Nat. Biotechnol.* **26**: 101–106 (2008).

57. J. B. Kim, H. Zaehres, G. Wu, L. Gentile, K. Ko, V. Sebastiano, M. J. Arauzo-Bravo, D. Ruau, D. W. Han, M. Zenke and H. R. Scholer, Pluripotent stem cells induced from adult neural stem cells by reprogramming with two factors, *Nature* **454**: 646–650 (2008).

58. H. Zhou, S. Wu, J. Y. Joo, S. Zhu, D. W. Han, T. Lin, S. Trauger, G. Bien, S. Yao, Y. Zhu, G. Siuzdak, H. R. Scholer, L. Duan and S. Ding, Generation of induced pluripotent stem cells using recombinant proteins, *Cell stem cell* **4**: 381–384 (2009).

59. M. Wernig, J. P. Zhao, J. Pruszak, E. Hedlund, D. Fu, F. Soldner, V. Broccoli, M. Constantine-Paton, O. Isacson and R. Jaenisch, Neurons derived from reprogrammed fibroblasts functionally integrate into the fetal brain and improve symptoms of rats with Parkinson's disease, *Proc Natl Acad Sci USA.* **105**: 5856–5861. Epub 2008 Apr 5857 (2008).

60. T. Vierbuchen, A. Ostermeier, Z. P. Pang, Y. Kokubu, T. C. Sudhof and M. Wernig, Direct conversion of fibroblasts to functional neurons by defined factors, *Nature* **463**: 1035–1041 (2010).

61. U. Pfisterer, A. Kirkeby, O. Torper, J. Wood, J. Nelander, A. Dufour, A. Bjorklund, O. Lindvall, J. Jakobsson and M. Parmar, Direct conversion of human fibroblasts to dopaminergic neurons, *Proc. Natl. Acad. Sci. USA* **108**: 10343–10348 (2011).

62. O. Torper, U. Pfisterer, D. A. Wolf, M. Pereira, S. Lau, J. Jakobsson, A. Bjorklund, S. Grealish and M. Parmar, Generation of induced neurons *via* direct conversion *in vivo*, *Proc. Natl. Acad. Sci. USA* **110**: 7038–7043 (2013).

63. E. Y. Son, J. K. Ichida, B. J. Wainger, J. S. Toma, V. F. Rafuse, C. J. Woolf and K. Eggan, Conversion of mouse and human fibroblasts into functional spinal motor neurons, *Cell stem cell* **9**: 205–218 (2011).

64. T. J. Heikkila, L. Yla-Outinen, J. M. Tanskanen, R. S. Lappalainen, H. Skottman, R. Suuronen, J. E. Mikkonen, J. A. Hyttinen and S. Narkilahti, Human embryonic stem cell-derived neuronal cells form spontaneously active neuronal networks *in vitro*, *Exp. Neurol.* **218**: 109–116 (2009).

65. K. A. Galvin and D. G. Jones, Adult human neural stem cells for autologous cell replacement therapies for neurodegenerative disorders, *Neuro-Rehabilitation* **21**: 255–265 (2006).

14

RECENT ADVANCES AND FUTURE PERSPECTIVES ON CELL REPROGRAMMING

Bilal Cakir, Kun-Yong Kim and In-Hyun Park

1. Introduction

Embryonic stem (ES) cells are cells derived from the inner mass of mammalian blastocysts. They have the capacity to self-renew while maintaining pluripotency (the ability to generate all cell types of the three germinal layers). Therefore, a variety of clinical applications has been proposed for ES cells, as well as *in vitro* studies of basic disease mechanisms, screens for drug discovery, and tissue engineering for degenerative diseases. However, ES cells are generic cells that are unrelated to the patient requiring treatment, and the usage of human embryonic tissue continues to be a contentious ethical issue.

Embryonic development and cellular differentiation are presumed to be unidirectional processes, and cells undergo a progressive loss of pluripotency during cell fate specification. In the 1950s, classical experiments demonstrated that differentiated cells retain the genetic information required to revert to pluripotency,[1] and attempts were made to generate cells functionally equivalent to ES cells from differentiated cells (reprogramming) — including by somatic cell nuclear transfer, fusion of differentiated cells with pluripotent cells, application of pluripotent stem cell extracts to somatic cells, and direct *in vitro* adaptation of germ cells.[2] Ultimately in 2006, Takahashi and Yamanaka showed that the enforced expression of four transcription factors — Oct4, Sox2, Klf4, and c-Myc — reprogrammed murine somatic cells to pluripotency.[3] After this discovery, reprogramming drew enormous attention not just from stem cell biologists, but also from clinicians, bioengineers, geneticists, and developmental biologists. In this chapter, we review recent advances in the reprogramming field and discuss the potential future applications of induced pluripotent stem (iPS) cells.

2. Nuclear Reprogramming

Briggs and King showed that nuclei from *Rana pipiens* blastula cells have the ability to undergo normal cleavage and develop into complete embryos when transplanted into enclosed oocytes.[1] This was the first demonstration that the oocyte cytoplasm contains factors that can reprogram the nuclei of differentiated cells back to a pluripotent state that allows donor cells to normally divide and develop into complete embryos. This finding was extended by studies by Gurdon and colleagues, who reported that even fully differentiated intestinal cells from *Xenopus* could be reprogrammed by frog oocytes.[4] Although the efficiency was very low, these oocytes could give rise to adult animals, indicating that the pluripotent state could be reacquired. Somatic cell nuclear transfer (SCNT) was not confirmed in other species until 1997, when Dolly the sheep was cloned.[5] The discovery of reprogramming by SCNT led to further advances in the understanding of the epigenetic processes governing self-renewal and differentiation in murine and human ES cells.

Several groups demonstrated that fusing a somatic cell with an ES cell could induce pluripotency.[6,7] This reprogramming can be accomplished when embryonal carcinoma (EC) cells, embryonic germ (EG) cells, or teratocarcinoma cells are fused with somatic cells. At the tetraploidy stage of the resulting cell, the molecular program of the original somatic cell can be erased and epigenetically reprogrammed to that of the pluripotent cells. With SCNT methods, reprogramming to a pluripotent state occurs within a few days, and can be greatly enhanced by co-expression of pluripotency-associated genes such as Oct4, Sox2, c-Myc, Klf4, Nanog, and Sall4. Yet, each of these methods has clinical limitations. SCNT has the advantage of using intrinsic reprogramming machinery in oocytes; it requires the donation of human oocytes that can be in limited supply and raise ethical concerns. In somatic cell fusion, there are no ethical problems and no limitations on the supply of cells, but the processes result in the formation of the undesired tetraploid that is genetically unstable, and the overexpression of transcription factors increases the frequency of tumorigenesis in the resulting progeny.

In recent years, nuclear and cytoplasmic extracts from undifferentiated cells have been tried to induce the pluripotency of differentiated somatic cells. Extracts from EC cells or ES cells that contain the regulatory components needed for reprogramming can induce the pluripotency of somatic cells.[8] The reprogramming extracts reversibly permeabilize the somatic cell and then induce transcriptional reprogramming, leading to epigenetic modification of the somatic cells. Even interspecies reprogramming via cell extracts showed some successes in dedifferentiating somatic cells. The extracts of regenerated new limbs induced partial dedifferentiation of C2C12-derived mouse myotubes into myoblasts and the extracts of Xenopus eggs induced the expression of the pluripotency genes Oct4 and Sox2 in mammalian somatic cells.[9,10] When fibroblast cells were reversibly permeabilized and transiently exposed to extracts from mouse EC cells, they lead to Oct4 biphasic activation. Incubation of extracts from mouse ES cells with reversibly permeabilized NIH3T3 cells induced partial reprogramming. This approach to inducing pluripotency using pluripotent cell extracts could be an excellent alternative to nuclear transfer reprogramming due to the fact that eggs are not required and the resultant cells are diploid. However, the continuous exposure of somatic cells to cell extracts is challenging, and deriving fully reprogrammed cells via cell extract has yet to be achieved.

Hence, a variety of studies have been proposed to better understand the reprogramming of pluripotent stem cells. However, the principal limitation to all the above technologies for cell reprogramming is that the generated pluripotent stem cells are unrelated to the patient who requires clinical treatment, and the use of human ES cells remains a contentious ethical issue.

3. Reprogramming by Defined Factors

Reprogramming somatic cells by using a defined set of factors is a novel concept and opens extraordinary possibilities for producing pluripotent stem cells without raising the ethical concerns associated with destroying human embryos, while allowing for patients to be treated with their own reprogrammed somatic cells. In their original report, Takahashi and Yamanaka retrovirally transduced different combinations of 24 candidate reprogramming genes into mouse embryonic fibroblasts derived from transgenic mice containing a neomycin resistance gene knocked into the Fbx15 locus.[3] Cells that recovered neomycin resistance from the reactivation of the reporter gene were selected. They found that the retrovirus-mediated introduction of transcription factors was sufficient to reprogram mouse fibroblasts back to a pluripotent state. Specifically, the combined introduction of the four transcription factors — Oct4, Sox2, Klf4, and c-Myc — were identified as sufficient to give rise to pluripotent cells, termed induced pluripotent stem (iPS) cells.

Selected iPS cells showed properties very similar to murine ES cells in morphology, proliferative characteristics, and expression of pluripotency markers. Moreover, iPS cells differentiated into all three germ layers *in vitro*, and were able to form teratomas when injected into nude mice, confirming their pluripotency. Although Fbx15-selected iPS cells formed chimeras, they did not result in germ line transmission, raising the concern that iPS cells are not equivalent to ES cells. However, several studies have reported that iPS cells selected using Nanog or Oct4 are more epigenetically related to ES cells and can produce chimeras capable of germ line transmission.[11,12] However, these iPS cells did not pass the most stringent pluripotency test: tetraploid complementation. Albeit negligible, the continuous expression of reprogramming genes seems to be responsible for the limited differentiation potential of the iPS cells, since the iPS cells produced by inducible lentiviral reprogramming factors succeeded to generate full-term mouse after tetracomplementation.[13,14]

After the identification of murine iPS cells was reported, using the same four transcription factors (Oct4, Sox2, Klf4, and c-Myc) or combined with novel factors(Nanog and Lin28), three different groups isolated human iPS cells.[15–17] Human iPS cells from either combination of factors shared defining characteristics with human ES cells, including gene expression profiles, morphology, proliferation, patterns of DNA methylation, and histone modification, as well as telomerase activities. Human iPS cells were also able to differentiate into three germ layers *in vitro* as embryonic bodies and to form teratomas after injection into immune-deficient mice.

In addition to generate iPS cells by identifying driving factors, the new concept of "direct reprogramming" has recently drawn considerable attention.[18] Direct reprogramming is converting differentiated cells into cells in different lineages or different cells in a same lineage without an intermediate iPS stage. The direct reprogramming strategy is considered more advantageous over iPS reprogramming by being a fast process due to less epigenetic remodeling and usage of minimal combination of factors, producing more mature cells. This approach was originally successful in *in vitro* conversion of fibroblasts to myocytes[19] and recently used to generate several somatic cells such as cardiomyocytes, neurons, and endothelial cells.[18,20] Ieda and his colleagues showed conversion of dermal fibroblasts into cardiomyocyte-like cells by using three transcription factors, Gata4, Mef2c, and Tbx5.[21] In addition, three neural specific transcription factors, Ascl1, Brn2, and Myt1l, were identified to readily convert fibroblasts into neurons.[22] Further studies also indicated the induction of different types of neurons via specified factors.[23,24] Last but not least, direct conversion of endothelial cells (ECs) mediated by SETSIP (SET similar protein) and vascular endothelial growth factor (VEGF) was generated.[25] Direct conversion of ECs could be utilized by vascular tissue engineering, and broaden regenerative medicine's horizon with potential clinical applications.

4. Recent Advances in Reprogramming Methods

When the first iPS cells were isolated, many questions regarding the techniques to generate iPS cells were raised, such as how to identify iPS cells, how to derive iPS cells without potentially tumorigenic oncogenes, and how to generate iPS cells without using retro/lentivirus. Over the last ten years, there have been many advances that have begun to overcome these hurdles.

Initially, Takahashi and Yamanaka created the first iPS cells using the neomycin-resistant gene regulated by the pluripotent cell-specific Fbx15 locus, and they claimed that reprogramming was a very inefficient process.[3] Selection tools seemed essential to identify the iPS cells, and Nanog or Oct4 promoter-based drug resistant genes were used to help isolate murine iPS cells.[11,12,26] Lately, murine and human iPS cells were derived by using the morphologies unique to ES cells.[15–17] Furthermore, silencing of the retrovirus and lentivirus could be retrospectively used to identify the reprogrammed cells from the partially reprogrammed or transformed cells.[27] The induction of the gene that mediates retroviral silencing in ES cells, Trim28, seems to be responsible for retroviral silencing during reprogramming in iPS cells.[28]

The use of oncogenenic c-Myc and Klf4 in generating iPS cells is a major hurdle for potential clinical applications. One of the four reprogramming factors, c-Myc, is a well-established oncogene, and its continued expression and/or potential reactivation in iPS cell derivatives is not suitable for future iPS cell therapies. Indeed, head and neck tumor formation has been observed in iPS cell-derived chimeric mic, possibly due to the reactivation of c-Myc.[11] However, c-Myc is dispensable for reprogramming murine and human somatic cells, although its absence significantly reduces the reprogramming efficiency.[29] Some somatic cells highly express endogenous reprogramming genes and may not need the expression of all reprogramming factors. Indeed, neural progenitor cells (NPCs) that highly express endogenous Sox2 can be reprogrammed without the ectopic expression of Sox2.[30] Furthermore, even introduction of Oct4 alone was shown to be sufficient to generate iPS cells from NPCs.[31,32] Similarly, meningiocytes and keratinocytes appear to be particularly prone to reprogramming due to their relatively high endogenous expression of Sox2, c-Myc, and Klf4.[33,34] These results indicate that the reprogramming cells like NPCs would require a minimal genetic modification in reprogramming and result in most desirably safe iPS cells.

A limitation to the current methods in generating patient-specific iPS cells is the presence of the reprogramming factors in chromosomes. Retro/lentiviral integration into the genome carries a risk of tumor formation when random integration activates pathways for cell proliferation or inhibits the tumor suppressor pathways. Using methods based on non-integrating vectors or direct exposure to reprogramming proteins is desirable. Stadtfeld and colleagues were able to generate iPS cells from adult mouse fibroblast and liver cells using non-integrating adenoviral vectors.[35] They demonstrated that adenoviral-mediated transient expression of the exogenous reprogramming factors eliminated the risk of insertional mutagenesis. Okita *et al.* also successfully generated iPS cells without using any viral vectors but multiple transient transfection of the reprogramming factors.[36] Lately, oriP/EBNA1-based episomal vector was successfully used to generate human iPS cells.[37] All the non-integrating vector methods provide safer iPS cells but suffer from the extremely low reprogramming efficiency. However, another non-integrating system, Sendai virus (SeV), was further developed to generate iPS cells with preserving higher reprogramming efficiency.[38,39] One drawback in using the SeV system for reprogramming is that around 10 passages are required to remove all virus RNA from the generated iPS cells.[38] To overcome this limitation of the SeV system, a temperature-sensitive mutant SeV vector was developed and could be utilized for complete removal of exogenous elements.[39]

Usage of three or four transcription factors for reprogramming results in the high incidence of multiple integration of reprogramming factors into chromosomes. Combining the reprogramming factors with picornaviral 2A sequences allowed the expression of multiple genes in one backbone.[40,41] This method generated iPS cells through less number of retroviral or lentiviral insertion into the genome, as opposed to using separate viral vectors for each reprogramming factor. Because the single viral copy may also be removed from the iPS cell genome after reprogramming (e.g. by loxP/Cre technology) the authors successfully generated safer iPS cells.[42] Similarly, Kaji *et al.* developed a piggyBac transposon-based vector expressing four genes enabling the creation of transgene-free iPS cells following removal of the transposon.[43] Despite these advancements, the concern lingers that all of these factors have links to tumorigenesis.

In addition to several vector-based systems, direct delivery of synthetic mRNAs of pluripotency factors, Oct4, Sox2, Klf4, and c-Myc, is a method to generate iPS cells without the exogenous elements.[44] Warren *et al.* also demonstrated the generation of iPS cells with enhanced conversion efficiency by adding Lin28 mRNA in combination of four factors to human fibroblasts, using a medium containing valproic acid and culturing cells at 5% O_2.[44] Even though this method is more efficient than the other non-integrating systems, it suffers from a few drawbacks. Firstly, this method requires extensive labor and daily mRNA transfection due to the nature of RNA. Additionally, modifications in mRNAs are critical to avoid cell death due to a strong innate immune system that may end up in only partial reprogramming. Lastly, the direct delivery of c-Myc mRNA and thus the high expression of c-Myc, may lead to genomic instability and pose an oncogenic threat to iPS cells. Thus, a similar approach without c-Myc could improve for reprogramming clinically useful iPS cells.

An alternative approach that allows the complete avoidance of the use of oncogenes altogether is to use small molecules instead. Shi *et al.* generated iPS cells from NPCs by transduction of Oct4 and Klf4, and found that simultaneous treatment with a G9a inhibitor, BIX-01294, remarkably increased the reprogramming efficiency.[45] Subsequent studies have confirmed that epigenetic modification or the activation of self-renewing signaling through small molecules can improve reprogramming. Additionally, the histone deacetylase (HDAC) inhibitor valproic acid (VPA) and Trichostatin A (TSA), as well as the DNA methyltransferase inhibitor 5-aza-cytidine, have been shown to increase reprogramming efficiency.[46,47] The Wnt signaling component WNT3a, or the L-channel calcium channel agonist Bayk8644 (BayK) also can increase the reprogramming efficiency.[30] In another study, the inhibition of mitogen-activated protein (MAP) kinase signaling and synthase kinase-3 (GSK3), allowed reprogramming to occur, even in the absence of Sox2 and c-Myc.[48] In addition, Yuan *et al.* demonstrated the Oct-4 induced reprogramming in the presence of transforming growth factor (TGF)-β inhibitor A-8301 and protein arginine methyltransferase inhibitor AMI-5 without Sox2, Klf4, and c-Myc.[49] Indeed, in a later study, iPS reprogramming was established by applying several small molecules, without any exogenous transcription factors.[50]

Hou *et al.* generated chemically induced iPS cells, showing similarity in expression pattern, methylation and histone profiles of four transcription factors compared to mouse ES and iPS cells, by using seven chemical molecules.[50] Recently, tumor suppressor pathways, including p53, Ink4a/Arf, and p21, were shown to play a role as a barrier to reprogramming.[51–54] Signaling pathways that are known to facilitate the ES cell self-renewal or cell proliferation seem to increase the reprogramming efficiency.

Ultimately, iPS cells have been generated without using vectors at all, but rather by directly introducing proteins into fibroblasts. Zhou *et al.* purified recombinant reprogramming factors fused with the poly-arginine (i.e. 11R) protein transduction domain of the C-termini and were able to generate iPS cells from mouse embryonic fibroblasts.[55] Kim *et al.* established 293T cell lines stably expressing reprogramming factors fused with the poly-arginine tag and demonstrated the formation of human iPS cells by exposure of the cell extract to fibroblasts.[56]

5. Future Perspectives on Reprogramming and iPS Cells

Dr. Yamanaka's reports of generating iPS cells using defined factors revolutionized the approach to manipulating the cellular identity. Initiated by the finding that fibroblasts became induced to myogenic lineage by expressing a given myogenic factor,[19] dedifferentiation or reprogramming to different cellular fates have been explored, including hematopoietic, pancreatic, and cardiac lineages.[57–59] Pluripotent iPS cells generated via reprogramming will have a broader applicability, since they can differentiate into any lineages of cells. They will be used to investigate the disease progression *in vitro*, and to provide a platform for screening chemicals for genetic disorders (Fig. 1). The recent advancement of high throughput sequencing, when combined with iPS cells, will allow us to investigate the early development of genetically defined, personalized cells *in vitro*.[60]

Since human ES cells were isolated, they have been used to make models of human diseases.[61,62] Human ES cells genetically modified by either overexpression or knock-down of genes of interest were proposed to mimic the early embryonic cells of human diseases *in vitro*. *In vitro* differentiation of ES cells was presumed to follow the early developmental program of normal embryonic development.[63] iPS cells derived directly

       B. Cakir, K.-Y. Kim and I.-H. Park

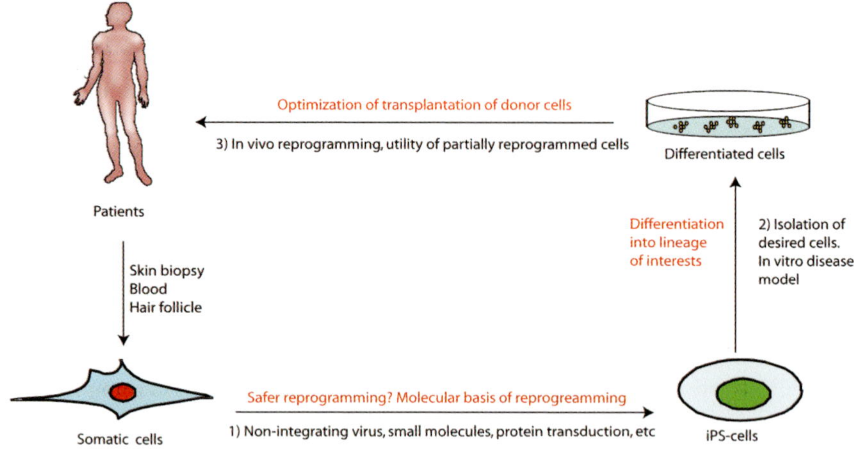

Fig. 1.   Recent advances and future perspectives of reprogramming. Since the Yamanaka lab reported the generation of iPS cells in murine fibroblasts, there was an explosion of research on factor-based reprogramming. (1) iPS cells were generated without using integrating retroviral and lentiviral vectors. Small molecules were identified that increased reprogramming efficiency. (2) Reprogrammed iPS cells will be used to generate *in vitro* disease model. (3) *In vivo* transdifferentiation will be a practical alternative to pluripotent stem cells for genetic or non-genetic degenerative diseases. Shown in red are challenges that will lead to successful utility of the iPS cells in regenerative medicine.

from patients provide an *in vitro* model even closer to the pathology of the diseases. Following are examples of the rigorous attempts to generate human disease models using murine and human iPS cells.

Murine iPS cells have been used to illustrate the therapeutic potential of iPS cells *in vivo*. For example, Hanna and colleagues showed that transgenic mice engineered to have human sickle cell anemia could be successfully treated with hematopoietic progenitor cells produced from autologous iPS cells and genetically repaired.[64] Wernig *et al.* showed that iPS cell-derived neuronal cells could integrate into fetal brains and ameliorate the symptoms of rats with Parkinson's disease.[65] Since then, more studies have reported the derivation of therapeutically relevant cell types from iPS cells, including human insulin-secreting cells, and functional mouse and human cardiomyocytes, as well as mouse endothelial cells that successfully treated mice with coagulation disorder hemophilia A.[66–68]

Human iPS cells from patients' fibroblasts were generated to investigate human diseases *in vitro*. Our group generated iPS cells derived from a variety of genetic disorders.[69] iPS cells generated from patients of neuromuscular diseases, including amyotrophic lateral sclerosis (ALS) and spinal muscular atrophy (SMA), were shown to differentiate into motor neurons, providing a novel *in vitro* motor neuron disease model.[70,71] A range of iPS cells from Parkinson's patients was also generated as an invaluable resource for studying the disease.[72] These cells do not contain transgenic reprogramming factors in their genomes as the factors have been excised by Cre recombinase from the integrated lentiviral vector. Recently, Lee *et al.* extended the iPS cell technology to model autosomal recessive familial dysautonomia (FD) *in vitro*, and demonstrated the FD-related mis-splicing of IKBKAP and concurrent defects in neurogenic differentiation and migration behavior.[73] Most of the iPS cells used to model human diseases still contained the retroviral vectors in their chromosomes. With the advent of reprogramming methods not relying on integrating virus, iPS cells are expected to be more close to ES cells and make a better disease model.[56] In addition, recent advancements in gene-editing technologies, based on the clustered regularly interspaced short palindromic repeats (CRISPR)/CRISPR-associated (CAS) system, enables the genetic modification in human ES or iPS cells with a previously unimaginable efficiency.[74] This facilitates the disease modeling and generating genetically correct human iPS cells for cellular therapeutics for future applications.

Human iPS cells hold great therapeutic potential in regenerative medicine with respect to organ generation. For instance, a recent study introduced clues to organ fabrication by generating a "heart on a chip" to investigate the mechanism of Barth Syndrome (BTHS) disease caused by mutation in tafazzin (*TAZ*) gene.[75] Generating patient-derived iPS cells indicated the role of *TAZ* gene in BTHS and also sarcomere assembly and myocardial contraction at the tissue level. With the advent of a tissue-engineering toolkit such as 3D printing technology[76] and the examples mentioned above, an iPS-driven organ would be applicable in the near future.

Although it is highly hopeful that iPS cells can be utilized in the various applications mentioned above, there are several issues needed to be resolved (Fig. 1). First of all, methods to generate clinically useful iPS cells should be improved. Non-integrating virus, direct protein transduction, and nuclear transfer will make genetically non-modified iPS cells. However, their efficiency is extremely low, and it will not be practical to produce iPS cells in a reliable manner. Improvement in viral vectors or screening small molecules will be essential to optimize the reprogramming methods.[77,78] Even though tremendous improvements have been achieved in non-integrating delivery systems (SeV and direct mRNA delivery), iPS cell reprogramming is still considered slow and inefficient. Additionally, each of the reprogramming methods suffers from a distinct limitation such as safety or efficiency. Identification of clinically more efficient, safe, and useful iPS protocols will be forthcoming regarding the rapid progress in this field.

In order to reach the final goal of regenerative medicine using pluripotent stem cells, it is crucial to make them to acquire the desired lineage *in vitro*. Directed differentiation of pluripotent stem cells relied on the previous knowledge of cell specification obtained from developmental biology.[63] The treatment of cytokines, growth factors, chemicals, and small molecules has been attempted to differentiate pluripotent stem cells. Ectopic expression of lineage-specific transcription factors is a reliable approach to direct the differentiating cells into a specific lineage.[79,80] Cell-surface markers were used to isolate the cells of interest. When mouse ES cell lines expressing fluorescent proteins of lineage-specific markers are available, they provide an effective way to isolate pure cell populations. The isolation of lineage-specific cells from human ES cells was facilitated by the transgenic human ES cell lines with lineage markers. Teratoma formed by the incompletely differentiated pluripotent stem cells continuously impedes the utility of cells differentiated from ES cells. Improvement of negative selection of partially differentiated or non-differentiated cells, or rigorous isolation of completely differentiated cells, is required. *In vivo* transdifferentiation will be an alternative to iPS cells of great practical importance. Most of the degenerative disorders show loss in parenchymal or functional cells in tissues: loss of endocrine beta cells in type I diabetes, dopaminergic neurons in Parkinson's diseases, and motor neurons in ALS. Despite the degeneration of these cell types, the stromal, adjacent, or surrounding cells are still intact. When expressed with a combination of genes important for the development of the same lineages of the degenerated cells, the surviving cells in the damaged tissue will change their cellular fate to the cells of interest and will help to recover the function. Given the great opportunity of manipulating cell fates, it will soon be possible to conquer devastating diseases, but there is much to overcome yet.

Acknowledgements

We thank the Park lab members for discussion and critical reading of the manuscript.

References

1. R. Briggs and T. J. King, Transplantation of living nuclei from blastula cells into enucleated frogs' eggs, *Proc Natl Acad Sci USA* **38**: 455–463 (1952).
2. K. Hochedlinger and R. Jaenisch, Nuclear reprogramming and pluripotency, *Nature* **441**: 1061–1067 (2006).
3. K. Takahashi and S. Yamanaka, Induction of pluripotent stem cells from mouse embryonic and adult fibroblast cultures by defined factors, *Cell* **126**: 663–676 (2006).
4. J. B. Gurdon, T. R. Elsdale and M. Fischberg, Sexually mature individuals of *Xenopus laevis* from the transplantation of single somatic nuclei, *Nature* **182**: 64–65 (1958).
5. K. H. Campbell, J. McWhir, W. A. Ritchie and I. Wilmut, Sheep cloned by nuclear transfer from a cultured cell line, *Nature* **380**: 64–66 (1996).
6. M. Tada, Y. Takahama, K. Abe, N. Nakatsuji and T. Tada, Nuclear reprogramming of somatic cells by *in vitro* hybridization with ES cells, *Curr Biol* **11**: 1553–1558 (2001).
7. C. A. Cowan, J. Atienza, D. A. Melton and K. Eggan, Nuclear reprogramming of somatic cells after fusion with human embryonic stem cells, *Science* **309**: 1369–1373 (2005).
8. C. T. Freberg, J. A. Dahl, S. Timoskainen and P. Collas, Epigenetic reprogramming of OCT4 and NANOG regulatory regions by embryonal carcinoma cell extract, *Mol Biol Cell* **18**: 1543–1553 (2007).
9. K. Miyamoto, T. Furusawa, M. Ohnuki, S. Goel, T. Tokunaga, N. Minami, M. Yamada, K. Ohsumi and H. Imai, Reprogramming events of mammalian somatic cells induced by *Xenopus laevis* egg extracts, *Mol Reprod Dev* **74**: 1268–1277 (2007).

10. C. J. McGann, S. J. Odelberg and M. T. Keating, Mammalian myotube dedifferentiation induced by newt regeneration extract, *Proc Natl Acad Sci USA* **98**: 13699–13704 (2001).

11. K. Okita, T. Ichisaka and S. Yamanaka, Generation of germline-competent induced pluripotent stem cells, *Nature* **448**: 313–317 (2007).

12. M. Wernig, A. Meissner, R. Foreman, T. Brambrink, M. Ku, K. Hochedlinger, B. E. Bernstein and R. Jaenisch, *In vitro* reprogramming of fibroblasts into a pluripotent ES-cell-like state. *Nature*, doi:10.1038/nature05944 (2007).

13. X. Y. Zhao, W. Li, Z. Lv, L. Liu, M. Tong, T. Hai, J. Hao, C. L. Guo, Q. W. Ma, L. Wang, F. Zeng and Q. Zhou, iPS cells produce viable mice through tetraploid complementation, *Nature* **461**: 86–90 (2009).

14. M. J. Boland, J. L. Hazen, K. L. Nazor, A. R. Rodriguez, W. Gifford, G. Martin, S. Kupriyanov and K. K. Baldwin, Adult mice generated from induced pluripotent stem cells, *Nature* **461**: 91–94 (2009).

15. K. Watanabe, M. Ueno, D. Kamiya, A. Nishiyama, M. Matsumura, T. Wataya, J. B. Takahashi, S. Nishikawa, S. Nishikawa, K. Muguruma and Y. Sasai, A ROCK inhibitor permits survival of dissociated human embryonic stem cells, *Nat Biotechnol* **25**: 681–686 (2007).

16. J. Yu, M. A. Vodyanik, K. Smuga-Otto, J. Antosiewicz-Bourget, J. L. Frane, S. Tian, J. Nie, G. A. Jonsdottir, V. Ruotti, R. Stewart, I. I. Slukvin and J. A. Thomson, Induced pluripotent stem cell lines derived from human somatic cells, *Science* **318**: 1917–1920 (2007).

17. I. H. Park, R. Zhao, J. A. West, A. Yabuuchi, H. Huo, T. A. Ince, P. H. Lerou, M. W. Lensch and G. Q. Daley, Reprogramming of human somatic cells to pluripotency with defined factors, *Nature* **451**: 141–146 (2008).

18. S. A. Morris, Direct lineage reprogramming *via* pioneer factors: a detour through developmental gene regulatory networks, *Development* **143**: 2696–2705 (2016).

19. R. L. Davis, H. Weintraub and A. B. Lassar, Expression of a single transfected cDNA converts fibroblasts to myoblasts, *Cell* **51**: 987–1000 (1987).

20. S. A. Morris and G. Q. Daley, A blueprint for engineering cell fate: current technologies to reprogram cell identity, *Cell Res* **23**: 33–48 (2013).

21. M. Ieda, J. D. Fu, P. Delgado-Olguin, V. Vedantham, Y. Hayashi, B. G. Bruneau and D. Srivastava, Direct reprogramming of fibroblasts into functional cardiomyocytes by defined factors, *Cell* **142**: 375–386 (2010).

22. T. Vierbuchen, A. Ostermeier, Z. P. Pang, Y. Kokubu, T. C. Südhof and M. Wernig, Direct conversion of fibroblasts to functional neurons by defined factors, *Nature* **463**: 1035–1041 (2010).

23. U. Pfisterer, A. Kirkeby, O. Torper, J. Wood, J. Nelander, A. Dufour, A. Björklund, O. Lindvall, J. Jakobsson and M. Parmar, Direct conversion of human fibroblasts to dopaminergic neurons, *Proc Natl Acad Sci USA* **108**: 10343–10348 (2011).

24. E. Lujan, S. Chanda, H. Ahlenius, T. C. Südhof and M. Wernig, Direct conversion of mouse fibroblasts to self-renewing, tripotent neural precursor cells, *Proc Natl Acad Sci USA* **109**: 2527–2532 (2012).

25. A. Margariti, B. Winkler, E. Karamariti, A. Zampetaki, T. N. Tsai, D. Baban, J. Ragoussis, Y. Huang, J. D. J. Han and L. Zeng, Direct reprogramming of fibroblasts into endothelial cells capable of angiogenesis and reendothelialization in tissue-engineered vessels. *Proc Natl Acad Sci USA* **109**: 13793–13798 (2012).

26. N. Maherali, R. Sridharan, W. Xie, J. Utikal, S. Eminli, K. Arnold, M. Stadtfeld, R. Yachechko, J. Tcheiu, R. Jaenisch, K. Plath and K. Hochedlinger, Directly reprogrammed fibroblasts show global epigenetic remodeling and widespread tissue contribution, *Cell Stem Cell* **1**: 55–70 (2007).

27. E. M. Chan, S. Ratanasirintrawoot, I. H. Park, P. D. Manos, Y. H. Loh, H. Huo, J. D. Miller, O. Hartung, J. Rho, T. A. Ince, G. Q. Daley and T. M. Schlaeger, Live cell imaging distinguishes bona fide human iPS cells from partially reprogrammed cells, *Nat Biotechnol* **27**: 1033–1037 (2009).

28. D. Wolf and S. P. Goff, TRIM28 mediates primer binding site-targeted silencing of murine leukemia virus in embryonic cells. *Cell* **131**: 46–57 (2007).

29. M. Nakagawa, M. Koyanagi, K. Tanabe, K. Takahashi, T. Ichisaka, T. Aoi, K. Okita, Y. Mochiduki, N. Takizawa and S. Yamanaka, Generation of induced pluripotent stem cells without Myc from mouse and human fibroblasts. *Nat Biotechnol* **26**: 101–106 (2008).

30. Y. Shi, C. Desponts, J. T. Do, H. S. Hahm, H. R. Scholer and S. Ding, Induction of pluripotent stem cells from mouse embryonic fibroblasts by Oct4 and Klf4 with small-molecule compounds, *Cell Stem Cell* **3**: 568–574 (2008).

31. J. B. Kim, H. Zaehres, G. Wu, L. Gentile, K. Ko, V. Sebastiano, M. J. Arauzo-Bravo, D. Ruau, D. W. Han, M. Zenke and H. R. Schöler, Pluripotent stem cells induced from adult neural stem cells by reprogramming with two factors, *Nature* **454**: 646–650 (2008).

32. J. B. Kim, B. Greber, M. J. Arauzo-Bravo, J. Meyer, K. I. Park, H. Zaehres and H. R. Schöler, Direct reprogramming of human neural stem cells by OCT4, *Nature* **461**: 649–653 (2009).

33. D. Qin, Y. Gan, K. Shao, H. Wang, W. Li, T. Wang, W. He, J. Xu, Y. Zhang, Z. Kou, L. Zeng, G. Sheng, M. A. Esteban, S. Gao and D. Pei, Mouse

meningiocytes express Sox2 and yield high efficiency of chimeras after nuclear reprogramming with exogenous factors, *J Biol Chem* **283**: 33730–33735 (2008).

34. T. Aoi, K. Yae, M. Nakagawa, T. Ichisaka, K. Okita, K. Takahashi, T. Chiba and S. Yamanaka, Generation of pluripotent stem cells from adult mouse liver and stomach cells, *Science* **321**: 699–702 (2008).

35. M. Stadtfeld, M. Nagaya, J. Utikal, G. Weir and K. Hochedlinger, Induced pluripotent stem cells generated without viral integration, *Science* **322**: 945–949 (2008).

36. K. Okita, M. Nakagawa, H. Hyenjong, T. Ichisaka and S. Yamanaka, Generation of mouse induced pluripotent stem cells without viral vectors, *Science* **322**: 949–953 (2008).

37. J. Yu, K. Hu, K. Smuga-Otto, S. Tian, R. Stewart, I. I. Slukvin and J. A. Thomson, Human induced pluripotent stem cells free of vector and transgene sequences, *Science* **324**: 797–801 (2009).

38. N. Fusaki, B. Hiroshi, A. Nishiyama, K. Saeki and M. Hasegawa, Efficient induction of transgene-free human pluripotent stem cells using a vector based on Sendai virus, an RNA virus that does not integrate into the host genome, *Proc Jpn Acad Ser B Phys Biol Sci* **85**: 348–362 (2009).

39. H. Ban, N. Nishishita, N. Fusaki, T. Tabata, K. Saeki, M. Shikamura, N. Takada, M. Inoue, M. Hasegawa and S. Kawamata, Efficient generation of transgene-free human induced pluripotent stem cells (iPSCs) by temperature-sensitive Sendai virus vectors, *Proc Natl Acad Sci USA* **108**: 14234–14239 (2011).

40. B. W. Carey, S. Markoulaki, J. Hanna, K. Saha, Q. Gao, M. Mitalipova and R. Jaenisch, Reprogramming of murine and human somatic cells using a single polycistronic vector, *Proc Natl Acad Sci USA* **106**: 157–162 (2009).

41. C. A. Sommer, M. Stadtfeld, G. J. Murphy, K. Hochedlinger, D. N. Kotton and G. Mostoslavsky, iPS cell generation using a single lentiviral stem cell cassette, *Stem Cells* **18**: 18 (2008).

42. C. A. Sommer, G. A. Sommer, T. A. Longmire, C. Christodoulou, D. D. Thomas, M. Gostissa, F. W. Alt, G. J. Murphy, D. N. Kotton and G. Mostoslavsky, Excision of reprogramming transgenes improves the differentiation potential of iPS cells generated with a single excisable vector, *Stem Cells* **10**: 10 (2009).

43. K. Kaji, K. Norrby, A. Paca, M. Mileikovsky, P. Mohseni and K. Woltjen, Virus-free induction of pluripotency and subsequent excision of reprogramming factors, *Nature* **458**: 771–775 (2009).

44. L. Warren, P. D. Manos, T. Ahfeldt, Y. H. Loh, H. Li, F. Lau, W. Ebina, P. K. Mandal, Z. D. Smith and A. Meissner, Highly efficient reprogramming to pluripotency and directed differentiation of human cells with synthetic modified mRNA, *Cell Stem Cell* **7**: 618–630 (2010).

45. Y. Shi, J. T. Do, C. Desponts, H. S. Hahm, H. R. Scholer and S. Ding, A combined chemical and genetic approach for the generation of induced pluripotent stem cells, *Cell Stem Cell* **2**: 525–528 (2008).

46. D. Huangfu, R. Maehr, W. Guo, A. Eijkelenboom, M. Snitow, A. E. Chen and D. A. Melton, Induction of pluripotent stem cells by defined factors is greatly improved by small-molecule compounds, *Nat Biotechnol* **26**: 795–797 (2008).

47. D. Huangfu, K. Osafune, R. Maehr, W. Guo, A. Eijkelenboom, S. Chen, W. Muhlestein and D. A. Melton, Induction of pluripotent stem cells from primary human fibroblasts with only Oct4 and Sox2, *Nat Biotechnol* **26**: 1269–1275 (2008).

48. J. Silva, O. Barrandon, J. Nichols, J. Kawaguchi, T. W. Theunissen and A. Smith, Promotion of reprogramming to ground state pluripotency by signal inhibition, *PLoS Biol* **6**: e253 (2008).

49. X. Yuan, H. Wan, X. Zhao, S. Zhu, Q. Zhou and S. Ding, Brief report: combined chemical treatment enables Oct4-induced reprogramming from mouse embryonic fibroblasts, *Stem Cells* **29**: 549–553 (2011).

50. P. Hou, Y. Li, X. Zhang, C. Liu, J. Guan, H. Li, T. Zhao, J. Ye, W. Yang and K. Liu, Pluripotent stem cells induced from mouse somatic cells by small-molecule compounds, *Science* **341**: 651–654 (2013).

51. H. Li, M. Collado, A. Villasante, K. Strati, S. Ortega, M. Canamero, M. A. Blasco and M. Serrano, The Ink4/Arf locus is a barrier for iPS cell reprogramming, *Nature* **460**: 1136–1139 (2009).

52. H. Hong, K. Takahashi, T. Ichisaka, T. Aoi, O. Kanagawa, M. Nakagawa, K. Okita and S. Yamanaka, Suppression of induced pluripotent stem cell generation by the p53-p21 pathway, *Nature* **460**: 1132–1135 (2009).

53. T. Kawamura, J. Suzuki, Y. V. Wang, S. Menendez, L. B. Morera, A. Raya, G. M. Wahl, and J. C. Belmonte, Linking the p53 tumour suppressor pathway to somatic cell reprogramming, *Nature* **460**: 1140–1144 (2009).

54. R. M. Marion, K. Strati, H. Li, M. Murga, R. Blanco, S. Ortega, O. Fernandez-Capetillo, M. Serrano and M. A. Blasco, A p53-mediated DNA damage response limits reprogramming to ensure iPS cell genomic integrity, *Nature* **460**: 1149–1153 (2009).

55. H. Zhou, S. Wu, J. Y. Joo, S. Zhu, D. W. Han, T. Lin, S. Trauger, G. Bien, S. Yao, Y. Zhu, G. Siuzdak, H. R. Schöler, L. Duan and S. Ding, Generation

of induced pluripotent stem cells using recombinant proteins, *Cell Stem Cell* **4**: 381–384 (2009).

56. D. Kim, C. H. Kim, J. I. Moon, Y. G. Chung, M. Y. Chang, B. S. Han, S. Ko, E. Yang, K. Y. Cha, R. Lanza and K. S. Kim, Generation of human induced pluripotent stem cells by direct delivery of reprogramming proteins, *Cell Stem Cell* **4**: 472–476 (2009).

57. H. Xie, M. Ye, R. Feng and T. Graf, Stepwise reprogramming of B cells into macrophages, *Cell* **117**: 663–676 (2004).

58. Q. Zhou, J. Brown, A. Kanarek, J. Rajagopal and D. A. Melton, *In vivo* reprogramming of adult pancreatic exocrine cells to beta-cells, *Nature* **455**: 627–632 (2008).

59. J. K. Takeuchi and B. G. Bruneau, Directed transdifferentiation of mouse mesoderm to heart tissue by defined factors, *Nature* **459**: 708–711 (2009).

60. J. H. Lee, I. H. Park, Y. Gao, J. B. Li, Z. Li, G. Q. Daley, K. Zhang and G. M. Church, A robust approach to identifying tissue-specific gene expression regulatory variants using personalized human induced pluripotent stem cells, *PLoS Genet* **5**: e1000718 (2009).

61. M. W. Lensch and G. Q. Daley, Scientific and clinical opportunities for modeling blood disorders with embryonic stem cells, *Blood* **107**: 2605–2612 (2006).

62. Y. Verlinsky, N. Strelchenko, V. Kukharenko, S. Rechitsky, O. Verlinsky, V. Galat and A. Kuliev, Human embryonic stem cell lines with genetic disorders, *Reprod Biomed Online* **10**: 105–110 (2005).

63. G. Keller, Embryonic stem cell differentiation: emergence of a new era in biology and medicine, *Genes Dev* **19**: 1129–1155 (2005).

64. J. Hanna, M. Wernig, S. Markoulaki, C. W. Sun, A. Meissner, J. P. Cassady, C. Beard, T. Brambrink, L. C. Wu, T. M. Townes and R. Jaenisch, Treatment of sickle cell anemia mouse model with iPS cells generated from autologous skin, *Science* **318**: 1920–1923 (2007).

65. M. Wernig, J. P. Zhao, J. Pruszak, E. Hedlund, D. Fu, F. Soldner, V. Broccoli, M. Constantine-Paton, O. Isacson and R. Jaenisch, Neurons derived from reprogrammed fibroblasts functionally integrate into the fetal brain and improve symptoms of rats with Parkinson's disease, *Proc Natl Acad Sci USA* **105**: 5856–5861 (2008).

66. L. Zwi, O. Caspi, G. Arbel, I. Huber, A. Gepstein, I. H. Park and L. Gepstein, Cardiomyocyte differentiation of human induced pluripotent stem cells, *Circulation* **120**: 1513–1523 (2009).

67. R. Maehr, S. Chen, M. Snitow, T. Ludwig, L. Yagasaki, R. Goland, R. L. Leibel and D. A. Melton, Generation of pluripotent stem cells from patients with type 1 diabetes, *Proc Natl Acad Sci USA* **106**: 15768–15773 (2009).

68. D. Xu, Z. Alipio, L. M. Fink, D. M. Adcock, J. Yang, D. C. Ward and Y. Ma, Phenotypic correction of murine hemophilia A using an iPS cell-based therapy, *Proc Natl Acad Sci USA* **106**: 808–813 (2009).

69. I. H. Park, N. Arora, H. Huo, N. Maherali, T. Ahfeldt, A. Shimamura, M. W. Lensch, C. Cowan, K. Hochedlinger and G. Q. Daley, Disease-specific induced pluripotent stem cells, *Cell* **134**: 877–886 (2008).

70. J. T. Dimos, K. T. Rodolfa, K. K. Niakan, L. M. Weisenthal, H. Mitsumoto, W. Chung, G. F. Croft, G. Saphier, R. Leibel, R. Goland, H. Wichterle, C. E. Henderson and K. Eggan, Induced pluripotent stem cells generated from patients with ALS can be differentiated into motor neurons, *Science* **321**: 1218–1221 (2008).

71. A. D. Ebert, J. Yu, F. F. Rose, Jr., V. B. Mattis, C. L. Lorson, J. A. Thomson and C. N. Svendsen, Induced pluripotent stem cells from a spinal muscular atrophy patient, *Nature* **457**: 277–280 (2009).

72. F. Soldner, D. Hockemeyer, C. Beard, Q. Gao, G. W. Bell, E. G. Cook, G. Hargus, A. Blak, O. Cooper, M. Mitalipova, O. Isacson and R. Jaenisch, Parkinson's disease patient-derived induced pluripotent stem cells free of viral reprogramming factors, *Cell* **136**: 964–977 (2009).

73. G. Lee, E. P. Papapetrou, H. Kim, S. M. Chambers, M. J. Tomishima, C. A. Fasano, Y. M. Ganat, J. Menon, F. Shimizu, A. Viale, V. Tabar, M. Sadelain and L. Studer, Modelling pathogenesis and treatment of familial dysautono-mia using patient-specific iPSCs, *Nature* **461**: 402–406 (2009).

74. S. N. Waddington, R. Privolizzi, R. Karda and H. C. O'Neill, A broad over-view and review of CRISPR-Cas technology and stem cells, *Curr Stem Cell Rep* **2**: 9–20 (2016).

75. G. Wang, M. L. McCain, L. Yang, A. He, F. S. Pasqualini, A. Agarwal, H. Yuan, D. Jiang, D. Zhang and L. Zangi, Modeling the mitochondrial car-diomyopathy of Barth syndrome with induced pluripotent stem cell and heart-on-chip technologies, *Nat Med* **20**: 616–623 (2014).

76. S. V. Murphy and A. Atala, 3D bioprinting of tissues and organs, *Nat Biotechnol* **32**: 773–785 (2014).

77. S. Chen, M. Borowiak, J. L. Fox, R. Maehr, K. Osafune, L. Davidow, K. Lam, L. F. Peng, S. L. Schreiber, L. L. Rubin and D. Melton, A small molecule that directs differentiation of human ESCs into the pancreatic line-age, *Nat Chem Biol* **5**: 258–265 (2009).

78. T. Lin, R. Ambasudhan, X. Yuan, W. Li, S. Hilcove, R. Abujarour, X. Lin, H. S. Hahm, E. Hao, A. Hayek and S. Ding, A chemical platform for improved induction of human iPSCs, *Nat Methods* **6**: 805–808 (2009).
79. M. Kyba, R. C. Perlingeiro and G. Q. Daley, HoxB4 confers definitive lymphoid-myeloid engraftment potential on embryonic stem cell and yolk sac hematopoietic progenitors, *Cell* **109**: 29–37 (2002).
80. R. Darabi, K. Gehlbach, R. M. Bachoo, S. Kamath, M. Osawa, K. E. Kamm, M. Kyba and R. C. Perlingeiro, Functional skeletal muscle regeneration from differentiating embryonic stem cells. *Nat Med* **14**: 134–143 (2008).

15

HIGH-THROUGHPUT SYSTEMS FOR STEM CELL ENGINEERING

David A. Brafman, Karl Willert and Shu Chien

1. Introduction

Stem cells have the unique ability to grow in number indefinitely (proliferate) or adopt new cellular fates (differentiate). As such, stem cells represent a unique cell-based system to study and model human development and diseases, to screen safety and mode of action of novel drugs, and to provide the raw material for cell-based therapies of presently incurable diseases. Human pluripotent stem cells (hPSCs), which are able to differentiate into all cell types in the human body, have the potential to revolutionize the treatment of many human disorders for which no effective therapies presently exist. Experimental manipulation of these cells to affect proliferation and differentiation is central to developing strategies for the production of defined and mature cell types that can be used to study development and disease progression, to perform drug screens, and to treat a variety of degenerative disorders such as Alzheimer's and heart disease.

Conventional cell culture methods are limited in their ability to screen the vast number of factors that can influence stem cell behavior. The establishment of high-throughput screen (HTS) technologies with stem cells is important for a broad range of applications from basic understanding of the role of certain signaling networks in self-renewal to the development of novel therapeutic approaches, such as cell replacement of damaged, diseased, or dead tissues. HTS technologies generally consist of three interrelated components: (1) platform fabrication, (2) data acquisition, and (3) data analysis and mining (Fig. 1). The behavior of stem cells can be controlled either by altering their extracellular environment (extrinsic manipulation) or by interfering with intracellular signaling pathways and transcriptional networks (intrinsic manipulation). This chapter discusses the emerging trend of using HTS technologies for the extrinsic and intrinsic manipulation and engineering of stem cells.

2. Sources of Stem Cells Suitable for High-Throughput Screening Approaches

A stem cell is defined by its ability to (1) maintain its undifferentiated state, or "blank slate," and (2) differentiate into mature and specialized cell types. Broadly speaking, there are two types of stem cell populations which vary significantly in their properties. Pluripotent stem cells can in principle be maintained indefinitely in an undifferentiated and highly proliferative state while retaining their pluripotency (i.e. the ability to

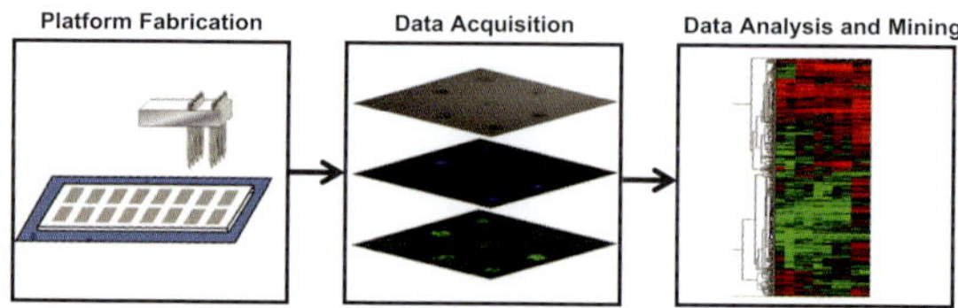

Fig. 1. Schematic of the three interrelated components of typical HTS approaches: (1) platform fabrication, (2) data acquisition, and (3) data analysis and mining.

differentiate into all mature cell types). In contrast, adult stem cells are restricted in their potency (multipotent or unipotent), giving rise to only a subset of specialized cell types. In addition, adult stem cells are extremely rare and hence difficult to isolate in pure form and expand in an undifferentiated state in the culture dish. Given these limitations, high-throughput screens are often not feasible with adult stem cells, and pluripotent stem cell lines provide a powerful alternative.

Until recently, human pluripotent stem cells were mainly derived by dissection of the inner cell mass from the blastocysts. The discoveries by Takahashi and Yamanaka ushered in a new era for human pluripotent stem cells: by introducing a combination of transcription factors into mature and specialized cell types, it is now possible to derive pluripotent stem cells, or induced pluripotent stem cells (iPSCs), that are nearly identical to embryonic stem cells (ESCs). This technology permits the derivation of patient-specific stem cell lines. Together, embryonic and induced pluripotent stem cell lines provide a virtually limitless supply of human cells for large-scale high-throughput screens of any kind.

3. The Stem Cell Niche: A Cellular Microenvironment That Controls Stem Cell Behavior

The *in vivo* cellular microenvironment is a complex mixture consisting of four distinct "components": (1) Immobilized protein factors such as extra-cellular matrix proteins (ECMPs) interact with the cell through integrin binding. (2) Soluble protein factors (such as growth factors, small molecules, and hormones) influence cell signaling pathways via their appropriate extracellular receptors. (3) Mechanical forcers (such as stretch or shear stress) modulate intracellular signaling processes by activating mechanoreceptors. (4) Neighboring cell types communicate with each other by cadherin-mediated signaling (Fig. 2).[1,2] Each of these four components include tens to thousands of "members" that influence numerous signaling pathways, which perturb gene and protein expressions, and ultimately affect cell fate and cell function.

Members of each of these components interact in complex manner to influence each other's signaling ability, a phenomenon known as cross-talk. For example, endothelial cell attachment to fibronectin *via* $\alpha 5\beta 1$

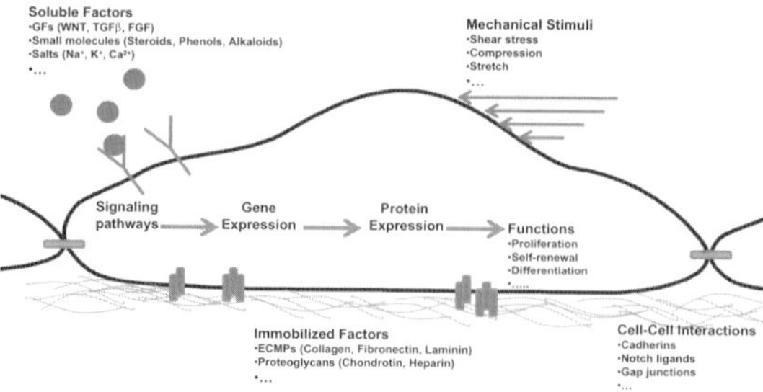

Fig. 2. Cellular microenvironment. Four distinct components of the cellular microenvironment that affect cell fate are as follows: (1) Immobilized proteins such as extracellular matrix proteins interact with the cell most commonly through integrin signaling. (2) Cell to cell interactions are mediated through cadherin signaling. (3) Chemical stimuli such as growth factors and hormones interact with the cell through their respective receptors. (4) Mechanical stimuli such as shear stress or flow signal the cell through mechanoreceptors. These distinct compartments have numerous members that interact in a complex manner, known as crosstalk, to affect a variety of signaling pathways which in turn affect gene and protein expression and ultimately cell fate and function.

integrin potentiates αvβ3-mediated migration on vitronectin.[3] Along similar lines, growth factors and ECMPs often have a reciprocal relationship — cell adhesion to ECMPs is required for activation of growth factor receptors, and growth factors are necessary to stimulate cell adhesion, migration, and the resulting integrin-dependent response. These interactions, which can be synergistic or antagonistic, are well documented in numerous biological systems.[4,5] For example, αvβ3 integrin associates with activated insulin growth factor (IGF) receptor and platelet-derived growth factor (PDGF) receptor, and potentiates the biological activity of IGF and PDGF, respectively.[6] Likewise, in human blood mononuclear cells, collagen-induced release of interleukin 1 (IL-1) through the binding of integrin α7β1 is potentiated by fibronectin binding to α5β1.[7] These examples demonstrate the importance in crosstalk in regulating cell fate. Since these interactions are complex and not predictable from studies on individual members of the various components, high-throughput systematic approaches to study crosstalk in stem cells are needed.

In vivo, stem cells reside in specialized microenvironments of organs and tissues called niches that regulate self-renewal and differentiation.[8,9]

The niche also serves to balance the choice of a stem cell to self-renew or differentiate: excessive self-renewal could lead to cancer,[10] while differentiation could lead to depletion of a tissue's regenerative potential. Thus, the niche must maintain a delicate balance between stem cell proliferation and differentiation.

Stem cell niches of various tissues with regenerative potential, such as skin, stomach, intestine, and blood, share certain structural and organizational properties. The niche cells provide cell-cell contacts and paracrine signaling that regulate the self-renewal of the neighboring stem cells. The extracellular matrix provides a scaffold for stem cell growth in the niche and can interact with soluble factors to regulate signal transduction. Additionally, glycoaminoglycans serve to locally concentrate and present soluble cytokines. The physiochemical environment, including oxygen gradients, pH, matrix stiffness, and topography, also contribute to the regulation of stem cell proliferation, self-renewal, and differentiation.[11]

The first stem cell niche identified in mammals was the hematopoietic stem cell (HSC) niche. Individual HSCs are multipotent and highly self-renewing, but yet proliferate quite slowly.[12] However, the manner in which HSCs interact with their niche to promote self-renewal has not been elucidated. Furthermore, the few existing culture systems that allow for maintenance and expansion of HSCs *in vitro* are not well defined.[13] More recently, niches have been identified in a wide range of tissues such as the skin, brain, gut, and liver.[14,15] Another well-characterized stem cell niche is that of the intestinal stem cells. Intestinal stem cells can be isolated to near homogeneity using the marker gene Lgr5, and these cells produce *in vitro* an intestinal-like structure that generates its own niche cells to maintain the Lgr5-marked stem cells. However, in most cases, expansion of stem/progenitor cell populations *in vitro* without loss of stem cell potential is often difficult or even impossible.

Human ESCs (hESCs) and iPSCs are able to generate all derivatives of the three primary germ layers — ectoderm, endoderm, and mesoderm — a property referred to as pluripotency.[16] Unlike tissue-specific adult stem cells, hESCs are only transiently present during development and do not have a stable niche *in vivo*. Nonetheless, various *in vitro* culture conditions for hESC proliferation and differentiation have been developed. A combination of components, including ECMPs, soluble factors, and other physiochemical factors, act to maintain and expand the stem cell population or promote particular differentiation programs. However, given the myriad of factors that may influence hESC proliferation and differentiation, the *in vitro* culture conditions and the cellular microenvironments that either promote hESC expansion or their specific differentiation have not been successfully developed and defined. Thus, the development of HTS methodologies is necessary in order to identify culture conditions for the maintenance, expansion, and directed differentiation of stem cells.

3.1 *Traditional HTS for indentifying modulators of stem cell fate*

Traditional HTS use microtiter plates along with robotics, liquid handling devices, and automated imagers to conduct rapidly thousands to millions of conditions (Fig. 3a). A highly sensitive, robust, and quick readout is critical, as is the case for any HTS method. Most HTS typically involve the use of libraries consisting of hundreds of thousands or even millions of small molecules that are created through combinatorial chemistry approaches, in order to identify compounds that produce the desired phenotype.[17,18] These screens are a useful tool to address questions in basic stem cell biology and chemistry.

Many groups have used HTS of chemical libraries to identify small molecules that regulate mouse ESC (mESC) self-renewal. For example, a transgenic reporter mESC line that expresses GFP under control of Oct4, a marker of pluripotency, was utilized to screen 50,000 compounds under differentiation conditions free of leukemia inhibitory factor (LIF) and feeder cells. This high-throughput cell-based screening approach was used to identify several small molecules that promote the self-renewal of mESCs in the absence of LIF and feeder cells. Specifically, these authors identified a previously uncharacterized heterocycle, SC1/pluripotin, that allowed for the propagation of mESCs in an undifferentiated, pluripotent state in the absence of feeder cells, serum, and LIF.[19]

These traditional HTS technologies have also been used to identify various small molecules that promote mESC differentiation into multiple lineages. A high-throughput phenotypic cell-based screen of kinase-directed libraries led to the identification of a synthetic small molecule inhibitor of glycogen synthase kinase-3β (GSK-3β), TWS119, that induces neurogenesis in mESCs.[20] Previously, a distinct GSK-3β inhibitor, called

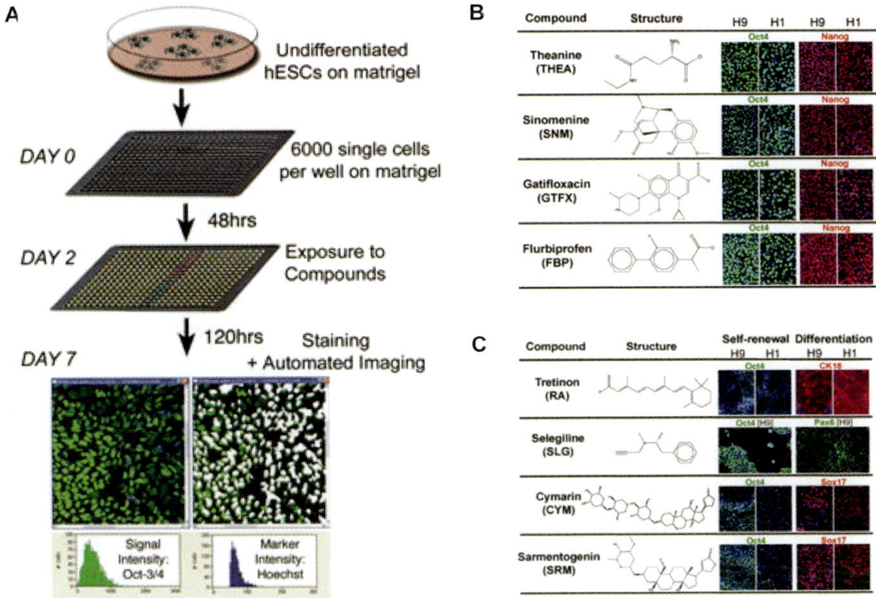

Fig. 3. Microtiter HTS for identifying modulators of human embryonic stem cell fate. (A) Undifferentiated hESCs were dissociated into single cells and plated onto 384 well-plates. After 48 hours, compounds were added and screened for their effects on hESC self-renewal or differentiation. After seven days, automated immunocytochemistry for Oct4 was performed and images were captured and analyzed using an automated laser-scanning confocal microscope. Several compounds were identified to promote (B) maintenance of pluripotency or (C) differentiation. Figure and legend adapted from Ref. 23 with permission.

BIO, was shown to promote the undifferentiated growth of ESCs. Likewise, using a combinatorial library of peroxisome proliferator-activated receptor (PPAR) ligands, novel molecules that promoted mesodermal differentiation of murine ESCs into beating cardiomyocytes were identified.[21] Along similar lines, a phenotypic cell-based screen of a large combinatorial chemical library was utilized to discover a class of diaminopyrimidine compounds that efficiently induce mESCs to differentiate into cardiomyocytes.[22]

The development of HTS for hESCs has been difficult because of challenges involved in establishing suitable growth conditions. More recently, though, a strategy was developed to adapt hESCs to HTS conditions, to screen 2880 compounds for their effects on hESC self-renewal or differentiation (Figs. 3b and 3c).[23] Use of this HTS system resulted in the identification of several drugs and natural compounds that promote short-term hESC maintenance and compounds that direct differentiation.[23] A similar approach was used to identify a small molecule, stauprimide, that downregulates c-Myc and thus increases the efficiency of the directed differentiation of hESCs.[24]

3.2 *Cellular microarray-based screening in stem cell research*

Although HTS have greatly advanced modern biology and drug discovery, they are not practical for all stem cell-related studies, given the cost and the large number of cells and reagents required. For example, typical HTS require $25–50 \times 10^4$ cells per condition screened.[23] Thus, HTS are not feasible for screens involving adult stem cells, which are rare (e.g. 1 in 200,000 blood cells is a hematopoietic stem cell) and difficult to isolate. HTS using conventional multi-well plates is cost-prohibitive and often does not provide adequate quantitative information on cell function. Additionally, most HTS approaches only have the capacity to investigate the effects of one factor at a time, often ignoring the complex crosstalks that typically occur in biological settings between combinations of molecules.

In order to overcome these obstacles, cellular microarrays have been used for screening the effects of large numbers of biological molecules on stem cell fate.[25] Typical cellular microarrays consist of a chip (e.g. glass microscope slide) where minute volumes (μL to nL) of various molecules (e.g. ECMPs, small molecules, cytokines, biopolymers) are deposited in defined locations and analyzed for their effect on cellular processes (e.g. changes in gene and protein expression levels). These arrays are fabricated using robotic spotting, photo-assisted, and soft-lithography approaches.

A major challenge in developing stem cell-based therapies is the identification of conditions that specifically regulate and influence their fate. Being able to identify components that mimic the stem cell niche will aid in the expansion and differentiation of stem cells *in vitro*. Considering the complexity of the microenvironments in which stem cells reside, it is impractical to test proliferation and differentiation conditions using current established HTS methods. Cellular microarrays have advantages over traditional well-based HTS in that they provide more information from smaller sample volumes in a rapid, efficient, and cost-effective manner.

3.2.1 *Combinatorial protein arrays for studying stem cell microenvironments*

Combinatorial protein arrays consist of immobilized biological signaling molecules on a surface onto which cells are seeded (Fig. 4). The first such platform was used to screen ECMPs and their effects on stem cells in a combinatorial fashion.[26] Specifically, with the use of a DNA robotic spotter, different ECMP combinations were spotted, and the commitment of mESCs toward an early hepatic fate was evaluated. Several combinations of ECMP components that influenced stem cell differentiation and hepatocyte function were identified. This platform was further developed to investigate the interactions between ECMPs and soluble growth factors on stem cell fate.[27] Several ECMPs and growth factors were found to influence mESC differentiation toward the cardiac lineage.

Similar robotic spotting techniques were used to print arrays of ECMPs along with growth factors and adhesion molecules to evaluate their effects on the proliferation and differentiation of human adult neural precursor cells,[28] and the results revealed significant effects of certain signaling molecules (such as BMPs, Wnts, and Notch) on the extent and direction of differentiation into neuronal or glial fate. For example, it was found that Wnt and Notch co-stimulation maintained the cells in an undifferentiated state. Along similar lines, a photo-assisted patterning process was used to create matrix-growth factor arrays to identify materials that direct neural stem cell growth and differentiation.[29]

Although these platforms were used to elucidate the role of certain microenvironmental components on stem cell fate, they have limitations. Specifically, these platforms either relied on the addition of signaling molecules to the surrounding media, thereby limiting the throughput and the complexity of the microenvironments that could be screened, or involved the covalent attachment of the signaling molecules to the chip, thus affecting their biological activity. Additionally, the throughput of these systems (< 100 conditions per chip) was significantly lower when compared to conventional HTS systems. In order to overcome these limitations, an integrated array platform was developed in which ECMPs, growth factors, and small molecules were non-covalently arrayed on acrylamide-coated slides, thereby creating comprehensive microenvironments

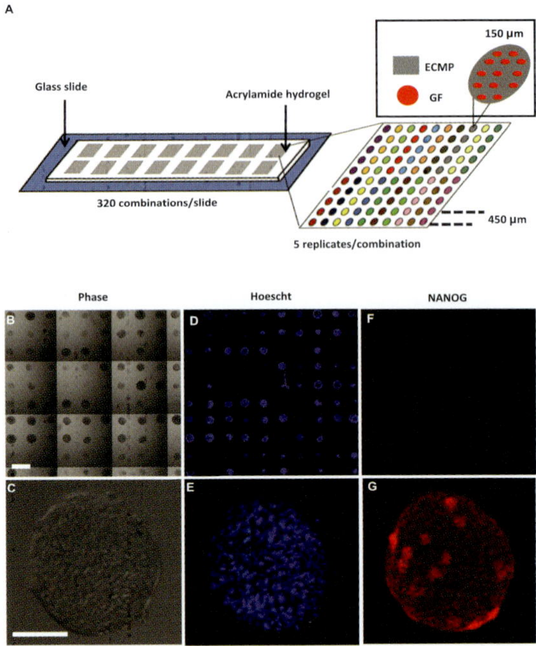

Fig. 4.   Combinatorial protein array technology for manipulating hESC fate. (A) Typical layout of arrayed cellular microenvironments. Arrays of pre-mixed combinations of extracellular matrix proteins (ECMPs) and signaling molecules are printed onto acrylamide-coated glass slides using a contact microarray printer. Human embryonic stem cells (hESCs) were cultured on the arrays for five days. Cells were imaged live (B–C) and then fixed and stained for DNA (D–E) and Nanog (F–G) to identify conditions of hESC attachment, proliferation, maintenance of pluripotency and differentiation. Figure and legend adapted from Ref. 30 with permission.

that closely resemble the *in vivo* microenvironment in which cells reside (Fig. 4).[30] This technology platform was used for the real-time simultaneous screening of thousands of physiochemical parameters on stem cell attachment, proliferation, differentiation, and gene expressions.[30,31] Specifically, through the systematic screening of ECMPs and other signaling molecules, a completely defined culture system for the long-term self-renewal of three independent hESC lines was developed.[30] In another study, this technology was used to investigate the effects of microenvironmental modulations on the fate of hepatic stellate cells (HeSCs), a progenitor cell that resides in the liver.[31] It was determined that different components of the microenvironment differentially influence other components to influence the HeSC phenotype. For example, it was found that the influences of Wnt signaling molecules on HeSC fate are dependent on the ECMP composition in which they are presented. Furthermore, this array platform technology was validated by the finding that data obtained from these experiments were indistinguishable from data obtained from traditional multi-well-based assays.

3.2.2 *Polymer arrays for screening of biomaterials that influence stem cell fate*

Biomaterials have been used for the expansion of many human adult stem cell and progenitor populations. Mesenchymal stem cells (MSCs), for example, were maintained and differentiated on several different classes of synthetic and natural biomaterials.[32–36] Unmodified and methyl-modified silane surfaces were used to enhance MSC proliferation.[32] Likewise, neural stem cells (NSCs) were maintained and expanded on three-dimensional scaffolds composed of amino polymers such as poly(D-lysine).[37] There has been some progress in the application of polymers for expansion and maintenance of ESCs. For example, aliphatic poly(α-hydroxy esters) such as poly(D,L-lactide) and poly(glycolide) were used to propagate mESCs.[38] Nonetheless, biomaterial-based expansion and differentiation of stem cells have been slow coming. This has been mainly due to the inefficiency of the iterative nature of biomaterials-based research in which materials are fabricated, tested, and redesigned.[39]

In order to establish a set of principles or properties that can assist in the prediction of which polymers would influence stem cell fate, high-throughput array-based approaches have been implemented. In a study using an array-based biomaterial screen to identify specific functional groups that promote human MSC (hMSC) differentiation,[40] it was found that phosphate surfaces promoted osteoblast formation, while t-butyl-modified surfaces promoted adipocyte formation. A similar approach was implemented to study the effects of 576 synthetic materials on stem cell differentiation (Fig. 5).[41] Using this method, several classes of polymers

     D. A. Brafman, K. Willert and S. Chien

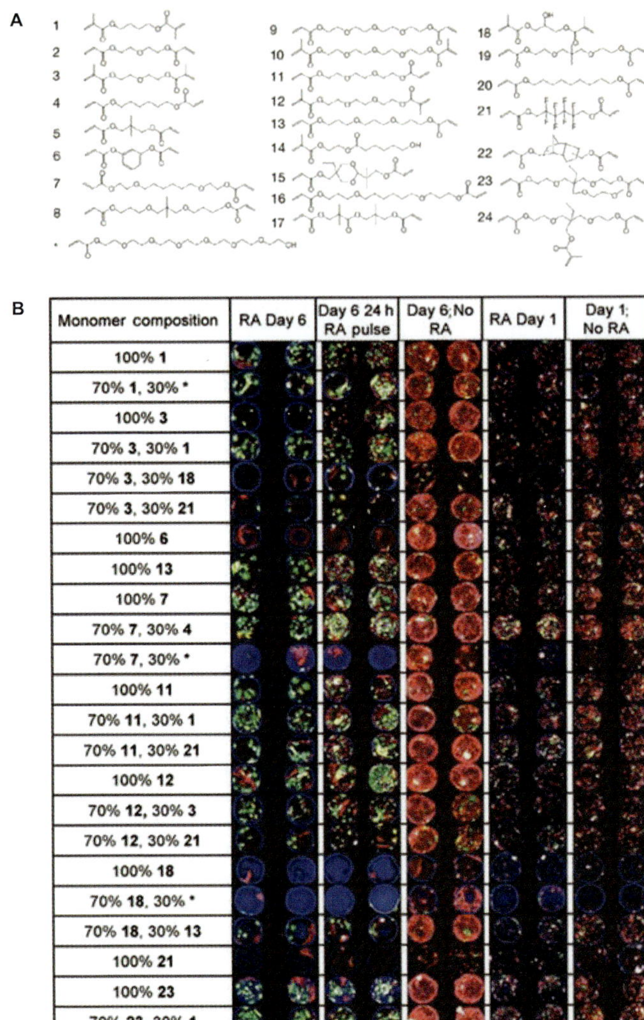

Fig. 5. Biomaterial microarrays for exploring polymer-stem cell interactions. (A) Monomers were mixed at a 70:30 ratio pairwise in all possible combinations. Monomer combinations were printed with a radical initiator onto a layer of poly(hydroxyethyl methacrylate) (pHEMA), on top of an epoxide-coated slide. (B) Day 6 embryoid bodies were dissociated and seeded onto polymer arrays in the presence of retinoic acid, the absence of retinoic acid, and with a 24-hour pulse of retinoic acid for one or six days. Cells were then stained for cytokeratin 7 (green), vimentin (red) and DNA (blue). Figure and legend reproduced from Ref. 41 with permission.

that promoted high levels of differentiation into epithelial cells were identified. More recently, this technology has been used to identify biodegradable polymers that support the growth and expansion of hMSCs and neural stem cells.[39]

3.2.3 *Microwell approaches to analyze physical cues regulating stem cell fate*

By using soft lithography methods, microwell arrays with defined dimensions can be fabricated (Fig. 6). For example, microfabrication was used to create an array of approximately 10,000 microwells on a glass coverslip for the parallel, quantitative analysis of single ESCs.[42] Furthermore, the well dimensions could be adjusted over ranges of 10–500 μm in height and 20–500 μm in diameter, and the platform was compatible with traditional light and fluorescent microscopy. The microwell platform was used to investigate the effect of cell density on the proliferation dynamics of rat neural stem cells (NSCs).

Directed differentiation of ESCs through formation of embryoid bodies (EBs) has been enhanced by using microwell array techniques.[43–47] EBs are typically formed using the hanging drop method[48] or in suspension culture.[49] The resulting EBs are heterogeneous in shape and size. As a result, cell populations obtained from EBs formed by these methods can vary considerably. To provide more uniform microenvironments to EBs and thus more uniformly direct EB differentiation, microwell approaches were implemented (Figs. 6a and 6b). For example, poly(ethylene glycol) (PEG) microwells were used to create EBs of homogenous size and shape (Figs. 6c and 6d).[45–47] More recently, such approaches were used to control hESC differentiation trajectories (Fig. 6e).[47] Specifically, it was observed that mesoderm and cardiac induction of hESCs was significantly higher at larger EB sizes.

Microwell arrays were utilized for the parallel manipulation and quantitative analysis of stem cells at the single-cell level. For example, combined biomimetic hydrogel matrix technology with microengineering was used to fabricate a microwell array which can used to control NSC fate and neurosphere formation.[50] Using this technology, the authors enhanced the viability and controlled the size of neurospheres formed from a single founding cell.

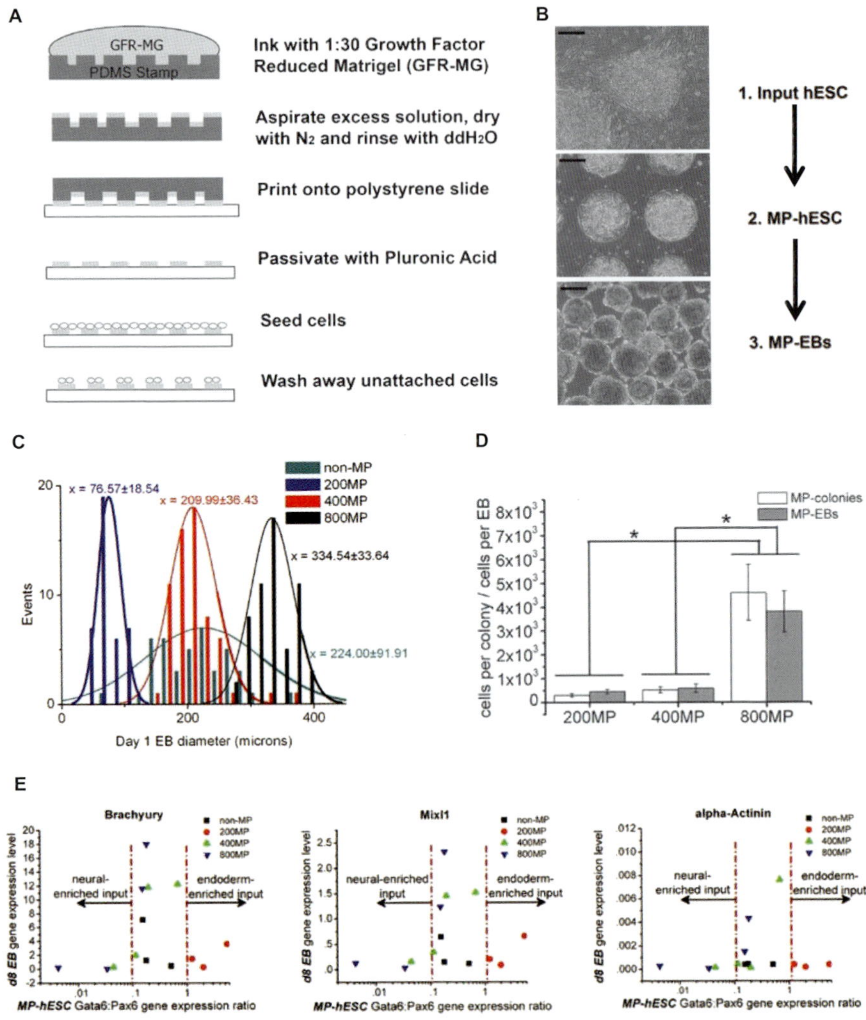

Fig. 6. Microwell platform for controlling the size of embryoid body (EB) formation. (A) Schematic of the process for generating microwell patterning Matrigel (MG) on a cell culture substrate, and subsequently seeding cells onto MG-patterned substrate. (B) To generate size-controlled micropatterened (MP) EBs, hESCs are dissociated to single cells and plated at high density onto patterned Matrigel islands and cultured to confluence. Intact colonies are then detached using a cell scraper and transferred to suspension in differentiation medium. Quantitative demonstration of EB size control in EBs generated from size-controlled human embryonic stem cell colonies (C) by EB diameter and (D) by comparing cell number in colonies to cell number in generated EBs. (E)

Fig. 6. (*Continued*) The effect of micropatterened human embryonic stem cell (MP-hESC) colony size and cell composition on mesoderm and cardiac induction in MP-embryoid bodies (EBs). Gene expression levels measured for mesoderm markers Brachyury (*Bry*) and *Mixl1*, and cardiac marker α-*Actin* in day 8 (d8) EBs plotted with respect to the *Gata6/Pax6* gene expression ratio in the corresponding MP-hESC starting population, and as a function of MP-hESC colony size demonstrate that mesoderm and cardiac induction of hESCs was significantly higher at larger EB sizes. Figure and legend reproduced from Ref. 47 with permission.

Microwell approaches provide precise control over microenvironmental parameters such as shape and size. However, in contrast to most multi-well formats, all wells in the microwell approach share the same culture media. Thus, it is difficult to vary other parameters of the microenvironment in a high-throughput manner.

3.2.4 *Microfluidic array approaches to study stem cell biology*

Microfluidic devices are powerful tools that have the ability to control the soluble and mechanical properties of the cell culture environment. Microfluidic devices are fabricated by casting poly-dimethylsiloxane (PDMS) over a prefabricated mold (Fig. 7a). The advantages of microfluidics include (1) decreased reaction rates and analysis times, (2) reduced consumption of reagents, (3) reduced production of harmful by-products, and (4) the ability to run multiple experiments on a single chip.[51]

Recently, microfluidic approaches were used for the analysis of signals that affect stem cell fate. A microfluidic device was developed for analyzing 16 unique mESC cultures in parallel.[52] Using this platform, the authors identified an optimal flow rate that enhanced mESC colony formation, proliferation, and maintenance of pluripotency (Figs. 7b and 7c). Along similar lines, a micro-bioreactor array was fabricated using soft lithography that contains 12 independent micro-bioreactors.[53] Each micro-bioreactor was perfused with independent culture media containing different biological molecules. Using this platform, a correlation between varying flow patterns and hESC differentiation into vascular lineages was established. Recently, an integrated microfluidic platform was designed that allows the screening of individual hESC colonies in real time using six individual cell-culture chambers.[54] Such approaches provided important information about the

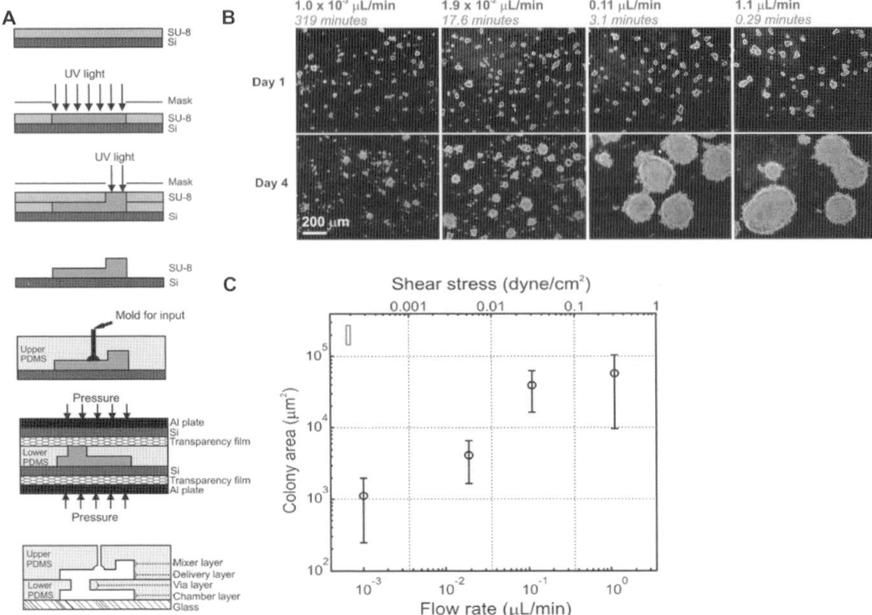

Fig. 7. Microfluidic arrays for culture of hESCs. (A) Fabrication of 4×4 microfluidic array device. (B) Culture of mESCs in microfluidic array under different flow rate conditions. Chambers with high flow rates display large, round colonies, suggesting a favorable growth environment. (C) Mean colony area versus flow rate mESCs grown under different flow rates. The mean colony areas are larger at higher flow rates while the mean colony areas are smaller at the lowest flow rate. Figure and legend reproduced from Ref. 52 with permission.

degree of hESC colony heterogeneity. Although these microfluidic platforms enable the multiplexing of experiments, they have the disadvantages of: (1) relative low throughput, (2) difficulty of fabrication, (3) inability to perform on chip immunocytochemistry, and (4) lack of compatibility with conventional, high-magnification light and fluorescent microscopy.

4. High-throughput Intrinsic Systems for Stem Cell Investigations

Each of the microenvironment factors acts individually and in combination to perturb intrinsic cellular signaling networks which in turn influence stem cell functions and ultimately stem cell fate. As a result, a variety of

HTS systems have been developed to manipulate the intrinsic signaling networks of stem cells. Such systems generally utilize large libraries comprised of small molecules, RNAi molecules or expression systems carrying shRNAs or cDNAs.

4.1 *High-throughput RNA interference studies to investigate gene functions in stem cells*

Elucidation of gene function is a critical aspect of advancing stem cell research. Functional genomics typically involve gain- and/or loss-of-function studies, which reveal the molecular mechanisms of a cellular phenotype, but these are difficult to implement at the genome-wide scale in cultured stem cells. Thus, most gene-silencing studies are restricted to knockout strains of model organisms such as yeast, flies, and mice.

Recent advances in RNA interference (RNAi) have aided the field of functional genomics by allowing for loss-of-function studies in mammalian cells without the need for germline inactivation of the gene being studied.[55] RNAi occurs through the effect of the ribonuclease (RNase) enzyme Dicer on double-stranded RNA (dsRNA). Dicer cleaves the dsRNA into double-stranded small interfering RNAs (siRNAs) which can act either through the RNA-induced silencing complex (RISC) to degrade complementary mRNA sequences or through the RNA-induced transcriptional silencing (RITS) complex to repress transcription and modify DNA and histone methylation.[56]

RNAi screens have been used to study the effect of genetic control elements on stem cell behavior. For example, a subtractive RNAi library approach was used to indentify multiple genes involved in the regulation of expression of Oct4 and of self-renewa.[57,58] However, large-scale cell-based RNAi screens have been hampered by the demands and inefficiency of traditional HTS. Recently, RNAi cell microarrays were used for effective gene knockdown in hMSCs (Fig. 8).[59] These technology platforms could offer an efficient approach for carrying out high-throughput loss-of-function studies.[60] Arrays can be fabricated by spotting either lentiviruses that express short hairpin RNA (shRNA) to silence gene expression through RNAi[61] or peptide transduction domain-double-stranded RNA binding domains (PTD-dRBDs) that carry siRNA across

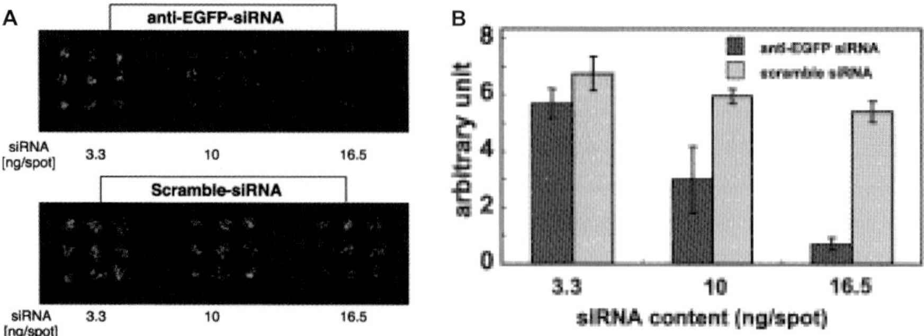

Fig. 8. siRNA microarray for high-throughput screening of gene function. (A) Co-transfection of an EGFP vector and anti-EGFP siRNA into hMSCs and co-transfection of an EGFP vector and scramble siRNA into hMSCs. (B) Quantification of the observed effects demonstrates the efficiency of the spotted siRNA gene knockdown. Figure and legend reproduced from Ref. 59 with permission.

the cell membrane and knockdown gene expression.[59,62] Such RNAi arrays will aid in the rapid functional annotation of stem cell genomes and in the identification of genes involved in stem cell self-renewal and differentiation.

4.2 *High-throughput generation of genetically modified stem-cell lines*

The ability to obtain information about gene function has been greatly aided by the generation of genetically modified stem-cell lines. Several techniques have been implemented to create genetically modified stem cells in a high-throughput manner. For example, gene-trap mutagenesis, a technique that randomly generates loss-of-function mutations, was used to create more than 8,000 mutagenized ES cell lines.[63,64] More recently, clonal microarrays were used to create genetically modified stem-cell lines (Fig. 9).[65] Clonal microarrays are produced by seeding stem cells on microfabricated surfaces generated using soft lithographic techniques. Stem cells grown on these arrays can be infected with DNA constructs and then isolated after assaying in parallel for various parameters, such as proliferation, signal transduction, and differentiation.

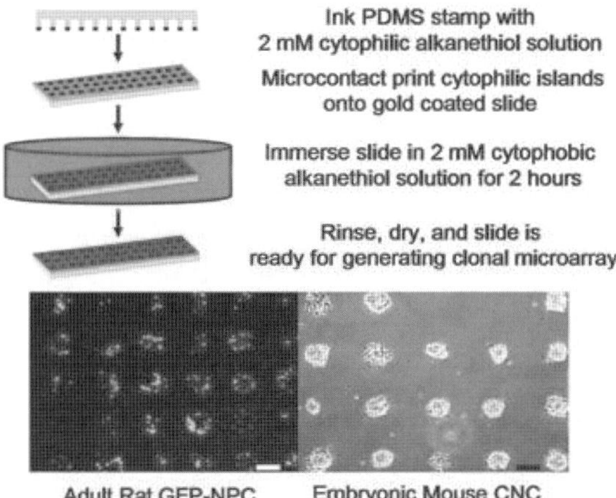

Fig. 9. Formation of neurosphere microarray. Schematic of procedure used to generate array. Patterned array of neurospheres of adult rat hippocampal neural progenitor cells expressing green fluorescent protein and mouse embryonic CNCs. Figure and legend adapted from Ref. 65 with permission.

5. Conclusions and Future Trends

The technologies presented here are capable of screening and perturbing several components of the stem cell microenvironment and can greatly advance the engineering of defined cell types from stem cells. However, additional factors, including oxygen and salt concentrations, mechanical forces, matrix stiffness, and dimensionality, also play critical roles in the microenvironment and determination of cell fate. *In vivo*, the complex network of signaling and matrix molecules is subject to mechanical forces (such as pressure, fluid shear stress, and stretch), which play important roles in specifying embryonic polarity[66] and tissue development.[67] Recent studies have shown that application of shear stress to mESCs induces cell proliferation and endothelial cell (EC) lineage differentiation with the expression of marker genes indicative of ECs and the enhancement of endothelial functions.[68–70] Application of cyclic stretch may induce differentiation toward smooth muscle linage,[70] and compression induces chondrogensis.[71] Substrate compliance is also known to influence cell fate decisions. For example, human mesenchymal stem cells (MSCs) showed higher rates of growth on stiffer substrates.[72] Recently, it was demonstrated that lineage-specific differentiation of MSCs is induced by matrix stiffness that matches the respective tissue — soft matrices are neurogenic, rigid matrices are osteogenic, and intermediate matrices are myogenic.[73] Finally, *in vivo*, cells often reside in three-dimensional as opposed to two-dimensional microenvironments. In fact, three-dimensionality has been shown to play a critical role in the microenvironment and can affect cell function.[74,75] In the future, these high-throughput tools can be enhanced to screen such additional components, thereby creating more *in vivo*-like screening conditions.

It has recently been demonstrated that stable genomic integration and high expression of four factors, Oct4/Sox2/Klf4/c-Myc or Oct4/Sox2/Nanog/LIN28, can reprogram fibroblast cells into iPSCs.[76,77] This ability to generate patient-specific pluripotent stem cells will have a major impact in regenerative medicine. However, the use of iPSCs in cell-based therapies is limited because of the presence of virally transduced transcription factors and oncogenes. Furthermore, current protocols to establish iPSCs are extremely inefficient (<0.1%). The technology platforms presented here could be used to identify microenvironment components, such as ECMPs, growth factors, and small molecules, not only for the purpose of enhancing reprogramming efficiency, but also identifying factors that could replace the need for virally transduced transcription factors and oncogenes.[78,79]

The HTS technologies described here could find usage in the pharmaceutical industry by aiding in crucial steps of drug development: toxicity screening, target identification, and lead assessment. The current method of target identification involves the use of 96- or 384-well microtiter plates with monolayer cell cultures.[25] However, the multi-well plate format suffers from several limitations, most notably the cost of the relatively large amounts of reagents and cells needed. The HTS technologies described here, by increasing the parallelism and efficiency of screening compound libraries, offer an attractive solution. Additionally, these technologies could be used to identify potentially toxic compounds earlier in the drug development process. For example, an array-based high-throughput system was recently developed that can be used to mimic the effects of human liver metabolism and simultaneously evaluate the cytotoxicity of small molecules and their metabolites.[80] Along the same line, the use of high-throughput technologies in conjunction with stem cells and their differentiated progenitors could provide more realistic *in vitro* models for predicting the effectiveness and toxicity of drug candidates and chemicals in humans. Most cell lines currently used in drug screening paradigms are virally transformed cells of tumor origin that are not representative of the disease to be investigated. The derivation of hPSCs has made it possible, for the first time, to study various aspects of human development and disease using cells representative of these conditions.

In summary, hPSCs represent an infinite supply of cellular "raw-material" that can be used to generate more realistic disease models for drug discovery. The ability to manipulate hPSCs and adult stem cells using high-throughput technologies will enable the production of large amounts of specialized cells needed for applications in regenerative medicine and drug discovery.

References

1. J. H. Jang and D. V. Schaffer, Microarraying the cellular microenvironment, *Mol Syst Biol* **2**: 39 (2006).
2. D. Schaffer, Exploring and engineering stem cells and their niches, *Curr Opin Chem Biol* **11**: 355–356 (2007).
3. S. Kim, M. Harris and J. A. Varner, Regulation of integrin alpha vbeta 3-mediated endothelial cell migration and angiogenesis by integrin alpha-5beta1 and protein kinase A, *J Biol Chem* **275**: 33920–33928 (2000).
4. S. J. Fashena and S. M. Thomas, Signalling by adhesion receptors, *Nat Cell Biol* **2**: E225–229 (2000).
5. E. A. Clark and J. S. Brugge, Integrins and signal transduction pathways: the road taken, *Science* **268**: 233–239 (1995).
6. M. Schneller, K. Vuori and E. Ruoslahti, Alphavbeta3 integrin associates with activated insulin and PDGFbeta receptors and potentiates the biological activity of PDGF, *EMBO J* **16**: 5600–5607 (1997).
7. R. Pacifici, C. Basilico, J. Roman, M. M. Zutter, S. A. Santoro and R. McCracken, Collagen-induced release of interleukin 1 from human blood mononuclear cells: potentiation by fibronectin binding to the alpha 5 beta 1 integrin, *J Clin Invest* **89**: 61–67 (1992).
8. E. Fuchs, T. Tumbar and G. Guasch, Socializing with the neighbors: stem cells and their niche, *Cell* **116**: 769–778 (2004).
9. S. J. Morrison and A. C. Spradling, Stem cells and niches: mechanisms that promote stem cell maintenance throughout life, *Cell* **132**: 598–611 (2008).

10. K. A. Moore and I. R. Lemischka, Stem cells and their niches, *Science* **311**: 1880–1885 (2006).

11. S. M. Dellatore, A. S. Garcia and W. M. Miller, Mimicking stem cell niches to increase stem cell expansion, *Curr Opin Biotechnol* **19**: 534–540 (2008).

12. I. L. Weissman, D. J. Anderson and F. Gage, Stem and progenitor cells: origins, phenotypes, lineage commitments, and transdifferentiations, *Annu Rev Cell Dev Biol* **17**: 387–403 (2001).

13. L. Li and T. Xie, Stem cell niche: structure and function, *Annu Rev Cell Dev Biol* **21**: 605–631 (2005).

14. M. R. Walker, K. K. Patel and T. S. Stappenbeck, The stem cell niche, *J Pathol* **217**: 169–180 (2009).

15. M. R. Walker and T. S. Stappenbeck, Deciphering the "black box" of the intestinal stem cell niche: taking direction from other systems, *Curr Opin Gastroenterol* **24**: 115–120 (2008).

16. J. A. Thomson, J. Itskovitz-Eldor, S. S. Shapiro, M. A. Waknitz, J. J. Swiergiel, V. S. Marshall and J. M. Jones, Embryonic stem cell lines derived from human blastocysts, *Science* **282**: 1145–1147 (1998).

17. N. Emre, R. Coleman and S. Ding, A chemical approach to stem cell biology, *Curr Opin Chem Biol* **11**: 252–258 (2007).

18. C. M. Dobson, Chemical space and biology, *Nature* **432**: 824–828 (2004).

19. S. Chen, J. T. Do, Q. Zhang, S. Yao, F. Yan, E. C. Peters, H. R. Scholer, P. G. Schultz and S. Ding, Self-renewal of embryonic stem cells by a small molecule, *Proc Natl Acad Sci USA* **103**: 17266–17271 (2006).

20. S. Ding, T. Y. Wu, A. Brinker, E. C. Peters, W. Hur, N. S. Gray and P. G. Schultz, Synthetic small molecules that control stem cell fate, *Proc Natl Acad Sci USA* **100**: 7632–7637 (2003).

21. Z. L. Wei, P. A. Petukhov, F. Bizik, J. C. Teixeira, M. Mercola, E. A. Volpe, R. I. Glazer, T. M. Willson and A. P. Kozikowski, Isoxazolyl-serine-based agonists of peroxisome proliferator-activated receptor: design, synthesis, and effects on cardiomyocyte differentiation, *J Am Chem Soc* **126**: 16714–16715 (2004).

22. X. Wu, S. Ding, Q. Ding, N. S. Gray and P. G. Schultz, Small molecules that induce cardiomyogenesis in embryonic stem cells, *J Am Chem Soc* **126**: 1590–1591 (2004).

23. S. C. Desbordes, D. G. Placantonakis, A. Ciro, N. D. Socci, G. Lee, H. Djaballah and L. Studer, High-throughput screening assay for the identification of compounds regulating self-renewal and differentiation in human embryonic stem cells, *Cell Stem Cell* **2**: 602–612 (2008).

24. S. Zhu, H. Wurdak, J. Wang, C. A. Lyssiotis, E. C. Peters, C. Y. Cho, X. Wu and P. G. Schultz, A small molecule primes embryonic stem cells for differentiation, *Cell Stem Cell* **4**: 416–426 (2009).

25. T. G. Fernandes, M. M. Diogo, D. S. Clark, J. S. Dordick and J. M. Cabral, High-throughput cellular microarray platforms: applications in drug discovery, toxicology and stem cell research, *Trends Biotechnol* **27**: 342–349 (2009).

26. C. J. Flaim, S. Chien and S. N. Bhatia, An extracellular matrix microarray for probing cellular differentiation, *Nat Methods* **2**: 119–125 (2005).

27. C. J. Flaim, D. Teng, S. Chien and S. N. Bhatia, Combinatorial signaling microenvironments for studying stem cell fate, *Stem Cells Dev* **17**: 29–39 (2008).

28. Y. Soen, A. Mori, T. D. Palmer and P. O. Brown, Exploring the regulation of human neural precursor cell differentiation using arrays of signaling microenvironments, *Mol Syst Biol* **2**: 37 (2006).

29. M. Nakajima, T. Ishimuro, K. Kato, I. K. Ko, I. Hirata, Y. Arima and H. Iwata, Combinatorial protein display for the cell-based screening of biomaterials that direct neural stem cell differentiation, *Biomaterials* **28**: 1048–1060 (2007).

30. D. Brafman, K. Shah, T. Fellner, S. Chien and K. Willert, Defining long-term maintenance conditions of human embryonic stem cells with arrayed cellular microenvironment technology, *Stem Cells Dev* **18**: 1141–1154 (2009).

31. D. A. Brafman, S. De Minicis, S. Seki, K. D. Shah, D. Teng, D. Brenner, K. Willert and S. Chien, Investigating the role of the extracellular environment in modulating hepatic stellate cell biology with arrayed combinatorial microenvironments, *Integr Biol (Camb)* **1**: 513–524 (2009).

32. J. M. Curran, R. Chen and J. A. Hunt, The guidance of human mesenchymal stem cell differentiation in vitro by controlled modifications to the cell substrate, *Biomaterials* **27**: 4783–4793 (2006).

33. N. Kotobuki, Y. Katsube, Y. Katou, M. Tadokoro, M. Hirose and H. Ohgushi, *In vivo* survival and osteogenic differentiation of allogeneic rat bone marrow mesenchymal stem cells (MSCs), *Cell Transplant* **17**: 705–712 (2008).

34. F. Zhao, W. L. Grayson, T. Ma, B. Bunnell and W. W. Lu, Effects of hydroxyapatite in 3-D chitosan-gelatin polymer network on human mesenchymal stem cell construct development, *Biomaterials* **27**: 1859–1867 (2006).

35. H. Fan, Y. Hu, C. Zhang, X. Li, R. Lv, L. Qin and R. Zhu, Cartilage regeneration using mesenchymal stem cells and a PLGA-gelatin/chondroitin/hyaluronate hybrid scaffold, *Biomaterials* **27**: 4573–4580 (2006).

36. S. M. Richardson, N. Hughes, J. A. Hunt, A. J. Freemont and J. A. Hoyland, Human mesenchymal stem cell differentiation to NP-like cells in chitosan-glycerophosphate hydrogels, *Biomaterials* **29**: 85–93 (2008).

37. M. W. Hayman, K. H. Smith, N. R. Cameron and S. A. Przyborski, Growth of human stem cell-derived neurons on solid three-dimensional polymers, *J Biochem Biophys Methods* **62**: 231–240 (2005).

38. J. Harrison, S. Pattanawong, J. S. Forsythe, K. A. Gross, D. R. Nisbet, H. Beh, T. F. Scott, A. O. Trounson and R. Mollard, Colonization and maintenance of murine embryonic stem cells on poly(alpha-hydroxy esters), *Biomaterials* **25**: 4963–4970 (2004).

39. D. G. Anderson, D. Putnam, E. B. Lavik, T. A. Mahmood and R. Langer, Biomaterial microarrays: rapid, microscale screening of polymer-cell interaction, *Biomaterials* **26**: 4892–4897 (2005).

40. D. S. Benoit, M. P. Schwartz, A. R. Durney and K. S. Anseth, Small functional groups for controlled differentiation of hydrogel-encapsulated human mesenchymal stem cells, *Nat Mater* **7**: 816–823 (2008).

41. D. G. Anderson, S. Levenberg and R. Langer, Nanoliter-scale synthesis of arrayed biomaterials and application to human embryonic stem cells, *Nat Biotechnol* **22**: 863–866 (2004).

42. V. I. Chin, P. Taupin, S. Sanga, J. Scheel, F. H. Gage and S. N. Bhatia, Microfabricated platform for studying stem cell fates, *Biotechnol Bioeng* **88**: 399–415 (2004).

43. H. C. Moeller, M. K. Mian, S. Shrivastava, B. G. Chung and A. Khademhosseini, A microwell array system for stem cell culture, *Biomaterials* **29**: 752–763 (2008).

44. A. Khademhosseini, L. Ferreira, J. Blumling, 3rd, J. Yeh, J. M. Karp, J. Fukuda and R. Langer, Co-culture of human embryonic stem cells with murine embryonic fibroblasts on microwell-patterned substrates, *Biomaterials* **27**: 5968–5977 (2006).

45. J. C. Mohr, J. J. de Pablo and S. P. Palecek, 3D microwell culture of human embryonic stem cells, *Biomaterials* **27**: 6032–6042 (2006).

46. J. M. Karp, J. Yeh, G. Eng, J. Fukuda, J. Blumling, K. Y. Suh, J. Cheng, A. Mahdavi, J. Borenstein, R. Langer and A. Khademhosseini, Controlling size, shape and homogeneity of embryoid bodies using poly(ethylene glycol) microwells, *Lab Chip* **7**: 786–794 (2007).

47. C. L. Bauwens, R. Peerani, S. Niebruegge, K. A. Woodhouse, E. Kumacheva, M. Husain and P. W. Zandstra, Control of human embryonic stem cell colony and aggregate size heterogeneity influences differentiation trajectories, *Stem Cells* **26**: 2300–2310 (2008).

48. G. M. Keller, *In vitro* differentiation of embryonic stem cells, *Curr Opin Cell Biol* **7**: 862–869 (1995).

49. J. Itskovitz-Eldor, M. Schuldiner, D. Karsenti, A. Eden, O. Yanuka, M. Amit, H. Soreq and N. Benvenisty, Differentiation of human embryonic stem cells into embryoid bodies compromising the three embryonic germ layers, *Mol Med* **6**: 88–95 (2000).

50. B. Buckner, J. Beck, K. Browning, A. Fritz, L. Grantham, E. Hoxha, Z. Kamvar, A. Lough, O. Nikolova, P. S. Schnable, M. J. Scanlon and D. Janick-Buckner, Involving undergraduates in the annotation and analysis of global gene expression studies: creation of a maize shoot apical meristem expression database, *Genetics* **176**: 741–747 (2007).

51. D. van Noort, S. M. Ong, C. Zhang, S. Zhang, T. Arooz and H. Yu, Stem cells in microfluidics, *Biotechnol Prog* **25**: 52–60 (2009).

52. L. Kim, M. D. Vahey, H. Y. Lee and J. Voldman, Microfluidic arrays for logarithmically perfused embryonic stem cell culture, *Lab Chip* **6**: 394–406 (2006).

53. E. Figallo, C. Cannizzaro, S. Gerecht, J. A. Burdick, R. Langer, N. Elvassore and G. Vunjak-Novakovic, Micro-bioreactor array for controlling cellular microenvironments, *Lab Chip* **7**: 710–719 (2007).

54. J. F. Zhong, Y. Chen, J. S. Marcus, A. Scherer, S. R. Quake, C. R. Taylor and L. P. Weiner, A microfluidic processor for gene expression profiling of single human embryonic stem cells, *Lab Chip* **8**: 68–74 (2008).

55. T. R. Brummelkamp and R. Bernards, New tools for functional mammalian cancer genetics, *Nat Rev Cancer* **3**: 781–789 (2003).

56. B. Alberts, A. Johnson, J. Lewis, M. Raff, K. Roberts and P. Walter, *Molecular Biology of the Cell* (Garland Science, New York, 2008).

57. J. Z. Zhang, W. Gao, H. B. Yang, B. Zhang, Z. Y. Zhu and Y. F. Xue, Screening for genes essential for mouse embryonic stem cell self-renewal using a subtractive RNA interference library, *Stem Cells* **24**: 2661–2668 (2006).

58. N. Ivanova, R. Dobrin, R. Lu, I. Kotenko, J. Levorse, C. DeCoste, X. Schafer, Y. Lun and I. R. Lemischka, Dissecting self-renewal in stem cells with RNA interference, *Nature* **442**: 533–538 (2006).

59. T. Yoshikawa, E. Uchimura, M. Kishi, D. P. Funeriu, M. Miyake and J. Miyake, Transfection microarray of human mesenchymal stem cells and on-chip siRNA gene knockdown, *J Control Release* **96**: 227–232 (2004).

60. D. B. Wheeler, A. E. Carpenter and D. M. Sabatini, Cell microarrays and RNA interference chip away at gene function, *Nat Genet* **37**: S25–30 (2005).

61. S. N. Bailey, S. M. Ali, A. E. Carpenter, C. O. Higgins and D. M. Sabatini, Microarrays of lentiviruses for gene function screens in immortalized and primary cells, *Nat Methods* **3**: 117–122 (2006).

62. A. Eguchi, B. R. Meade, Y. C. Chang, C. T. Fredrickson, K. Willert, N. Puri and S. F. Dowdy, Efficient siRNA delivery into primary cells by a peptide transduction domain-dsRNA binding domain fusion protein, *Nat Biotechnol* **27**: 567–571 (2009).

63. W. L. Stanford, J. B. Cohn and S. P. Cordes, Gene-trap mutagenesis: past, present and beyond, *Nat Rev Genet* **2**: 756–768 (2001).

64. E. Medico, G. Gambarotta, A. Gentile, P. M. Comoglio and P. Soriano, A gene trap vector system for identifying transcriptionally responsive genes, *Nat Biotechnol* **19**: 579–582 (2001).

65. R. S. Ashton, J. Peltier, C. A. Fasano, A. O'Neill, J. Leonard, S. Temple, D. V. Schaffer and R. S. Kane, High-throughput screening of gene function in stem cells using clonal microarrays, *Stem Cells* **25**: 2928–2935 (2007).

66. R. Keller, Mechanisms of elongation in embryogenesis, *Development* **133**: 2291–2302 (2006).

67. M. Benjamin and B. Hillen, Mechanical influences on cells, tissues and organs: "mechanical morphogenesis", *Eur J Morphol* **41**: 3–7 (2003).

68. K. Yamamoto, T. Sokabe, T. Watabe, K. Miyazono, J. K. Yamashita, S. Obi, N. Ohura, A. Matsushita, A. Kamiya and J. Ando, Fluid shear stress induces differentiation of Flk-1-positive embryonic stem cells into vascular endothelial cells *in vitro*, *Am J Physiol Heart Circ Physiol* **288**: H1915–1924 (2005).

69. B. Illi, A. Scopece, S. Nanni, A. Farsetti, L. Morgante, P. Biglioli, M. C. Capogrossi and C. Gaetano, Epigenetic histone modification and cardio-vascular lineage programming in mouse embryonic stem cells exposed to laminar shear stress, *Circ Res* **96**: 501–508 (2005).

70. H. Huang, Y. Nakayama, K. Qin, K. Yamamoto, J. Ando, J. Yamashita, H. Itoh, K. Kanda, H. Yaku, Y. Okamoto and Y. Nemoto, Differentiation from embryonic stem cells to vascular wall cells under *in vitro* pulsatile flow loading, *J Artif Organs* **8**: 110–118 (2005).

71. P. Angele, D. Schumann, M. Angele, B. Kinner, C. Englert, R. Hente, B. Fuchtmeier, M. Nerlich, C. Neumann and R. Kujat, Cyclic, mechanical compression enhances chondrogenesis of mesenchymal progenitor cells in tissue engineering scaffolds, *Biorheology* **41**: 335–346 (2004).

72. S. X. Hsiong, P. Carampin, H. J. Kong, K. Y. Lee and D. J. Mooney, Differentiation stage alters matrix control of stem cells, *J Biomed Mater Res A* **85**: 145–156 (2008).

73. A. J. Engler, S. Sen, H. L. Sweeney and D. E. Discher, Matrix elasticity directs stem cell lineage specification, *Cell* **126**: 677–689 (2006).

74. C. Fischbach, R. Chen, T. Matsumoto, T. Schmelzle, J. S. Brugge, P. J. Polverini and D. J. Mooney, Engineering tumors with 3D scaffolds, *Nat Methods* **4**: 855–860 (2007).

75. L. G. Griffith and M. A. Swartz, Capturing complex 3D tissue physiology *in vitro*, *Nat Rev Mol Cell Biol* **7**: 211–224 (2006).

76. K. Takahashi, K. Tanabe, M. Ohnuki, M. Narita, T. Ichisaka, K. Tomoda and S. Yamanaka, Induction of pluripotent stem cells from adult human fibroblasts by defined factors, *Cell* **131**: 861–872 (2007).

77. J. Yu, M. A. Vodyanik, K. Smuga-Otto, J. Antosiewicz-Bourget, J. L. Frane, S. Tian, J. Nie, G. A. Jonsdottir, V. Ruotti, R. Stewart, I. I. Slukvin and J. A. Thomson, Induced pluripotent stem cell lines derived from human somatic cells, *Science* **318**: 1917–1920 (2007).

78. Y. Shi, J. T. Do, C. Desponts, H. S. Hahm, H. R. Scholer and S. Ding, A combined chemical and genetic approach for the generation of induced pluripotent stem cells, *Cell Stem Cell* **2**: 525–528 (2008).

79. H. Zhou, S. Wu, J. Y. Joo, S. Zhu, D. W. Han, T. Lin, S. Trauger, G. Bien, S. Yao, Y. Zhu, G. Siuzdak, H. R. Scholer, L. Duan and S. Ding, Generation of induced pluripotent stem cells using recombinant proteins, *Cell Stem Cell* **4**: 381–384 (2009).

80. M. Y. Lee, R. A. Kumar, S. M. Sukumaran, M. G. Hogg, D. S. Clark and J. S. Dordick, Three-dimensional cellular microarray for high-throughput toxicology assays, *Proc Natl Acad Sci USA* **105**: 59–63 (2008).

16

NOVEL METHODS FOR CHARACTERIZING AND SORTING SINGLE STEM CELLS FROM THEIR TISSUE NICHES

Ju Li, Eric Jabart, Sachin Rangarajan and Irina Conboy

1. Introduction

Both embryonic organogenesis and adult tissue regeneration rely on multiple subsets of stem cells. Understanding and deliberately controlling the activity of stem cells are of great therapeutic value since these cells build and repair every tissue in the human body. However, stem cells are difficult to study because they constitute minute populations in adult organs, express multiple and not-entirely-characterized cell-surface markers, and change their behavior based on the biochemical composition of their tissue niches. In skeletal muscle, for example, muscle stem cells (also known as satellite cells) represent only 4% of total myonuclei in adult mammals and are heterogeneous with respect to the expression of different combinations and different levels of CD34,[1,2] CXCR4,[3] β1-integrin,[3]

397

M-cadherin,[1] α-7-integrin,[4] and Syndecan-4.[5] Furthermore, while some studies exclude Sca-1 as a marker of muscle stem cells, other publications suggest that this cell-surface molecule marks a small subset of myogenic satellite cells.[6] Adding to this complexity of stem cell populations, the body of recent work suggests that the regenerative capacity and the gene expression of muscle stem cells are regulated by signals from their niches and change with age in ways which inhibit tissue maintenance and repair.[1,7–11]

This high level of heterogeneity compounded by the microniche-directed changes reinforces the importance of applying single-cell methods in modern approaches to studying organ stem cells.[12] Numerous techniques have been developed for characterizing and sorting single stem cells from their organ environments, including flow cytometry, fluorescence microscopy, and microfluidics. In this chapter, we will focus on both traditional and newly developed methods for single-stem cell analysis, and their implication in biomedical research.

2. Techniques for Single-Cell Analysis

Traditional methods such as gel electrophoresis, chromatography, and mass spectrometry are used to analyze large cell populations, and consequently yield population-averaged quantitative data. Several techniques, which were initially developed for large cell populations, have been more recently adapted and optimized for single-cell analysis, such as fluorescence-activated cell sorting (FACS), two-photon microscopy (TPM), laser scanning cytometry,[13] capillary electrophoresis,[14] and laser capture microdissection.[15] Among these methods, flow cytometry and fluorescence microscopy are the most common in stem cell research.

2.1 *Flow cytometry*

Since its invention in the 1960s, flow cytometry has been widely used in studying numerous cell types.[16–18] Traditional flow cytometry combines fluorescence labeling and high content analysis, which allows quantitative and objective determination of cell size, granularity, and the levels of macromolecules (e.g. protein, nucleic acid) in each cell of a given cell population. In recent years flow cytometry has been further developed from originally measuring 1–2 wavelengths to now measuring up to 10–15 wavelengths at the same time, thus allowing one to assay many different combinations of cell markers and to uniquely identify rare cells in heterogeneous populations.[19] This platform is particularly suited for studying stem cells, which are characterized not by one marker, but by a specific combination of proteins.[20,21]

The behavior of stem cells is tightly regulated by various signals emanating from their tissue niches, which modulate the intensity of numerous intracellular signal transduction pathways.[8,22,23] Intracellular multi-parameter flow cytometry was hence developed to measure not only the levels of different intracellular proteins, but also to determine the basal versus active states of the key effectors of signaling pathways.[24–26] Moreover, this new technique allows simultaneous profiling of the multiple signaling pathways in the rare populations of stem cells.[26–28] To further improve the sensitivity of these methods, an enzyme amplification approach called tyramide signal amplification (TSA) has been applied to intracellular flow cytometry, thereby increasing the detection capabilities ten-fold.[29] The abovementioned developments have made this technology even more applicable to comprehensive analysis of rare/single stem cells.

While commercial flow cytometry is capable of screening 10–15 different fluorophores, reliable analysis of rare stem cells (low starting material) and the detection of more than ~10 fluorophores are complicated by the (1) high noise to signal ratios, (2) spectral overlap, and (3) difficulty with controlling non-specific fluorescence. To overcome these limitations, a new technique called fluorescent cell barcoding (FCB) has been developed, where each sample is labeled with a different barcode of fluorescence intensity and wavelength. With a single fluorophore, FCB can analyze four to six different samples simultaneously, and with a combination of three fluorophores, 64–216 samples can be barcoded and analyzed with high-throughput screening.[30] FCB and phospho-specific flow cytometry have been currently applied for single cell analysis of primary mouse splenocytes, and the results reveal differences in sensitivity and kinetics between B cells, T cells, and neutrophils.[30]

The ability to measure multiple molecular markers and signal transduction pathways in rare stem cell populations has made flow cytometry the most powerful current tool for drug discovery and a popular cell-based diagnostic instrument for a variety of diseases.[31,32] To further apply flow cytometry for single-cell analysis, low-volume machines have emerged in recent years[33] that enable the study of small samples with few (100–1000) starting cells. Miniaturization of this method is likely to bring about theoretical and clinical breakthroughs.

2.2 *Fluorescence microscopy*

Microscopy has been used for biological research since the mid-17th century and is perfectly suited for single-cell analysis, including the study of stem cells in their tissue niches. However, cells are typically not alive during examination by microscopy, and cellular as well as subcellular phenomena are not easily quantifiable.[34,35] In traditional fluorescence microscopy, linear absorption of light of a specific wavelength generates the contrast between the signal and noise. This detection is limited to the surface of cells and tissues and is not effective at high magnification, because of strong and multiple light scattering, which reduces the contrast and blurs the images. To solve this problem, the first confocal microscope was developed in the 1950s, which uses point illumination and a pinhole to eliminate out-of-focus light in order to increase optical resolution and contrast. By changing the focal plane, images at different depths can be collected and used to reconstruct the three-dimensional structure of the specimen.[36] Nevertheless, the samples are fixed for confocal microscopy and hence, this method does not allow one to correlate the expression of a particular stem cell marker with cell fate or regenerative behavior.

While the development of *in vivo* fluorescent labels and fluorescent genes/proteins has moved microscopy toward the single-cell analysis of live stem cells, one major limitation of this technique is the photo-bleaching and phototoxicity caused by the damaging illumination.[37] To solve this problem and improve live-cell imaging, the new technique of two-photon laser-scanning microscopy was developed. In this method, a laser beam scans a focal plane within a specimen (penetrating up to 2 inches), and the fluorescence label is excited only at the focus of the beam. As a result, there is no out-of-focus fluorescence, and photobleaching and phototoxicity are limited in space and time.[37] Using this technology, it is possible to directly visualize the position and behavior of hematopoietic stem cells within their bone marrow niches.[38] Moreover, with a similar technique, behavior of epithelial stem cells and their progeny can be traced in live mice during the entire process of hair-follicle regeneration.[39] In more recent research, two-photon microscopy has been combined with other techniques, such as synthetic dye injection and transgenic mouse strains, in which particular stem cell fates are fluorescently detected, and with *ex vivo* labeled cells, which are used in transplantations; as a result, single stem cells can now be traced inside the animal tissues both during embryonic development and in the process of adult organ maintenance and repair.[40]

Most biological processes, such as the immune response or tissue repair, are dynamic in nature; however, earlier studies employed a static method of fluorescence microscopy in fixed tissue sections.[41] Recently, efforts have been made to assay the biological phenomena (including the stem cell responses) in real time by two-photon microscopy or by conventional confocal microscopy. For example, Stoll *et al.* used traditional confocal microscopy to visualize antigen-specific T cells interacting with dendritic cells within lymph nodes, including immunological synapse formation, crosstalk between these two types of cells, and subsequent activation, dissociation, and migration of T cells.[42] Instead of one-photon microscopy, several groups used two-photon laser microscopy to image the dynamic behavior of individual cells deeper in live organs, in their native niches, such as lymphocytes in lymph nodes,[43] and thymocytes in thymic stroma.[44] A great promise of two-photon real-time microscopy is enabling a better understanding of dynamic stem cell behavior in the context of the tissue niche. This method is likely to yield novel findings that are based on single stem cell analysis.

2.3 *Other techniques for single-cell analysis and their biomedical applications*

Flow cytometry is perfectly suited for large numbers of cells, and although a new wave of cytometers is being currently developed for analysis and sorting of low-starting cell populations,[15] microfluidics is much better suited for assaying minute numbers of stem and other cells. While flow cytometry is highly quantitative and a well-controlled method, it cannot be easily applied to niche-specific characterization of single stem cells. For example, in the case of muscle satellite cells, FACS cannot discriminate between cells that have been isolated from different myofibers or even separate hind-leg muscles. On the other hand, microscopy is capable of imaging stem cells in their niches; however, it is not easy to quantify gene expression levels. Furthermore, both flow cytometry and fluorescence microscopy utilize irreversible antibody binding to stem-cell proteins, thus potentially altering cell properties during analysis.[33]

To address these challenges, microfluidic approaches for single-cell analysis have been developed and used for a number of years. Microfluidics-based cell sorting and analysis are nowadays commonplace as cheap, effective, and versatile platforms can be rapidly produced for a variety of specific applications that are ill-suited for traditional sorting techniques such as FACS and magnetic-activated cell sorting (MACS).[45] Particularly appropriate in situations where obtaining large numbers of cells or volumes, dealing with expensive reagents and costs, or portability are issues, microfluidic systems can often outperform classic techniques.[46] Cell separations in microfluidic devices can be achieved using a variety of modalities: physical properties such as hydrodynamic separation, dielectrophoresis (separation through application of a non-uniform electric field),[47] acoustophoretic sorting using high intensity sound waves, or magnetophoresis (separation through application of a magnetic field)[48] and binding to antibodies that are specific to cell-surface molecules and are functionalized in the micro-channels have all been successfully used for single-cell analysis.[49]

An additional feature of a recently developed microfluidic platform is a combination of quantitative screening and sorting of live tissue stem cells with the clonogenic functional analysis of cell-fate and regenerative capacity at single-cell levels.[50] Such a method allows one to isolate distinct, single stem cells that reside in the same tissue microniche and correlate the heterogeneous cell-surface marker expression of these cells with their functional myogenic performance.[50] While, thus far, such microfluidics have been applied only to skeletal muscle fibers and their associated satellite cells, in the future the functional cell-fate mapping can be applied broadly to adult and embryonic single stem cells residing in their tissue microniches. These types of approaches to single-cell analysis aim to enhance our understanding of the molecular causes and evolutionarily conserved reasons for the phenomenon of stem cell heterogeneity.

In addition to individual techniques, several new, integrative methods have been designed in order to combine the benefits and overcome the limitations of each separate methodology for stem cell analysis (Table 1). With respect to flow cytometry, to improve the sampling and amplification schemes, the first mass cytometer has been developed to combine conventional tagged antibody screening and time-of-flight parameters with mass spectrometry.[51] With the high sensitivity and accuracy of this new technique, up to 34 low-abundance proteins can be profiled simultaneously in each single cell.[52] Another novel platform, Flow Chip, has been developed to integrate microfluidics for sample handling with flow cytometry for cell screening and sorting; this allows automatic high-throughput cell culture, capture, imaging, analysis, and sorting.[53,54] The Flow Chip assay has been used to monitor the mitogen-activated protein kinases ERK1/2 and p38 activation in response to Escherichia coli lipopolysaccharide (LPS) stimulation in murine macrophage cells.[53]

Similarly to the abovementioned biomedical technologies, microfluidics can also be modular, such that multiple approaches can be integrated in order to increase efficiency and accuracy. For example, several devices use the low Reynolds number flow properties of small-scale fluid flow in conjunction with antibody-coated channels, or antibody-functionalized magnetic beads, to increase the separation of individual cell types in blood.[55, 56] Alternatively, combining the dielectrophoretic method to direct cells toward antibody functionalized capture sites led to the creation of a device with a relatively high efficiency and purity when used to isolate prostate circulating cancer cells.[57]

The increased effectiveness of integrated devices for single-cell analysis is likely to yield brand-new technological platforms for disease diagnostics, as well as a better understanding of stem cell properties.

3. Necessity for Single-Cell Analysis

Stem cells are rare; their genetic and epigenetic landscapes are complex; and their behavior is strictly dependent on their microenvironment. While

Table 1.　Summary of Techniques for Single-Cell Analysis.

Technique	Advantages	Limitations	Example References
Flow cytometry	Up to 15 parameters Sorting live cells	Manual sampling Spectral overlap	19
Fluorescence cell barcoding	Up to 216 barcoded samples Sorting live cells	Manual sampling Separation of fluorophore intensity	30
Fluorescence microscopy	Analyzing cells within niches Detecting cellular localization	Photobleaching and phototoxicity Not applicable for imaging in deep body Not applicable for live cell sorting	34, 35
Confocal microscopy	Analyzing cells within niches Deep cell imaging	Photobleaching and phototoxicity Not applicable for live cell sorting	36
Two-photon microscopy	Analyzing cells within niches Tracing live cells in animals	Not applicable for live cell sorting Not high-throughput Not quantitative	37, 40
Microfluidic channel	Quantitative analysis Label-free	Currently only one parameter	46
Mass cytometry	Currently 42 parameters Accuracy for protein identification	Requires multiple chemical tags Not applicable for live cell sorting	47, 48
Flow chip	Automatic system High throughput	Needs alteration for different cells	49, 50

the heterogeneity of stem cells is undisputed, the reasons and molecular mechanisms regulating stem cell diversity are poorly understood, yet are thought to be responsible for such important events as targeting of stem cells to their niches and cell-fate determination during embryonic development and adult tissue repair.[58,59] When the stem cell population is analyzed as a whole, the causes and consequences of stem cell heterogeneity cannot be fully uncovered. With advances of the technologies for single-cell analysis, we are beginning to be able to profile the gene expression, sort, and functionally study single stem cells, which provide great insight into the fundamental laws governing the heterogeneity of stem cells (even when they reside in the same tissue niche), as well as have the potential to improve stem cell-based transplantation therapies for combating tissue degenerative disorders.

3.1 *Stem cell heterogeneity*

Significant heterogeneity in cell-surface markers, in intensity of signaling pathways, as well as in the transcriptome, proteome, and epigenome is a typical property of stem cells;[58–61] such heterogeneity plays an important role in the commitment of a stem cell to a particular cell fate.[20] As a result, understanding the heterogeneity of stem cells at the single-cell level presents a great challenge as well as a unique opportunity to uncover the key mechanisms regulating embryonic development and adult tissue regeneration. In this regard, one important advance in single-cell analysis is our ability to characterize the gene expression and functional properties of single stem cells in their native environment. For example, a combination of high-resolution confocal microscopy and two-photon video imaging is used to track individual transplanted hematopoietic stem cells (HSCs) in the bone marrow of living mice, and the results show the interactions between the HSCs and the surrounding tissues as they home and engraft in recipient mice.[38] The understanding of stem cell heterogeneity also enables better approaches for prospective isolation of the stem cell of interest: for example, mesenchymal stem cells can be purified from bone marrow through lineage negative selection followed by positive enrichment of cells that express the Sca-1 and c-Kit surface markers.[62]

3.2 *Significance of heterogeneity for the success of stem-cell therapy*

Stem cell-based regenerative therapy is a promising new approach for treating many diseases, such as heart failure, neurological disorders, muscular dystrophy, immune dysfunction, Alzheimer's disease, and other tissue degenerative conditions. A variety of stem cells including embryonic stem cells, induced pluripotent stem cells, and adult stem cells have been explored as therapeutic stem cell sources. However, the method of stem cell transplantation is limited by numerous side effects, such as poor survival and insufficient engraftment of donor cells, immune rejection, and low to no regenerative efficiency of the transplant.[63] Interestingly, even within the same type of stem cells from the same donor, there are huge variations in cell performance, which are due to the stem cell heterogeneity.[64] As a result, there is a tempting opportunity to compare the stem cells of a population for their regenerative capacity (at single-cell levels) and link the regenerative response to expression of particular genes/markers, thereby selecting the most efficient subpopulation of stem cells for organ repair and maintenance. Identifying and selecting stem cells with the best ability to self-renew and differentiate *in vivo* would enable continuous regeneration of the transplant, which would with time replace genetically defective and/or damaged tissue with the healthy progeny of transplanted cells. All these translational approaches depend on better technologies for the analysis of stem cells at single levels, and such new experimental platforms are being developed at a rapid pace.

For example, HSCs, which can be found in cord blood, peripheral blood, and bone marrow, are able to differentiate into various blood cell types such as red blood cells, white blood cells, and platelets, and therefore garner an important therapeutic potential.[65,66] HSCs are typically isolated from whole blood and selected for cell-based transplantation based on their surface marker expression. The most well-defined population, $CD34^-c\text{-}Kit^{\pm}Sca\text{-}1^{\pm}Lin^-$, shows the greatest regenerative and self-renewal potential; one cell can rescue a lethally irradiated mouse from the hematopoietic decline.[67,68] Nevertheless, more effective transplantation methods could be developed upon deciphering the significance of HSCs' heterogeneity, since currently, only 20% of recipients show successful engraftment and long-term constitution after the single-cell transplantations.[67,69] HSC heterogeneity can often be missed when looking at bulk populations, but can be detected by observing growth of single HSCs or by assaying single-HSC gene expression.[70–73]

Purification of HSCs from blood or bone marrow is a key process for enhancing the therapeutic outcomes of cell transplantation; however, there is room for improvement, as efficiencies can often be low.[74] While FACS and MACS are routinely and effectively used for the bulk sorting and analysis of these rare tissue stem cells, microfluidic approaches are in nascent form, although a few recent studies have been promising in being able to characterize single HSCs. For example, a polydimethylsiloxane (PDMS) membrane microdevice was used to purify HSCs from bone marrow.[56] Unlike simpler devices with filters that often clogged, the membrane allowed for both valving and filtering, enabling wash steps and preventing clogging. The sorting, which took advantage of the fact that HSCs are larger than most of the cells in bone marrow, was rapid (up to 17.2 ml/min) and efficient (up to 98% enrichment), and resulted in low cytotoxity (cell survival above 90%). In addition, the entire process is label-free, although cells could be labeled prior to sorting if downstream analysis was desired. Another recent paper showcased a microfluidic device that could isolate, count, and sort HSCs.[49] Also made of PDMS, this device combined fluorescent and magnetic bead cell labeling, and operated using a four-step system. Cells were first labeled with magnetic beads, then transported, magnetically isolated, and then finally counted (*via* fluorescence detection) and sorted. Close to 90% efficiency was achieved using cord blood. Although the processing time (100 μL in 40 minutes) was high, cell purity was better than the one obtained in the standard larger scale system, and the scalability or ability to do parallel processing with such microdevices could easily increase the future efficiency of single-cell sorting and characterization of the HSCs.

Similarly to the HSCs, in skeletal muscle, phenotypic heterogeneity of muscle stem cells reflects their functional heterogeneity, although the consensus on the best combination of cell-surface markers that define all muscle stem cells has not yet been reached. Purified subpopulations of muscle stem cells show differences in cell-surface marker expression[75–77] and differ in lineage markers,[4,78] as well as in cell shape and proliferation rates.[79,80] To identify the subset of muscle stem cells with robust regenerative potential, different research groups use different combinations of cell-surface markers. A combination of five markers (CD45$^-$Sca-1$^-$Mac-1$^-$ CXCR4$^+$β1-integrin$^+$) is thought to define a population of muscle stem cells that exhibit an efficient functional regeneration of dystrophic or injured muscles upon transplantation.[3,75] However, CXCR4 is also expressed by differentiated muscle progenitors (and not stem) cells and by many immune cells, as well as by other cell types, and β1-integrin is expressed by a majority of the cells in the mammalian organism. Another subpopulation of muscle stem cells with high regenerative capacity is enriched by a combination of α7-integrin/CD34 double-positive selection and Sca-1, CD45, CD31, and Lin-negative selection.[4,81,82] However, the common consensus and recent data suggest that not every myogenic muscle stem cell expresses every cell-surface marker that is used for the satellite cell isolation, and furthermore, that the levels of particular proteins vary significantly in satellite cell populations. Thus, many important subpopulations are typically missed in these bulk cell sorting approaches. Therefore, so far, all/most true muscle stem cells are best defined by their micro-anatomical location: beneath the basement membrane and on top of the plasma membrane of muscle fibers.

Summarily, future work is required to identify and understand in cellular and molecular terms the heterogeneity of tissue stem cells, and such knowledge will improve the outcomes of cell transplantation therapies.

3.3 *Blood cell analysis and sorting*

Detection, sorting, and analysis of cells (or other elements such as proteins, viruses, etc.) from blood have important ramifications in disease diagnosis, prevention, and treatment. While advanced equipment found in hospitals or clinics can often reliably detect such analytes, these instruments suffer from several drawbacks. They are often bulky (i.e. non-portable), expensive, require specialized training to operate, and are often restricted to people with the financial means to both afford and access them in their permanent location. The advent of microfluidic devices has allowed access to similar, if not equally effective, means of detection and analysis, combined with portability and accessibility, as microfluidic devices can typically be delivered at the point-of-care, manufactured at low cost, and often require little training to operate.

Blood, which is approximately three parts plasma and two parts cells by volume (red blood cells make up 99% of cells in blood in addition to leukocytes, platelets, and some rare cells[83]), is one of the most important fluids in the body, and ripe with applications for microfluidic diagnostics and single-cell analysis. The ability to selectively isolate the different cell types and components in blood is the focus of many microfluidic research groups.[84] A review by Han and Frazier sums up many of the recent techniques for plasma, leukocyte, platelet, and some rare cell separations from blood.[85]

Advanced microfluidic purification methods using dielectrophoresis or acoustophoresis, or electrohydrodynamic methods have also achieved high purity of the negative selection of blood cells, by trapping or concentrating the blood cells away from the plasma.[86–92]

White blood cell (leukocyte) isolation from blood is essential for diagnosis and prevention in certain disease.[93] A good review highlights the effects of a series of silicon microdevices in trapping leukocytes;[94] weirs, pillars, cross-flow, and membranes were showcased using this platform. As with whole-blood isolation, many early leukocyte isolation methods experienced clogging or throughput issues,[85] and novel studies have looked into hydrodynamic filtration methods to deal with clogging, improving cell concentrations up to 50-fold.[95,96] Novel microfluidic devices using dielectrophoresis or magnetic sorting akin to MACS have also appeared to further advance the methods of analysis of rare/single cells.[85, 97–100]

As for red and white blood cells, centrifugation is commonly used to extract platelets from blood.[85] Apheresis, the platelet-rich plasma preparation, or the buffy-coat preparation are common alternatives[101] to the more recently developed microfluidic methods: acoustophoresis[85] and dielectrophoresis;[102,103] many possibilities still remain to be explored.

3.4 *Detection of circulating tumor cells*

Circulating tumor cells (CTCs) are among the most commonly searched-for rare cells in the blood, and microfluidic devices have recently focused heavily on isolating them in order to improve disease diagnosis and assess progression. The presence and concentration of CTCs in blood is a valuable diagnostic tool. Studies have shown that measuring CTC concentration as a marker for metastasis and tumor progression has comparable reliability with established methods, while providing the benefits of reducing the incidence of inter-observer variability (in radiographic studies) or patient variability (in tumor marker expression).[104–106] This diagnostic tool is also more immediately useful in terms of monitoring the tumor's response to treatment, with measurements of CTC levels in response to treatment being diagnostically relevant within three to four weeks, as opposed to the standard 8–12 weeks required before radiographic studies can show any diagnostically relevant changes in tumor growth; this allows doctors to provide a more fine-tuned therapeutic regimen for patients, and to more quickly inform doctors when a patient does not respond to a treatment.[102,107] However, the diagnostic utility of CTC concentration is limited due to the difficulty in making an accurate measurement, primarily due to the low concentration of CTCs. A very high degree of accuracy and efficiency is required; distinguishing, for example, between 1 cell per 7.5 mL and 5 cells per 7.5 mL of blood is necessary since these levels correlate with a favorable prognosis and an unfavorable prognosis, respectively.[107]

There exist several methods of measuring CTC concentration. The first diagnostic tool to obtain FDA approval, called CellSearch, uses antibody-functionalized magnetic beads, fluorescent nuclear staining, and immunofluorescence to isolate and identify CTCs.[107] Other methods in development include negative selection by removing erythrocytes and white blood cells, either by lysis and immunohistochemistry, or by size-based straining of blood samples.[108,109] However, these methods face limitations due to cell loss, low throughput, and user intensiveness. These approaches can, however, be optimized though microfluidics, whereby higher throughput, increased efficiency and accuracy, and label-free methods can be achieved.

Identification of CTCs can be mediated, for example, through the presence of the EpCAM surface marker on a large percentage of CTCs. Indeed, CellSearch functionalizes magnetic beads with anti-EpCAM antibodies for isolation of CTCs.[107] Several devices have been developed whereby microfluidic channels are built with microposts functionalized with anti-EpCAM antibodies, to which CTCs preferentially adhere to, and which are shown to have an efficiency ranging from 65% to 85%.[110, 111] While such an approach has been generally effective, reliance on EpCAM to differentiate between CTCs and blood cells is problematic in cases of certain types of melanoma, brain cancers, sarcomas, and other cancers, in which EpCAM expression is minimized.[112,113]

An alternative approach involves utilizing the unique size and structural properties of CTCs. Specifically, CTCs are both larger (10–25 μm), as well as less deformable, than white and red blood cells.[104,109,114,115] These properties are useful as devices can be designed with microcavities and/or narrow channels on the order of 5-μm diameters in order to selectively isolate CTCs; blood cells can pass through the 5-μm gaps (by size, with red blood cells, or through deformation, with white blood cells), while the larger and sturdier CTCs remain trapped within the device.[110,111,116] Such an approach is unique in that it does not require antibodies or labeling, and instead utilizes only physical properties of CTCs for their isolation.

A more nuanced size-based technique takes advantage of the unique properties of low Reynolds number flow that occur at the miniature scale of microfluidic devices. Namely, particles of different sizes that undergo laminar flow at these small scales tend to concentrate on different streamlines based on inertial forces. Several devices utilize a curvilinear channel result in separation of particles (and cells) of different sizes, into distinct streamlines along the flow profile; by including outlet ports for the streamlines corresponding to smaller and larger cells, continuous high throughput screening can be performed on large volumes of blood in order to isolate CTCs.[109,117]

Another unique property of CTCs is the dielectrophoretic potentials of CTCs that contrast with the potentials of other blood cells. The ApoStream device takes advantage of the dissimilarities in frequency-dependent dielectric properties of CTCs and erythrocytes, monocytes, lymphocytes, and granulocytes, in order to select for rare/single CTCs in a mixed population.[108] While this device circumvents the issue of labeling cells, and is generally more accurate than purely sorting by size, its efficiency remains below 70% which is problematic due to the low starting concentrations of CTCs.

4. Conclusions and Future Directions

Single-cell analysis has become one of the main thrusts of stem cell science in the 21st century. It is not only the frontier in academic research, but is also recognized for a great potential to develop novel strategies for combating currently incurable diseases and for increasing a healthy human life span. Numerous new techniques have been developed and each advance improves multiple aspects of single-cell analysis, such as sensitivity, throughput, resolution, and accuracy. In the future the ongoing integration of different fields — life science, medicine, engineering, physics, and chemistry — will yield an ideal quantitative and automated single-cell analysis platform that is applicable to all stem cell types and allows one to profile and sort rare stem cells at single levels. Deciphering the meaning and regulation of the fundamental phenomenon of stem-cell heterogeneity will without a doubt enable prospective isolation of the most regenerative cells that are tailor-made for specific applications, including effective tissue repair, speed of differentiation into the tissue of need, and efficient self-renewal and repopulation of pathological organs with the healthy transplant.

References

1. I. M. Conboy, M. J. Conboy, G. M. Smythe and T. A. Rando, Notch-mediated restoration of regenerative potential to aged muscle, *Science* **302**: 1575–1577 (2003).
2. P. S. Zammit, Y. Nagata, A. P. Ruiz, C. A. Collins, T. A. Partridge and J. R. Beauchamp, Pax7 and myogenic progression in skeletal muscle satellite cells, *J Cell Sci* **119**: 1824–1832 (2006).
3. R. I. Sherwood, J. L. Christensen, I. M. Conboy, M. J. Conboy, T. A. Rando, I. L. Weissman and A. J. Wagers, Isolation of adult mouse myogenic progenitors: functional heterogeneity of cells within and engrafting skeletal muscle, *Cell* **119**: 543–554 (2004).
4. D. Montarras, J. Morgan, C. Collins, F. Relaix, S. Zaffran, A. Cumano, T. Partridge and M. Buckingham, Direct isolation of satellite cells for skeletal muscle regeneration, *Science* **309**: 2064–2067 (2005).
5. K. K. Tanaka, J. K. Hall, A. A. Troy, D. D. Cornelison, S. M. Majka and B. B. Olwin. Syndecan-4-expressing muscle progenitor cells in the SP engraft as satellite cells during muscle regeneration, *Cell Stem Cell* **4**: 217–225 (2009).

6. P. O. Mitchell, T. Mills, R. S. O'Connor, E. R. Kline, T. Graubert, E. Dzierzak and G. K. Pavlath, Sca-1 negatively regulates proliferation and differentiation of muscle cells, *Dev Biol* **283**: 240–252 (2005).

7. M. E. Carlson and I. M. Conboy, Loss of stem cell regenerative capacity within aged niches, *Aging Cell* **6**: 371–382 (2007).

8. I. M. Conboy, M. J. Conboy, A. J. Wagers, E. R. Girma, I. L. Weissman and T. A. Rando, Rejuvenation of aged progenitor cells by exposure to a young systemic environment, *Nature* **433**: 760–764 (2005).

9. A. J. Wagers and I. M. Conboy, Cellular and molecular signatures of muscle regeneration: current concepts and controversies in adult myogenesis, *Cell* **122**: 659–667 (2005).

10. M. J. Conboy, I. M. Conboy and T. A. Rando, Heterochronic parabiosis: historical perspective and methodological considerations for studies of aging and longevity, *Aging Cell* **12**: 525–530 (2013).

11. M. E. Carlson, C. Suetta, M. J. Conboy, P. Aagaard, A. Mackey, M. Kjaer and I. Conboy, Molecular aging and rejuvenation of human muscle stem cells, *EMBO Mol Med* **1**: 381–391 (2009).

12. H. H. Chang, M. Hemberg, M. Barahona, D. E. Ingber and S. Huang, Transcriptome-wide noise controls lineage choice in mammalian progenitor cells, *Nature* **453**: 544–547 (2008).

13. M. M. Harnett, Laser scanning cytometry: understanding the immune system *in situ*, *Nat Rev Immunol* **7**: 897–904 (2007).

14. I. G. Arcibal, M. F. Santillo and A. G. Ewing, Recent advances in capillary electrophoretic analysis of individual cells, *Anal Bioanal Chem* **387**: 51–57 (2007).

15. J. Kehr, Single cell technology, *Curr Opin Plant Biol* **6**: 617–621 (2003).

16. H. R. Hulett, W. A. Bonner, J. Barrett and L. A. Herzenberg, Cell sorting: automated separation of mammalian cells as a function of intracellular fluorescence, *Science* **166**: 747–749 (1969).

17. W. A. Bonner, H. R. Hulett, R. G. Sweet and L. A. Herzenberg, Fluorescence activated cell sorting, *Rev Sci Instrum* **43**: 404–409 (1972).

18. L. A. Herzenberg and R. G. Sweet, Fluorescence-activated cell sorting, *Sci Am* **234**: 108–117 (1976).

19. Roederer, M. *et al.* 8 color, 10-parameter flow cytometry to elucidate complex leukocyte heterogeneity, *Cytometry* **29**: 328–339 (1997).

20. S. C. Bendall and G. P. Nolan, From single cells to deep phenotypes in cancer, *Nat Biotechnol* **30**: 639–647 (2012).

21. M. J. Conboy, M. Cerletti, A. J. Wagers and I. M. Conboy, Immuno-analysis and FACS sorting of adult muscle fiber-associated stem/precursor cells, *Methods Mol Biol* **621**: 165–173 (2010).

22. S. J. Morrison, A. M. Wandycz, K. Akashi, A. Globerson and I. L. Weissman, The aging of hematopoietic stem cells, *Nat Med* **2**: 1011–1016 (1996).

23. T. A. Rando, Stem cells, ageing and the quest for immortality, *Nature* **441**: 1080–1086 (2006).

24. K. D. Bauer and J. W. Jacobberger, Analysis of intracellular proteins, *Methods Cell Biol* **41**: 351–376 (1994).

25. P. O. Krutzik and G. P. Nolan, Intracellular phospho-protein staining techniques for flow cytometry: monitoring single cell signaling events, *Cytometry A* **55**: 61–70 (2003).

26. O. D. Perez and G. P. Nolan, Simultaneous measurement of multiple active kinase states using polychromatic flow cytometry, *Nat Biotechnol* **20**: 155–162 (2002).

27. S.C. de Rosa, L. A. Herzenberg and M. Roederer, 11-color, 13-parameter flow cytometry: identification of human naive T cells by phenotype, function, and T-cell receptor diversity, *Nat Med* **7**: 245–248 (2001).

28. S. P. Perfetto, P. K. Chattopadhyay and M. Roederer, Seventeen-colour flow cytometry: unravelling the immune system, *Nat Rev Immunol* **4**: 648–655 (2004).

29. M. R. Clutter, G. C. Heffner, P. O. Krutzik, K. L. Sachen and G. P. Nolan, Tyramide signal amplification for analysis of kinase activity by intracellular flow cytometry, *Cytometry A* **77**: 1020–1031 (2010).

30. P. O. Krutzik and G. P. Nolan, Fluorescent cell barcoding in flow cytometry allows high-throughput drug screening and signaling profiling, *Nat Methods* **3**: 361–368 (2006).

31. K. Sachs, O. Perez, D. Pe'er, D. A. Lauffenburger and G. P. Nolan, Causal protein-signaling networks derived from multiparameter single-cell data, *Science* **308**: 523–529 (2005).

32. J. M. Irish, R. Hovland, P. O Krutzik, O. D. Perez, Ø. Bruserud, B. T. Gjertsen and G. P. Nolan, Single cell profiling of potentiated phospho-protein networks in cancer cells, *Cell* **118**: 217–228 (2004).

33. K. C. Cheung, M. Di Berardino, G. Schade-Kampmann, M. Hebeisen, A. Pierzchalski, J. Bocsi, A. Mittag and A. Tárnok, Microfluidic impedance-based flow cytometry, *Cytometry A* **77**: 648–666 (2010).

34. M. Oheim, High-throughput microscopy must re-invent the microscope rather than speed up its functions, *Br J Pharmacol* **152**: 1–4 (2007).

35. R. Pepperkok, and J. Ellenberg, High-throughput fluorescence microscopy for systems biology, *Nat Rev Mol Cell Biol* **7**: 690–696 (2006).

36. M. Minsky, Memoir on inventing the confocal scanning microscope, *Scanning* **10**: 128–138 (1988).

37. W. Denk, J. H. Strickler and W. W. Webb, Two-photon laser scanning fluorescence microscopy, *Science* **248**: 73–76 (1990).

38. C. L. Celso, H. E. Fleming, J. W. Wu, C. X. Zhao, S. Miake-Lye1, J. Fujisaki., D. Côté, D. W. Rowe, C. P. Lin and D. T. Scadde, Live-animal tracking of individual haematopoietic stem/progenitor cells in their niche, *Nature* **457**: 92–96 (2009).

39. P. Rompolas, E. R. Deschene, G. Zito, D. G. Gonzalez, I. Saotome, A. M. Haberman and V. Greco, Live imaging of stem cell and progeny behaviour in physiological hair-follicle regeneration, *Nature* **487**: 496–499 (2012).

40. F. Helmchen and W. Denk, Deep tissue two-photon microscopy, *Nat Methods* **2**: 932–940 (2005).

41. M. K. Jenkins, A. Khoruts, E. Ingulli, D. L. Mueller, S. J. McSorley, R. L. Reinhardt, A. Itano and K. A. Pape, *In vivo* activation of antigen-specific CD4 T cells, *Annu Rev Immunol* **19**: 23–45 (2001).

42. S. Stoll, J. Delon, T. M. Brotz and R. N. Germain, Dynamic imaging of T cell-dendritic cell interactions in lymph nodes, *Science* **296**: 1873–1876 (2002).

43. M. J. Miller, S. H. Wei, I. Parker and M. D. Cahalan, Two-photon imaging of lymphocyte motility and antigen response in intact lymph node, *Science* **296**: 1869–1873 (2002).

44. P. Bousso, N. R. Bhakta, R. S. Lewis and E. Robey, Dynamics of thymocyte-stromal cell interactions visualized by two-photon microscopy, *Science* **296**: 1876–1880 (2002).

45. J. Autebert, B. Coudert, F. Bidard, J. Pierga, S. Descroix, L. Malaquin and J. Viovy, Microfluidic: an innovative tool for efficient cell sorting, *Methods* **57**: 297–307 (2012).

46. V. Lecault, A. K. White, A. Singhal and C. L. Hansen, Microfluidic single cell analysis: from promise to practice, *Curr Opin Chem Biol* **16**: 381–390 (2012).

47. B. J. Kirby, *Micro- and Nanoscale Fluid Mechanics: Transport in Microfluidic Devices* (Cambridge University Press, Cambridge, 2010).

48. M. Zborowski, G. R. Ostera, L. R. Moore, S. Milliron, J. J. Chalmers and A. N. Schechter, Red blood cell magnetophoresis, *Biophys J* **84**: 2638–2645 (2003).

49. Y. Gao, W. Li and D. Pappas, Recent advances in microfluidic cell separations, *Analyst* (2013).

50. M. R. Chapman, K. R. Balakrishnan, J. Li, M. J. Conboy, H. Huang, S. K. Mohanty, E. Jabart, J. Hack, I. M. Conboy and L. L. Sohn, Sorting single satellite cells from individual myofibers reveals heterogeneity in cell-surface markers and myogenic capacity, *Integr Biol (Camb)* **5**: 692–702 (2013).

51. D. R. Bandura, V. I. Baranov, O. I. Ornatsky, A. Antonov, R. Kinach, X. Lou, S. Pavlov, S. Vorobiev, J. E. Dick and S. D. Tanner, Mass cytometry: technique for real time single cell multitarget immunoassay based on inductively coupled plasma time-of-flight mass spectrometry, *Anal Chem* **81**: 6813–6822 (2009).

52. S. C. Bendall, E. F. Simonds, P. Qiu, A. D. Amir, P. O. Krutzik, R. Finck, R. V. Bruggner, R. Melamed, A. Trejo, O. I. Ornatsky, R. S. Balderas, S. K. Plevritis, K. Sachs, D. Pe'er, S. D. Tanner and G. P. Nolan, Single-cell mass cytometry of differential immune and drug responses across a human hematopoietic continuum, *Science* **332**: 687–696 (2011).

53. N. Srivastava, J. S. Brennan, R. F. Renzi, M. Wu, S. S. Branda, A. K. Singh and A. E. Herr, Fully integrated microfluidic platform enabling automated phosphoprofiling of macrophage response, *Anal Chem* **81**: 3261–3269 (2009).

54. B. K. McKenna, J. G. Evans, M. C. Cheung and D. J. Ehrlich, A parallel microfluidic flow cytometer for high-content screening, *Nat Methods* **8**: 401–403 (2011).

55. J. W. M. Visser, S. J. L. Bol and G. Vandenengh, Characterization and enrichment of murine hematopoietic stem-cells by fluorescence activated cell sorting, *Exp Hematol* **9**: 644–655 (1981).

56. E. A. Ozkumur, M. Shah, J. C. Ciciliano, B. L. Emmink, D. T. Miyamoto, E. Brachtel, M. Yu, P. I. Chen, B. Morgan, J. Trautwein, A. Kimura, S. Sengupta, S. L. Stott, N. M. Karabacak, T. A. Barber, J. R. Walsh, K. Smith, P. S. Spuhler, J. P. Sullivan, R. J. Lee, D. T. Ting, X. Luo, A. T. Shaw, A. Bardia, L. V. Sequist, D. N. Louis, S. Maheswaran, R. Kapur, D. A. Haber and M. Toner, Inertial focusing for tumor antigen-dependent and -independent sorting of rare circulating tumor cells, *Sci Transl Med* **5**: 179ra147 (2013).

57. C. Huang, H. Liu, N. H. Bander and B. J. Kirby, Enrichment of prostate cancer cells from blood cells with a hybrid dielectrophoresis and immuno-capture microfluidic system, *Biomed Microdevices* (2013).

58. A. W. Bruce and M. Zernicka-Goetz, Developmental control of the early mammalian embryo: competition among heterogeneous cells that biases cell fate, *Curr Opin Genet Dev* **20**: 485–491 (2010).

59. I. M. Conboy and T. A. Rando, Aging, stem cells and tissue regeneration: lessons from muscle, *Cell Cycle* **4**: 407–410 (2005).

60. M. L. Suva, N. Riggi and B. E. Bernstein, Epigenetic reprogramming in cancer, *Science* **339**: 1567–1570 (2013).

61. H. Baharvand, A. Fathi, D. van Hoof and G. H. Salekdeh, Concise review: trends in stem cell proteomics, *Stem Cells* **25**: 1888–1903 (2007).

62. S. A. Jacobs, V. D. Roobrouck, C. M. Verfaillie and S. W. Van Gool, Immunological characteristics of human mesenchymal stem cells and multipotent adult progenitor cells, *Immunol Cell Biol* **91**: 32–39 (2013).

63. T. F. Lee-Pullen and M. D. Grounds, Muscle-derived stem cells: implications for effective myoblast transfer therapy, *IUBMB Life* **57**: 731–736 (2005).

64. M. Prinz, J. Priller, S. S. Sisodia and R. M. Ransohoff, Heterogeneity of CNS myeloid cells and their roles in neurodegeneration, *Nat Neurosci* **14**: 1227–1235 (2011).

65. H. W. Wu, R. C. Hsu, C. C. Lin, S. M. Hwang and G. B. Lee, An integrated microfluidic system for isolation, counting, and sorting of hematopoietic stem cells, *Biomicrofluidics* **4** (2010).

66. P. S. Dhot, V. Nair, D. Swarup, D. Sirohi and P. Ganguli, Cord blood stem cell banking and transplantation, *Indian J Pediatr* **70**: 989–992 (2003).

67. M. Osawa, K. Hanada, H. Hamada and H. Nakauchi, Long-term lympho-hematopoietic reconstitution by a single CD34-low/negative hematopoietic stem cell, *Science* **273**: 242–245 (1996).

68. G. J. Spangrude, S. Heimfeld and I. L. Weissman, Purification and characterization of mouse hematopoietic stem cells, *Science* **241**: 58–62 (1988).

69. G. J. Spangrude, D. M. Brooks and D. B. Tumas, Long-term repopulation of irradiated mice with limiting numbers of purified hematopoietic stem cells: *in vivo* expansion of stem cell phenotype but not function, *Blood* **85**: 1006–1016 (1995).

70. Q. Zhang and R. H. Austin, Applications of microfluidics in stem cell biology, *Bionanoscience* **2**: 277–286 (2012).

71. V. Lecault, M. Vaninsberghe, S. Sekulovic, D. J. Knapp, S. Wohrer, W. Bowden, F. Viel, T. McLaughlin, A. Jarandehei, M. Miller, D. Falconnet, A. K. White, DG Kent, M. R. Copley, F. Taghipour, C. J. Eaves, R. K. Humphries, J. M. Piret and C. L. Hansen, High-throughput analysis of single hematopoietic stem cell proliferation in microfluidic cell culture arrays, *Nat Methods* **8**: 581–586 (2011).

72. J. P. Glotzbach, M. Januszyk, I. N. Vial, V. W. Wong, A. Gelbard, T. Kalisky, H. Thangarajah, M. T. Longaker, S. R. Quake, G. Chu and G. C. Gurtner, An information theoretic, microfluidic-based single cell analysis permits identification of subpopulations among putatively homogeneous stem cells, *PLoS One* **6**: e21211 (2011).

73. B. Dykstra, J. Ramunas, D. Kent, L. McCaffrey, E. Szumsky, L. Kelly, K. Farn, A. Blaylock, C. Eaves and E. Jervis, High-resolution video monitoring of hematopoietic stem cells cultured in single-cell arrays identifies new features of self-renewal, *P Natl Acad Sci USA* **103**, 8185–8190 (2006).

74. D. G. Kent, M. R. Copley, C. Benz, S. Wöhrer, B. J. Dykstra, E. Ma, J. Cheyne, Y. Zhao, M. B. Bowie, Y. Zhao, M. Gasparetto, A. Delaney, C. Smith, M. Marra, C. J. Eaves, Prospective isolation and molecular characterization of hematopoietic stem cells with durable self-renewal potential, *Blood* **113**: 6342–6350 (2009).

75. M. Cerletti, S. Jurga, C. A. Witczak, M. F. Hirshman, J. L. Shadrach, L. J. Goodyear and A. J. Wagers, Highly efficient, functional engraftment of skeletal muscle stem cells in dystrophic muscles, *Cell* **134**: 37–47 (2008).

76. S. Fukada, S. Higuchi, M. Segawa, K. Koda, Y. Yamamoto, K. Tsujikawa, Y. Kohama, A. Uezumi, M. Imamura, Y. Miyagoe-Suzuki, S. Takeda and H. Yamamoto, Purification and cell-surface marker characterization of quiescent satellite cells from murine skeletal muscle by a novel monoclonal antibody, *Exp Cell Res* **296**: 245–255 (2004).

77. B. Pawlikowski, L. Lee, J. Zuo and R. H. Kramer, Analysis of human muscle stem cells reveals a differentiation-resistant progenitor cell population expressing Pax7 capable of self-renewal, *Dev Dyn* **238**: 138–149 (2009).

78. S. Kuang, K. Kuroda, F. Le Grand and M. A. Rudnicki, Asymmetric self-renewal and commitment of satellite stem cells in muscle, *Cell* **129**: 999–1010 (2007).

79. N. Hashimoto, T. Murase, S. Kondo, A. Okuda and M. Inagawa-Ogashiwa, Muscle reconstitution by muscle satellite cell descendants with stem cell-like properties, *Development* **131**: 5481–5490 (2004).

80. Rouger, K. *et al.* Muscle satellite cell heterogeneity: *in vitro* and *in vivo* evidences for populations that fuse differently. *Cell Tissue Res* **317**, 319–326 (2004).

81. A. Sacco, R. Doyonnas, P. Kraft, S. Vitorovic, and H. M. Blau, Self-renewal and expansion of single transplanted muscle stem cells, *Nature* **456**: 502–506 (2008).

82. A. Pasut, P. Oleynik, and M. A. Rudnicki, Isolation of muscle stem cells by fluorescence activated cell sorting cytometry, *Methods Mol Biol* **798**: 53–64 (2012).

83. W. Han, A. Ali, C. Wong, H. Sha, H. Jongyoon and T. Chwee, Microfluidic devices for blood fractionation, *Micromachines* **2**: 319–343 (2011).

84. M. Toner and D. Irimia, Blood-on-a-chip, *Annu Rev Biomed Eng* **7**: 77–103 (2005).

85. K. H. Han and A. B. Frazier, Lateral-driven continuous dielectrophoretic microseparators for blood cells suspended in a highly conductive medium, *Lab Chip* **8**: 1079–1086 (2008).

86. A. Lenshof, A. Ahmad-Tajudin, K. Järås, A. M. Swärd-Nilsson, L. Aberg, G. Marko-Varga, J. Malm, H. Lilja and T. Laurell, Acoustic whole blood plasmapheresis chip for prostate specific antigen microarray diagnostics, *Anal Chem* **81**: 6030–6037 (2009).

87. F. Petersson, A. Nilsson, C. Holm, H. Jonsson and T. Laurell, Separation of lipids from blood utilizing ultrasonic standing waves in microfluidic channels, *Analyst* **129**: 938–943 (2004).

88. F. Petersson, A. Nilsson, C. Holm, H. Jonsson and T. Laurell, Continuous separation of lipid particles from erythrocytes by means of laminar flow and acoustic standing wave forces, *Lab Chip* **5**: 20–22 (2005).

89. L. Y. Yeo, J. R. Friend and D. R. Arifin, Electric tempest in a teacup: the tea leaf analogy to microfluidic blood plasma separation, *Appl Phys Lett* **89** (2006).

90. D. R. Arifin, L. Y. Yeo and J. R. Friend, Microfluidic blood plasma separation *via* bulk electrohydrodynamic flows, *Biomicrofluidics* **1**: 14103 (2007).

91. C. Xu, Y. Wang, M. Cao and Z. Lu, Dielectrophoresis of human red cells in microchips, *Electrophoresis* **20**: 1829–1831 (1999).

92. Y. Nakashima, S. Hata and T. Yasuda, Blood plasma separation and extraction from a minute amount of blood using dielectrophoretic and capillary forces, *Sensor Actuat B-Chem* **145**: 561–569 (2010).

93. E. D. Pratt, C. Huang, B. G. Hawkins, J. P. Gleghorn and B. J. Kirby, Rare cell capture in microfluidic devices, *Chem Eng Sci* **66**: 1508–1522 (2011).

94. H.M. Ji, V. Samper, Y. Chen, C. K. Heng, T. M. Lim and L. Yobas, Silicon-based microfilters for whole blood cell separation, *Biomed Microdevices* **10**: 251–257 (2008).

95. M. Yamada, and M. Seki, Hydrodynamic filtration for on-chip particle concentration and classification utilizing microfluidics, *Lab Chip* **5**: 1233–1239 (2005).

96. S. Y. Zheng, J. Q. Liu and Y. C. Tai, Streamline-based microfluidic devices for erythrocytes and leukocytes separation, *J Microelectromech Syst* **17**: 1029–1038 (2008).

97. D. W. Inglis, R. Riehn, R. H. Austin and J. C. Sturm, Continuous microfluidic immunomagnetic cell separation, *Appl Phys Lett* **85**: 5093–5095 (2004).

98. D. W. Inglis, R. Riehn, J. C. Sturm and R. H. Austin, Microfluidic high gradient magnetic cell separation, *J Appl Phys* **99** (2006).

99. K. H. Han and A. B. Frazier, Paramagnetic capture mode magnetophoretic microseparator for high efficiency blood cell separations, *Lab Chip* **6**: 265–273 (2006).

100. B. Y. Qu, Z. Y. Wu, F. Fang, Z. M. Bai, D. Z. Yang and S. K. Xu, A glass microfluidic chip for continuous blood cell sorting by a magnetic gradient without labeling, *Anal Bioanal Chem* **392**: 1317–1324 (2008).

101. J. Kaufman, S. L. Spinelli, E. Schultz, N. Blumberg and R. P. Phipps, Release of biologically active CD154 during collection and storage of platelet concentrates prepared for transfusion, *J Thromb Haemost* **5**: 788–796 (2007).

102. J. Voldman, Electrical forces for microscale cell manipulation, *Annu Rev Biomed Eng* **8**: 425–454 (2006).

103. M.S. Pommer, Y. T. Zhang, N. Keerthi, D. Chen, J. A. Thomson, C. D. Meinhart and H. T. Soh, Dielectrophoretic separation of platelets from diluted whole blood in microfluidic channels, *Electrophoresis* **29**: 1213–1218 (2008).

104. G. T. Budd, M. Cristofanilli, M. J. Ellis, A. Stopeck, E. Borden, M. C. Miller, J. Matera, M. Repollet, G. V. Doyle, L. W. Terstappen and D. F. Hayes, Circulating tumor cells versus imaging: predicting overall survival in metastatic breast cancer, *Clin Cancer Res* **12**: 6403–6409 (2006).

105. M. G. Cristofanilli, T. Budd, M. J. Ellis, A. Stopeck, J. Matera, M. C. Miller, J. M. Reuben, G. V. Doyle, W. J. Allard, L. W. M. M. Terstappen and D. F. Hayes, Circulating tumor cells, disease progression, and survival in metastatic breast cancer. *New Engl J Med* **351**: 781–791 (2004).

106. D. F. Hayes, M. Cristofanilli, G. T. Budd, M. J. Ellis, A. Stopeck, M. C. Miller, J. Matera, W. J. Allard, G. V. Doyle and L. W. Terstappen, Circulating tumor cells at each follow-up time point during therapy of metastatic breast cancer patients predict progression-free and overall survival, *Clin Cancer Res* **12**: 4218–4224 (2006).

107. F. Petersson, L. Aberg, A. M. Sward-Nilsson and T. Laurell, Free flow acoustophoresis: microfluidic-based mode of particle and cell separation, *Anal Chem* **79**: 5117–5123 (2007).

108. B. P. Casavant, R. Mosher, J. W. Warrick, L. J. Maccoux, S. M. Berry, J. Y. Becker, A negative selection methodology using a microfluidic platform for the isolation and enumeration of circulating tumor cells, *Methods* (2013) [Epub ahead of print].

109. G. Vona, A. Sabile, M. Louha, V. Sitruk, S. Romana, K. Schütze, F. Capron, D. Franco, M. Pazzagli, M. Vekemans, B. Lacour, C. Bréchot and P. Paterlini-Brécho, Isolation by size of epithelial tumor cells: a new method for the immunomorphological and molecular characterization of circulatingtumor cells, *Am J Pathol* **156**: 57–63 (2000).

110. S. Nagrath, L. V. Sequist, S. Maheswaran, D. W. Bell, D. Irimia, L. Ulkus, M. R. Smith, E. L. Kwak, S. Digumarthy, A. Muzikansky,P. Ryan, U. J. Balis, R. G. Tompkins, D. A. Haber and M. Toner, Isolation of rare circulating tumour cells in cancer patients by microchip technology, *Nature* **450**: 1235–1239 (2007).

111. T. Ohnaga, Y. Shimada, M. Moriyama, H. Kishi, T. Obata, K. Takata, T. Okumura, T. Nagata, A. Muraguchi and K. Tsukada, Polymeric microfluidic devices exhibiting sufficient capture of cancer cell line for isolation of circulating tumor cells, *Biomed Microdevices* **15**: 611–616 (2013).

112. V. Gupta, I. Jafferji, M. Garza, V. O. Melnikova, D. K. Hasegawa, R. Pethig and D. W. Davis, ApoStream™, a new dielectrophoretic device for antibody independent isolation and recovery of viable cancer cells from blood, *Biomicrofluidics* **6**: 24133 (2012).

113. H. W. Hou, M. E. Warkiani, B. L. Khoo, Z. R. Li, R. A. Soo, D. S. W. Tan, W. T. Lim, J. Y. Han, A. A. S. Bhagat and C. T. Lim, Isolation and retrieval of circulating tumor cells using centrifugal forces, *Sci Rep* **3**: 1259 (2013).

114. M. Hosokawa, T. Hayata, Y. Fukuda, A. Arakaki, T. Yoshino, T. Tanaka and T. Matsunaga, Size-selective microcavity array for rapid and efficient detection of circulating tumor cells, *Anal Chem* **82**: 6629–6635 (2010).

115. H. Mohamed, M. Murray, J. N. Turner and M. Caggana, Isolation of tumor cells using size and deformation, *J Chromatogr A* **1216**: 8289–8295 (2009).

116. S. J. Tan, L. Yobas, G. Y. H. Lee, C. N. Ong and C. T. Lim, Microdevice for the isolation and enumeration of cancer cells from blood, *Biomed Microdevices* **11**: 883–892 (2009).

117. S. Zheng, H. Lin, J. Q. Liu, M. Balic, R. Datar, R. J. Cote and Y. C. Tai, Membrane microfilter device for selective capture, electrolysis and genomic analysis of human circulating tumor cells, *J Chromatogr A* **1162**: 154–161 (2007).

17

LABEL-FREE MICROFLUIDIC TECHNIQUES TO ISOLATE AND SCREEN SINGLE STEM CELLS

Eric Jabart, Karthik Balakrishnan and Lydia L. Sohn

1. Introduction

Embryonic stem cells (ESCs) are pluripotent,[1] having the potential to become every type of cell in the body. While some pluripotent cells have been discovered in adults,[2,3] it is the resident adult stem cells — multipotent cells that can give rise to a number of different specific cell types — in organs and tissues that are responsible for maintaining tissue homeostasis and repairing tissues after injury.[4] Successful and reliable identification of these important stem cells, e.g. *via* surface-marker expression, is invaluable for the fields of tissue engineering and regeneration. Yet, many stem-cell types are often identifiable only through expression of multiple markers and, in some cases, such as that of skeletal muscle stem cells (known as satellite cells), no actual consensus exists on which combination of markers identify the stem cells.[5-7] Of paramount importance, therefore, are methods that can identify, sort, and characterize stem cells systematically and accurately. Traditional methods of sorting stem cells, including fluorescence-activated cell sorting (FACS)[8] and magnetic-activated cell sorting (MACS),[9] are best suited for dealing with large numbers of cells. They are not capable of accounting for the significant heterogeneity present in small stem-cell populations from specific niches.[5] Screening and sorting methods that can satisfy the above requirements will play a pivotal role in harnessing the full regenerative potential of stem cells.

Microfluidic devices are exquisitely poised to provide the capabilities required to identify and characterize stem cells. Easy to fabricate, inexpensive, versatile, and customizable, microfluidic devices can be rapidly tailor-made for a variety of applications,[10] including cell culture,[11,12] western blotting,[13,14] PCR,[15-17] and *in vitro* tissue reconstruction.[12,18] For microfluidic cell sorting and analysis, modalities based on FACS[10, 19-21] and MACS[10] have already been developed. A number of other microfluidic sorting techniques exist as well: hydrodynamic sorting based on particle/cell size,[22-25] acoustophoretic sorting using high-intensity sound waves,[22,26,27] and dielectrophoretic sorting based on cell polarity,[22,28,29] to name only a few.

In the case of skeletal muscle stem cells, a recent microfluidic platform explored the heterogeneity of single satellite cells from single muscle fibers using an antibody-functionalized pore and a detection technique called resistive-pulse sensing (RPS).[5] The study showed that the heterogeneity of satellite cells is in fact greater than what was once suspected and that cell-surface marker expression could be correlated with the myogenicity of the cells, thereby enabling selection of the most regenerative cells. In this chapter, we will describe this study in greater detail, as well as discuss how this microfluidic platform technology and method could be easily applied to other stem-cell systems. We will also discuss how the method has recently been extended to perform multimarker analysis and sorting.

2. Resistive-Pulse Sensing

Resistive-pulse sensing (RPS), or the Coulter technique of particle sizing,[30] is based on measuring a current (or resistance) pulse as a particle passes through a microchannel (i.e. "pore") connected to two fluid-filled reservoirs (Fig. 1). In simplest terms, when a particle transits a pore, it partially blocks the flow of current, leading to a transient increase in the pore's electrical resistance and a defined current pulse. RPS has long been used to both characterize cells several microns in diameter[31] and detect nanoscale particles[32–35] such as viruses.[36] Furthermore, this method has been used to detect single molecules[37–39] and study their interaction.[39] In our own group, we have used RPS to: detect the nanometer increase in diameter of a submicron latex colloid upon binding to an unlabeled specific antibody,[40,41] sense single molecules of unlabeled lambda (λ) phage DNA,[42] and screen for specific surface markers on cells.[5,43,44]

The sensitivity of RPS relies upon the relative size of the pore and the particle to be measured (Fig. 2), and the magnitude of the current pulse created by a particle transiting a pore can be estimated as the volume ratio of particle to pore, i.e. $V_{particle}/V_{pore}$. In greater detail, the resistance of a pore, R_p, increases by δR_p when a particle enters since the particle displaces conducting fluid. δR_p can be estimated for a pore aligned along the z-axis by

$$\delta R_p = \rho \int \frac{dz}{A(z)} - R_p \tag{1}$$

where $A(z)$ represents the successive cross-sections of the pore containing a particle, and ρ is the resistivity of the solution. For a spherical particle of diameter d in a pore of diameter D and length L, the relative change in resistance is

$$\frac{\delta R_p}{R_p} = \frac{D}{L} \left| \frac{\arcsin(d/D)}{(1-(d/D)^2)^{1/2}} - \frac{d}{D} \right| \tag{2}$$

Eqs. (1) and (2) assume that the current density is uniform across the channel, and thus is not applicable for cases when $d \ll D$. For that

     E. Jabart, K. Balakrishnan and L. L. Sohn

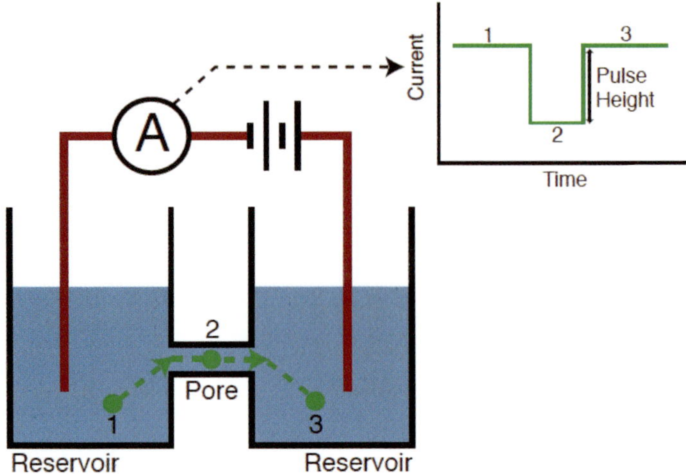

Fig. 1. A resistive-pulse sensor is made up of two reservoirs separated by a small pore or sensing aperture. When a particle transits the pore, the increase in resistance is measured as a drop (pulse) in the current.[46]

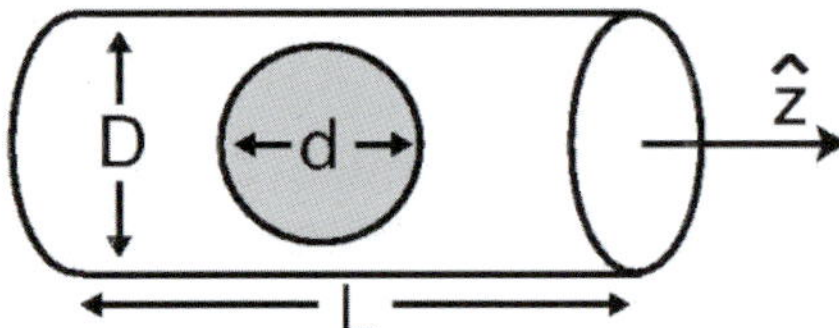

Fig. 2. A diagram of a particle of diameter d in a pore of length L and diameter D.[45]

particular case, Deblois and Bean[32] formulated an equation for δR_{p} based on an approximate solution to the Laplace equation,

$$\frac{\delta R_p}{R_p} = \frac{d^3}{LD^2}\left[\frac{D^2}{2L^2} + \frac{1}{\sqrt{1+(D/L)^2}}\right]F\left(\frac{d^3}{D^3}\right) \tag{3}$$

where $F(d^3/D^3)$ is a numerical factor that accounts for the bulging of the electric-field lines into the pore wall. When employing Eq. (3) to predict resistance changes, one finds an effective value for D by equating the cross-sectional area of a rectangular channel with that of a cylindrical one.

If R_p is the dominant resistance of the measurement circuit, then relative changes in the current I are equal in magnitude to the relative changes in the resistance, i.e. $|\delta I/I| = |\delta R_p/R_p|$, and Eqs. (2) and (3) can be both directly compared to measure current changes. This comparison is disallowed if R_p is similar in magnitude to other series resistances, such as the electrode/fluid interfacial resistance, $R_{e/f}$. One can completely remove $R_{e/f}$ from the electrical circuit by performing a *four-point* measurement of the current.[45]

3. RPS-Based Cytometry

While most work with microfluidic RPS to date has been with colloids or microparticles, there is significant interest in using microfluidic cytometry to study live cells. The vast majority of microfluidics-based cytometry has focused on optical detection of labeled cells, i.e. a true miniaturization of flow cytometry.[20,21] Very little work, however, has been done using microfluidic RPS to screen cells. Advantages of RPS screening of cells include label-free detection, which eliminates the need for sample preparation and potential modification of cells,[5,43,46,47] standard microfabrication techniques that enable scaling up of design for high throughput, and simple, straightforward electronics. Because of these advantages, microfluidic RPS has the potential to screen samples that have low cell numbers (such as satellite cells), are in low volume, and have the potential to be lost during sample preparation.

While microfluidic cytometry would appear to be ideal, it does present a number of challenges such as clogging or throughput issues. In 2004, Nieuwenhuis *et al.* used a "liquid aperture" to overcome the issue of channel clogging, all the while achieving the sensitivity of smaller apertures.[48] Use of a liquid aperture for RPS helps prevent channel blockage without compromising sensitivity, as the physical channel is much larger than the sensing aperture. In addition to a high probability of clogging, throughput does decrease with decreasing pore cross-section. In order to improve throughput, one can either fabricate a device that can detect smaller particles in a larger channel, or integrate additional channels in parallel into the device. In 2008, Wu *et al.* used two parallel channels with a two-stage differential amplification detection scheme to achieve higher throughput, increased sensitivity, and greater dynamic range.[49] Although throughput on a single chip can be increased by fabricating multiple distinct devices on a single substrate,[41] or by using a single input with multiple sensing apertures branching off into separate output reservoirs,[50] commercial viability of high-throughput RPS devices requires the simplicity of a single input and output reservoir connected by multiple channels in parallel. The major challenge toward this realization has been isolating the resistance of adjacent channels to minimize crosstalk between channels. In 2007, Zhe *et al.* demonstrated that by positioning the electrodes in the center of each channel, they could decouple four parallel channels sufficiently, and count colloids and juniper pollen spores with all four, simultaneously.[51]

Our group's contribution to microfluidic cytometry has been to develop the first microchip Coulter counter that, in a single, label-free measurement, accurately determines not only the size and shape of a cell, but also specific cell-surface markers and their levels of expression.[43,46,47] As individual cells pass through our device's pore, the magnitude of the current pulse generated corresponds to cell size; as discussed in the previous sections, this is the basis of the Coulter technique of particle sizing. The current-pulse width corresponds to the time needed for the cell to pass through the pore, and the current-pulse shape indicates how the cell transited through the pore (e.g. rolling or translating). When the pore is functionalized with proteins that have a high affinity to a particular surface epitope of the cell, specific interactions between a cell-surface marker and those proteins retard the cell passing through the pore. This correspondingly leads to an increase in pulse width, thereby indicating the presence of that specific surface marker. Cells do not have to be labeled with an exogenous label, as the label is incorporated in the device itself. Our device is fabricated with great ease and control using common soft-lithography techniques.[52] The chip-based fabrication outlined below provides flexibility in device design, thus greatly extending the capabilities and cell-based applications of the microchip Coulter counter.

4. Device Fabrication and Experimental Methods

Figure 3 shows a photographic image of a typical pore device we now fabricate in our laboratory: a PDMS mold sealed to a glass substrate with pre-defined platinum (Pt) electrodes. The PDMS mold consists of reservoirs,

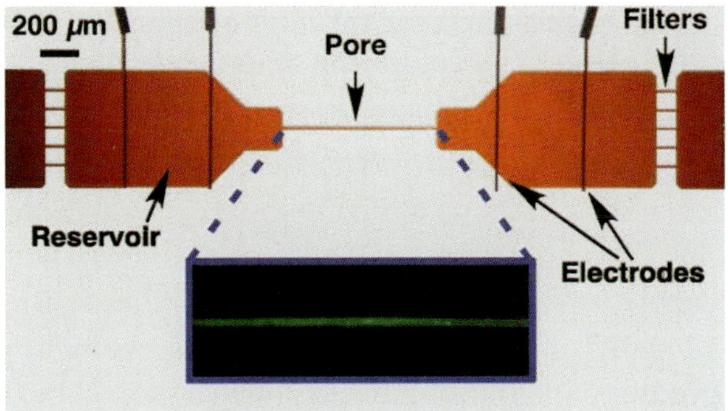

Fig. 3.   Image of our 15 μm × 15 μm × 800-μm pore linking two 30 μm-deep reservoirs. The entire device is embedded in PDMS bound to a glass substrate. The inset shows complete coverage of the pore after functionalizing with 1500 μg/mL of FITC-conjugated anti-CD34 antibody.[43]

a set of filters, and a pore whose diameter is comparable to that of the cells to be interrogated and whose length is typically several hundred microns. We lithographically create a negative master of the reservoirs, filters, and pore, which is subsequently cast into a PDMS slab. We create the master in two steps, each involving patterning SU-8 resist (MicroChem Corp.) on a polished silicon substrate to form the negatives of the pore, filters, and reservoirs. SU-8 resist is exceptionally durable once crosslinked, thereby allowing us to reuse a master indefinitely. Following standard micromolding techniques,[52] we pour PDMS (Sylgard 184, Dow Corning) over the master and cure it. We subsequently cut a slab surrounding the PDMS mold, core the reservoirs to provide inlet and outlet ports, align the mold with the glass substrate under a microscope, and permanently seal it to a glass substrate.

4.1 *Functionalization of the pore*

Prior to sealing the PDMS slab onto the glass substrate, we functionalize the glass substrate. After exposing the glass substrate with the pre-defined Pt electrodes to an oxygen plasma, we coat the region between the electrodes via microcontact printing[53] with amino-silane groups using a solution of aminopropyltriethoxysilane (APTES) (Sigma Aldrich) in anhydrous toluene (10% w/w) (Sigma Aldrich) at room temperature. To ensure a stable APTES layer, we bake the substrate in an oven at 80°C for 4–5 hours, thereby crosslinking the APTES. Following baking, we soak the cured substrate in toluene (10 minutes) and then soak and rinse in 18 MΩ deionized (DI) water (twice, 10 minutes each) to remove any unbound APTES. Once the APTES is patterned onto the glass substrate, we use a micropipette to deliver a droplet (50 μL) of a 5 mM solution of N-5-Azido-nitrobenzoyloxysuccinimide (ANB-NOS) (Pierce Biotechnology) dissolved in dimethylsulphoxide (DMSO) (Sigma Aldrich) (10% of the final reaction volume) in 20 mM HEPES (Invitrogen) (pH 7.3) to the area between the electrodes.[54] Use of this flexible crosslinker ensures that the functionalized proteins extend beyond the substrate surface.[55] After incubating the substrate in the dark, we remove the excess ANB-NOS by first washing the substrates with HEPES (10 minutes) and then rinsing with 18 MΩ DI water. We subsequently thermally bond the PDMS mold, embedded with the pore, filters, and reservoir, to the prepared substrate.

We next inject a protein solution (5–1500 μg/mL) into the pore and incubate for 3–4 hours. For the covalent binding of proteins with the ANB-NOS-treated glass substrate, we place the sealed device directly under a UV-light source. UV light activates the ANB-NOS aryl-azide groups and subsequently binds the protein covalently to the substrate. We remove any excess protein by flushing the device thoroughly with buffer solution (1× PBS at 25°C). In general, antibodies functionalized via this procedure may have reduced activity; however, previous work has shown that they are still able to interact specifically with protein receptors.[56] Fig. 3 (inset) is an epifluorescent image of a pore functionalized with 1500 μg/mL of FITC-conjugated anti-CD34 antibody (eBioscience).

4.2 *Estimate of the density of functionalized antibodies*

Following Cozens-Roberts *et al.*,[56] who investigated the kinetics of receptor-mediated cell adhesion to a ligand-coated surface, we estimate the functionalized-antibody density, N_{AB}, within our pores as

$$N_{AB} = 0.7[Ab]V_p \left(\frac{A_V}{A_p M_W} \right) \tag{4}$$

Here, [Ab] is the antibody concentration used to incubate the pore, V_p is the pore volume, A_v is Avogadro's number, A_p is the area of the functionalized glass, and M_w is the molecular weight of the antibody. Like Cozens-Roberts *et al.*[56] and Clausen,[57] we assume that only 70% of the antibodies bind to the surface. Thus, for the pores described here, the estimated functionalized-antibody densities are $N_{Ab} \sim 2.6 \times 10^2$ Ab/μm^2, when using an antibody concentration of [Ab] = 5 μg/mL, and $N_{Ab} \sim 7.8 \times 10^4$ Ab/μm^2, when using a saturating antibody concentration of [Ab] = 1500 μg/mL. If only steric limits in a single layer of antibodies [$\sim$80 nm^2/antibody][40] were to be considered, then the upper limit to the functionalized antibody density would be $\sim 1.2 \times 10^4$ antibodies/μm^2. The quantitative discrepancy between the antibody density derived from Eq. (4) using [Ab] = 1500 μg/mL and that from steric considerations is due to the aforementioned assumption that 70% of the antibodies bind to the surface. This assumption is highly dependent on the particular functionalization protocol, its efficiency, and substrate type, to name only a few factors. Nonetheless, we do have agreement regarding the order of magnitude of antibody concentration, i.e. $\sim 10^4$ Ab/μm^2.

4.3 *Fluid handling, data acquisition, and analysis*

Upon device completion, we are able to inject a sample of cells ($\sim 10^5$ to 10^6 cells/mL) into one of the outer reservoirs and apply a non-pulsatile pressure (typically 10.5 kPa, although pressures as low as 10.0 kPa and as high as 34.0 kPa have been employed) using a commercial microfluidic pump (Fluidigm Inc.) to drive individual cells ($\sim$100 cells/min) through the filters, the inner reservoir, and finally through the pore. Cellular debris and clumps are excluded from the pore by the filters (Fig. 3).

We use RPS to measure the current pulse caused by a single cell transiting the pore. In all performed measurements, the current is low-pass filtered below 0.3 ms in the rise time and is sampled at 50 kHz. This sampling frequency corresponds to a time resolution of 20 μs and ensures a high-resolution pulse width. Data are recorded and analyzed using custom-written software. Pulses generated by the rare occurrence of two cells flowing simultaneously across the pore have a characteristic shape (Fig. 4) that is easily recognized and discarded.

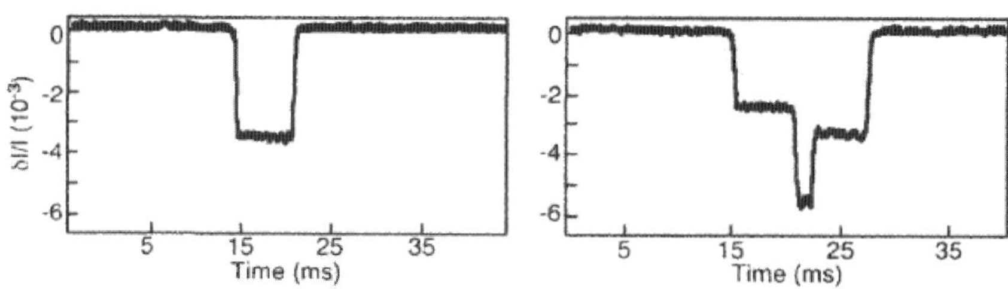

Fig. 4. Characteristic current pulses for a single cell (*left*) or two cells (*right*) transiting a pore.[46]

5. Single Surface-Marker Detection Using RPS

As mentioned earlier, in RPS, the current-pulse *width* corresponds to the transit time, τ, of a cell passing through a pore. A cell's transit time through a functionalized pore (τ_2 or τ_3 corresponding to isotype control or specific antibody) is greater than that through an unfunctionalized (i.e. "blank") pore (τ_1), due to interactions between the cell-surface marker of interest and the functionalized proteins. If interactions are non-specific, then $\tau_2 \geq \tau_1$. If interactions are specific, then $\tau_3 \gg \tau_1$. Thus, by comparing τ of a cell passing through a functionalized or a blank pore, we are able to screen for a specific cell-surface marker.[43] Of great importance is that at no time are cells strongly bound to the functionalized proteins, as they may be with cell-affinity chromatography or even MACS. Cells are flowing at a high enough shear force such that they do not have time to bind to the proteins although the shear force is orders of magnitude less than that found in MACS or FACS, thereby leaving the cells undamaged. Finally, while transient binding does occur, we have performed numerous tests to show that the cells are not activated.[5]

As proof-of-principle, we detected the CD34 receptors expressed on the surface of murine erythroleukemia (MEL) cells. A suspension of MEL cells ($\sim 10^5$ cells/mL) was injected into (1) a blank pore; (2) a pore functionalized with rat IgG_{2a} isotype control antibody (eBioscience); and (3) a pore functionalized with anti-CD34 monoclonal antibody (eBioscience) (clone: RAM34; isotype IgG_{2a}, k). Fig. 5a shows typical traces of the normalized current, $\delta I/I$, vs. time for cells passing through the three different pores: the pulses are well resolved in width. A comparison among the

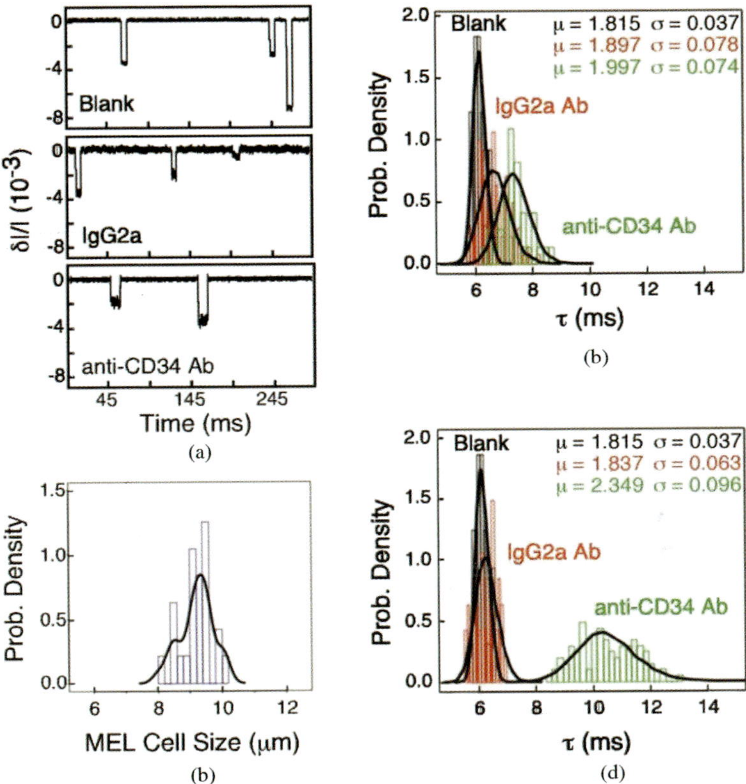

Fig. 5. CD34 detection on murine erythroleukemia (MEL) cells using RPS. (a) Representative current pulse data of cells transiting a blank (*top*), isotype control (*middle*), and anti-CD4 Ab functionalized (*bottom*). (b) Cell-size distribution for MEL cells screened in a blank pore. (c) and (d) Transit-time τ-distributions of MEL cells flowing through a blank, isotype control, anti-CD34 Ab pore.[43]

different τ-distributions shows that MEL cells traveled slowest through the anti-CD34 pore. Further, as shown in Figs. 5b–5d, τ significantly depends on the antibody concentration used to functionalize the pore. When the pores were functionalized with 100 μg/mL of antibody, the average transit time of MEL cells passing through an anti-CD34 pore was $\tau_{avg} = 7.35$ ms. This transit time was 1.21 ms and 0.67 ms more than through the blank and the isotype-control pores, respectively. More importantly, when a 15-fold higher concentration of antibody was used, τ_{avg} significantly increased only for the anti-CD34 pore. Equally important, the variance of the τ-distribution increased with antibody concentration only for the anti-CD34 pore (Figs. 5b–c). We attribute this increase to the number of binding events that occur between cells with varying CD34 densities and the anti-CD34 pore.

6. Stem-Cell Applications

While microfluidic cytometry may not be able to achieve the high throughput of flow cytometry, there are a number of applications for which current RPS is already well suited. For example, there has been significant interest in studying biological phenomena at the single-cell level in diverse areas ranging from circulating tumor cells[58–61] and meta-static potential to stem-cell biology.[62] Often when studying rare cells, sample size can be limited to as few as 5–10 cells. While traditional cytometry systems typically require samples in excess of 100,000 cells, our group has used our microfluidic Coulter counter to perform quantitative analysis of single muscle satellite cells that have been directly isolated from single to a few myofibers.[5] As we will show, we have been able to screen and isolate single satellite cells based on surface-marker expression.

6.1 *Single marker screening of skeletal muscle stem cells*

We used our label-free microfluidic technique to screen and sort single satellite cells from uninjured, single myofibers.[5] We also quantified the heterogeneous expression of surface markers and correlated these findings with myogenic capacity to reveal insight into the functional importance of marker expression. Using our platform, we analyzed the expression of CXCR4, B1-integrin, Sca-1, M-cadherin, Syndecan-4, and Notch-1.

Devices were prepared as described in the previous section, with the exception that we modified our antibody functionalization procedure to include Protein G, thereby ensuring that the antibodies were all oriented parallel to one another. To prepare the satellite cells for testing, extensor digitorum longus (EDL) muscle was dissected from the hind leg of a three-month-old C57/black-6 mouse and incubated at 37°C in digestion medium for one hour with gentle agitation. Digested muscle was gently triturated and a microscope was used to handpick single fibers. Satellite cells were liberated from three single fibers by digestion with 1 U mL^{-1} Dispase and 40 U mL^{-1} Collagenase type II in medium for one hour before screening.

As one of the first demonstrations that we could successfully screen primary cells with RPS, we screened cultured myoblasts for Sca-1 and M-cadherin and found that 2.7% were Sca-1^{+} and 93.0% were M-cadherin^{+}, confirming results from previous studies.[63-66] We also compared results of screening a bulk population of satellite cells for Sca-1 with the RPS system and FACS, and both methods yielded similar results — RPS found 67.7% to be Sca-1^{+} while FACS found 66.1% to be Sca-1^{+}. While both FACS and our device can screen cells from bulk-tissue preparations, our method uniquely allows for the analysis of rare cells from a few single myofibers.

To illustrate the applicability of our platform for screening small stem-cell populations, we examined the fiber-to-fiber heterogeneity of Sca-1, CXCR4, β1-integrin, M-cadherin, Syndecan-4, and Notch-1 expression in freshly isolated muscle satellite cells from different sets of myofibers. To determine which cells exhibited positive expression of each marker, we statistically analyzed the transit-time data and recorded if particular cells had slower transit times as compared to cells passing through the isotype control pore (Fig. 6a). Cells with significantly slower transit times in comparison to control runs were considered to have high levels of expression (Fig. 6, red), while less stringent transit time cutoffs were used to determine cells with medium (Fig. 6, green) and low (Fig. 6, blue) expression levels. In summary of the experiments performed (Fig. 6b), fibers had almost no β1-integrin, M-cadherin, Notch-1, or Sca-1 expressing satellite cells, while other fibers had a significant number of satellite cells expressing Syndecan-4, β1-integrin, CXCR4, or Notch-1. Significantly, almost no fibers had satellite cells with medium or low expression of CXCR4, Notch-1, or M-Cadherin. Notch-1 expression was highly variable between fibers, while Syndecan-4 showed the greatest variance in expression level in cells positive for this marker. We also compared freshly isolated and cultured satellite cells and found that the heterogeneity and variance in expression levels of the studied markers increased upon *in vitro* culture, as expected (Fig. 6b).

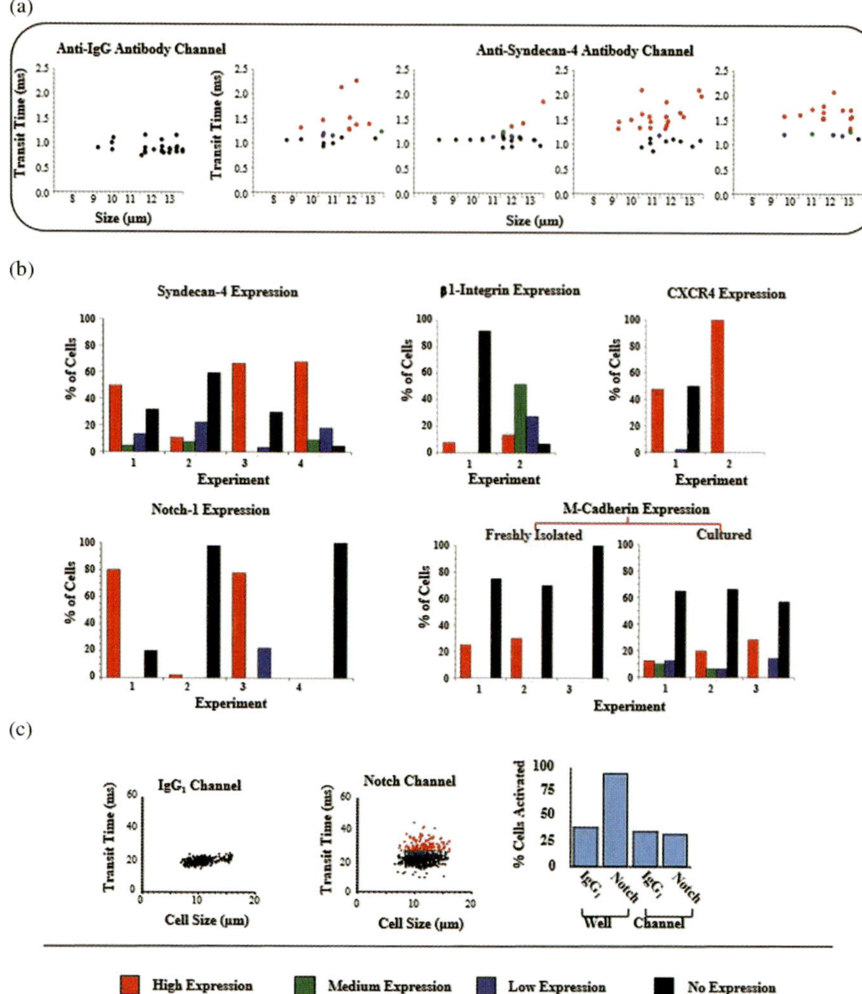

Fig. 6. RPS data from screening satellite cells. (a) Size and transit time plots of anti-IgG and anti-Syndecan-4 functionalized pore screening. Based on False Discovery Rate analysis, cells were determined to have high (red), medium (green), low (blue), or no (black) expression. (b) Summary of expression levels for experiments using anti-Syndecan-4 Ab, anti-β1-integrin Ab, anti-CXCR4 Ab, anti-Notch-1 Ab, anti-M-cadherin Ab, and anti-Sca-1 Ab functionalized pores. (c) Transit time and cell-size plots of isotype-control and anti-Notch-1-functionalized pores (*left*) and percentage of cells showing levels of nuclear Notch-1 after screening vs. control cultured conditions (*right*).[5]

Following the discovery of microniche-specific heterogeneity of marker expression in satellite cells freshly isolated from single fibers, we investigated the possibility that transient binding between the functionalized antibody and cell-surface receptors activated receptor signaling and could ultimately change cell properties.[67] Anti-Notch-1 antibody, specific for the external part of Notch-1 receptor, has been shown to mimic native ligand binding and activate Notch-1 in satellite cells, resulting in high levels of the truncated intracellular portion of Notch that is localized to the cell nucleus.[68] We therefore analyzed whether Notch would become activated when satellite cells were screened in our microfluidic device (Fig. 6c). As additional controls, unscreened satellite cells were plated on IgG_1 and anti-Notch-1 antibody-coated culture wells overnight. Immunofluorescence was used to determine whether the transient interactions between the extracellular portion of the receptor and the functionalized antibodies activated the Notch pathway, using an antibody that specifically recognizes the truncated-activated form of Notch-1. In contrast to the anti-Notch-1-antibody-coated wells, cells from the IgG_1 and anti-Notch-1-antibody microchannels and the IgG_1-coated wells showed low levels of Notch activation. Thus, the transient binding between functionalized antibody in our microchannel and specific receptors does not contribute to Notch-pathway activation in satellite cells. Overall, we conclude that changes in signal transduction and cell behavior should not be expected when stem-cell populations are screened with our method.

Our results demonstrate that our technology is very applicable to the analysis of stem cells, and offers advantages such as the ability to correlate cell surface-marker expression with myogenicity. With our simple method, we revealed the variation of satellite-cell subsets among single fibers in the same muscle and characterized this heterogeneity based on multiple surface markers, further showcasing the potential of our device for rare-cell analysis. The label-free nature of our multi-parametric measurement enables further propagation and secondary analysis of the screened cells, since they are not damaged and there are no residual reagents left to interfere. We believe that the ability to handle small samples in a label-free manner is one of the greatest strengths of our method, and we are very excited by the possibility of using this technology to access biological phenomena that have been elusive using traditional bulk cytometry. More importantly, the generality of our method lends itself to other stem-cell systems for screening given its simplicity — by varying the selection of functionalization antibodies, we can easily verify the presence of, or uncover, new surface markers in an effort to identify stem-cell populations specifically and uniquely.

6.2 *Toward multi-marker screening of stem cells*

Having demonstrated the strength of our method for single surface-marker screening, we have also leveraged the versatility of microfluidics to develop a multi-marker screening platform which employs node-pore sensing (NPS).[69] NPS, which involves using pores that have a sequence of nodes along them, produces distinct electronic signals (Fig. 7, bottom), unlike those generated through traditional RPS (Fig. 7, top). As a cell enters a node-pore, the current initially drops from the baseline as expected. When the cell enters the first node, however, the current rises transiently, only to drop again once it exits. This rise and fall in current is repeated as the cell enters and exits the next node. When the cell exits the node-pore, the current returns to the baseline value. Because of the unique segmentation of generated signals, the particle transit times

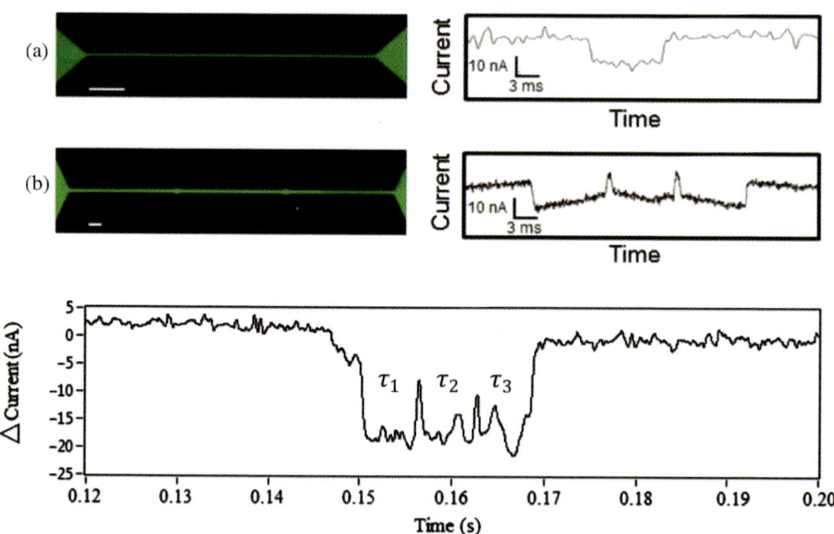

Fig. 7. (*top*) Fluorescent image of a traditional pore and a representative current pulse of a cell transiting the pore. (*bottom*) Fluorescent image of a pore segmented into two regions by a single node and the current signal generated as a cell travels through the single-node pore.

through each region of the node-pore can be determined. Recently, we successfully utilized NPS to screen single cells for multiple surface markers, simultaneously.[76] Using temporary channels, we functionalize the individual segments between the nodes with different antibodies, each corresponding to a distinct cell-surface receptor (at least one segment is functionalized with an isotype control antibody). By aligning and clamping the node channel to the functionalized substrate, we can subsequently screen cells for multiple markers, simultaneously. A cell transiting the NPS device produces a current pulse whose sub-pulses correspond to the different antibody segments the cell traverses. Measuring the duration of the sub-pulses and comparing to that of the isotype control allows for multi-surface-marker, or phenotypic, identification. We have successfully applied our innovative technology to screen simultaneously CD13, CD14, CD15, CD33, CD34, HLA-DR, and CD45 on cells from established human leukemic cell lines and on blast cells from patient bone-marrow samples.[76] Applying NPS multi-marker screening to phenotype single stem cells can thus be achieved.

7. Conclusion

The exquisite sensitivity and simplicity of our RPS-based device impart a number of general advantages over other current cell characterization methods, such as flow cytometry,[70] fluorescence microscopy, MACS,[71] cell-affinity chromatography,[72–74] and impedance spectroscopy.[34,75] First, exogenous labeling is not required. This significantly reduces both cost and sample preparation time and is especially attractive in cases where it is not known whether labeling affects cell physiology and function. Second, accuracy is not lost as a result of the small number of cells interrogated. Thus, our method is suitable for cases in which only a few hundred cells may be available for interrogation. Third, separation (using microfluidic valves) and recovery of cells for further propagation are possible. Although the throughput of a single pore is not comparable to that of flow cytometry (100 cells/min vs. 10,000 cells/s), our microfluidic platform enables the construction of arrays of pores[41,50,51] for performing many measurements or assays in parallel. Finally, our measurements are sensitive to the density of receptors on the cell surface. For those cells that express low levels of receptors on the cell surface, the flow rate in our pore could be readily reduced to increase sensitivity.

Acknowledgements

The authors would like to thank the numerous people who have contributed to the development of the RPS-based cytometry method described here: O. A. Saleh, A. Carbonaro, L. Godley, H. Huang, S. K. Mohanty, and M. Chapman. We thank M. Conboy and J. Li for isolating the satellite cells we screened. This work has been partially funded by the W. M. Keck Foundation Medical Research Program, the National Institutes of Health, Siemens-Berkeley Center of Knowledge Interchange, National Defense Science and Engineering Graduate Fellowship (K. B.), and the Siebel Scholars Foundation (E. J.).

References

1.. J. A. Thomson, J. Itskovitz-Eldor, S. S. Shapiro, M. A. Waknitz, J. J. Swiergiel, V. S. Marshall and J. M. Jones, Embryonic stem cell lines derived from human blastocysts, *Science,* **282**: 1145–1147 (1998).

2. S. Heneidi, A. Simerman, E. Keller, P. Singh, X. Li, D. A. Dumesic and G. Chazenbalk, Awakened by cellular stress: isolation and characterization of a novel population of pluripotent stem cells derived from human adipose tissue, *PLoS One* **8**: e64752 (2013).

3. S. Roy, P. Gascard, N. Dumont, J. Zhao, D. Pan, S. Petrie, M. Margeta and T. P. Tisty, Rare somatic cells from human breast tissue exhibit extensive lineage plasticity, *Proc Natl Acad Sci USA* **110**: 4598–4603 (2013).

4. M. Korbling and Z. Estrov, Adult stem cells for tissue repair: a new therapeutic concept? *N Engl J Med* **349**: 570–582 (2003).

5. M. R. Chapman, K. R. Balakrishnan, J. Li, M. J. Conboy, H. Huang, S. K. Mohanty, E. Jabart, J. Hack, I. M. Conboy and L. L. Sohn, Sorting single satellite cells from individual myofibers reveals heterogeneity in cell-surface markers and myogenic capacity, *Integr Biol* **5**: 692–702 (2013).

6. P. S. Zammit, J. P. Golding, Y. Nagata, V. Hudon, T. A. Partridge and J. R. Beauchamp, Muscle satellite cells adopt divergent fates: a mechanism for self-renewal? *J Cell Biol* **166**: 347–357 (2004).

7. Y. Ono, L. Boldrin, P. Knopp, J. E. Morgan and P. S. Zammit, Muscle satellite cells are a functionally heterogeneous population in both somite-derived and branchiomeric muscles, *Dev Biol* **337**: 29–41 (2010).

8. L. A. Herzenberg and S. C. De Rosa, Monoclonal antibodies and the FACS: complementary tools for immunobiology and medicine, *Immunol Today* **21**: 383–390 (2000).

9. B. Schmitz, A. Radbruch, T. Kummel, C. Wickenhauser, H. Korb, M. L. Hansmann, J. Thiele and R. Fischer, Magnetic activated cell sorting (MACS): a new immunomagnetic method for megakaryocytic cell isolation: comparison of different separation techniques, *Eur J Haematol* **52**: 267–275 (1994).

10. S. Lindstrom and H. Andersson-Svahn, Overview of single-cell analyses: microdevices and applications, *Lab Chip* **10**: 3363–3372 (2010).

11. G. H. Underhill, P. Galie, C. S. Chen and S. N. Bhatia, Bioengineering methods for analysis of cells *in vitro*, *Annu Rev Cell Dev Biol* **28**: 385–410 (2012).

12. X. J. Li, A. V. Valadez, P. Zuo and Z. Nie, Microfluidic 3D cell culture: potential application for tissue-based bioassays, *Bioanalysis* **4**: 1509–1525 (2012).

13. S. Q. Tia, M. He, D. Kim and A. E. Herr, Multianalyte on-chip native Western blotting, *Anal Chem* **83**: 3581–3588 (2011).

14. A. J. Hughes and A. E. Herr, Microfluidic Western blotting, *Proc Natl Acad Sci USA* **109**: 21450–21455 (2012).

15. J. Burn, Company profile: QuantuMDx Group Limited, *Pharmacogenomics* **14**: 1011–1015 (2013).

16. F. Ahmad and S. A. Hashsham, Miniaturized nucleic acid amplification systems for rapid and point-of-care diagnostics: a review, *Anal Chim Acta* **733**: 1–15 (2012).

17. S. Park, Y. Zhang, S. Lin, T. H. Wang and S. Yang, Advances in microfluidic PCR for point-of-care infectious disease diagnostics, *Biotechnol Adv* **29**: 830–839 (2011).

18. D. Huh, Y. S. Torisawa, G. A. Hamilton, H. J. Kim and D. E. Ingber, Microengineered physiological biomimicry: organs-on-chips, *Lab Chip* **12**: 2156–2164 (2012).

19. D. J. Ehrlich, B. K. McKenna, J. G. Evans, A. C. Belkina, G. V. Denis, D. Sherr and C. C. Man, Parallel imaging microfluidic cytometer, *Methods Cell Biol* **102**: 49–75 (2011).

20. B. K. McKenna, A. A. Selim, R. F. Bringhurst and D. J. Ehrlich, 384-channel parallel microfluidic cytometer for rare-cell screening, *Lab Chip* **9**: 305–310 (2009).

21. A. Kummrow, J. Theisen, M. Frankowski, A. Tuchscheerer, H. Yildrim, K. Brattke, M. Schmidt and J. Neukammer, Microfluidic structures for flow cytometric analysis of hydrodynamically focused blood cells fabricated by ultraprecision micromachining, *Lab Chip* **9**: 972–981 (2009).

22. G. J. Kirby, *Micro- and Nanoscale Fluid Mechanics: Transport in Microfluidic Devices* (Cambridge University Press, 2009).

23. C. Blattert, R. Jurischka, I. Tahhan, A. Schoth, P. Kerth and W. Menz, Separation of blood in microchannel bends, *Conf Proc IEEE Eng Med Biol Soc* **4**: 2627–2630 (2004).

24. C. Blattert, R. Jurischka, A. Schoth, P. Kerth and W. Menz, Separation of blood cells and plasma in microchannel bend structures. *Proc SPIE* **5591**: 143–151 (2004).

25. R. D. Jaggi, R. Sandoz and C. S. Effenhauser, Microfluidic depletion of red blood cells from whole blood in high-aspect-ratio microchannels, *Microfluid Nanofluid* **3**: 47–53 (2007).

26. M. Zborowski, G. R. Ostera, L. R. Moore, S. Milliron, J. J. Chalmers and A. N. Schechter, Red blood cell magnetophoresis, *Biophys J* **84**: 2638–2645 (2003).

27. Y. Gao, W. Li and D. Pappas, Recent advances in microfluidic cell separations. *Analyst* (2013).

28. M. S. Pommer, Y. T. Zhang, N. Keerthi, P. Chen, J. A. Thomson, C. D. Meinhart and H. T. Soh, Dielectrophoretic separation of platelets from diluted whole blood in microfluidic channels, *Electrophoresis* **29**: 1213–1218 (2008).

29. J. Voldman, Electrical forces for microscale cell manipulation, *Annu Rev Biomed Eng* **8**: 425–454 (2006).

30. W. H. Coulter, Means for counting particles suspended in a fluid, United States of America patent (1953).

31. E. C. Gregg and K. D. Steidley, Electrical counting and sizing of mammalian cells in suspension, *Biophys J* **5**: 393–405 (1965).

32. R. W. Deblois and C. P. Bean, Counting and sizing of submicron particles by the resistive pulse technique, *Rev Sci Instrum* **41**: 909–917 (1970).

33. M. Koch, G. R. Evans and A. Brunnschweiler, Design and fabrication of a micromachined coulter counter. *J Micromech Microeng* **9**: 159–161 (1999).

34. L. Sun and R. M. Crooks, Fabrication and characterization of single pores for modeling mass transport across porous membranes, *Langmuir* **15**: 738–741 (1999).

35. Y. Kobayashi and C. R. Martin, Toward a molecular coulter counter type device, *J Electroanal Chem* **431**: 29–33 (1997).

36. R. W. Deblois, C. P. Bean and R. K. A. Wesley, Electrokinetic measurements with submicron particles and pores by the resistive pulse technique, *J Colloid Interface Sci* **61**: 323–335 (1977).

37. H. Bayley and P. S. Cremer, Stochastic sensors inspired by biology, *Nature* **413**: 226–230 (2001).

38. S. M. Bezrukov, I. Vodyanoy and V. A. Parsegian, Counting polymers moving through a single ion channel, *Nature* **370**: 279–281 (1994).

39. L. Q. Gu, S. Cheley, and H. Bayley, Capture of a single molecule in a nano-cavity, *Science* **291**: 636–640 (2001).

40. O. A. Saleh and L. L. Sohn, Direct detection of antibody-antigen binding using an on-chip artificial pore, *Proc Natl Acad Sci USA* **100**: 820–824 (2003).

41. A. Carbonaro and L. L. Sohn, A resistive-pulse sensor chip for multianalyte immunoassays, *Lab Chip* **5**: 1155–1160 (2005).

42. O. A. Saleh and L. L. Sohn, An artificial nanopore for molecular sensing, *Nano Lett* **3**: 37–38 (2003).

43. A. Carbonaro, S. K. Mohanty, H. Huang, L. A. Godley and L. L. Sohn, Cell characterization using a protein-functionalized pore, *Lab Chip* **8**: 1478–1485 (2008).

44. K. R. Balakrishnan, M. R. Chapman, A. Kesavaraju and L. L. Sohn, in *Nanopores for Bioanalytical Applications: Proceedings of the International Conference*, Joshua Edel and Tim Albrecht (eds.) (Royal Society of Chemistry, 2012).

45. O. A. Saleh and L. L. Sohn, Quantitative sensing of nanoscale colloids using a microchip Coulter counter, *Rev Sci Instrum* **72**: 4449–4451 (2001).

46. M. R. Chapman and L. L. Sohn, Label-free resistive-pulse cytometry, *Methods Cell Biol* **102**: 127–157 (2011).

47. K. R. Balakrishnan and L. L.Sohn, in *Methods in Cell Biology,* P. Michael Conn (ed.) (Elsevier, 2012).

48. J. H. Nieuwenhuis, F. Kohl, J. Bastemeijer, P. M. Sarro and M. J. Vellekoop, Integrated Coulter counter based on 2-dimensional liquid aperture control, *Sens Actuators B* **102**: 44–50 (2004).

49. X. Wu, Y. Kang, Y. N. Wang, D. Xu, D. Li and D. Li, Microfluidic differential resistive pulse sensors, *Electrophoresis* **29**: 2754–2759 (2008).

50. A. V. Jagtiani, R. Sawant and J. A. Zhe, Label-free high throughput resistive-pulse sensor for simultaneous differentiation and measurement of multiple particle-laden analytes, *J Micromech Microeng* **16**: 1530–1539 (2006).

51. J. Zhe, A. V. Jagtiani, P. J. H. Dutta and J. Carletta, A micromachined high throughput Coulter counter for bioparticle detection and counting, *J Micromech Microeng* **17**: 304–313 (2007).

52. Y. N. Xia and G. M. Whitesides, Soft lithography, *Angew Chem Int Ed*, 551–575 (1998).

53. J. L. Wilbur, A. Kumar, E. Kim and G. M. Whitesides, Microfabrication by microcontact printing of self-assembled monolayers, *Adv Mater* **6**: 600–604 (1994).

54. S. Karrasch, M. Dolder, F. Schabert, J. Ramsden and A. Engel, Covalent binding of biological samples to solid supports for scanning probe microscopy in buffer solution, *Biophys J* **65**: 2437–2446 (1993).

55. K. D. Patel, M. U. Nollert and R. P. McEver, P-selectin must extend a sufficient length from the plasma membrane to mediate rolling of neutrophils. *J Cell Biol* **131**: 1893–1902 (1995).

56. C. Cozens-Roberts, D. A. Lauffenburger and J. A. Quinn, Receptor-mediated cell attachment and detachment kinetics: I. Probabilistic model and analysis. *Biophys J* **58**: 841–856 (1990).

57. J. Clausen, *Immunochemical Techniques for the Identification and Estimation of Macromolecules* (North-Holland Publishing Co., 1981).

58. G. T. Budd, M. Cristofanilli, M. J. Ellis, A. Stopeck, E. Borden, M. C. Miller, J. Matera, M. Repollet, G. V. Doyle, L. W. Terstappen and D. F. Hayes, Circulating tumor cells versus imaging: predicting overall survival in metastatic breast cancer, *Clin Cancer Res* **12**: 6403–6409 (2006).

59. Circulating tumor cells, disease progression, and survival in metastatic breast cancer. *N Engl J Med* **351**: 781–791 (2004).

60. D. F. Hayes, M. Cristofanilli, G. T. Budd, M. J. Ellis, A. Stopeck, M. C. Miller, J. Matera, W. J. Allard, G. V. Doyle and L. W. Terstappen, Circulating tumor cells at each follow-up time point during therapy of metastatic breast cancer patients predict progression-free and overall survival, *Clin Cancer Res* **12**: 4218–4224 (2006).

61. F. Petersson, L. Aberg, A. M. Sward-Nilsson and T. Laurell, Free flow acoustophoresis: microfluidic-based mode of particle and cell separation, *Anal Chem* **79**: 5117–5123 (2007).

62. S. A. Kobel, O. Burri, A. Griffa, M. Girotra, A. Seitz and M. P. Lutolf, Automated analysis of single stem cells in microfluidic traps. *Lab Chip* **12**: 2843–2849 (2012).

63. R. Moore and F. S. Walsh, The cell adhesion molecule M-cadherin is specifically expressed in developing and regenerating, but not denervated skeletal muscle, *Development* **117**: 1409–1420 (1993).

64. P. O. Mitchell, T. Mills, R. S. O'Connor, T. Graubert, E. Dzierzak and G. K. Pavlath, Sca-1 negatively regulates proliferation and differentiation of muscle cells, *Dev Biol* **283**: 240–252 (2005).

65. A. Irintchev, M. Zeschnigk, A. Starzinski-Powitz and A. Wernig, Expression pattern of M-cadherin in normal, denervated, and regenerating mouse muscles. *Dev Dyn* **199**: 326–337 (1994).

66. U. Kaufmann, B. Martin, D. Link, K. Witt, R. Zeitler, S. Reinhard and A. Starzinski-Powitz, M-cadherin and its sisters in development of striated muscle. *Cell Tissue Res* **296**: 191–198 (1999).

67. A. Tarnok, H. Ulrich and J. Bocsi, Phenotypes of stem cells from diverse origin, *Cytometry A* **77**: 6–10 (2010).

68. I. M. Conboy, M. J. Conboy, G. M. Smythe and T. A. Rando, Notch-mediated restoration of regenerative potential to aged muscle, *Science* **302**: 1575–1577 (2003).

69. K. R. Balakrishnan, G. Anwar, M. R. Chapman, T. Nguyen, A. Kesavaraju, L. L. Sohn, Node-pore sensing: a robust, high-dynamic range method for detecting biological species, *Lab Chip* **13**: 1302–1307 (2013).

70. H. M. Shapiro, *Practical Flow Cytometry 4th Ed.* (Wiley, 2003).

71. P Lang, M. Pfeiffer, R. Handgretinger, M. Schumm, B. Demirdelen, S. Stanojevic, Th Klingebiel, U. Kohl, S Kuci and D. Niethammer, Clinical scale isolation of T cell-depleted CD56+ donor lymphocytes in children, *Bone Marrow Transplant* **29**: 497–502 (2002).

72. S. Katoh, Scaling-up affinity chromatography, *Trends Biotechnol* **5**: 328–331 (1987).

73. J. J. Killion and G. M. Kollmorgen, Isolation of immunogenic tumor cells by cell-affinity chromatography, *Nature* **259**: 674–676 (1976).

74. D. S. Hage, Affinity chromatography: a review of clinical applications, *Clin Chem* **45**: 593–615 (1999).

75. D. Holmes and H. Morgan, Single cell impedance cytometry for identification and counting of CD4 T-cells in human blood using impedance labels, *Anal Chem* **82**: 1455–1461 (2010).

76. K. Balakrishnan, J. C. Whang, R. Hwang, J. H. Hack, L. A. Godley and L. L. Sohn, Node-pore sensing enables label-free surface-marker profiling of single cells. *Anal Chem* **87**: 2988–2995 (2015).

18

MICROSCALE TECHNOLOGIES FOR TISSUE ENGINEERING AND STEM CELL DIFFERENTIATION

Jason W. Nichol, Hojae Bae, Nezamoddin N. Kachouie,
Behnam Zamanian, Mahdokht Masaeli and Ali Khademhosseini

1. Introduction

There is a severe shortage of tissues and organs for use in repair of diseased or otherwise defective tissues. Tissue engineering has emerged as a field to fabricate tissues in sufficient quantities to repair damaged tissues.[1] While tissues such as skin[2] or bone[3] can repair a small injury in a reasonable amount of time, tissues such as cartilage[4] and myocardium[5] cannot regenerate, and will continue to degenerate and decrease in function without intervention.

Early "top-down" approaches comprised a majority of tissue engineering approaches, whereby biodegradable polymer scaffolds, such as poly (glycolic acid) (PGA),[6] were seeded with the different cell types. In this approach, it was expected that the cells would proliferate and migrate throughout the scaffold and secrete the appropriate extracellular matrix (ECM) often aided by growth factors,[7] perfusion,[6] and mechanical[8] or other stimulation.[9] However, despite advances in scaffold technology, such as surface patterning[10] or decellularization techniques for native ECM structures,[11] it is still difficult to recreate the intricate tissue microstructure necessary to create functional tissues.

The "bottom-up" approach aims to recreate the tissue microarchitecture by designing individual building blocks with specific microstructural features and assembling these blocks into larger engineered tissues. There are a number of techniques to fabricate these building blocks, such as fabrication of cell-laden microgels,[12,13] self-assembled aggregation,[14] creation of cell sheets[15,16] or tissue printing technologies.[17] After creating the building blocks, larger engineered tissues can be fabricated from these modules by a variety of approaches such as stacking,[18,19] random packing,[20] or self assembly.[14] The bottom-up approach attempts to mimic nature as many tissues are largely composed of repeated units with similar functions and architectures, including the lobule of the liver.[21] Recapitulation of the native microarchitecture of naturally repeated units of tissues could lead to improved function at the microscale, creating biomimetic engineered tissues with improved functional properties.

One primary objective of bottom-up approaches is control of the cellular microarchitecture to better direct cellular function and ultimately tissue morphogenesis.[22] In this respect, many microscale technologies and techniques used to create cell-laden building blocks can also be used as investigative models for determining and controlling cell behavior.[23] These techniques have been employed recently for driving stem cell differentiation and function down specific lineages,[24] which could be beneficial for use in a variety of tissue engineering and regenerative medicine applications. As many tissues are comprised of cells that typically do not undergo extensive self-replication, such as cardiomyocytes in cardiac tissue, the ability to control and dictate stem cell behavior is crucial to the future success of engineered tissues. In addition, recent research has confirmed the existence of stem cells residing in niches that are unique to the tissues and organs, which contain highly ordered microarchitectures and cellular compartmentalization and arrangement.[25] Microscale technologies have the ability to recreate many of these complex features, giving great promise to creating *in vitro* microenvironments with the ability to quickly and effectively direct both adult and embryonic stem cell (ESC) behavior bringing us closer to the ultimate goal of clinical regenerative medicine applications.

The following chapter will highlight the current techniques for using microscale technologies both for creating engineered tissues using the bottom-up approach and for investigating and directing stem cell behavior. These two applications of microscale engineering are crucial to the future of regenerative medicine. In addition, as one field progresses and experiences new breakthroughs, similar analogous techniques can drive research in the other field. The future clinical success of regenerative medicine is highly dependent on successes in both tissue engineering and stem cell differentiation, suggesting that the more these fields can interact and use similar techniques the faster development will occur hopefully shortening the time needed to bring these techniques to the clinic.

2. Control of Cellular and Tissue Microarchitecture

The ability to control the cellular and tissue microarchitecture is key to both stem cell differentiation and bottom-up tissue engineering techniques. Through restriction of the cellular geometry, controlled environments can be created to investigate cell behavior.[26] These geometrical restrictions can be performed by a number of methods, such as cell seeding in microscale channels[27] or microwells[28] or creating cell-laden hydrogels using micromolding.[12]

2.1 *Hydrogels*

Engineering an environment in which cell proliferation, differentiation and function can be tightly controlled is of great importance to re-generative medicine.[29] As the cellular microenvironment has a profound effect on stem cell physiology, including proliferation and differentiation, the ability to precisely control the cellular microenvironment could allow for better control over directing stem cell behavior.[30] Similarly, the cellular microenvironment, such as cell-cell and cell-ECM and cell-soluble factor interactions, have a substantial effect on the function and tissue morphogenesis of engineered tissues.[31] Due to their characteristics, including biocompatibility, flexible methods of synthesis and a wide variety of physical characteristics, hydrogels have long been employed as scaffolding materials for tissue engineering.[32]

Hydrogels are composed of networks of hydrophilic polymer chains, which are the product of chemical or thermal interactions between prepolymer chains. These chemical bonds could be permanent (such as covalent and ionic bonds) or temporary bonds (such as hydrogen bonds).[33] Such structures make it feasible to control the crosslinking and network formation through manipulation of the reaction process, ultimately leading to better control over the microenvironment of the hydrogel.[32] Adjusting the temperature of the hydrogel or manipulating the reaction energy by tuning the ultraviolet light (UV) exposure time and power are examples of such manipulations. In terms of ionic charges, based on the groups incorporated into the hydrogel backbone, hydrogels could be neutral, cationic, anionic or ampholytic.[34] However, it is observed that charged hydrogels tend to have a greater potential for cell attachment as compared to neutrally-charged hydrogels. As a result, charged hydrogels are potentially better candidates for engineering tissue scaffolds.[35] Many of the currently available hydrogels are synthesized from monomers which are either ionized or ionizable, improving the possibility of cell attachment and proliferation within cell-laden hydrogel scaffolds. Furthermore, incorporation of various cell-binding peptide domains, such as arginine-glycine-aspartic acid (RGD), into the hydrogel can dramatically increase cell binding and spreading within the gel.[36]

The physical properties of hydrogels are another key factor making hydrogels attractive for tissue engineering applications. To optimize cell viability, having a microenvironment in which cells can easily exchange nutrients and waste products, oxygen and other soluble factors is of great importance. Diffusion is a key factor in solute transport in hydrogels, however in hydrogels with microscale pore structures or forced flow conditions, convection can be a critical factor as well.[37] Experimental results have demonstrated that the hydrogels pH, temperature, mesh size and environmental conditions could significantly affect diffusion in ionic hydrogels.[38] In addition, as will be detailed more thoroughly later in the text, hydrogels have been demonstrated to be highly amenable to a number of techniques to create structures with controllable features on the microscale. Microfabrication technologies have been applied to hydrogels to mimic the complex spatiotemporal *in vivo* ECM environment leading to engineered tissues with biomimetic properties and microarchitectures. The ability to exert tighter control over the cellular microenvironment through microfabrication of hydrogels could greatly enhance the ability to control cell and tissue behavior ultimately leading to functional tissue structures.

2.2 *Cell seeding in microwells*

Control of cell aggregation through seeding in fixed geometrical spaces is an effective means of restricting aggregate size and dimensions to direct tissue morphogenesis. Seeding cells in non-adherent microwells formed spheroid structures,[28] while larger organoid structures were created by culturing cells in linear channels[39] leading to improved cell alignment and resulting tissue function. A major advantage of these techniques, for compatible cell types, is the ability to allow the cells to remodel and recreate the microarchitecture using only natural materials while allowing the cells to dictate the pace and organization of the tissue morphogenesis.

Taking this one step further, researchers have used micromolds to generate shape-controlled cellular aggregates, in geometries such as spheroids,[40] individual and connected tori or honeycomb structures.[41] Cells were then seeded onto the molded hydrogels, leading to self assembled cellular organization within the shape restricted templates.

2.3 *Cell-laden hydrogels*

To create microtissues researchers combine cells with hydrogels to create tissue-like structures of specific mechanical properties and geometries. Some common techniques mix cells within polymer hydrogels, such as poly (ethylene glycol) (PEG)[21] or other photopolymerizable materials,[42] self-assembling peptide gels[43] or temperature sensitive hydrogels.[5] In this approach, cells are combined with the hydrogel precursors, pipetted into micromolds and polymerized using incubation[5] or UV light.[28] An alternative approach creates cell-laden hydrogels by directly passing UV light through a patterned mask containing specific shapes, which, when optimized, only polymerizes the hydrogels where the UV light is able to penetrate the mask.[14]

One example of using cell-laden hydrogels for tissue engineering created rings of primary cardiac cells within molded collagen or Matrigel. These cell-laden rings were arranged in contact with other rings and cultured under cyclic mechanical stretch forming a composite tissue, with improved contractile function and cell alignment. *In vivo* implantation of these tissues in a rat myocardial infarct model displayed significant improvements in cardiac function demonstrating the ability to create functional tissues using assembled macroscale cell-laden hydrogels.[5]

2.4 *Microarray systems*

Microscale technologies have emerged as a powerful tool for high-throughput cellular and biological studies as well as a means for manipulating biological systems and miniaturizing experiments.[44] A microarray is a fabricated device composed of a specified number of microwells of defined shape and size which can be used to culture cells and/or to deposit combinations of different materials such as collagen, laminin, and fibronectin to enable the study of cell behavior in a high-throughput manner. These miniaturized arrays can be used to segregate cells for single cell analysis and multiple cells studies, such as forming embryoid bodies.[45] One major advantage of microarray systems is that the throughput is significantly higher than is possible using traditional cell culture techniques as these arrays can contain hundreds or thousands of microwells. In addition, due to the microscale size, expensive reagents and soluble factors, such as growth factors, cytokines or drugs, are conserved making large-scale combinatorial experiments feasible. Using time-lapse microscopy and immunostaining, the fate of several hundred single stem cells can be tracked simultaneously making this a powerful tool for *in vitro* determination of the impact of exogenous factors on cell and tissue behavior.

3. Microscale Technologies to Investigate and Control Stem Cell Behavior

The derivation of ESCs from both mouse and human[46] has opened up new possibilities for cell-based therapies. However, the use of human embryos to generate ESC lines is controversial and therefore recognized as ethically problematic.[47] A breakthrough to this problem is in the form of reprogrammed somatic cells.[48] These induced pluripotent stem (iPS) cells have been shown to be functionally and molecularly similar to ESCs[49] and offer new opportunities in regenerative cell therapy applications.[50] The ability to reliably direct stem cell differentiation to ectodermal, meso-dermal, and endodermal lineages is a potentially powerful method for generating functional tissues, as tissues emerge from well-organized sequences of cell renewal, differentiation, and assembly under normal physiological conditions.[51]

Stem cells are sensitive to a variety of microenvironmental stimuli that regulate both self-renewal and differentiation.[52] Thus, through microscale engineering, cell differentiation can be guided by control-ling the interplay between regulatory factors down to the individual cellular level.

3.1 *Microarray analysis of ESC differentiation*

ESC differentiation can be directed through different combinations of factors that influence the microenvironment. Spatiotemporal microenvi-ronment signals may be influenced by the size of embryoid bodies (EBs), co-culture of different cell lineages, small molecules present in the microenvironment and ECM materials in combination with other unknown parameters. To study the effect of all stem cell differentiation factors is an enormous combinatorial problem that would be extremely difficult to analyze without the facilitation of high-throughput analysis, screening, and imaging. This goal can be achieved much more readily through the use of microarray systems for analysis and screening of multiple factors individually and in combination. Because of the significant number of possible combinations, high-throughput approaches can be used to rapidly test and discover important combinations of parameters in ESC differentiation. Parameters such as ECM molecules, soluble factors (e.g. cytokines, growth factors), and biomaterial interactions can influence various biological pathways that influence ESC differentiation. To improve our understanding of ESC differentiation, studying the combined influence of parameters that may regulate stem cell expansion and specialization in an efficient and cost effective manner using high-throughput analysis, screening, and imaging may prove beneficial.

In addition to growth factors and cell-secreted factors, synthetic small molecules can direct ESC differentiation. Small permeable, naturally occurring molecules such as vitamin C, sodium pyruvate, dexamethasone, thyroid hormones, and retinoic acid have been used to regulate stem cell fate.[53] Similarly, new synthetic, heterocyclic small molecules that can alter stem cell fate have recently been studied to control stem cell differentiation. Ding *et al.* reported ESC differentiation into various cell types employing small molecules.[54] One manner by which synthetic molecules can be used to selectively control and regulate stem cell differentiation and proliferation is through adjusting the activities of proteins. Such processes have been successfully employed to cause controllable neurogenetic and cardiomyogenetic induction in murine ESCs, osteogenesis induction in mesenchymal stem cells (MSC), and skeletal muscle cell differentiation.[55] Further investigation into small molecule discovery could play a substantial role in furthering the ability to reliably differentiate stem cells down specific pathways.

The interactions between ESCs and the surrounding matrix environment can profoundly influence stem cell behavior. To assess this parameter Anderson *et al.* used a biomaterial library comprised of 576 different acrylate-based polymers to analyze stem cell behavior.[56] Combinations of the different polymers were mixed in 384-well plates and were robotically printed on coated glass slides. After printing, the slides were exposed to long wave UV to initiate polymerization, dried, sterilized, and washed with PBS and cell culture media. Embryonic stem cells were then seeded onto the slides and the influence of the materials on their differentiation was analyzed. Based on these experiments, it was concluded that ESC differentiation may be induced toward epithelial cell lineage through interactions with specific combinations of materials. This study demonstrated the potential for creating a library of materials to study ESC-material interactions which could be used to better design scaffolds or culture conditions for controlling the behavior of ESCs, or other stem or differentiated cell types.

3.2 *Microwell fabrication*

As briefly described above, microengineered wells can be a useful tool to control the geometry and homogeneity of cell aggregates. To fabricate microwell arrays, UV-photocrosslinkable PEG prepolymer containing photoinitiator can be placed on a glass slide that has been treated to be adhesive to the hydrogel. The precursor solution was crosslinked between the surface-treated glass slide and a coverslip with a photomask on top to control the spatial geometry of the crosslinking (Fig. 1). Microscopy cover slips were used as spacers between the glass support and the coverglass to define the depth of the microwells. The PEG precursor solution was then irradiated through a bright field photomask with UV light of 350–500 nm to generate the microwells. The PEG precursor solution only underwent polymerization in the areas where UV light was able to pass through the photomask, while all other areas remained in the liquid prepolymer state. Following polymerization of the polymers, the coverglass was carefully removed and the uncrosslinked PEG macromers were removed with deionized water.

3.3 *Microfluidic systems*

The field of microfluidics could be useful for controlling the interaction of stem cells with their surrounding soluble environment as it allows manipulation of microliter volumes of fluids within microchannels.[57] Microfluidic systems have been used in a number of different applications for controlling the cellular microenvironment such as micropatterning of cells, subcellular localization of media components, high-throughput drug screening, introduction of a large range of laminar-flow rates, and creation of soluble factor gradients.[58,59]

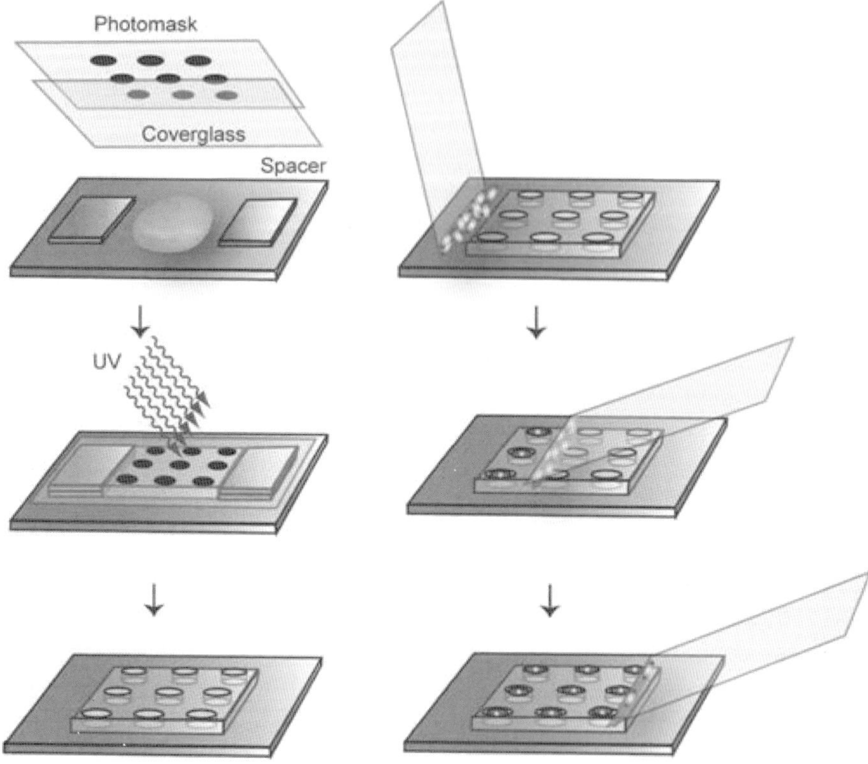

Fig. 1. (**Left**) PEG microwell fabrication through UV crosslinking. (**Right**) Localizing cells within arrays of microwells using a wiping technique. This method produces cell seeding densities that vary consistently with microwell geometry and cell concentration. (Copyright (2009) Wiley. Used with permission from Ref.63.)

A unique aspect of microfluidic systems is the ability to control mixing and shear stress in which the microfluidic devices can produce a logarithmic scale of flow rates and a logarithmic concentration gradient.[60] This ability has been utilized to study ESCs where high flow rates resulted in increased proliferation with minimized usage of media as compared to traditional macroscale experiments.[61] This important advantage may be applied for the screening of media components, conditioned media, or different chemical formulation with minimized consumption and optimized cell behavior. The use of microfluidic systems in controlled flow rates or concentration gradients have been shown to be important for regulating cell proliferation and differentiation.[58]

3.4 *Controlled microbioreactors*

One of the main challenges in regenerative medicine is obtaining ample amount of specific cell types for capable transplantation. However, these efforts have been hindered by limited proliferative capacity of the desired cells. Incorporation of bioreactor systems with ESCs is an active area of investigation since ESCs can be differentiated into cell lineages of all three primary germ layers.[51] ESCs can greatly benefit from bioreactors as biological processes can be carried out under tightly controlled (oxygen, nutrients, or other molecular and physical regulatory factors) environmental conditions to account for controlled nutrient transfer. These conditions will help minimize the batch to batch variability, making the process sufficiently reproducible in regenerative medicine. Bioreactor systems may better facilitate the transition from laboratory to clinical scale production due to high level of control, while providing a means to quantitatively study cell behavior in response to various stimuli. Some of the essential requirements to be considered during design are: (1) rapid and controllable expansion of cells; (2) enhanced cell seeding of 3D scaffolds (at a desired cell density, high yield, high kinetic rate, and spatial uniformity); (3) efficient local exchange of oxygen, nutrients, and metabolites; and (4) implementation of physiological stimuli.[52]

4. Assembly Techniques for Creating Engineered Tissues from Microscale Building Blocks

Just as ESC behavior can be directed and manipulated through microscale technologies, engineered microtissues can be engineered by microgels with specific microarchitectural features. The field of bottom-up tissue engineering aims to create macroscale engineered tissues with tightly controlled microarchitectural structures to better recapitulate the native structure and function of the tissues we are aiming to repair or replace. The major challenge of tissue engineering using bottom-up techniques is assembling macroscale engineered tissues from microscale building blocks. For optimal function *in vivo*, engineered tissues must recreate the native microarchitecture, while also possessing robust mechanical properties for appropriate interactions with the surrounding tissues and to withstand the mechanical environment. Some specific challenges to overcome for this field are: determination and recapitulation of native mechanical properties, building block integration to form robust tissues, creation of integrated microvasculature, scale-up technologies to move from the laboratory to the clinic at the appropriate length scale and successful demonstration of *in vivo* functional improvement in diseased tissues.

4.1 *Layer by layer production using cell-laden hydrogels*

Layer by layer assembly of tissue sheets can be used to generate larger 3D structures that mimic tissues. In one approach to create engineered tissues using this layer by layer approach, arrays of cells and ECM/polymer were photocrosslinked in an additive manner.[62] Through additive photopolymerization, multilayer engineered tissues were fabricated for hepatic tissue engineering using cell-laden hydrogels. Hepatocytes from adult Lewis rats were mixed with PEG prepolymer that had been functionalized with cell-adhesive RGD motifs. The cell-PEG mixture was pipetted into a custom designed sterile chamber and polymerized through the application of UV light exposed through one of three photomasks, the first of which was in the shape of three pointed stars. Next, the spacer height was increased to create a second subsequent layer which extended the full spacer height partially resting on top of the initial layer (Fig. 2). Finally, the spacer height was increased a final time, and a honeycomb enclosure was created around each two-layer unit, creating engineered microtissues with microarchitecture and function similar to native hepatic tissues. Using this technique Bhatia and colleagues were able to demonstrate many benchmarks of functional hepatic tissue, such as urea production, while also greatly improving hepatocyte adhesion both by inclusion of RGD molecules as well as through recapitulation of the native microarchitecture of the liver lobule. While there are many clear advantages of this technique, such as recreation of the

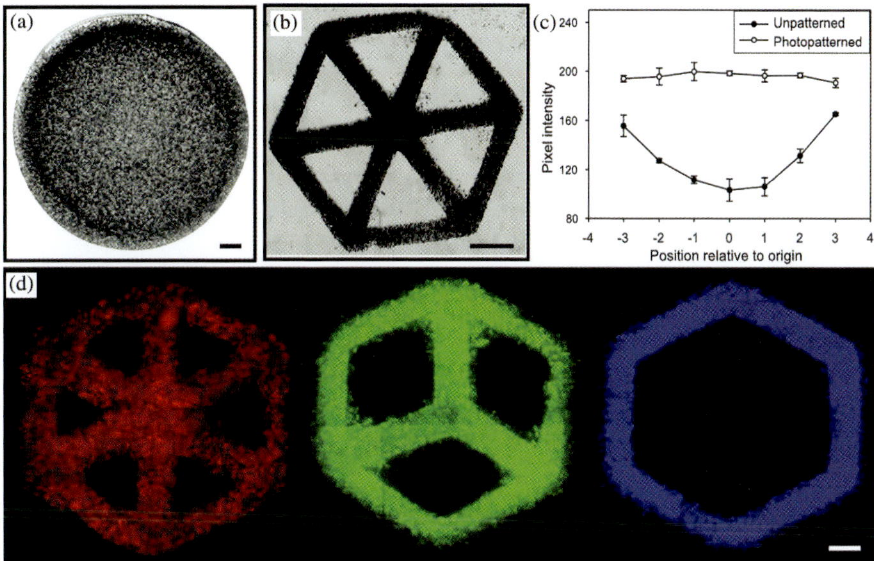

Fig. 2. Photopatterning of hepatocytes to improve nutrient transport and cell viability. (a) Control — hepatocytes encapsulated in 1.5 mm thick by 10 mm diameter hydrogel. MTT stain (black) demonstrated mitochondrial activity that was stronger nearer the outside edge. Scale bar = 500 μm. (b) Hepatocytes micropatterned at the same density and overall hydrogel dimensions demonstrated good viability with no diffusive limitation. Scale bar = 1 mm. (c) Quantification of MTT demonstrates the reduced viability in the center of the control construct that is alleviated by micropatterning. (d) Layer by layer demonstration of micropatterning technique. Scale bar = 500 μm. (Copyright (2007) by *Federation of American Societies for Experimental Biology* (*FASEB*). Reproduced with permission of *FASEB*.[21])

native microarchitecture combined with positive functional properties similar to the native tissues, one potential problem with this technique is the inability to tightly control the layer thickness. There is no simple method for restricting subsequent cell-PEG mixtures from filling the chamber, making it difficult to have multiple layers that only occupy one plane. For some tissues using different cell types on each layer, this technique would have difficulty recapitulating the native cellular arrangement. Clearly the positive aspects outweigh the negative, however in an effort to create more complex patterns and mimic more complex tissues, addressing these shortcomings could make this technique applicable to more tissues and applications.

4.2 *Packing of co-cultured building blocks for capillary filtration*

A major challenge in creating engineered tissues that can function and integrate *in vivo*, is the ability to create tissues containing a functional microvasculature to integrate with and ensure perfusion through and around the surrounding tissues. Recently an approach has been demonstrated that aims to create a perfusable tissue without creating microvasculature.

To create perfusable tissues analogous to capillary filtration, a random hydrogel packing technique was developed to create tissues with tortuous, perfusable networks of capillaries.[20] Cylindrical tissue units were created through packing of cell-laden collagen hydrogel building blocks within perfused silicon tubing. To create the building blocks, collagen was mixed either with HepG2 cells, a model hepatic cell line, or without. Following gelation in cylindrical tubing, cylindrical cell-laden building blocks were created by chopping the long cylindrical tissue into pieces. The cell-laden gels were collected in media, then were combined with HUVEC cells to form a confluent endothelial layer around the perimeter of the hydrogel blocks. The HepG2-HUVEC modules were perfused into tubing containing a porous plug to both trap the gels and also to allow the modules to aggregate and compact together. Over time the co-cultured constructs compacted further and remodeled together to form a porous, perfusable tissue. The engineered filtration devices were perfused with blood to demonstrate both high viability of the encapsulated HepG2 and the surface coated HUVEC cells, and the ability to perfuse without clotting, for use as clinical blood filtration devices (Fig. 3). While, this relatively simple technique for creating perfusable tissues could potentially be effective in replacing tissues which primary function as filtration systems, such as the kidney or liver, other tissues would potentially not be possible due to the lack of a self-contained enclosure as well as poor mechanical properties. In addition,

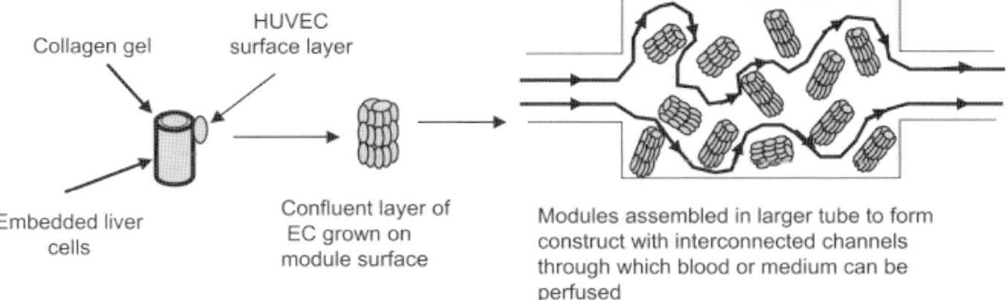

Fig. 3. Packing of cell-containing building blocks in a perfusable reactor. HepG2 liver cells were encapsulated in collagen, cast into modular units then coated with a confluent layer of HUVECs and packed inside a perfused tube to generate a perfusable filtration unit. (Reprinted with permission from the *Proceedings of the National Academy of Sciences USA*.[20])

the encapsulated cells would have to retain their viability and function while being surrounded by endothelial cells, potentially limiting the available cell types appropriate for encapsulation.

4.3 *Directed assembly of cell-laden hydrogels*

Recent work in our laboratory has used directed assembly to demonstrate the ability to create ordered tissue structures using cell-laden microgels as building blocks.[14] This technique demonstrates one method by which higher order structures can be created when the building block materials are fragile or difficult to handle. In addition, this technique also demonstrates the potential for scale-up or automated techniques, as the engineered tissues can be created almost entirely without intervention.

This technique was made possible by harnessing the properties of surface tension with regard to hydrophilic hydrogels to assemble cell-laden building blocks into tissue-like structures (Fig. 4). For the first step, rectangular PEG hydrogels of varying aspect ratios were created containing encapsulated cells through direct UV photopolymerization using specially designed photomasks. These building blocks were then placed into mineral oil, which is hydrophobic, to cause the hydrophilic microgels to aggregate. These aggregates were then exposed to a brief, second UV exposure to crosslink the assembled microgels. The ultimate shape

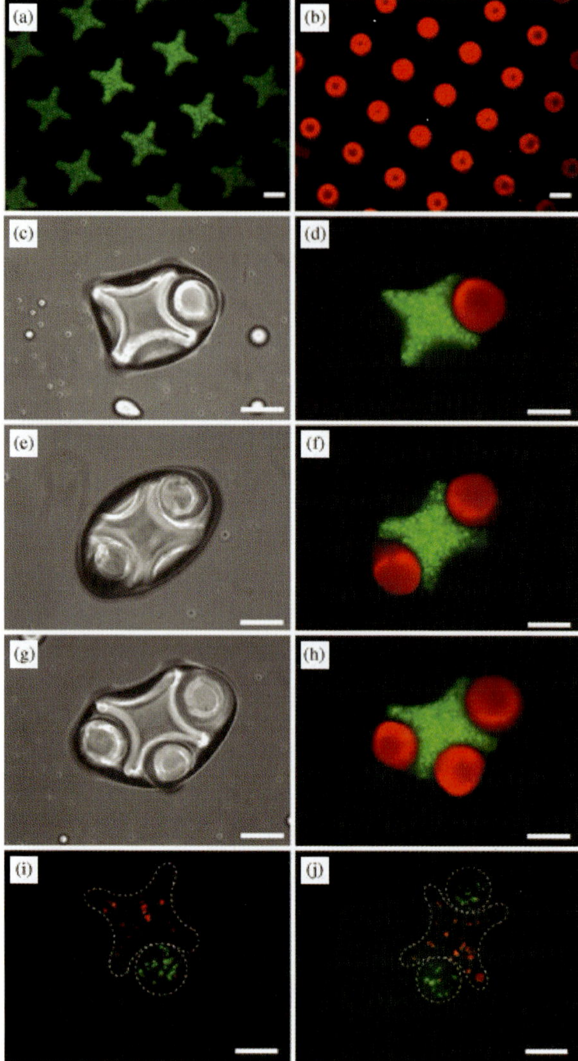

Fig. 4. Directed assembly of shape controlled hydrogel building blocks using a two-phase reactor. Lock (a) and key (b) shaped hydrogel modules were created directly by photopolymerization using UV through a photomask, then allowed to aggregate and self-assemble in a hydrophobic media (mineral oil), into single (c, d), double (e, f) and triple (g, h) arrangements demonstrating the possible control of assembled co-cultured structures. Scale: 200 μm. After assembly a second UV polymerization solidified the structures. (Reprinted with permission from the *Proceedings of the National Academy of Sciences USA.*[14])

and overall dimensions of the tissues were demonstrated to be controllable to a certain extent based on the aspect ratio of the building blocks used, with the total number of microgels increasing with the aspect ratio. Even greater control of the process, and the potential for making more complex structures, was demonstrated through use of spatially controllable lock and key geometries, suggesting a potential way to further increase the size and intricacy of tissues created using this technique. While there are clear advantages to this technique, such as the ability to create tissues of specific sizes based on the geometries of the building block materials and without complicated manipulation or assembly techniques, there are also potential drawbacks to this technique. For instance, the ultimate size of the assembled tissues was on the order of millimeters, making a clear clinical application difficult to envision, unless these tissues were used as the basis for a secondary assembly process. In addition, a large fraction of the microgels assembled randomly instead of in a controllable manner, suggesting that optimization must be performed to reduce the variability in the system. Current research into adapting this technique to be performed on a surface, rather than as a 3D immersion, are beginning to overcome many of these challenges and show great promise towards creating engineered tissues with controlled co-culture on a clinically relevant length scale.

5. Conclusions and Future Directions

Some major challenges affecting both the fields of microscale tissue engineering and stem cell differentiation in the future are improving the overall spatial resolution leading to smaller size structures with enhanced biomimetic functionalities. Control on an even smaller scale of structures and stem cell will allow the investigation of interactions on a broader spectrum than what is possible with current techniques. The major challenge for bottom-up tissue engineering approaches will be the improving the ability to create tissues with controllable microarchitectures, yet, on a clinically, and biologically, relevant length scale for *in vivo* implantations. One of the greatest challenges of microscale stem cell technologies is the ability to direct proliferation and differentiation more reliably. The combination of these fields contains great potential for the advancement of the field of regenerative medicine. The improvements in the fields mentioned above will make the possibility of regenerating degenerated or defective tissues much closer to reality, providing hope for millions of current and future patients with no other alternatives for cure.

References

1. R. Langer and J. P. Vacanti, Tissue engineering, *Science* **260**: 920–926 (1993).
2. Y. M. Bello, A. F. Falabella and W. H. Eaglstein, Tissue-engineered skin. Current status in wound healing, *Am J Clin Dermatol* **2**: 305–313 (2001).
3. M. Hillsley and J. Frangos, Review — bone tissue engineering: the role of interstitial fluid flow, *Biotech Bioeng* **43**: 575–581 (1994).
4. A. I. Caplan, M. Elyaderani, Y. Mochizuki, S. Wakitani and V. M. Goldberg, Principles of cartilage repair and regeneration, *Clin Orthop*: **342**: 254–269 (1997).
5. W. H. Zimmermann, I. Melnychenko, G. Wasmeier, M. Didie, H. Naito, U. Nixdorff, A. Hess, A. Budinsky, K. Brune, B. Michaelis, S. Dhein, A. Schwoerer, H. Ehmke and T. Eschenhagen, Engineered heart tissue grafts improve systolic and diastolic function in infarcted rat hearts, *Nat Med* Apr: 452–458 (2006).
6. L. E. Niklason, J. Gao, W. M. Abbott, K. K. Hirschi, S. Houser, R. Marini and R. Langer, Functional arteries grown *in vitro*, *Science* **284**: 489–493 (1999).
7. K. J. Gooch, T. Blunk, D. L. Courter, A. L. Sieminski, G. Vunjak-Novakovic and L. E. Freed, Bone morphogenetic proteins-2, -12, and -13 modulate *in vitro* development of engineered cartilage, *Tissue Eng* **8**: 591–601 (2002).
8. J. Boublik, H. Park, M. Radisic, E. Tognana, F. Chen, M. Pei, G. Vunjak-Novakovic and L. E. Freed, Mechanical properties and remodeling of hybrid cardiac constructs made from heart cells, fibrin, and biodegradable, elastomeric knitted fabric, *Tissue Eng* Jul–Aug: 1122–1132 (2005).
9. M. Radisic, H. Park, H. Shing, T. Consi, F. J. Schoen, R. Langer, L. E. Freed and G. Vunjak-Novakovic, Functional assembly of engineered myocardium by electrical stimulation of cardiac myocytes cultured on scaffolds, *Proc Natl Acad Sci USA* **101**: 18129–18134 (2004).

10. S. C. Daxini, J. W. Nichol, A. L. Sieminski, G. Smith, K. J. Gooch and V. P. Shastri, Micropatterned polymer surfaces improve retention of endothelial cells exposed to flow-induced shear stress, *Biorheology* **43**: 45–55 (2005).

11. P. J. Schaner, N. D. Martin, T. N. Tulenko, I. M. Shapiro, N. Tarola, R. F. Leichter, R. A. Carabasi and P. J. Dimuzio, Decellularized vein as a potential scaffold for vascular tissue engineering, *J Vasc Surg* **40**: 146–153 (2004).

12. A. Khademhosseini, G. Eng, J. Yeh, J. Fukuda, J. Blumling, R. Langer and J. A. Burdick, Micromolding of photocrosslinkable hyaluronic acid for cell encapsulation and entrapment, *J Biomed Mater Res A* **79**: 522–532 (2006).

13. A. Khademhosseini and R. Langer, Microengineered hydrogels for tissue engineering, *Biomaterials* **28**: 5087–5092 (2007).

14. Y. Du, E. Lo, S. Ali and A. Khademhosseini, Directed assembly of cell-laden microgels for fabrication of 3D tissue constructs, *Proc Natl Acad Sci USA* **105**: 9522–9527 (2008).

15. K. Nishida, M. Yamato, Y. Hayashida, K. Watanabe, N. Maeda, H. Watanabe, K. Yamamoto, S. Nagai, A. Kikuchi, Y. Tano and T. Okano, Functional bioengineered corneal epithelial sheet grafts from corneal stem cells expanded *ex vivo* on a temperature-responsive cell culture surface, *Transplantation* **77**: 379–385 (2004).

16. M. Yamato and T. Okano, Cell sheet engineering, *Materials Today* May: 42–47 (2004).

17. N. E. Federovich, J. Alblas, J. R. DeWijn, W. E. Hennink, A. J. Verbout and W. J. A. Dhert, Hydrogels as extracellular matrices for skeletal tissue engineering: state-of-the-art and novel application in organ printing, *Tissue Eng* **13**: 1905–1925 (2007).

18. N. L'Heureux, S. Paquet, R. Labbe, L. Germain and F. A. Auger, A completely biological tissue-engineered human blood vessel, *FASEB J* **12**: 47–56 (1998).

19. Y. Shiroyanagi, M. Yamato, Y. Yamazaki, H. Toma and T. Okano, Transplantable urothelial cell sheets harvested noninvasively from temperature-responsive culture surfaces by reducing temperature, *Tissue Eng* **9**:1005–1012 (2003).

20. A. P. McGuigan and M. V. Sefton, Vascularized organoid engineered by modular assembly enables blood perfusion, *Proc Natl Acad Sci USA* **103**: 11461–11466 (2006).

21. V. L. Tsang, A. A. Chen, L. M. Cho, K. D. Jadin, R. L. Sah, S. DeLong, J. L. West and S. N. Bhatia, Fabrication of 3D hepatic tissues by additive photopatterning of cellular hydrogels, *FASEB J* **21**: 790-801 (2007).

22. J. W. Nichol and A. Khademhosseini, Modular tissue engineering: engineering biological tissues from the bottom up, *Soft Matter* **5**: 1312–1319 (2009).

23. C. M. Nelson and C. S. Chen, Cell-cell signaling by direct contact increases cell proliferation via a PI3K-dependent signal, *FEBS Lett* **514**: 238–242 (2002).

24. A. J. Engler, S. Sen, H. L. Sweeney and D. E. Discher, Matrix elasticity directs stem cell lineage specification, *Cell* **126**: 677–689 (2006).

25. B. Murtuza, J. W. Nichol and A. Khademhosseini, Micro- and nanoscale control of the cardiac stem cell niche for tissue fabrication, *Tissue Eng Part B Rev* **15**: 443–454 (2009).

26. J. Tien and C. S. Chen, Patterning the cellular microenvironment, *IEEE Eng Med Biol Mag* **21**: 95–98 (2002).

27. A. Khademhosseini, J. Yeh, S. Jon, G. Eng, K. Y. Suh, J. A. Burdick and R. Langer, Molded polyethylene glycol microstructures for capturing cells within microfluidic channels, *Lab Chip* **4**: 425–430 (2004).

28. J. M. Karp, J. Yeh, G. Eng, J. Fukuda, J. Blumling, K.-Y. Suh, J. Cheng, A. Mahdavi, J. Borenstein, R. Langer and A. Khademhosseini, Controlling size, shape and homogeneity of embryoid bodies using poly(ethylene glycol) microwells, *Lab Chip* **7**: 786–794 (2007).

29. R. Langer and J. Vacanti, Tissue engineering, *Science* **260**: 920–926 (1993).

30. J. A. Burdick and G. Vunjak-Novakovic, Review: Engineered microenvironments for controlled stem cell differentiation, *Tissue Eng Part A* **15**: 205–219 (2009).

31. A. Khademhosseini, J. P. Vacanti and R. Langer, Progress in tissue engineering, *Sci Am* **300**: 64–71 (2009).

32. N. A. Peppas, J. Z. Hilt, A. Khademhosseini and R. Langer, Hydrogels in biology and medicine: from molecular principles to bionanotechnology, *Adv Mater* **18**: 1345–1360 (2006).

33. X. Qu, A. Wirsen and A. C. Albertsson, Structural change and swelling mechanism of pH-sensitive hydrogels based on chitosan and D,L-lactic acid, *J Appl Polym Sci* **74**: 3186–3192 (1999).

34. A. S. Hoffman, T. A. Horbett and B. D. Ratner, Interactions of blood and blood components at hydrogel interfaces*, *Ann NY Acad Sci* **283**: 372–382 (1977).

35. G. B. Schneider, A. English, M. Abraham, R. Zaharias, C. Stanford and J. Keller, The effect of hydrogel charge density on cell attachment, *Biomaterials* **25**: 3023–3028 (2003).

36. C. G. W. Fan Yang, D.-A. Wang, H. Lee, P. N. Manson and J. Elisseeff, The effect of incorporating RGD adhesive peptide in polyethylene glycol diacrylate hydrogel on osteogenesis of bone marrow stromal cells, *Biomaterials* **26**: 5991–5998 (2005).

37. M. T. A. Ende and N. A. Peppas, Transport of ionizable drugs and proteins in crosslinked poly(acrylic acid) and poly(acrylic acid-co-2-hydroxyethyl methacrylate) hydrogels. 2. Diffusion and release studies, *J Control Release* **48**: 47–56 (1997).

38. M. T. Amende, D. Hariharan and N. A. Peppas, Factors influencing drug and protein-transport and release from ionic hydrogels, *Reactive Polym* **25**: 127–137 (1995).

39. A. Khademhosseini, G. Eng, J. Yeh, P. A. Kucharczyk, R. Langer, G. Vunjak-Novakovic and M. Radisic, Microfluidic patterning for fabrication of contractile cardiac organoids, *Biomed Microdevices* **9**: 149–157 (2007).

40. A. P. McGuigan, D. A. Bruzewicz, A. Glavan, M. Butte and G. M. Whitesides, Cell encapsulation in sub-mm sized gel modules using replica molding, *PLoS One* **3**: 1–11 (2008).

41. D. M. Dean, A. P. Napolitano, J. Youssef and J. R. Morgan, Rods, tori, and honeycombs: the directed self-assembly of microtissues with prescribed microscale geometries, *FASEB J* **21**: 4005–4012 (2007).

42. M. D. Brigham, A. Bick, E. Lo, A. Bendali, J. A. Burdick and A. Khademhosseini, Mechanically robust and bioadhesive collagen and photocrosslinkable hyaluronic acid semi-interpenetrating networks, *Tissue Eng* **15**: 1645–1653 (2009).

43. J. W. Nichol, G. C. Engelmayr Jr, M. Cheng and L. E. Freed, Co-culture induces alignment in engineered cardiac constructs via MMP-2 expression, *Biochem Biophys Res Commun* **373**: 360–365 (2008).

44. A. Khademhosseini, R. Langer, J. Borenstein and J. P. Vacanti, Microscale technologies for tissue engineering and biology, *Proc Natl Acad Sci USA* **103**: 2480–2487 (2006).

45. J. M. Karp, J. Yeh, G. Eng, J. Fukuda, J. Blumling, K. Y. Suh, J. Cheng, A. Mahdavi, J. Borenstein, R. Langer and A. Khademhosseini, Controlling size, shape and homogeneity of embryoid bodies using poly(ethylene glycol) microwells, *Lab Chip* **7**: 786–794 (2007).

46. J. A. Thomson, J. Itskovitz-Eldor, S. S. Shapiro, M. A. Waknitz, J. J. Swiergiel, V. S. Marshall and J. M. Jones, Embryonic stem cell lines derived from human blastocysts, *Science* **282**: 1145–1147 (1998).

47. A. Rolletschek and A. M. Wobus, Induced human pluripotent stem cells: promises and open questions, *Biol Chem* **390**: 845–849 (2009).

48. K. Takahashi and S. Yamanaka, Induction of pluripotent stem cells from mouse embryonic and adult fibroblast cultures by defined factors, *Cell* **126**: 663–676 (2006).

49. G. Amabile and A. Meissner, Induced pluripotent stem cells: current progress and potential for regenerative medicine, *Trends Mol Med* **15**: 59–68 (2009).

50. I. H. Park, N. Arora, H. Huo, N. Maherali, T. Ahfeldt, A. Shimamura, M. W. Lensch, C. Cowan, K. Hochedlinger and G. Q. Daley, Disease-specific induced pluripotent stem cells, *Cell* **134**: 877–886 (2008).

51. A. M. Wobus and K. R. Boheler, Embryonic stem cells: prospects for developmental biology and cell therapy, *Physiol Rev* **85**: 635–678 (2005).

52. J. A. Burdick and G. Vunjak-Novakovic, Engineered microenvironments for controlled stem cell differentiation, *Tissue Eng Part A* **15**: 205–219 (2009).

53. X. Wu, S. Ding, G. Ding, N. S. Gray and P. G. Schultz, Small molecules that induce cardiomyogenesis in embryonic stem cells, *J Am Chem Soc* **126**: 1590–1591 (2004).

54. Y. Xu, Y. Shi and S. Ding, A chemical approach to stem-cell biology and regenerative medicine, *Nature* **453**: 338–344 (2008).

55. S. Ding and P. G. Schultz, A role for chemistry in stem cell biology, *Nat Biotechnol* **22**: 833–840 (2004).

56. A. J. Urquhart, M. Taylor, D. G. Anderson, R. Langer, M. C. Davies and M. R. Alexander, TOF-SIMS analysis of a 576 micropatterned copolymer array to reveal surface moieties that control wettability, *Anal Chem* **80**: 135–142 (2008).

57. G. M. Whitesides, The origins and the future of microfluidics, *Nature* **442**: 368–373 (2006).

58. S. W. Rhee, A. M. Taylor, C. H. Tu, D. H. Cribbs, C. W. Cotman and N. L. Jeon, Patterned cell culture inside microfluidic devices, *Lab Chip* **5**: 102–107 (2005).

59. S. Takayama, E. Ostuni, P. LeDuc, K. Naruse, D. E. Ingber and G. M. Whitesides, Subcellular positioning of small molecules, *Nature* **411**: 1016 (2001).

60. L. Kim, M. D. Vahey, H. Y. Lee and J. Voldman, Microfluidic arrays for logarithmically perfused embryonic stem cell culture, *Lab Chip* **6**: 394–406 (2006).

61. L. Meinel, V. Karageorgiou, R. Fajardo, B. Snyder, V. Shinde-Patil, L. Zichner, D. Kaplan, R. Langer and G. Vunjak-Novakovic, Bone tissue engineering using human mesenchymal stem cells: effects of scaffold material and medium flow, *Ann Biomed Eng* **32**: 112–122 (2004).

62. V. L. Tsang and S. N. Bhatia, Fabrication of three-dimensional tissues, *Adv Biochem Eng/Biotechnol* **103**: 189–205 (2007).
63. L. Kang, M. J. Hancock, M. D. Brigham and A. Khademhosseini, Cell confinement in patterned nanoliter droplets in a microwell array by wiping, *J Biomed Mater Res Part A* **93**: 547–557 (2010).

19

DESIGNING PROTEIN-ENGINEERED BIOMATERIALS FOR STEM CELL THERAPY

Lei Cai and Sarah C. Heilshorn

1. Introduction

Stem cell therapy by transplanting cells to the site of repair has emerged as a promising therapeutic strategy for a variety of regenerative medicine applications.[1] While minimally invasive, numerous recent studies have shown that the stem cell therapy typically results in low cell retention and engraftment post-delivery, which is a major obstacle to long-term therapeutic benefits and clinical translation.[2,3] In the past decade, researchers have attempted to use a variety of biomaterials as cell carriers to improve stem cell delivery efficiency and to provide more relevant microenvironmental cues to promote cell function.[4]

Among these biomaterials, synthetic polymeric materials have well-defined chemical structures and tunable material properties. However, many of these materials are bioinert and have poor biocompatibility

471

without further modification.[5] Short peptide sequences have often been used to modify these materials, but most of them are cell-adhesive ligands that may not be able to emulate the full biochemical signals of ECMs.[5] Naturally derived extracellular matrix (ECM) materials are generally cell-compatible and present better biochemical signaling. However, they are often subject to large batch-to-batch variations, risk of immunogenicity, and lack of tunability of material properties, rendering difficult clinical translation.[6]

Protein-engineered biomaterials have evolved as a promising class of biomaterials that are comprised of well-defined, modular peptide domains to recapitulate several stem cell niche properties.[7–9] These peptide domains include structural domains that control the physical properties of the scaffold and bioactive domains that provide biological functionality to tailor specific stem cell-material interactions. As with other synthetically designed systems, protein-engineered biomaterials enable the design of stem cell delivery vehicles with independently customizable properties including structure, mechanics, and biochemical function that are normally improbable to achieve using naturally derived materials. Protein-engineered biomaterials are commonly generated using recombinant protein synthesis, which is the templated synthesis of a pre-designed amino acid sequence from a genetically encoded DNA sequence. This strategy ensures molecular-level control of monodispersed polypeptides that can be consistently reproduced for clinical usage.

In this chapter, we will discuss the design and synthesis strategies to create modular peptide domains to build up novel protein-engineered biomaterials. We will focus on the use of these biomaterials to enable better stem cell therapies by delivering stem cells to the site of repair and optimizing cell-material interactions to promote therapeutic benefits for regenerative medicine applications.

2. Protein Engineering Strategies

2.1 *Design strategies of modular domains*

Protein-engineered biomaterials are created by stringing together multiple modular peptide domains along a single protein polymer chain with molecular-level precision. These modular peptide domains can be derived from native sequences with site-specific mutations, directly evolved by high-throughput combinatorial screening, or rationally developed by computational modeling.[7] For example, most of the structural peptide domains used in protein-engineered biomaterials to date are derived from native sequences, including collagen,[10] elastin,[11] resilin,[12] and silk.[13] High-throughput combinatorial screening has been used to generate mutants with high specificity, such as calmodulin binding domains to calcium ions.[14] Computational modeling utilizes different algorithms to detect and predict protein-protein interactions, which have been used to identify target sites for mutation.[15] As an example, several synthetic WW domains have been computationally designed to undergo heteroassembly with proline-rich sequences via molecular recognition.[16]

While small peptides can be synthesized using solid-phase or liquid-phase peptide synthesis techniques up to ~150 amino acids,[17] protein-engineered biomaterials typically consist of a series of peptide domains with much longer sequences and therefore are produced using recombinant protein synthesis. This strategy utilizes a DNA template that genetically encodes the target peptide domains spliced together for each engineered protein. This engineered DNA sequence is then inserted into a plasmid vector using molecular biology protocols. The resulting plasmid vector is transformed into a host organism, most commonly *Escherichia coli*. The *E. coli* bacteria grow rapidly with relatively high expression efficiency, although they are not able to perform post-translational modifications, such as glycosylation. These processes are often achieved using yeast or mammalian cell expression systems, but with lower efficiency. Therefore, the selection of a protein expression system is mainly determined by the quantity needed and the complexity of the protein sequence. After expression, the target engineered protein is extracted from the host cell lysate and then purified from other proteins by differences in size, charge, hydrophobicity, or protein interactions. For use in animal studies or clinical trials, the engineered proteins need to be further purified from lipopolysaccharides, or endotoxin, which are present on the cell surface of gram negative bacterial cells such as *E. coli*.

2.2 *Structural domains*

2.2.1 *Naturally derived structural domains*

The structural domains of protein-engineered biomaterials are critical for creating three-dimensional (3D) scaffolds to encapsulate and deliver cells to the site of repair. These domains are often selected based on the sequences of native structural proteins. Elastin-like polypeptide (ELP) is derived from elastin, which is one of the extracellular matrix proteins that offers elasticity to many types of tissues including skin, blood vessels, and lungs.[18] ELP is often composed of classic pentapeptide repeat units $(VPGXG)_n$, where X is any amino acid except for proline.[18] Hydrogels can be formed by chemically crosslinking amine-reactive ELP chains in the presence of live cells for cell-compatible encapsulation and culture using a variety of bioconjugate chemistry schemes.[19–23] ELP hydrogels have been demonstrated to be suitable for encapsulating embryonic stem cells (ESCs)[21] and mesenchymal stem cells (MSCs).[22] ESCs have shown the ability to proliferate within the 3D ELP hydrogels with high viability and to differentiate into cardiomyocytes.[21] MSCs encapsulated within ELP hydrogels could be guided down a chondrocytic differentiation pathway by upregulating chondrocytic-associated genes such as SOX9 and collagen II in low oxygen tension conditions.[23] Besides forming hydrogels, ELP can be used as surface coatings with fibril formation to promote MSC adhesion, proliferation, and differentiation.[24] MSCs seeded on these ELP coatings were found to differentiate into osteoblast-like cells in the absence of osteogenic media.[24] Tropoelastin, the soluble precursor of elastin, was genetically engineered to be a soft and elastomeric protein.[25] To enhance its mechanical strength and integrity, a structural protein blend system was created by blending a crystallizable silk fibroin with beta-sheet crystals into tropoelastin at different ratios.[25] The mechanical properties were tuned with resilience from 68% to 97%, and elastic modulus between 2 and 9 MPa, depending on the ratio of the two protein polymers. Interestingly, enhanced proliferation and osteogenic differentiation of MSCs were found by increasing the content of human tropoelastin in the silk-tropoelastin blends with higher surface roughness and stronger micro/nano-scale surface patterns.[26]

Silk-like polypeptide is derived from silk protein, which consists of repetitive peptide sequences that organize into hydrophobic beta-sheet crystals and amorphous hydrophilic regions.[27] It has high tensile strength, toughness, and extensibility, owing to the precise control of the size, number, distribution, orientation, and spatial arrangement of crystalline and amorphous domains at the nanometer scale.[27] Silk-like polypeptide can be genetically engineered with other structural domains to create new copolymer proteins with combined structural features and improved mechanical properties. One of the early examples is the synthesis of recombinant silk-elastin-like polypeptide (SELP).[28] The addition of elastin to silk decreased crystallinity and increased hydrophobicity, leading to a range of different physicochemical properties.[28] Recently, high-throughput screening of dynamic SELP was implemented to build up a library of proteins with unique stimuli-responsive features, including tensile strength and adhesion.[29] Sixty-four SELPs with a variety of sequences and molecular weights were selected out of over 2000 recombinant *E. coli* colonies to obtain fundamental understanding of sequence-function relationships.[29] As another example, silk-collagen-like polypeptide, which consists of hydrophilic collagen-inspired and histidine-rich silk-inspired blocks, was synthesized by the yeast *Pichia pastoris*, with the ability to produce highly repetitive sequences and to perform most post-translational modifications.[30] This polypeptide can self-assemble into a nanofibrous hydrogel with structural and functional properties similar to those of the ECM of connective tissue.[30] MSCs cultured on these hydrogels remained fully viable; however, in the absence of cell-adhesive motifs, MSC proliferation and mineralization were relatively lower compared to collagen hydrogel controls.[30] In a different design, a silk-like polypeptide was functionalized with a collagen-like sequence fused to a consensus sequence that allows the two polypeptide domains to self-assembly into chimeras.[31] The collagen-silk chimeric polypeptides substantially promoted MSC spreading and proliferation in comparison to the silk scaffold controls, leading to improved regulation of cell growth and differentiation.[31] In another elegant example, a recombinant silk polypeptide has been genetically engineered with a mineralizable domain to create a silk-silica fusion polypeptide, resulting in the formation of controlled mineralized silk films with different silica morphologies and distributions to generate 3D porous networks with clustered silica nanoparticles.[32] MSCs adhered, proliferated, and differentiated toward osteogenic lineages on the silk/silica films, upregulated gene expression of alkaline phosphatase, bone sialoprotein, and collagen type I markers, as well as formed calcium-containing deposits, indicative of early bone formation.[32]

Resilin-like polypeptide (RLP) is derived from resilin, a rubber-like protein that exists in specialized regions of most arthropods, possessing excellent mechanical properties including high resilience and elasticity.[33] RLP has been engineered with 12 repeats of the same 15 amino acid resilin consensus sequence, followed by a single and distinct biological domain (cell-binding (RGD), RDG, MMP-sensitive, and heparin-binding domains).[33] A phosphine-based crosslinker was used to prepare RLP hydrogels with comparable mechanical properties independent of the identity and concentration of the biological domains in the hydrogel.[33] MMP-sensitive domains were found susceptible to MMP-1 enzymatic degradation within 48 hours, in contrast to the completely stable RLP hydrogels lacking the MMP-sensitive domain.[33] An RGD-containing RLP hydrogel supported the adhesion and spreading of MSCs as well as 3D encapsulation with high cell viability and hydrogel integrity for at least seven days.[33] A continuation of this study showed that the RLP hydrogels did not activate cultured macrophages.[34] Subcutaneous transplantation of RLP hydrogels in a rat model further demonstrated that these materials did not elicit a significant inflammatory response, suggesting their potential for mechanically demanding tissue engineering applications including vocal fold and cardiovascular tissues.[34] In another study, a combination of structural domains of resilin-like, elastin-like, and collagen-like sequences was modularly designed and strung together to form a multifunctional polypeptide hydrogel with excellent elasticity and fiber formation.[35] MSCs cultured on these polypeptide films showed preferential adhesion and proliferation compared to the tissue culture polystyrene surface.[35] These studies suggest that we could select multiple modular domains from a repertoire of available structural domains and mix and match them to create a versatile family of biomaterials with properties mimicking their original structural function.

2.2.2 *Repurposed structural domains*

In addition to naturally derived structural domains, polypeptide domains can be repurposed to have structural properties distinct from its original function. These domains are often designed to create multi-component hydrogels that undergo gelation through specific interactions and rationally controlled binding kinetics for supporting 3D stem cell function. For example, hydrogels of protein-engineered biomaterials can be achieved by molecular recognition between two complementary protein binding domains.[36] In one design, multiple repeats of the WW domain have been incorporated into a protein polymer that specifically recognize proline-rich polypeptide domains in another protein polymer to form a 3D hydrogel. This hydrogel, termed mixing-induced two-component hydrogel (MITCH), undergoes gelation in about 10 seconds upon mixing of the two components and cells in physiological conditions, without the need for environmental triggers including pH, temperature, or ionic strength.[36] The resulting hydrogel is shear-thinning during injection and self-healing after injection, facilitating the injectable cell therapy for a variety of regenerative medicine applications. Neural stem cells (NSCs)[36] and adipose-derived stem cells (ASCs)[37] have been encapsulated within this hydrogel with good cell viability for at least two weeks. Proliferation and differentiation of these encapsulated stem cells have also been supported within the 3D MITCH.[36]

Another example of 3D protein-engineered hydrogel formation is the self-assembly of leucine zipper domains through coiled-coil interactions.[38] These peptide domains contain hydrophobic leucine residues and charged residues to form amphiphilic alpha-helical structures that can be triggered by external stimuli such as temperature, pH, and ionic strength. The aggregations of these structures, most often in the form of coiled-coil dimers, serve as junction points to create a hydrogel network.[38] The properties of the leucine zipper hydrogels can be regulated over a wide range through controlling the chain lengths and flanking amino acid sequences that alter aggregate structure, aggregation number, stability, and dimerization specificity.[38] The viability of NSCs seeded on 2D leucine zipper hydrogels was found to be sustained.[39] Because of the weak interactions between coiled coils, hydrogels based solely on leucine zipper interactions were not sufficiently stable to support 3D cell encapsulation. A second generation of leucine zipper hydrogels were recombinantly synthesized with additional cysteine residues to form covalent disulfide bonds and to support 3D hydrogel structures with improved stability (Fig. 1).[40] This hydrogel fostered good MSC viability after encapsulation as well as cell spreading, proliferation, and migration.[40]

Self-assembling protein-engineered hydrogels can also be created through many other protein-protein interactions for cell encapsulation. For example, the Dock-and-Lock system was used to achieve hydrogels via interactions between a docking and dimerization domain and an A-kinase anchoring protein domain (Fig. 2).[41] The rapid association and dissociation between these two domains allow for the formation of a shear-thinning and self-healing hydrogel. MSCs were encapsulated within this hydrogel and remained highly viable and homogeneously dispersed after injection through a 28 G needle.[41] As another example, two hydrogel components each containing tax-interactive protein-1 (TIP1) domains or PDZ domains formed a physical hydrogel upon mixing at physiological conditions.[42] Different PDZ-binding sequences resulted in a range of affinities to TIP1, enabling good tunability of hydrogel mechanical properties and injectability. MSCs were encapsulated within these hydrogels and showed high cell viability and proliferation over seven days.[43] Another type of self-assembling peptide hydrogel is (RADA),[16] which undergoes spontaneous assembly into well-ordered interwoven nanofibers in water due to its bilayer β-sheet structure. These structures can further undergo gelation to rapidly form hydrogels with ~10 nm fiber diameter, 5–200 nm pore size and >99% water content under physiological conditions with mechanical stiffness similar to brain tissue.[44] NSCs have been encapsulated within this hydrogel and differentiated toward a neuronal lineage.[44] ASCs have also been cultured within this 3D hydrogel with enhanced cell function including cell migration, proliferation, and growth factor-secreting capability.[45] As a final example, calmodulin not only undergoes Ca^{2+}-stimulated assembly, but it also undergoes a hinge-like conformational change from an extended dumbbell to a collapsed conformation upon binding to its ligands.[14] This interesting structural transformation has been used to design hydrogels with reversible and dynamic properties for a variety of tissue engineering applications.[14]

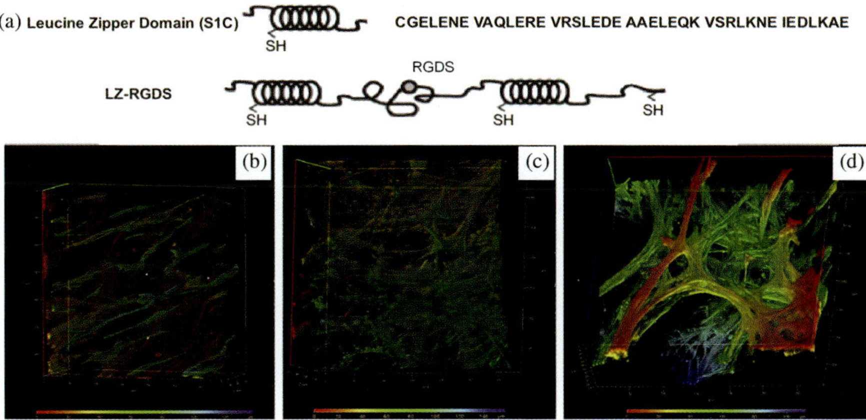

Fig. 1. (a) Schematic representation showing the peptide sequences used in the synthesis of a second-generation, chimeric leucine zipper protein with the cell-adhesive RGDS domain. 3D reconstructed images of MSC cultured within the hydrogel after (b) 7, (c) 14, and (d) 21 days. Cellular f-actin is stained with TRITC-Phalloidin and imaged using Z-stack confocal microscopy. Color-coding represents the depth profile from the surface (red) to a depth of 150 μm (blue). Adapted from Ref. [40], copyright 2014, reprinted with permission from Elsevier Ltd.

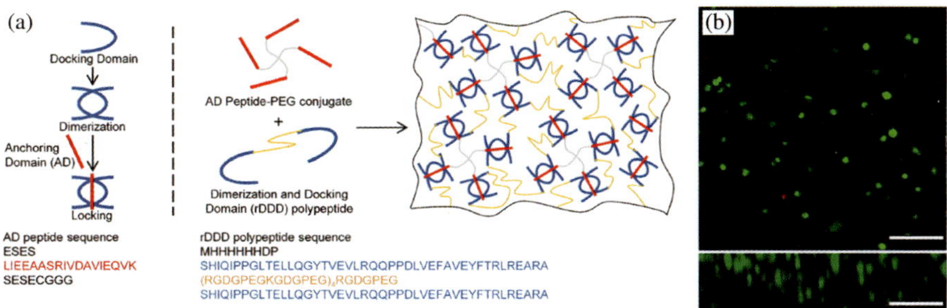

Fig. 2. (a) Dock-and-Lock self-assembly mechanism. Docking domains dimerize and lock with the anchoring domain (*left*). Engineered anchoring domains and docking domains self-assemble to form dynamic, shear-thinning, self-healing hydrogels when mixed (*right*). (b) Top-down and side views of MSCs encapsulated in the gels and injected into a collagen gel after three days of culture (live/dead staining imaged with confocal microscopy). Adapted from Ref. [41], copyright 2011, reprinted with permission from Elsevier Ltd.

2.3 *Bioactive domains*

2.3.1 *Cell-adhesive domains*

Besides structural domains, a wide range of bioactive domains can be customized and incorporated into the protein sequences to offer important biofunctionality. Cell-adhesive domains are critical in promoting adhesion, spreading, and subsequent cellular function of the encapsulated cells within the engineered protein hydrogel. A variety of cell-adhesive domains derived from ECM components have been designed into engineered protein sequences and exploited to regulate stem cell behavior, including RGD, IKVAV, YIGSR, DGEA, PHSRN, REDV, IKLLI, and GFOGER.[46] For example, the RGD domain, which is found in multiple ECM proteins, has been designed into a leucine zipper hydrogel to support MSC adhesion, spreading, proliferation, migration, and secretion of angiogenic factors.[40] As another example, an RGD-modified ELP hydrogel through chemical conjugation of maleimide-thiol coupling significantly enhanced adhesion, spreading, migration, and proliferation of MSCs compared to a hydrogel the RGD peptide passively absorbed.[47] The RGD sequence conjugated to the surface of the protein could bind to multiple integrins that mediated cellular functions and directed cell fates via diverse signaling pathways.[47] The DGEA domain, derived from collagen type I, has been encoded into the sequence of a self-assembling nanofibrous peptide to support adhesion, spreading, proliferation, and early commitment of osteogenic differentiation of MSCs.[48] DGEA was found to facilitate direct contact between the cells and surface-bound nanofibers, because DGEA ligand promoted cell binding through its integrin $\alpha2\beta1$ receptor.[48] The laminin-derived IKVAV domain was included in a self-assembling peptide and was found to decrease cell proliferation rate and to promote neuronal differentiation of pluripotent embryonic carcinoma cells with upregulated βIII-tubulin and MAP2 expression.[49] In another elegant study, an engineered protein sequence, which spans fragments 7–10 of fibronectin module III and fragments 1–2 of cadherin-11 (rFN/Cad-11), was used to modify the interface of collagen type II sponges.[50] The protein-engineered collagen sponges stimulated cell proliferation and promoted better chondrogenic differentiation of MSCs than fibronectin-modified surfaces.[50] These results indicate that the bioactive domains take into account the roles of both the core RGD sequence and the auxiliary PHSRN motif. The cell-cell interactions are hypothesized to be enhanced by heterophilic interactions mediated by FNIII7-10 and the homophilic interactions mediated by Cad-11. The cells were thus crosslinked to the more hydrophilic surfaces of the collagen sponges, suggesting that linking motifs with binding functions can have a significant synergistic effect.[50]

2.3.2 *Growth factor mimetic domains*

In addition to cell adhesive-domains, bioactive sequences that mimic the function of growth factors have been included in protein-engineered biomaterials to impart additional biofunctionality. Although many studies directly adsorb or immobilize growth factors to the surface of the biomaterial, incorporation of growth factor sequences into the engineered protein chains may enable better spatial distribution and presentation to induce growth factor activity. As an example, CCN1, a heparin-binding protein that displays pro-angiogenic activity, has been multimerized and cloned into an ELP sequence.[51] This protein contains four bioactive domains with sequences similar to insulin-like growth factor binding protein. The CCN1-engineered ELP hydrogel promoted cell adhesion, migration, focal adhesion assembly, and spreading. The enhanced cell responses are hypothesized to function through integrin $\alpha_v\beta_3$-mediated signaling pathways.[51] In another study, vascular endothelial growth factor (VEGF) mimetic sequences were introduced to the sequence of self-assembling nanofibers to induce angiogenic signaling at the site of delivery.[52] MSC adhesion and spreading were significantly enhanced, whereas endothelial cell migration, VEGF receptor activation, and infiltration rate were significantly elevated, as compared to unmodified nanofibers.[52] In a subcutaneous injection rat model, rapid angiogenesis was found with the development of mature vessels in just seven days. It is hypothesized that the mechanism of angiogenesis works as follows: the hydrophobic and ionically bound nanofibers at the periphery of scaffolds recruit endothelial cells to infiltrate scaffolds to form lumens prior to implantation. Then the subcutaneously implanted scaffolds could rapidly form CD31+, vWF+, Nestin+, and α-SMA+ microvessels that have CD45+ cells to flow through their lumen to induce maturation of endothelial cells and neovessels.[52] As a final example, short peptide motifs SKPPGTSS (SKP, bone marrow homing motif), FHRRIKA (FHR, heparin-binding motif), and PRGDSGYRGDS (two-unit RGD cell adhesion motif) were used to extend the C-terminus of self-assembling peptide (RADA)$_{16}$ to obtain functionalized peptide scaffolds.[45] The heparin-binding domain interacts with transmembrane proteoglycans to enhance cell responses including attachment, spreading, and formation of discrete focal contacts and organized cytoskeletal assembly.[45] The SKP motif has partial sequence homology with a region of CD84, and plays a role in stem cell homing.[45] The functionalized peptide hydrogels were found to be suitable 3D scaffolds for promoting ASC function with higher cell proliferation, migration, and the secretion of angiogenic growth factors compared with tissue culture plates and pure (RADA)$_{16}$ scaffolds. Interestingly, incorporation of the two-unit RGD motif showed the highest proliferation rate of ASCs within the scaffolds, while the FHR motif supported the highest growth factor-secreting capability of ASCs.[45]

2.3.3 *Enzyme-responsive domains*

To mimic the functionality of the native ECM that undergoes dynamic remodeling in response to stem cells, researchers have shown great interest in exploring and incorporating protein domains that respond to stem cell-secreted enzymes. The major cell responsive domains used in the design of protein-engineered biomaterials have been matrix metalloproteinase (MMP) degradable domains, including sequences sensitive to MMP-1, -2, -8, -9, -13, and -14.[53] These cell-dictated enzymatic activities lead to the cleavage of structural domains or bioactive domains along the engineered protein sequence, resulting in material degradation or release of bioactive peptides. MMP-1 sensitive domains were engineered into RLP that underwent enzymatic degradation within two days, while RLP construct lacking the MMP-sensitive domain remained intact during the entire incubation with MMP-1.[33] The cell-degradable hydrogel supported the adhesion, spreading, and 3D encapsulation of MSCs.[33] In a similar study, a peptide sequence responsive to MMP-2 and MMP-9 was encoded in SELPs.[54] The MMP degradable SELP was subject to fast degradation with complete cleavage in 48 hours, while no observable degradation was found for unmodified SELP.[54] As another example, MMP-2 cleavable domains and RGD cell-adhesive domains were incorporated into amphiphilic self-assembling peptides.[55] This multi-domain polypeptide hydrogel significantly increased MSC viability and spreading and encouraged cell migration into the hydrogel matrix, as compared to unmodified hydrogel, rendering it a suitable cell delivery vehicle for a variety of tissue engineering applications.[55] In another elegant design, streptococcal collagen-like 2 protein was modified with hyaluronic acid or chondroitin sulfate-binding peptides and then crosslinked with an MMP-7-sensitive peptide to form biodegradable hydrogels.[56] MSCs encapsulated within these hydrogels exhibited improved viability and significantly enhanced chondrogenic differentiation compared to controls within no modification, suggesting the potential of this novel degradable hydrogel to modulate cell-mediated material properties.[56] In a final example, MMP degradable sequence was engineered into the recombinant fusion protein of LAP-MMP-mTGF-β3 to trigger the release of mTGF-β3 from the self-assembled peptide scaffolds, thus stimulating the differentiation of the encapsulated ASCs into chondrogenesis.[57] This study demonstrates that controlled local delivery of the degradable LAP-MMP-mTGF-β3 constructs can accelerate differentiation of ASCs into cartilage *in vivo.*[57]

3. Stem Cell Delivery Using Engineered Proteins

There have recently been a growing number of studies using *in vivo* models to assess protein-engineered materials as stem cell carriers for a variety of tissue engineering applications, including dermal, bone, cartilage, and nerve repair and regeneration. These protein-based biomaterials, with highly tunable structural and bioactive components, have been evaluated as stem cell carriers *in vivo* in terms of stem cell survival, proliferation, and secretome, as well as other cell functions and therapeutic outcomes, in comparison to current standards.

3.1 *Subcutaneous cell delivery*

Subcutaneous injection/implantation is one of the simplest *in vivo* models to assess biomaterial carriers for stem cell delivery efficiency and subsequent cell behavior. MSCs (5×10^5 cells/20 µL) were injected within leucine-zipper hydrogels subcutaneously in mice.[40] The hydrogels did not trigger unfavorable foreign body response with the absence of macrophages after two weeks of implantation. Although the survival of MSCs was not determined, neovascularization was observed originating from the surface of the hydrogel to the interior. Higher vessel density was also seen with protein sequences containing the RGD motif.[40] As another example, ASCs (1×10^6 cells/50 μL) were injected within MITCH in a subcutaneous mouse model.[58] As monitored by bioluminescence, significantly higher cell survival and retention were found for the MITCH-delivered cells over seven days compared to other commonly used hydrogels including collagen and alginate.[58] In a subsequent study, MITCH was designed with an additional thermoresponsive domain to enhance its stability and mechanical properties at body temperature. This novel hydrogel, termed shear-thinning hydrogel for injectable encapsulation and long-term delivery (SHIELD), was employed to deliver ASCs (1×10^6 cells/30 μL) subcutaneously in mice and achieved significantly enhanced hydrogel retention and cell retention over two weeks.[59] These results suggest that these protein-engineered hydrogel carriers supporting good cell survival and retention have great potential in a variety of regenerative medicine applications.

3.2 *Dermal regeneration*

In the context of dermal repair, numerous animal studies and clinical trials have indicated the benefit of using MSCs to inhibit scar formation and attenuate skin fibrosis.[60] However, GFP$^+$ MSCs (1×10^6 cells/100 μL saline) delivered subcutaneously in a skin fibrosis mouse model showed that all cells were no longer detectable within 24 hours post-delivery.[60] Therefore, using a protein-engineered hydrogel carrier to prolong cell retention time is hypothesized to significantly promote effectiveness of the cell therapy. As an example, ASCs were delivered using a recombinant, electrospun, tropoelastin scaffold for dermal repair.[61] The physiological response of ASC-seeded electrospun tropoelastin membranes was studied in a murine excisional wound model and compared to treatment with sterile petrolatum gauze and Tegaderm as a control. Significantly faster wound closure and thicker regenerated epithelium at day 6 was found for the ASC-seeded tropoelastin treatment group, as compared to wound healing by the scaffold only and controls.[61] As another example, highly elastic

ELP hydrogels have demonstrated long-term structural stability *in vivo*, and early and progressive host integration with no immune response, suggesting their potential for supporting wound repair.[62] Combining the ELPs with colloidal solutions of silica nanoparticles showed enhanced clotting ability and hemostasis, leading to effective treatment of a lethal bleeding liver wound.[62]

3.3 *Bone and cartilage regeneration*

For bone regeneration, silk protein scaffolds mimicking mechanical features of native bone have been developed with a high compressive strength (~13 MPa hydrated state) based on silk protein-protein interfacial bonding. Micron-sized silk fibers (10–600 µm) obtained by alkali hydrolysis were used to reinforce the scaffolds in a compact fiber composite with tunable compressive strength, surface roughness, and porosity.[63] Bone marrow stem cell (BMSC) proliferation and differentiation toward osteogenesis were significantly promoted with upregulated biochemical and gene expression of bone markers and formation of bone-like tissue *in vitro*, as compared to control silk sponges.[63] The 3D scaffolds were implanted subcutaneously and exhibited dense tissue ingrowth with vascularization and minimal immunomodulatory responses.[63] In another recent study, three types of ELP sequences containing RGD and statherin-derived bioactive domains were designed to deliver MSCs in an orthotopic critical-size rat calvarial defect model (Fig. 3).[64] The statherin-derived domain is known to promote mineralization *in vitro*.[64] MSC adhesion, mineralization, and differentiation toward an osteogenic lineage were significantly promoted, as compared to control groups of ELP without these bioactive domains. Animals implanted with these scaffolds exhibited the highest volume of ossified tissue, with active osteoblasts within the defect.[64] ELP has also been used to prepare thermally triggered coacervates as 3D scaffolds that maintained chondrocytic phenotypes for cartilage repair.[65] When implanted *in vivo*, the ELP scaffolds displayed minimal cytotoxicity and immune response and supported cell infiltration and matrix synthesis.[66] In another example, 3D scaffolds of self-assembled peptides were used to culture ASCs *in vitro* for three weeks prior to transplantation *in vivo*. Incorporation of the

L. Cai and S. C. Heilshorn

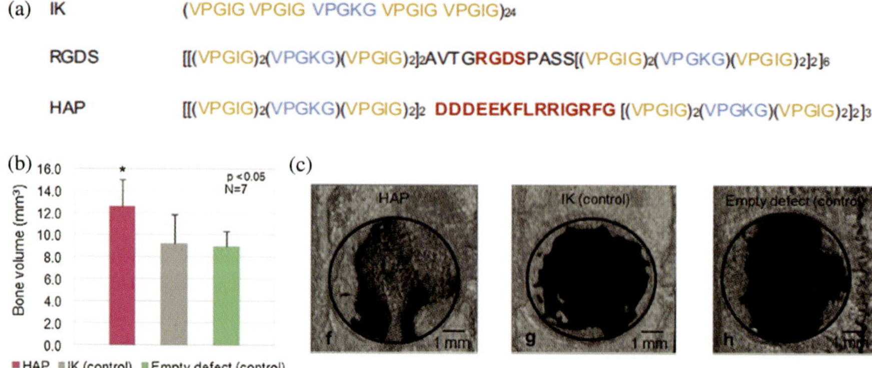

Fig. 3. (a) Sequences used in the synthesis of elastin-like polypeptides containing no biofunctional domains (IK), a cell-adhesive domain (RGDS), and a mineralization domain (HAP). The structural domains are shown in yellow, crosslinking sites in blue, and bioactive domains in red. (b) HAP scaffolds presented the highest mean volume of ossified tissue in a clavarial defect model. (c) The ossified tissue within the defect compared to animals receiving the HAP (left), non-bioactive IK (middle), and untreated group (right) imaged using high-resolution micro-computed tomography. Adapted from Ref. [64], copyright 2014, reprinted with permission from Elsevier Ltd.

recombinant fusion protein of LAP-MMP-mTGF-β3 significantly upregulated the expression of chondrocytic markers of ASCs compared to the one without MMP-sensitive sequence. After three weeks of culture *in vitro*, the cell-laden peptide scaffolds with ASC differentiated chondrocytes were then injected subcutaneously into nude mice and retrieved after four weeks *in vivo*. The formation of cartilage matrix was significantly more distinct than the control groups of peptides modified with TGF-β3 alone or without LAP-MMP-mTGF-β3, suggesting enhanced chondrogenesis upon controlled delivery of TGF-β3 *via* degradable MMP-sensitive sequence.[57]

3.4 *Nerve regeneration*

With regard to nerve regeneration, ELP was fabricated into nerve guidance conduits to support Schwann cell adhesion and proliferation, with consistent expression of Schwann cell-specific phenotypic marker S100.[67] ELP containing laminin-derived neurite outgrowth-promoting sequence was also synthesized as thin films to coat the surface of nerve guidance conduits and to promote neurite outgrowth of neuronal cells.[68] As another example, self-assembling $RADA_{16}$-IKVAV scaffolds were used to encapsulate and deliver GFP[+] NSCs in a rat traumatic brain wound model.[44] The GFP[+] NSC survival was significantly enhanced and cells could be traced over three weeks post-transplantation. More interestingly, the polypeptide scaffolds incorporating the IKVAV motif supported NSC differentiation into mature neuronal cells and significantly improved brain tissue regeneration after six weeks post-transplantation, as compared to $RADA_{16}$-only scaffolds, where more astroglial cells were observed.[44] In another recent study, NSC transplantation into mouse spinal cord injury lesions was conducted using diblock copolypeptide hydrogels that were liquid at room temperature and formed semi-rigid gels at body temperature.[69] Cell retention was significantly improved at the lesion site, with 18% of GFP[+] NSC surviving within the hydrogel at three weeks post-transplantation, as compared to 6% of cells delivered in medium. The transplanted NSCs within the hydrogel were also found to integrate with healthy neural cells at the lesion perimeter and to support the regrowth of host nerve fibers.[69]

4. Conclusions and Future Directions

In summary, protein-engineered biomaterials have been rationally designed and reproduced with a range of structural and bioactive modular peptide domains to construct 3D scaffolds that mimic several aspects of native stem cell microenvironments. These biomaterials are highly tunable and can be designed to have decoupled material properties. Furthermore, given the expansive diversity of proteins, these modular biomaterials encompass virtually limitless possibilities to incorporate novel peptide domains for multifunctionality. The successful examples highlighted in this chapter demonstrate the capability of these biomaterials to serve as versatile platforms for stem cell cultures *in vitro* and cell transplantation *in vivo*. As the diversity of peptide domains continues to expand, it is expected that more bioactive domains that address specific cell-material interactions will be included in a single material formulation to regulate complex stem cell signaling toward better regenerative function and therapeutic outcomes. Besides molecular-level composition, the 3D microstructures of the protein-engineered scaffolds could be better regulated through advancing material fabrication techniques. Further investigation is warranted to optimize protein design strategies and production yields, to carefully control immunogenicity, and to evaluate the potential for FDA approval and clinical translation. Given the remarkable progress made in the past two decades, protein-engineered biomaterials serve as a powerful tool that opens up tremendous opportunities to enable stem cell therapy for treating a variety of injuries and diseases.

Acknowledgements

This work was supported by NIH NRSA (F32HL128094) and Interdisciplinary Scholar Award from Stanford Neuroscience Institute.

References

1. M. A. Fischbach, J. A. Bluestone and W. A. Lim, Cell-based therapeutics: the next pillar of medicine, *Science translational medicine* **5**: 179ps7 (2013).
2. W. Chang, B. W. Song and K. C. Hwang, in *Stem Cells and Cancer Stem Cells, Volume 10*, edited by M. A. Hayat (Springer Netherlands, 2013), Vol. 10, pp. 35–43.
3. C. A. Herberts, M. S. Kwa and H. P. Hermsen, Risk factors in the development of stem cell therapy, *J Transl Med* **9**: 29 (2011).
4. M. W. Tibbitt, C. B. Rodell, J. A. Burdick and K. S. Anseth, Progress in material design for biomedical applications, *Proc Natl Acad Sci USA* **112**: 14444–14451 (2015).
5. J. Zhu, Bioactive modification of poly(ethylene glycol) hydrogels for tissue engineering, *Biomaterials* **31**: 4639–4656 (2010).
6. S. F. Badylak, D. O. Freytes and T. W. Gilbert, Extracellular matrix as a biological scaffold material: structure and function, *Acta Biomater* **5**: 1–13 (2009).
7. R. L. DiMarco and S. C. Heilshorn, Multifunctional materials through modular protein engineering, *Adv Mater* **24**: 3923–3940 (2012).
8. J. A. Werkmeister and J. A. M. Ramshaw, Recombinant protein scaffolds for tissue engineering, *Biomed Mater* **7**: 012002 (2012).
9. J. E. Gagner, W. Kim and E. L. Chaikof, Designing protein-based biomaterials for medical applications, *Acta Biomater* **10**: 1542–1557 (2014).

10. M. Parvizi, J. A. Plantinga, C. A. van Speuwel-Goossens, E. M. van Dongen, S. G. Kluijtmans and M. C. Harmsen, Development of recombinant collagen-peptide-based vehicles for delivery of adipose-derived stromal cells, *J Biomed Mater Res A* **104**: 503–516 (2016).

11. D. L. Nettles, A. Chilkoti and L. A. Setton, Applications of elastin-like polypeptides in tissue engineering, *Adv Drug Deliv Rev* **62**: 1479–1485 (2010).

12. C. M. Elvin, A. G. Carr, M. G. Huson, J. M. Maxwell, R. D. Pearson, T. Vuocolo, N. E. Liyou, D. C. C. Wong, D. J. Merritt and N. E. Dixon, Synthesis and properties of crosslinked recombinant pro-resilin, *Nature* **437**: 999–1002 (2005).

13. Z. Megeed, J. Cappello and H. Ghandehari, Genetically engineered silk-elastinlike protein polymers for controlled drug delivery, *Adv Drug Deliv Rev* **54**: 1075–1091 (2002).

14. Z. Sui, W. J. King and W. L. Murphy, Protein-based hydrogels with tunable dynamic responses, *Adv Funct Mater* **18**: 1824–1831 (2008).

15. J. Zahiri, J. H. Bozorgmehr and A. Masoudi-Nejad, Computational prediction of protein–protein interaction networks: algorithms and resources, *Curr Genomics* **14**: 397–414 (2013).

16. W. P. Russ, D. M. Lowery, P. Mishra, M. B. Yaffe and R. Ranganathan, Natural-like function in artificial WW domains, *Nature* **437**: 579–583 (2005).

17. L. Raibaut, N. Ollivier and O. Melnyk, Sequential native peptide ligation strategies for total chemical protein synthesis, *Chem Soc Rev* **41**: 7001–7015 (2012).

18. S. G. Wise, G. C. Yeo, M. A. Hiob, J. Rnjak-Kovacina, D. L. Kaplan, M. K. C. Ng and A. S. Weiss, Tropoelastin: a versatile, bioactive assembly module, *Acta Biomater* **10**: 1532–1541 (2014).

19. K. J. Lampe, A. L. Antaris and S. C. Heilshorn, Design of 3D engineered protein hydrogels for tailored control of neurite growth, *Acta Biomater* **9**: 5590–5599 (2013).

20. L. Cai, C. B. Dinh and S. C. Heilshorn, One-pot synthesis of elastin-like polypeptide hydrogels with grafted VEGF-mimetic peptides, *Biomater Sci* **2**: 757–765 (2014).

21. C. Chung, K. J. Lampe and S. C. Heilshorn, Tetrakis(hydroxymethyl) phosphonium chloride as a covalent cross-linking agent for cell encapsulation within protein-based hydrogels, *Biomacromolecules* **13**: 3912–3916 (2012).

22. J. Ozsvar, S. M. Mithieux, R. Wang and A. S. Weiss, Elastin-based biomaterials and mesenchymal stem cells, *Biomater Sci* **3**: 800–809 (2015).

23. H. Betre, S. R. Ong, F. Guilak, A. Chilkoti, B. Fermor and L. A. Setton, Chondrocytic differentiation of human adipose-derived adult stem cells in elastin-like polypeptide, *Biomaterials* **27**: 91–99 (2006).
24. B. Celebi, M. Cloutier, R. B. Rabelo, D. Mantovani and A. Bandiera, Human elastin-based recombinant biopolymers improve mesenchymal stem cell differentiation, *Macromol Biosci* **12**: 1546–1554 (2012).
25. X. Hu, X. Wang, J. Rnjak, A. S. Weiss and D. L. Kaplan, Biomaterials derived from silk-tropoelastin protein systems, *Biomaterials* **31**: 8121–8131 (2010).
26. X. Hu, S. H. Park, E. S. Gil, X. X. Xia, A. S. Weiss and D. L. Kaplan, The influence of elasticity and surface roughness on myogenic and osteogenic-differentiation of cells on silk-elastin biomaterials, *Biomaterials* **32**: 8979–8989 (2011).
27. B. Kundu, R. Rajkhowa, S. C. Kundu and X. Wang, Silk fibroin biomaterials for tissue regenerations, *Adv Drug Deliv Rev* **65**: 457–470 (2013).
28. W. Huang, A. Rollett and D. L. Kaplan, Silk-elastin-like protein biomaterials for the controlled delivery of therapeutics, *Expert Opin Drug Deliv* **12**: 779–791 (2015).
29. Q. Wang, X. Xia, W. Huang, Y. Lin, Q. Xu and D. L. Kaplan, High throughput screening of dynamic silk-elastin-like protein biomaterials, *Adv Funct Mater* **24**: 4303–4310 (2014).
30. M. K. Włodarczyk-Biegun, M. W. T. Werten, F. A. de Wolf, J. J. J. P. van den Beucken, S. C. G. Leeuwenburgh, M. Kamperman and M. A. Cohen Stuart, Genetically engineered silk-collagen-like copolymer for biomedical applications: production, characterization and evaluation of cellular response, *Acta Biomater* **10**: 3620–3629 (2014).
31. B. An, T. M. DesRochers, G. Qin, X. Xia, G. Thiagarajan, B. Brodsky and D. L. Kaplan, The influence of specific binding of collagen-silk chimeras to silk biomaterials on hMSC behavior, *Biomaterials* **34**: 402–412 (2013).
32. A. J. Mieszawska, L. D. Nadkarni, C. C. Perry and D. L. Kaplan, Nanoscale control of silica particle formation via silk-silica fusion proteins for bone regeneration, *Chem Mater* **22**: 5780–5785 (2010).
33. L. Li, Z. Tong, X. Jia and K. L. Kiick, Resilin-like polypeptide hydrogels engineered for versatile biological function, *Soft Matter* **9**: 665–673 (2013).
34. L. Li, A. Mahara, Z. Tong, E. A. Levenson, C. L. McGann, X. Jia, T. Yamaoka and K. L. Kiick, Recombinant resilin-based bioelastomers for regenerative medicine applications, *Adv Healthc Mater* **5**: 266–275 (2016).

35. F. Sbrana, C. Fotia, A. Bracalello, N. Baldini, G. Marletta, G. Ciapetti, B. Bochicchio and M. Vassalli, Multiscale characterization of a chimeric biomimetic polypeptide for stem cell culture, *Bioinspir Biomim* **7**: 046007 (2012).

36. C. T. S. W. P. Foo, J. S. Lee, W. Mulyasasmita, A. Parisi-Amon and S. C. Heilshorn, Two-component protein-engineered physical hydrogels for cell encapsulation, *Proc Natl Acad Sci USA* **106**: 22067–22072 (2009).

37. M. Greenwood-Goodwin, E. S. Teasley and S. C. Heilshorn, Dual-stage growth factor release within 3D protein-engineered hydrogel niches promotes adipogenesis, *Biomater Sci* **2**: 1627–1639 (2014).

38. W. A. Petka, J. L. Harden, K. P. McGrath, D. Wirtz and D. A. Tirrell, Reversible hydrogels from self-assembling artificial proteins, *Science* **281**: 389–392 (1998).

39. S. E. Fischer, X. Liu, H. Q. Mao and J. L. Harden, Controlling cell adhesion to surfaces via associating bioactive triblock proteins, *Biomaterials* **28**: 3325–3337 (2007).

40. C. C. Huang, S. Ravindran, Z. Yin and A. George, 3-D self-assembling leucine zipper hydrogel with tunable properties for tissue engineering, *Biomaterials* **35**: 5316–5326 (2014).

41. H. D. Lu, M. B. Charati, I. L. Kim and J. A. Burdick, Injectable shear-thinning hydrogels engineered with a self-assembling Dock-and-Lock mechanism, *Biomaterials* **33**: 2145–2153 (2012).

42. F. Ito, K. Usui, D. Kawahara, A. Suenaga, T. Maki, S. Kidoaki, H. Suzuki, M. Taiji, M. Itoh, Y. Hayashizaki and T. Matsuda, Reversible hydrogel formation driven by protein-peptide-specific interaction and chondrocyte entrapment, *Biomaterials* **31**: 58–66 (2010).

43. J. Wang, J. Zhang, X. Zhang and H. Zhou, A protein-based hydrogel for *in vitro* expansion of mesenchymal stem cells, *PLoS One* **8**: e75727 (2013).

44. T. Y. Cheng, M. H. Chen, W. H. Chang, M. Y. Huang and T. W. Wang, Neural stem cells encapsulated in a functionalized self-assembling peptide hydrogel for brain tissue engineering, *Biomaterials* **34**: 2005–2016 (2013).

45. X. Liu, X. Wang, X. Wang, H. Ren, J. He, L. Qiao and F. Z. Cui, Functionalized self-assembling peptide nanofiber hydrogels mimic stem cell niche to control human adipose stem cell behavior *in vitro*, *Acta Biomater* **9**: 6798–6805 (2013).

46. M. B. Rahmany and M. Van Dyke, Biomimetic approaches to modulate cellular adhesion in biomaterials: a review, *Acta Biomater* **9**: 5431–5437 (2013).

47. S. Ravi, V. R. Krishnamurthy, J. M. Caves, C. A. Haller and E. L. Chaikof, Maleimide-thiol coupling of a bioactive peptide to an elastin-like protein polymer, *Acta Biomater* **8**: 627–635 (2012).

48. H. Ceylan, S. Kocabey, H. Unal Gulsuner, O. S. Balcik, M. O. Guler and A. B. Tekinay, Bone-like mineral nucleating peptide nanofibers induce differentiation of human mesenchymal stem cells into mature osteoblasts, *Biomacromolecules* **15**: 2407–2418 (2014).

49. Q. Li, W. H. Cheung, K. L. Chow, R. G. Ellis-Behnke and Y. Chau, Factorial analysis of adaptable properties of self-assembling peptide matrix on cellular proliferation and neuronal differentiation of pluripotent embryonic carcinoma, *Nanomedicine* **8**: 748–756 (2012).

50. S. Dong, H. Guo, Y. Zhang, Z. Li, F. Kang, B. Yang, X. Kang, C. Wen, Y. Yan, B. Jiang and Y. Fan, rFN/Cad-11-modified collagen type II biomimetic interface promotes the adhesion and chondrogenic differentiation of mesenchymal stem cells, *Tissue Eng Part A* **19**: 2464–2477 (2013).

51. S. Ravi, C. A. Haller, R. E. Sallach and E. L. Chaikof, Cell behavior on a CCN1 functionalized elastin-mimetic protein polymer, *Biomaterials* **33**: 2431–2438 (2012).

52. V. A. Kumar, N. L. Taylor, S. Shi, B. K. Wang, A. A. Jalan, M. K. Kang, N. C. Wickremasinghe and J. D. Hartgerink, Highly angiogenic peptide nanofibers, *ACS Nano* **9**: 860–868 (2015).

53. K. B. Fonseca, P. L. Granja and C. C. Barrias, Engineering proteolytically-degradable artificial extracellular matrices, *Prog Polym Sci* **39**: 2010–2029 (2014).

54. J. A. Gustafson, R. A. Price, J. Frandsen, C. R. Henak, J. Cappello and H. Ghandehari, Synthesis and characterization of a matrix-metalloproteinase responsive silk-elastinlike protein polymer, *Biomacromolecules* **14**: 618–625 (2013).

55. K. M. Galler, L. Aulisa, K. R. Regan, R. N. D'Souza and J. D. Hartgerink, Self-assembling multidomain peptide hydrogels: designed susceptibility to enzymatic cleavage allows enhanced cell migration and spreading, *J Am Chem Soc* **132**: 3217–3223 (2010).

56. P. A. Parmar, L. W. Chow, J. P. St Pierre, C. M. Horejs, Y. Y. Peng, J. A. Werkmeister, J. A. M. Ramshaw and M. M. Stevens, Collagen-mimetic peptide-modifiable hydrogels for articular cartilage regeneration, *Biomaterials* **54**: 213–225 (2015).

57. D. Zheng, Y. Dan, S. H. Yang, G. H. Liu, Z. W. Shao, C. Yang, B. J. Xiao, X. Liu, S. Wu, T. Zhang and P. K. Chu, Controlled chondrogenesis from adipose-derived stem cells by recombinant transforming growth factor-β3 fusion protein in peptide scaffolds, *Acta Biomater* **11**: 191–203 (2015).

58. A. Parisi-Amon, W. Mulyasasmita, C. Chung and S. C. Heilshorn, Protein-engineered injectable hydrogel to improve retention of transplanted adipose-derived stem cells, *Adv Healthc Mater* **2**: 428–432 (2013).

59. L. Cai, R. E. Dewi and S. C. Heilshorn, Injectable hydrogels with in situ double network formation enhance retention of transplanted stem cells, *Adv Funct Mater* **25**: 1344–1351 (2015).

60. Y. Wu, S. Huang, J. Enhe, K. Ma, S. Yang, T. Sun and X. Fu, Bone marrow-derived mesenchymal stem cell attenuates skin fibrosis development in mice, *Int Wound J* **11**: 701–710 (2014).

61. H. Machula, B. Ensley and R. Kellar, Electrospun tropoelastin for delivery of therapeutic adipose-derived stem cells to full-thickness dermal wounds, *Adv Wound Care* **3**: 367–375 (2014).

62. Y. N. Zhang, R. K. Avery, Q. Vallmajo-Martin, A. Assmann, A. Vegh, A. Memic, B. D. Olsen, N. Annabi and A. Khademhosseini, A highly elastic and rapidly crosslinkable elastin-like polypeptide-based hydrogel for biomedical applications, *Adv Funct Mater* **25**: 4814–4826 (2015).

63. B. B. Mandal, A. Grinberg, E. Seok Gil, B. Panilaitis and D. L. Kaplan, High-strength silk protein scaffolds for bone repair, *Proc Natl Acad Sci USA* **109**: 7699–7704 (2012).

64. E. Tejeda-Montes, A. Klymov, M. R. Nejadnik, M. Alonso, J. C. Rodriguez-Cabello, X. F. Walboomers and A. Mata, Mineralization and bone regeneration using a bioactive elastin-like recombinamer membrane, *Biomaterials* **35**: 8339–8347 (2014).

65. H. Betre, L. A. Setton, D. E. Meyer and A. Chilkoti, Characterization of a genetically engineered elastin-like polypeptide for cartilaginous tissue repair, *Biomacromolecules* **3**: 910–916 (2002).

66. D. L. Nettles, K. Kitaoka, N. A. Hanson, C. M. Flahiff, B. A. Mata, E. W. Hsu, A. Chilkoti and L. A. Setton, In situ crosslinking elastin-like polypeptide gels for application to articular cartilage repair in a goat osteochondral defect model, *Tissue Eng Part A* **14**: 1133–1140 (2008).

67. Y. S. Hsueh, S. Savitha, S. Sadhasivam, F. H. Lin and M. J. Shieh, Design and synthesis of elastin-like polypeptides for an ideal nerve conduit in peripheral nerve regeneration, *Mater Sci Eng C* **38**: 119–126 (2014).

68. S. Kakinoki and T. Yamaoka, Thermoresponsive elastin/laminin mimicking artificial protein for modifying PLLA scaffolds in nerve regeneration, *J Mater Chem B* **2**: 5061–5067 (2014).

69. S. Zhang, J. E. Burda, M. A. Anderson, Z. Zhao, Y. Ao, Y. Cheng, Y. Sun, T. J. Deming and M. V. Sofroniew, Thermoresponsive copolypeptide hydrogel vehicles for central nervous system cell delivery, *ACS Biomater Sci Eng* **1**: 705–717 (2015).

20

QUALITY CONTROL OF AUTOLOGOUS CELL- AND TISSUE-BASED THERAPIES

Nathalie Dusserre, Todd McAllister and Nicolas L'Heureux

1. Introduction

Cell- and tissue-based therapies can be broadly defined as the treatment of human diseases using human or animal cells. Original examples of such therapies were introduced with the intent to restore the blood and immune system of patients. In its simplest version, this approach dates back to the first successful blood transfusion in 1818 by James Blundel. Later, Joseph E. Murray performed the first syngeneic (1956) and then allogeneic (1969) bone marrow transplants, pioneering work for which he received the Nobel Prize of Medicine in 1990.[1,2] More recently, stem cells of a different origin have been successfully used for similar purposes.[3] Varying the nature of the starting cells or tissue and the method or extent of any subsequent processing has generated a stunning range of applications for cell- and tissue-based therapies. Some of these applications currently include diabetes regulation, osteoarthritis treatment, immune system modulation, regeneration of neural cells through injection of stem cells into the spinal cord, and replacement of whole organs (cornea, skin, blood vessels) through tissue engineering.

Cell sources for cell- and tissue-based therapies include differentiated cells extracted from the tissues of living adult humans or cadavers, adult stem and progenitor cells, and embryonic and fetal stem cells. For most applications, the cells or tissues used require some level of *ex vivo* processing. Some applications require only minimal manipulation for cell purification and storage. For others, a more extensive manufacturing process may include the proliferation, differentiation, selection, pharmacological treatment, or genetic modification of the cells. *Ex vivo* modifications are made to expand a specific cell type, restore a specific cell function, or build a specific tissue structure from the cells (tissue-engineered products). Personalized cell- and tissue-based therapies add to this level of complexity by customizing the cell-based treatment to individual patient needs.

Regulations are in place to ensure the safety of the patients by controlling the quality of the cell- and tissue-based product. They aim to control all aspects of the manufacturing process from the selection of the source of the cell and tissue, to its extraction and subsequent processing, and ultimately its implantation or re-injection into the patient. The spectrum of possible cell- and tissue-based therapeutic applications is wide enough that regulations applying to the quality control of one application might not apply to another. This chapter will focus on the quality control of therapeutic products derived from autologous cells or tissues and requiring extensive *ex vivo* manipulation. Such products meet the FDA's definition of autologous somatic cell therapy products.[4]

2. Regulations Pertaining to Quality Control of Cell- and Tissue-Based Products

Regulations pertaining to quality control of autologous cell- and tissue-based product manufacturing are based on pre-existing regulations designed for drugs and pharmaceutical products. These pre-existing regulations were adjusted to meet the specific challenges associated with the manufacture of therapeutic products derived from cells or tissues manipulated *ex vivo*. Not surprisingly, the regulatory burden varies with the perceived risk associated with each product.

In the current Good Tissue Practices (cGTP) Final Rule (21 CFR 1271), the FDA categorized cell- and tissue-based products into three classes based on their relative safety risk.[5] This risk assessment is based on the extent to which the required processing alters the initial biological characteristics of the cells or tissues used in the product. Products requiring only minimal manipulation and where the cells or tissues perform the function for which they were biologically intended are considered low-risk and are not subject to 21 CFR 1271 rules.

In contrast, cell- and tissue-based products that either

- require significant *ex vivo* manipulation such as cell expansion through tissue culture,
- are combined with another article,
- are intended for a use that is different from their initial biological purpose, or
- have a systemic effect and are dependent upon the metabolic activity of live cells for their primary function,

are considered high-risk and are subject to more stringent regulations (see Table 1 for more details on three-tiered cGTP classification). Products derived from autologous cells or tissues and requiring extensive *ex vivo* manipulations belong to this high-risk category. As such, these products are regulated as biological drugs under the Federal Food, Drug, and Cosmetic Act and Section 351 of the Public Health Services (PHS) Act (42 USC 262, "351 products").[6] They must be evaluated through a Biologic License Application (BLA) pathway before they reach the market, an Investigational New Drug (IND) application must be granted by the FDA before they are clinically tested, and their manufacturing process is subject to current Good Manufacturing Practices (cGMP) (21 CFR 210 and 211) and applicable parts of 21 CFR 1271 (subparts A, C, and D).[4,7,8] They are also subject to the various regulations listed in Tables 2 and 3.

Table 1.   Three-tiered approach of cGTP Final Rule (21 CFR 1271).

Risk Assessment	Criteria Met by Products
Low risk, not subject to 21 CFR 1271 rules.	Requiring only minimal manipulation and are meant to replace a function for which the cells or tissues were biologically intended (homologous use)

Higher risk, subject to 21 CFR 1271 rules and Section 361of PHS Act*	All of the following:
	• Minimally manipulated
	• Intended for homologous use only
	• Manufacture does not involve combining cells or tissue with another article, except for water, crystalloids, sterilizing, preserving, or storage agents that do not raise new clinical safety concerns
	• Either:
	— Does not have a systemic effect and is not dependent upon the metabolic activity of live cells for their primary function; or,
	— Has a systemic effect and is dependent upon the metabolic activity of live cells for its primary function, and is for
	  a. Autologous use
	  b. Allogeneic use in a first- or second-degree blood relative, or
	  c. Reproductive use

High risk, subject to 21 CFR 1271 rules and Section 351 of PHS Act*	Any of the following:
	• More than minimally manipulated: expanded, activated, genetically modified
	• Combined with another article (except some preserving and storage reagents)
	• Intended for a use that is different than its initial biological purpose
	• Has a systemic effect and is dependent upon the metabolic activity of live cells for its primary function, and is not for
	  a. Autologous use
	  b. Allogeneic use in a first- or second-degree blood relative, or
	  c. Reproductive use

*Public Health Services Act.

Table 2. Regulations that apply to "351" products while under IND.

Regulation	Title
21 CFR 312	Investigational New Drug (IND) Application
21 CFR 210/211	Current Good Manufacturing Practices
21 CFR 50	Protection of Human Subjects
21 CFR 56	Institutional Review Boards
21 CFR 1271	Subparts A, C, and D

Table 3. Regulations that apply to "351" products once licensed as Biologics (BLA).

Regulation	Title
21 CFR 201	Labeling
21 CFR 202	Advertising
21 CFR 210/211	Current Good Manufacturing Practices
21 CFR 600	Biological Products; General (includes Reporting of Adverse Experiences and Biological Deviations)
21 CFR 610	General Biologics Standards
21 CFR 1271	Subparts A, B, C, and D

Similar pathways and risk assessment-based classifications exist in Europe where autologous cells or tissues manipulated *ex vivo* and intended for medical applications are labeled as Advanced Therapy Medicinal Products.[9–13]

3. cGMP, cGTP, and Quality System

The objective of cGMP and cGTP is to ensure patient safety by controlling the identity, purity, stability, viability, and consistency of both the intermediate and final products. In order to ensure compliance with cGMP and cGTP requirements, every establishment processing "351 products" must devise its own quality system. This system is a comprehensive program developed by the investigator or manufacturer and is tailored to monitor specific cell- and tissue-based products. According to 21 CFR 1271 (1271.3 (hh)), this program should be designed, among other things, to prevent, detect, and correct deficiencies that might increase the risk of introducing, transmitting, and spreading communicable diseases. While there are multiple organizational variations, the quality program typically includes the following:

- A set of **standard operating procedures** (SOPs): written documents that describe all activities performed.
- A **quality control program**: a program that monitors certain characteristics of the process and product.
- A **quality assurance program**: a program that utilizes audits of records, revision of SOPs, personnel training, deviation monitoring and investigation, implementation of corrective and preventive actions (CAPA), and review of the final-products release process to ensure that all activities are performed according to SOPs and that all quality control results meet pre-set criteria.

Fig. 1 succinctly illustrates how these three basic elements interact to control the manufacturing and clinical testing of cell- and tissue-based products.

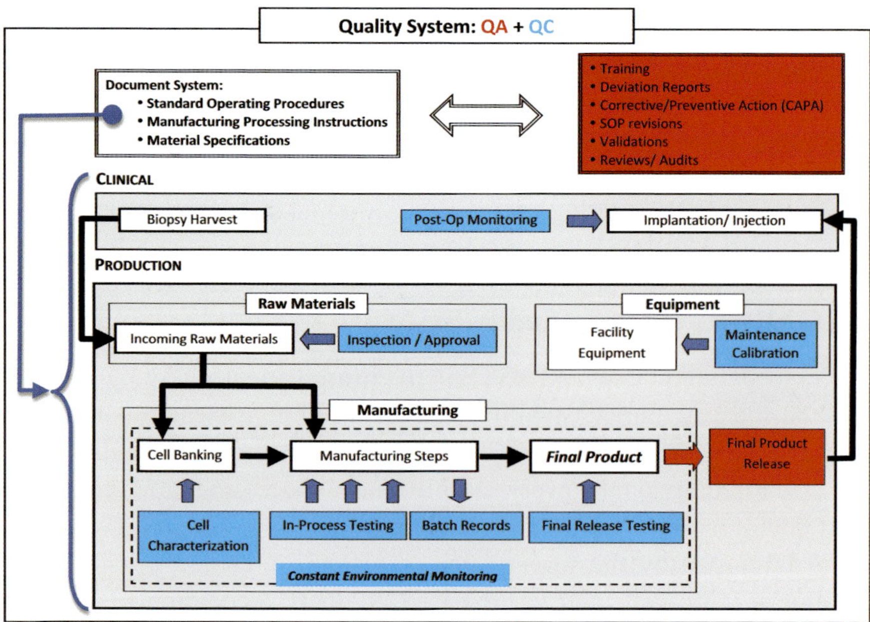

Fig. 1.

4. Core Requirements of a Quality Program

4.1 *Facilities*

(§ 1271.190(a) and (b)): Any facility used in the manufacture of cell therapy products must be of suitable size, construction, and location to prevent contamination of the product with communicable disease agents and to ensure orderly processing without mix-up. The facility must be designed so that each manufacturing step has a dedicated area, whose cleanliness matches the safety risk associated with that step. A cleaning program, supported by appropriate environmental monitoring, must be developed, implemented, and carefully documented. The flow of supplies, personnel, products, and wastes in the facility must be conceived so as to minimize the risk of contamination or cross-contamination of any of the products manufactured in the facility.

4.2 *Environmental control and monitoring*

(§ 1271.195(a)): Where environmental conditions could reasonably be expected to cause contamination or cross-contamination of the product or equipment, they must be adequately controlled. The following control activities or systems should be provided as needed: temperature and humidity controls, proper ventilation and air filtration, adequate cleaning and disinfection of rooms and equipment to ensure aseptic processing, and maintenance of equipment used to control conditions necessary for aseptic processing operations. Documents such as FDA's Guidance for Industry: Sterile Drug Products Produced by Aseptic Processing-cGMP[14] and the EU's Annex 1 (from Guidelines to GMP for Medicinal Products for Human Veterinary Use)[15] can be used to determine the appropriate level of control needed in each production area. This level will depend on factors such as the manufacturing steps involved, whether they are performed in an open or closed system, etc.

Monitoring of controlled environments must be performed regularly, but the appropriate frequency is defined by the manufacturer or investigator. Monitoring can include non-viable and viable particulate air monitoring, clean area positive pressure levels, as well as surface and personnel monitoring. USP <1116>[16] and ISO 14644-2[17] contain useful information to help determine the type, method, and frequency of environmental monitoring needed for any specific controlled environment.

Records of environmental control and monitoring activities must be maintained. It is important that relevant procedures define alert and action levels for environmental monitoring results and specify relevant corrective actions when these levels are exceeded.

4.3 *Equipment*

(§ 1271.200(a)): Equipment design, location, and installation should facilitate all operations, including those pertaining to maintenance and cleaning. Cleaning, sanitizing, and maintenance of all equipment should be performed according to well-defined procedures and schedules. All equipment used for inspecting, measuring, or testing must be demonstrated to be capable of producing valid results and calibrated according to established procedures and schedules. This includes, for example, equipment used to monitor the temperature of a storage unit or equipment used to measure air particle counts inside a controlled environment. Calibration accuracy should be in accordance with accepted standards (e.g. National Institute of Standards and Technology). Although § 1271.200 does not specifically require equipment qualification and certification, doing so is strongly recommended, particularly as part of process validation.[18–20] Equipment in which aseptic operations are performed, e.g. laminar flow hoods and biological safety cabinets, should be certified according to a strict schedule. Records of maintenance, cleaning, sanitization, calibration, certification, and use for each piece of equipment used during the manufacture of any batch of product should be kept on file.

4.4 *Supplies and reagents*

(§1271.210 (a) and (b)): All materials used during manufacturing, whether or not they come in direct contact with the processed product, are considered supplies or reagents. Examples of supplies include sterile drapes, gloves, pipettes, cell culture flasks, and shipping containers. Examples of reagents include culture medium, saline, antibiotic solutions, cleaning agents, and chemicals used in processing.

A system must be implemented that will ensure that supplies and reagents are not used until they have been verified to meet predetermined specifications. These specifications are designed to diminish the risk of introducing, transmitting, or spreading communicable diseases, and to prevent toxicity. The establishment using the supply or reagent can either verify that it meets the requisite specifications through relevant testing or by obtaining proper documentation from the manufacturer (typically a Certificate of Analysis for reagents or a Certificate of Compliance for supplies). Each supply or reagent must go though an approval process during which accompanying documentation is checked to verify that it corresponds to the predetermined specifications. Prior to inspection, items should be stored in a quarantined location according to manufacturer instructions. Accepted items must be moved to areas dedicated to the storage of approved items (1271.26(a–d)). They should then be stored and used according to the manufacturer's instructions. Reagents produced in-house should also undergo a verification process to ensure they meet the predetermined specifications (e.g. solutions are sterile, have the proper concentrations, and fall within a specific pH range). The water used throughout the manufacturing process (to rinse equipment or parts or prepare reagents) must be of appropriate cleanliness and, if produced in the facility, properly monitored. To limit the possibility of endotoxin contamination, water used for preparing reagents is usually Water for Injection, USP grade. Processes used to produce these reagents should be validated.

Keeping information data sheets listing specifications for all supplies and reagents is a good way to ensure consistency during the approval process. These sheets should be updated when products change and an archive of previously used data sheets should be kept. Under §1271.210 (d), records must be maintained for receipt of each supply or reagent, including the type, quantity, source, lot number, date of receipt, and expiration date. Records of the approval process for each supply or reagent, including tests results or Certificates of Analysis from the vendor, should be kept on file. Records of all lots of supplies or reagents used in the manufacture of each batch of product should also be kept.

4.5 *Manufacturing process validation*

The ability of an established manufacturing process to effectively and reproducibly provide intermediate and final products of predetermined specifications must be documented through adequate validations.[18,19] There are three possible validation types: prospective, concurrent, or retrospective. Prospective validation is the preferred approach and should be completed prior to commercial distribution of the final product. The number of process batches to be included in a validation is a function of the complexity of the manufacturing process. For prospective and concurrent validations, three consecutive, successful production batches should typically be used. More batches may be necessary to prove the consistency of a manufacturing process if it is particularly complex or requires prolonged completion times. For retrospective validations, data from 10 to 30 consecutive batches should generally be reviewed.

Critical parameters should be controlled and monitored during process validation studies. Such studies should include all critical process steps and be performed using qualified equipment and facilities according to a written validation protocol. The quality assurance team should regularly review validated processes. If the reviews confirm that the process consistently produces products meeting specifications, no re-validation is normally needed.

Additionally, the aseptic status of a manufacturing process should be validated through a medium fill simulation.[15,16,20] This is done by simulating each of the critical manufacturing steps using a growth medium (broth) that promotes the growth of bacteria and fungi instead of any liquid reagent regularly used. The broth is subsequently incubated and observed to detect any microbial growth. Whenever feasible, simulations should be performed at the end of a production shift to capture the worst-case scenario conditions in terms of microbial contamination risk. For the initial (start-up) validation, three medium fill simulations should typically be performed per manufacturing step. One simulation per manufacturing step can be performed during each subsequent validation. Subsequent validations should be performed once every six months.

5. Tailoring Quality Control to the Manufacturing Process

5.1 *Quality control: safety, consistency, and traceability*

A proper quality program should address critical safety issues specific to the various steps of the manufacturing process. This is easiest to achieve when the quality system framework is developed in parallel with the manufacturing process during the product development (a concept called "quality by design").[21] This is realized by thoroughly assessing the safety risk associated with each manufacturing step and by implementing specific risk control measures to reduce the identified risk to acceptable levels.[13] The adequacy of the risk control measures should be regularly reviewed as part of risk control management.[22] Importantly, the level of risk associated with the manufacturing process and the resulting final product must be communicated to all interested parties (including regulatory authorities and patients, when relevant), throughout the product lifecycle.[23]

To ensure consistency, proper ways to execute and control each manufacturing step must be described in written documents, called standard operating procedures (SOPs). It is also critical that personnel be regularly trained to follow SOPs. Furthermore, every step of the manufacturing process, as well as all results from quality control tests, should be documented in manufacturing batch records that undergo regular scrutiny by the quality assurance staff (21 CFR 211.188) and must be thoroughly reviewed as part of the process leading to the final product release (21 CFR 211.192). According to 21 CFR 211.188, batch records should include full traceability of patient material through all steps of the manufacturing process, including initial biopsy, final-product shipping, and eventual disposal. They should also contain the identification of all equipment, reagents, and supplies used during the manufacturing process. Personnel performing, supervising, or controlling any significant step of the process should be listed in the batch record. Any quality assurance investigation made during the process (according to 211.192) should also be documented.

To thoroughly control the safety and consistency of a manufacturing process, a quality control program should at least assess the following:

- starting material (cell or tissue) sourcing and raw materials qualifications,
- characteristics of pre-production cells,
- critical process steps through in-process control testing, and
- final product properties through final release testing.

5.2 *Cell and tissue sourcing*

Quality control of the manufacturing process starts with cell and tissue sourcing, also known as "collection." To prevent the transmission of communicable diseases from the cell donor to the recipient of the therapy product, determining donor eligibility, and donor screening and testing are all required by cGTP regulations (§ 1271.215). This is especially important for allogeneic cell therapy applications. For autologous applications, where the donor and the recipient are the same individuals, donor eligibility determination or donor screenings are not requested (21 CFR 1271.90(a)). However, if manufacturing procedures have the potential to increase the risk of propagating pathogenic agents that may be present in the donor, then the donor should be tested for these pathogens and the results should be documented. Typically, autologous cells and tissues are considered potentially infectious and manipulated as such.

Cross-contamination between cells of different patients is a key safety issue. To address this issue, a proper quality control program begins with specific, documented, and controlled labeling of the specimens harvested from the patient. For privacy issues, a unique patient identifier is assigned to each patient and recorded with his or her name on a log accessible only to the clinical staff. The labeling follows the patient material whenever it changes containers. Forms requiring a witness's signature or the use of a bar code labeling system can ensure stringent traceability from the initial biopsy to the final product. At the patient's bedside, the final product labeling is reconciled with the patient identity before implantation or injection. Very strict segregation is also required for different batches of cells or tissues being processed simultaneously. This can be achieved through dedication of specific equipment to specific batches as well as by thorough and validated cleaning of shared equipment before each use.

Unless processing occurs in a closed unit at the patient's bedside, which is unlikely for applications that involve extensive *ex vivo* manipulation, collected cells or tissues must be transported to the processing site, usually by shipping. There are many requirements to properly control the shipping of the samples. Each patient sample must be packaged in a dedicated, properly labeled, shipping container. Shipping container specifications must prevent the contamination or cross-contamination of the material shipped. The containers should be sterile, endotoxin-free, and leak-proof. In addition, the surrounding shipping package should provide environmental conditions (e.g. temperatures ranges) that ensure that the biological characteristics and viability of the material shipped will be properly preserved. Both the shipping containers and the package should comply with the Pressure Differential and Thermal Shock requirements outlined in the Department of Transportation's Title 49 CFR (§173.196 (a) (6) and (a)(7))[24] and International Air Transport Association (IATA) regulations, and be validated to maintain their specifications in any shipping environment. Optimally, environmental conditions inside the package should be recorded during the whole shipping process and documented.

5.3 *Cell culture*

Cell extraction is usually the next step of the manufacturing process. Extraction typically requires the use of enzymes of animal origin, like collagenase or trypsin. Generally speaking, all raw and ancillary biological material of animal or human origin used in any manufacturing step, including the enzymes necessary for the extraction phase, should be demonstrated to be free of adventitious agents, including bacteria and fungi, mycoplasmas, mycobacteria, and viruses.[25]

Processes used to remove or inactivate potential infectious contaminants from biological raw materials should be validated. Sourcing, donor screening data (for human sourced-reagents[26]), and the results of adventitious agents testing must be fully documented. Useful recommendations regarding the testing of all materials of human and animal origin (e.g. enzymes, adhesion factors, albumin, growth factors, etc.) can be found in references 25 and 26. The type of testing necessary depends on the animal species from which the materials are sourced. Because of the risk of bovine spongiform encephalopathy transmission, reagents of bovine origin are of special concern.[27,28] Interacting with regulatory agencies early during the development of the manufacturing process is valuable for identifying questionable reagents and either finding alternative reagents or determining appropriate actions to minimize the risks associated with those reagents.

The composition and purity of raw materials of non-biological origin should also be thoroughly documented. It is very advantageous to use materials that are already licensed for human use.

Whether extracted cells are expanded, banked (stored frozen), or immediately used for further production, it is important and required that pre-production cells (or bulk cells) be characterized (see 21 CFR 610 for relevant required tests).[29] The objective of this characterization is to verify that the cells are suitable for further production. Proper bulk-cell characteristics increase the likelihood of manufacturing a final product with adequate biological properties.

Generally, bulk cell identity, purity, viability, and microbial safety must be assessed.[30,31] However, the consequences that a drift in the bulk cells' characteristics might have on the properties of the final product must be evaluated. Results from this risk evaluation are necessary to determine the extent of necessary cell characterization and will dictate whether additional testing might be necessary (e.g. tumorogenicity, etc). Therefore, tests should be established during the development of the product to help identify which cell characteristics are necessary for the production of an adequate product. Based upon historical data, a set of acceptance criteria can then be determined for pre-production cells. Those criteria are subsequently used as a means to control the quality of these cells.

In many instances, extracted and expanded cells are banked. Most often, banking cells for cell therapy applications involves establishing a master bank. Working banks are then created by expanding cells from the master bank and then using them to manufacture the product. For autologous applications, however, the number of cells can be limited, and it may not be feasible to create a master bank. Regardless of the cell banking strategy, viability and phenotypic stability of the cells during cryostorage should be established.

Assessing microbial safety of pre-production (or bulk) cells entails testing for sterility (absence of bacteria, yeast, and fungi per USP <71>),[32] endotoxin amount (per USP <85>),[33] and mycoplasma (per FDA's "Points to Consider in the Characterization of Cell Lines Used to Produce Biologicals").[34] To prevent spreading microbial contaminations to other patient cells, sterility and mycoplasma should be tested before the cells are transferred to liquid nitrogen storage containers for cryopreservation. In liquid nitrogen storage, viruses (hepatitis viruses) can spread from one sample to another.[35,36] Therefore, vapor-phase nitrogen storage is preferred. Note that testing for the presence of specific human viruses, including CMV, HIV-1 and -2, HTLV-1 and -2, EBV, B19, HBV, and HCV, is not required for autologous cell banks. However, for cells that have been exposed to bovine or porcine components (e.g. serum, serum components, trypsin), it is appropriate to test for bovine and/or porcine adventitious agent.[37]

5.4 *In-process testing*

Quality control tests should be performed along each critical step of the manufacturing process through in-process control testing. During the manufacturing of cell-based products, microorganisms and impurities can be introduced from multiple sources, including components of biological and non-biological sources, poorly controlled process steps, or failures in aseptic processing techniques. To thoroughly control the manufacturing-process quality, each step of the process must be assessed to identify any potential routes of microbial contamination or impurities. This type of risk assessment should take into consideration different parameters, such as the following:

- materials sources,
- level of manipulation (e.g. multiple or few manual processes, automated processes, no manipulation), and the processing path (e.g. open, partially closed, closed, sealed),
- cell culture or incubation duration (longer times are riskier),
- cell culture vessel (e.g. open (flask, tube, dish), partially closed (sterile collection bag), or closed (bioreactor) system).

Results from this risk assessment should be used to determine critical time points for testing, the most appropriate samples to harvest, and the tests to be performed at these times. For example, in-process sterility testing might be performed during an extended period of culture or after manufacturing steps involving manual processing.

Similarly, it is advisable to test an intermediary-step product for potency (i.e. measuring the product characteristics and biological activity that contribute to its function) or identity immediately after a manufacturing step that is designed to (or that could) modify those parameters, e.g. cell expansion, differentiation, gene modification, etc.

Typically, in-process materials should be tested for sterility, endotoxins, identity, strength (concentration and potency), quality, and purity, as appropriate. The test methods used for in-process control are at the discretion of the manufacturer. However, regulatory authorities will review the testing procedures, the justification of their choice, their limits of acceptance, and the rationale for these limits. Most importantly, valid in-process specifications for such characteristics should be consistent with the final product specifications. They should be derived from previous process averages and variability estimates, where possible, and determined by the application of suitable statistical procedures, where appropriate. In-process control tests are the responsibility of the quality control team and their results are part of the batch records. The quality assurance team will audit them as part of the final-product release.[8]

As specified in 21 CFR 211.10, written procedures (SOPs) detailing each manufacturing step must be established and followed carefully at all times. They should include the description of the in-process controls, the tests or examinations to be conducted, and the appropriate samples of in-process materials to be tested for each batch. In-process control procedures are useful in monitoring the output and validating the performance of the manufacturing steps that may be responsible for causing variability in the characteristics of in-process materials and the final product. Performing such tests during the manufacturing process allows for the early identification and rejection of compromised batches of products downstream and immediate investigation of the integrity of the manufacturing step.

5.5 *Final release testing*

At the end of the manufacturing process, according to 21 CFR 211.165, each batch of products to be released should be appropriately tested to determine satisfactory conformance to final specifications, including identity, strength (or potency), purity, and freedom from contaminating microorganisms. For aseptic products, sterility and absence of endotoxins are essential qualities (21 CFR 211.167). For products whose biological properties depend on the presence of specific viable cells, testing for viability and purity of the cell population is also critical. Sampling and testing plans for final release should be established and described in written procedures that should include the method of sampling and the number of units per batch that will be tested. For each test, or each specification, the acceptable quality level must be set based on a well-documented record of developmental data. The selection of these levels, also called acceptance criteria, is very important since they will be used in making the decision to accept or reject batches of final products. Typically, acceptance criteria are revised and refined as a historical data record is established; however, they cannot be changed to justify the acceptance of a batch already produced (e.g. during a clinical trial).

The quality control team ensures that batches of products meet each acceptance criteria necessary for their approval and release. The accuracy, sensitivity, specificity, and reproducibility of the test methods employed must be established and documented. Such validation and documentation must be in accordance with 21 CFR 211.194 (a)(2). For cell- and tissue-based therapy products, the tests necessary for final release are listed in Table 4. Note that cell therapy products are exempt from the General Safety Test requirement (per 21 CFR 610.11).

Significant challenges are associated with the final release of cell- and tissue-based products. One of the challenges for small-sized autologous cell therapy batches is that the sampling size might be different from those specified in the regulations, which were originally developed for large-sized batches of pharmaceutical drugs. Acceptable sampling size and strategy will have to be negotiated on a case-by-case basis with the appropriate regulatory authorities. Other challenges stem from the fact that most cell-based products are manufactured using aseptic manipulations because they cannot undergo sterile filtration or terminal sterilization. This makes the assessment of sterility, absence of endotoxins, and freedom from mycoplasma paramount. However, many cell-based products cannot be cryopreserved or otherwise stored without affecting viability and potency. Thus, their shelf life is usually shorter than the

Table 4.   Tests required for final release.

Required Test	Relevant Biologics Standard	Test Method
Potency	21 CFR 610.10	Not specified
Sterility	21 CFR 610.12	Not specified
Purity (pyrogenicity)	21 CFR 610.13	Specified
Identity	21 CFR 610.14	Not specified
Constituents materials	21 CFR 610.15	Not specified
Mycoplasma*	21 CFR 610.30	Specified
Communicable diseases	21 CFR 610.40	Not specified
Viability		Not specified
Phenotype		Not specified

*Only required for cells that are cultured.

standard test read-out times. Test durations for sterility and mycoplasma testing (per CFR 610.12) had previously been standardized as 14 and 28 days, respectively.[29,34]

Until recently, in the absence of rapid and effective microbiological testing that avoided administrating the final product before the final sterility test results became available, the FDA recommended a program tailored specifically for these products.[38] This program combined standard microbiological testing (sterility, endotoxin, and mycoplasma) initiated a few days (48 to 72 hours) before the final release of the products with a Gram staining performed at the time of shipping or implantation. Another microbiological sample was also harvested at the time the final product was shipped. No-growth results from the 48- to 72-hour sterility test and the negative Gram stain were used for pre-release criteria, and results from the last harvested sample were used for final release.

During the last decade, alternate microbiological tests have been developed and are now available for cell therapy products with a limited shelf life. These rapid microbiological methods are designed to yield accurate and reliable results in less time than the previously standardized method. Recent amendments to sterility test requirements for biological products[39] promote the use of alternative tests with rapid and advanced detection capabilities, but specify that these tests must be validated, in accordance with an established protocol, as required for most methods of final product testing under 21 CFR 211.165(e) and 211.94(a)(2). The validation should demonstrate that the test is appropriate for the product and capable of consistently detecting the presence of viable microorganisms. Principles of validation specific to rapid growth-based microbiological methods were described in guidelines that the FDA withdrew in 2013. Helpful guidance for conducting such validation studies can still be found in USP General Chapter 1223,[40] as well as in the ICH guideline Validation of Analytical Procedures: Text and Methodology Q2(R1).[41] Rapid mycoplasma detection techniques using polymerase chain reaction (PCR) have been developed as well. Prior to licensing, data demonstrating that the PCR test selected has a sensitivity and specificity similar or greater than the standard test must be provided.

Guaranteeing final product safety by assessing the potency of the final product constitutes another huge challenge for autologous biological products. Potency of an autologous product often varies from one patient to another. Live cell- and tissue-based therapies frequently rely on the activity of cell populations that are not 100% pure or on the combined activities of several cell populations that can have complex mechanisms of action. The characteristics of autologous cell therapy products that make the development of potency tests difficult are described in Table 5.[42] Historical developmental data, relevant pre-clinical information, and early clinical study results can be used to determine which product attributes are most relevant for measuring potency.

For products that contain more than one active ingredient, potency measurements should be designed to assess the biological activity of all active ingredients. The potential for interference and synergy between active ingredients should also be considered. A collection of assays (biological and/or analytical assays) can be developed for products with complex mechanisms of action or presenting multiple biological activities. At least one quantitative assay must be included in the combination. Analytical assays can evaluate immunochemical, biomechanical, and molecular attributes of the products. Their data should be correlated with a relevant product-specific biological activity.

To market a biological product, a validated potency assay with defined acceptance criteria must be described in the BLA (21 CFR 211.165(e)). The acceptance criteria, intended for subsequent final release testing,

Table 5. Autologous cell- and tissue-based product characteristics that make the development of potency test difficult.

Challenges to Potency Assay Development	Examples
Inherent variability of starting materials	• Donor variability • Cell line heterogeneity
Limited lot size and limited material for testing	• Single-dose therapy using cells suspended in a small volume • One single lot/patient
Limited stability	• Viability of cellular products
Lack of appropriate reference standards	• Autologous cellular materials
Multiple active ingredients	• Multiple cell lines combined in final product • Heterogeneous mixtures of immune-modulating cells
Potential for interference or synergy between active ingredients	• Multiple cell types in cell pre paration
Complex mechanism of action(s)	• Multiple characteristics required for appropriate functions (e.g. biomechanical resistance and anti-thrombogenic properties for an autologous tissue-engineered blood vessel) • Multiple potential effector functions of cells
In vivo fate of product	• Migration from site of administration • Cellular differentiation • Remodeling of tissue-engineered construct • Invasion by recipient's own cells • Immune response to components introduced during manufacturing process

Modified from Guidance for Industry. Potency Tests for Cellular and Gene Therapy Products, CBER, 2008.

should be based on knowledge gained through manufacturing experience and data collected during all phases of product development and clinical investigation. Reference material used as controls for the assays should be available and can include well-characterized clinical lots. Several resources are available for analytical method validations.[40,41,43,44]

5.6 *Shipping the final product*

After final release, biological products are either stored or shipped to the patient's bedside. Requirements similar to the ones pertaining to biopsy shipping, labeling, and packaging, apply to the final product shipping. Supplementary labeling (e.g. "For autologous use only," "Not evaluated for infectious diseases," "Warning: advise recipient of communicable disease risk," "Biohazard," etc.) might also be required.[45] In addition, under 21 CFR 211.166, an adequate number of batches must be tested to determine the product stability under specific storage and shipping conditions. The results of such stability testing are used in determining appropriate storage and shipping conditions as well as in establishing the product shelf life. Stability of such products should be assessed in the worst shipping conditions possible (duration, temperature) in the same container system in which the product will be marketed. For further information, refer to ICH 5C, Q1A(R2) and Q1E.[46–48] Typically, living autologous cell- and tissue-based therapy products will have a short shelf life.

6. Conclusion

Cell- and tissue-based therapies, especially those requiring extensive, and sometimes lengthy, *ex vivo* cell manipulation, present an additional challenge for both regulatory authorities and the companies developing them. The challenge lies in the necessity to strike a balance between the requirements needed by the QC officials to ensure product safety and the company's ability to comply. More specifically, the extent and number of the quality control steps, which tend to multiply with the complexity and length of the manufacturing process, need to guarantee patient safety, but should not prevent a much-needed treatment from becoming clinically available. This is especially true for autologous products where each manufacturing batch is produced for a specific patient. In these cases, the QC samples-to-patient ratio become extremely high compared to what is demanded of conventional medicinal product processes. The segregation of batches is also a key and costly issue specific of autologous therapies. Although many autologous somatic cell therapies have been developed so far, only a few are commercially available. In the United States, only two autologous cell therapy products requiring extensive *ex vivo* manipulation and regulated through the BLA pathway, have reached the market in 2010. Carticel®, produced by Genzyme Tissue Repair, became available in August 1997 and Provenge®, from Dendreon Corporation, in April 2010. The rate of approval has only slightly accelerated since then. Azficel, from Fibrocell Science Inc., was approved in June 2011, and MACI™, by Vericel Corporation, in December 2016. Obviously, over the last two decades, some significant discrepancy has existed between the rapid development of autologous cell- and tissue-based therapies and their slow access to the market. This has pushed various companies to develop and test new applications outside of the United States. Hopefully, initiatives such as the new Regenerative Medicine Advanced Therapy Designation, an expedited FDA program included in the 21st Century Cures Act, signed into law by the US Congress in December 2016, will help increase the rate of approval for such therapies so that more US patients can safely benefit from them.

Acknowledgements

The authors thank Marissa Peck and Corey Iyican for their help in revising and editing this chapter.

References

1. E. D. Thomas, H. L. Lochte, J. H. Cannon, O. D. Sahler and J.W. Ferrebee, Supralethal whole body irradiation and isologous marrow transplantation in man, *J Clin Invest* **38**: 1709–1716 (1959).
2. C. D. Buckner, R. B. Epstein, R. H. Rudolph, R. A. Clift, R. Storb and E. D. Thomas, Allogeneic marrow engraftment following whole body irradiation in a patient with leukemia, *Blood* **35**: 741–750 (1970).
3. E. Gluckman, H. A Broxmeyer, A. D. Auerbach, H. S. Friedman, G. W. Douglas, A. Devergie, H. Esperou, D. Thierry, G. Socie, P. Lehn, et al. Hematopoietic reconstitution in a patient with Fanconi's anemia by means of umbilical-cord blood from an HLA-identical sibling, *N Engl J Med* **321**: 1174–1178 (1989).
4. CBER/FDA. Guidance for Industry: Guidance for Human Somatic Cell Therapy and Gene Therapy (1998).
5. 21 CFR 1271: Human Cells, Tissues, and Cellular and Tissue-Based Products (2016).

6. 42 USC § 262: Regulation of Biological Products.

7. 21 CFR 210: Current Good Manufacturing Practice in Manufacturing, Processing, Packing or Holding of Drugs.

8. 21 CFR 211: General and Current Good Manufacturing Practice for Finished Pharmaceuticals.

9. Regulation (EC) No 1394/2007 of the European Parliament and of the Council on Advanced Therapy Medicinal Products and amending Directive 2001/83/EC and Regulation (EC) No 726/2004. *Official Journal of the European Union*, L 324/121–137 (2007).

10. Commission Directive 2003/94/EC laying down the principles and guidelines of good manufacturing practice in respect of medicinal products for human use and investigational medicinal products for human use. *Official Journal of the European Union*, L 262/22–26 (2003).

11. Directive 2001/83/EC of the European Parliament and of the Council of Six on the Community Code Relating to Medicinal Products for Human Use. *Official Journal of the European Union*, L 311/67–128 (2001). Consolidated version published in 2012.

12. European Commission. EU Guidelines to Good Manufacturing Practices for Medical Products for Human and Veterinary Use. Eudralex: The Rules Governing Medicinal Products in the European Union, Volume 4.

13. EMEA/CHMP. Guideline on Human Cell-Based Medicinal Products (Doc. Ref EMEA/CHMP/410869/2006) (2008).

14. CBER/CDER/FDA. Guidance for Industry: Sterile Drug Products Produced by Aseptic Processing — Current Good Manufacturing Practices (2004).

15. European Commission. Annex 1 of the EU Guidelines to Good Manufacturing Practices for Medical Products for Human and Veterinary Use. Eudralex: The Rules Governing Medicinal Products in the European Union, Volume 4 (2008).

16. United States Pharmocopeia <1116>: Microbiological Evaluation of Clean Rooms and other Controlled Environments. In: *USP 36-NF 31* (2013).

17. ISO. International Standard ISO 14644-2: Cleanrooms and Associated Controlled Environments: Part 2: Specifications for Testing and Monitoring to Prove Continued Compliance with ISO 14644-1 (2015).

18. CBER/FDA. Guidance for Industry: Analytical Procedures and Methods Validation Chemistry, Manufacturing, and Controls Documentation (2015).

19. ICH Q7A: Good Manufacturing Practice Guide for Active Pharmaceutical Ingredients (2016).

20. Pharmaceutical Inspection Convention and Pharmaceutical Inspection Co-Operation Scheme (PIC/S). Recommendation on the Validation of Aseptic Processes (2017).

21. ANSI/ISO/ASQ Q9000-2000: Quality Management Systems: Fundamentals and Vocabulary (2000).
22. EMEA/CHMP Guideline on Risk Management System for Medicinal Products for Human Use. EMEA/CHMP/96268/2005 (2005).
23. ICH Q9: Quality Risk Management (2006).
24. 49 CFR 173.196: General Requirements for Shipments and Packages: Infectious Substances (2009).
25. CBER/FDA. Guidance for Industry: Characterization and Qualification of Cell Substrates and Other Biological Materials Used in the Production of Viral Vaccines for Infectious Diseases (2010).
26. 21 CFR 640: Additional Standards for Human Blood and Blood Derivatives.
27. FDA. 72 FR 1581: Use of Materials Derived from Cattle in Medical Products Intended for Use in Humans and Drugs Intended for Use in Ruminants (2007).
28. EMEA/CHMP Guideline on the Use of Bovine Serum in the Manufacture of Human Biological Medicinal Products. EMA/CHMP/BWP/457920/2012 rev 1 (2013).
29. 21 CFR 610: General Biological Products Standards.
30. CBER/FDA. 61 CFR 26523: Guidance on Applications for Products Comprised of Living Autologous Cells Manipulated *Ex Vivo* and Intended for Structural Repair or Reconstruction (1996).
31. ICH Q5D: Derivation and Characterization of Cell Substrates Used for Production of Biotechnological/Biological products (1998).
32. United States Pharmacopoeia <71>: Sterility Testing. In: *USP 32-NF 27* (2009).
33. United States Pharmacopoeia <85>: Bacterial Endotoxin Testing. In: *USP 32-NF 27* (2009).
34. CBER/FDA. Points to Consider in the Characterization of Cell Lines Used to Produce Biologics (1993).
35. R. S. Tedder, M. A. Zuckerman, A. H. Goldstone, A. E. Hawkins, A. Fielding, E. M. Briggs, D. Irwin, S. Blair, A. M. Gorman, K. G. Patterson, *et al.* Hepatitis B transmission from contaminated cryopreservation tank, *Lancet* **346**: 137–140 (1995).
36. A. E. Hawkins, M. A. Zuckerman, M. Briggs, R. J. Gilson, A. H. Goldstone, N. S. Brink, *et al.* Hepatitis B nucleotide sequence analysis: linking an outbreak of acute hepatitis B to contamination of a cryopreservation tank, *J Virol Methods* **60**: 81–88 (1995).
37. 9 CFR 113: Detection of Extraneous Viruses by the Fluorescent Antibody Technique: Animals and Animal Products.

38. CBER/FDA. Guidance for FDA Reviewers and Sponsors. Content and Review of Chemistry, Manufacturing, and Control Information for Human Somatic Cell Therapy Investigational New Drug Applications (2008).

39. 21 CFR Parts 600, 610, and 680: Amendments to Sterility Tests Requirements for Biological Products (2012).

40. United States Pharmacopoeia <1223>: Validation of Alternative Microbiological Methods. In: *USP 38-NF 33* (2015).

41. ICH Q2 (R1): Validation of Analytical Procedures: Text and Methodology (2005).

42. CBER/FDA. Guidance for Industry: Potency Tests for Cellular and Gene Therapy Products (2011).

43. CBER/FDA. Guidance for Industry: Analytical Procedures and Methods Validation Chemistry, Manufacturing, and Controls Documentation (2015).

44. United States Pharmacopoeia <1225>: Validation of Compendial Procedures. In: *USP 32-NF 27* (2009).

45. CBER/FDA. Guidance for Industry. Eligibility Determination for Donors of Human Cells, Tissues, and Cellular and Tissue-Based Products (HCT/Ps) (2007).

46. ICH Q5C: Quality of Biotechnological Products: Stability Testing of Biotechnological/Biological Products (1996).

47. ICH Q1A (R2): Stability Testing of New Drug Substances and Products (2003).

48. ICH Q1E: Evaluation of Stability Data (2004).

21

REGULATORY CHALLENGES FOR CELL-BASED THERAPEUTICS

Todd McAllister, Corey Iyican and Nicolas L'Heureux

1. Introduction

Interactions with the Food and Drug Administration (FDA) or other regulatory agencies typically conjure images of a bureaucratic quagmire at best, or an adversarial fight at worst. In all fairness, these regulatory bodies are tasked with an incredibly challenging situation: trying to balance the mutually exclusive demands for a "least burdensome process" from industry with a "zero risk" expectation from the patients and overly eager litigators behind them. Industry is quick to criticize the FDA for its lengthy approach to review and approval, and the lay press is quick to chastise the agency over rare clinical debacles such as those associated with Vioxx (Merck) or TBN 1412 (TeGenero). Few people, however, laud these same regulatory agencies for presiding over nearly 50 years of medical innovations in the modern era of drug and device regulation.

Indeed, in the post-thalidomide era, the average life expectancy has increased from roughly 70 years in 1962 to nearly 80 years in 2003.[1] Moreover, the quality of life has increased dramatically for the aged population. In effect, we can thank the FDA for safely adding more than two hours to each 24-hour day that we live.

While there have been some dramatic failures of drugs and devices, none in the modern era of regulation have approached the scope of the thalidomide disaster. When viewed in light of the billions of devices and drug doses that have been safely administered, the overall performance of the worldwide regulatory bodies is commendable. Safely managing the transition of medical innovations from bench-top research to clinical use has driven this improvement in longevity and quality of life. This is even more remarkable in light of the funding disparity between the regulatory agencies and the scientific communities. Fueled by an annual budget of approximately \$30 billion from the National Institutes of Health (NIH) and billions more from industry, medical innovation has leapt forward at unfathomable rates. The FDA is understandably hard pressed to keep pace with this innovation on a budget of just over \$2 billion annually. Consider that in the early 1960s, which many consider to be the dawn of the modern era of drug and device regulation, the structure of DNA was still unknown, catheter-based cardiovascular interventions were essentially non-existent, and drug therapies for cancer and cardiovascular disease were in their infancies. Thus, a realistic and sympathetic approach should be sustained when setting expectations for regulatory agencies to keep pace with innovations such as gene therapies or regenerative medicine.

While it is appropriate to acknowledge the unenviable tasks assigned to the FDA and other regulatory agencies, it is clear that there are unnecessary burdens in each system, particularly with respect to new therapies like regenerative medicine and cell-based therapeutics in particular. Indeed, there are dramatic regional disparities in the data required to support submission, the tolerance of risk, and the subsequent time required to receive the study approvals. Additionally, there are significant differences in costs associated with clinical studies and access to patient populations. In this chapter we will explore some of these regional differences and provide insights into more efficient management of the regulatory processes irrespective of location. As evidenced by the dramatic increase in the number of clinical trials performed outside countries like the United States, France, Germany, and the United Kingdom, it is clear that developing a cohesive global clinical trials strategy has become a critical developmental priority for a burgeoning regenerative medicine company.[2]

Specifically, in this chapter we will address the following:

- Key regulatory considerations for each clinical trial phase.
- Strategies that target certain geographical regions for specific developmental advantages.
- Use of a clinical research organization (CRO).
- Universal regulatory considerations for cell-based therapeutics.

2. Regulatory Challenges at Each Phase of Clinical Development

2.1 *Clinical trials design considerations*

While it may be obvious, it should be stated at the outset that a well-thought-out clinical trials strategy is perhaps the most important element to consider in the process of transitioning to human use. It is difficult to make sweeping recommendations relating to protocol design, as each trial will be driven by its own particular requirements. It is worth emphasizing, however, that the appropriate selection of endpoints can make the difference between approval and rejection of the protocol from a regulatory perspective, and more broadly, can make the difference between perceived success and perceived failure in a clinical study. While the perception of success or failure from a clinical perspective is beyond the scope of this chapter, one should realize that setting aggressive endpoints can result in adverse interpretations from the Data Safety Monitoring Board (DSMB) or the regulatory agencies once the data is reviewed. For example, in our own Phase I/II trials with hemodialysis access, a cell-based vascular graft was used in a subset of "worst-case scenario" patients. This type of protocol is quite common with radically new technologies. Data relating to the standard of care, typically reported across a much broader patient population, predicts a life expectancy of nearly six years. So in the context of the standard of care, it is tempting to set Phase II endpoints at 12 or even 18 months, as is typically done with the standard patient population. However, with the short life expectancy unique to our subset of patients (one to two years), we would have been unlikely to meet these endpoints. This same argument can apply to Phase I studies in regenerative medicine, as often there are secondary endpoints that relate to longer-term efficacy. Regenerative medicine trials rely upon complex biological mechanisms, and it is difficult to accurately predict the outcomes and the most appropriate endpoints. To minimize the likelihood of studies being suspended due to outcomes that are unexpected or difficult to interpret, one strategy we have successfully employed is to define clear failure criteria and clear continuation criteria in the original protocol. By negotiating failure and continuation criteria upfront, there is less pressure on the DSMB or regulatory agencies to suspend a study that has an unexpected result.

While there is no question that the regulatory challenges facing the clinical development of cell-based therapeutics are onerous, there are many shortcuts that can, and should, be made to streamline the overall approval process. It is important to understand that philosophically, the FDA is completely risk averse and depends largely upon a strategy of exhaustive data collection to demonstrate a thorough review. While sponsors tend to shy away from direct confrontation with the FDA or similarly established regulatory agencies, these agencies, particularly in Europe, are actually quite responsive to rational arguments against unnecessary data collection. So while the guidance documents may encourage detailed, standardized testing pulled from every conceivable precedent, many of these tests can be eliminated from lot release criteria (or at a minimum be done a single time as a validation study) if there is supporting rationale. Moreover, as detailed below, some experimental "requirements" can be delayed until Phase II or even Phase III studies are to be performed.

The classic lines defining Phase I, II, and III trials are often blurred in regenerative medicine. While some simple cell-based injections (myocardial regeneration, for example) may require a classic Phase I dosing study, even these initial safety studies have efficacy endpoints. Thus, we tend to call our initial studies Phase I/II studies, which are clearly distinct from controlled Phase III studies. So while the study phases may be defined differently for specific applications, the recommendations below may accelerate clinical studies. These strategies, however, should be considered as a part of a detailed risk analysis to ensure that these "shortcuts" do not appreciably compromise patient safety.

2.2 *Phase I/II clinical testing*

Most companies do not fully appreciate or take advantage of the relaxed manufacturing requirements for Phase I trials. While the FDA does not encourage "garage" manufacturing during pre-IND (Investigational New Drug) meetings, the guidance documents very clearly establish that there is no requirement for Good Manufacturing Principles (GMP)/Good Tissue Practices (GTP) manufacturing at this stage of clinical development.[3] This is not the case for Europe, as GMPs are required for products manufactured for human clinical trials.[4] However, even in Europe, the regulatory bodies tend to offer some leeway in co-developing a company's various GMP processes early on, especially if it is a small and medium-sized enterprise (SME) and/or the company does not yet have a product on the market. Given the fact that early-stage trials are extraordinarily expensive (that is, early valuations are low, making early-stage expenditures disproportionately "expensive" from a capitalization perspective), it makes little sense to implement overly extensive quality assurance/quality control (QA/QC) procedures for Phase I trials. For most Phase I cell-based therapeutics trials, the key manufacturing elements that affect patient safety (functionality, uniformity, sterility, avoidance of cross-contamination, etc.) can be consistently performed in any reasonably organized university lab, as long as the appropriate release criteria are employed. Relatively basic documentation of the key manufacturing steps usually suffices for initial studies. These basic systems can then be filled in as the study progresses with more detailed procedures and hardware. Indeed, our own approach to Phase I clinical trials was to establish GMP-level manufacturing controls for the key, high-risk steps in the manufacturing process, as well as for those repeated frequently, such as change of culture medium. This more focused GMP effort significantly augments product safety without unnecessary early-stage expenditures. Additional safeguards, such as stringent functional release criteria, conservative enrollment rates, and long post-treatment surveillance periods, were employed. We felt that these safeguards probably did more for overall patient safety in Phase I trials than the more costly efforts later required to bring the full production process up to GMP standards. Our experience with eventually building a GMP-compliant manufacturing facility was that 80% of the cost was applied to the final 20% of the facility. This final 20% focused on relatively obscure details that had little impact on the overall risk analysis for our small, early-stage trials. This recommendation of a streamlined QA/QC process should not be confused with a poorly defined manufacturing process. It should be noted that even minor changes to the manufacturing process itself can result in the necessity to repeat Phase I studies (or similarly perform a validation study to demonstrate equivalence between the processes). Therefore, an important strategy in Phase I trials is therefore to make every attempt to finalize and document the details of the manufacturing process. Similarly, *in vitro* functional assays should be developed that can be used in lieu of clinical validation studies to demonstrate equivalence in the likely event that process changes are made as the technology evolves through Phase II and Phase III studies.

Some aspects of manufacturing controls or lot release criteria are deemphasized in regions with emerging clinical programs. It is tempting to take advantage of these relaxed requirements in a globalized clinical trials strategy. However, since FDA or Western European reviewers will analyze the enrollment and manufacturing safeguards in place for earlier studies before accepting human data, it is essential, to design the clinical study as if it were being conducted in Western Europe or the United States. The principal advantage to a globalized clinical trials approach is primarily associated with a quicker process for receiving study approval and decreased study costs, as discussed later in this chapter.

A final point worthy of mention for Phase I studies is the use of clinical research organizations (CROs). The common perception is that regulators require the use of CROs. These entities may seem omnipresent in all stages of the clinical trials process, but careful review of the guidance documents in most countries reveals that while outside, independent monitoring is encouraged, it is not required. While the CROs do provide convenience and some guidance, ultimately, the use of a CRO becomes an important issue when cost is factored in. Even in the relatively inexpensive regions, CROs seem to have uniformly elevated their fees to American levels. Most clinical cost estimates we have seen show that more than 50%

of the *overall* clinical costs for all studies are associated with regulatory submissions, legal representation, and clinical monitoring performed by CROs. Moreover, budget overruns associated with trials outside the United States and Western Europe are almost always associated with the CRO and not the clinical study itself. While additional discussion on this topic can be found later in the chapter, overall, we recommend careful cost analysis and vigilant contractual navigation when partnering with a CRO.

2.3 *Phase II/III clinical testing*

Transitioning to Phase II/III trials brings with it the responsibility of full GMP and GTP manufacturing compliance and all of the associated costs.[3] While the requirements for GMP manufacturing are discussed elsewhere in this book, there are two key elements to Phase II/III clinical trials that should be carefully considered: endpoints that demonstrate geographic distribution and cost-effectiveness.

2.3.1 *Geographic distribution*

There is clear pressure from investors to gain early access to the large US and EU markets, regions that are also the most likely to support the price premiums required from cell-based therapeutics. While this makes sense for classic drug and device business models, the model is less relevant for cell-based therapies, where the overall sales volumes are expected to be low for the near future. Even the most successful commercial regenerative medicine products, such as Organogenesis' Apligraf®, Advanced BioHealing's Dermagraft®, or Dendreon's Provenge sold to a very small fraction of the wound-healing population. There are no cell-based therapies that sell in sufficient volumes to claim a penetration rate of more than a few percentage points. Said another way, even these icons of industrial success, which are both clinically efficacious and profitable, sell little more than a few tens of thousands of units per year. It should also be noted that all three of these examples of early commercial success have failed to scale to blockbuster status. Changes in the reimbursement status for woundcare and the subsequent arrival of amniotic membranes essentially killed the tissue-engineered woundcare field. Similarly, high production costs coupled with reduced reimbursement effectively bankrupted Dendreon. Taken together, the sales volumes afforded by the US marketplace have not materialized for any cell-based product. Thus it would seem that early access to the US market might be a dated concept. Moreover, current manufacturing capacities by most regenerative medicine companies cannot service significant penetration in these largest markets. Thus, one can make the argument that a slow-growth business model, even in a smaller market, could provide the basis for a successful outcome. Our own clinical trials strategy is to continue with Phase II/III clinical studies in Eastern Europe and South America to lower the overall cost, and expand to a single Western EU country to facilitate eventual commercialization in Europe. For our clinical indication, which addresses an extremely large patient population, even modest penetration into a single country like Germany would represent a major commercial success for a cell-based therapy and would likely exceed our current manufacturing capacity. Thus, the typical strategy of concurrent regulatory and commercial activities in both the United States and Western Europe should be carefully evaluated.

2.3.2 *Cost-effectiveness*

While clearly outside the scope of regulatory considerations, clinical trials design should also be mindful of the increasing pressure to demonstrate cost-effectiveness. While Medicare policy does not acknowledge the requirement for demonstrating cost-effectiveness, reimbursement for cell-based therapeutics will *absolutely* require convincing cost-effectiveness data. Indirect arguments associated with technology assessment, quality adjusted life years, etc., are meaningless unless there is a clear cost benefit relative to the standard of care. Treatments for life- or limb-threatening diseases or orphan diseases, where few or no clinical options exist, may be immune from this rule (certain cancer vaccines, for example) but in general, our industry will be required to demonstrate that our treatments can be cost-effective. This realization is critical for clinical trials design, as requirements from the Centers for Medicare and Medicaid Services (CMS), and not FDA requirements, may be the real drivers of enrollment numbers and endpoints. In other words, while a trial can demonstrate safety and efficacy to the FDA, a more successful trial will have been designed to provide key elements to argue cost-effectiveness to CMS.

3. Regional Considerations for Clinical Trials

Historically, smaller medical device companies and regenerative medicine companies have focused on early clinical trial approvals in the United States and Europe. These approvals are seen as key milestones as they are associated with access to larger markets and can lead to a subsequent increase in valuation. While this is more relevant for Phase III trials, Phase I and II trials, particularly for an SME in the field of regenerative medicine, should be driven more by cost and speed considerations. The key valuation inflection point for fledgling regenerative medicine companies with a limited funding horizon is first-in-man studies. As long as Good Clinical Practices are followed (suggesting the data would ultimately be admissible as a part of later-stage EU or US regulatory applications), it would seem appropriate to avoid expensive early-stage capital by choosing a region with lower costs and more rapid approvals. Indeed, the appeal of lower cost and faster study approval has driven a significant globalization of clinical trials. That is, there has been a shift toward performing early-stage clinical trials in such countries such as Russia, Brazil, China, and India. Interestingly, the United States provides an unparalleled entrepreneurial and scientific environment for innovation, yet, the US's complete aversion to clinical trial risks and related litigation problems are driving access to cutting-edge clinical treatment out of the country.

In deciding where clinical trials should be performed, a balance must be struck between the overall cost, speed of study approval, access to an appropriate patient population, quality of care, and the perceived quality of the data produced (e.g. does clinical data from India impart the perception of sufficient quality to a regulatory agency in Germany?). Trumping all of these questions, however, should be the selection of a principal investigator (PI) and a CRO. Ultimately, the value of the clinical study is directly related to the quality of the data collected, and it is the PI and the CRO that will create, collect, and manage this data. Note that Table 1 summarizes by region the difficulty ranking of the various regulatory hurdles discussed in this section.

Table 1. Ranking by region of difficulty of completing various regulatory processes for cell-based therapeutics.

	Clinical Trial Approval Process	Relative Cost to Complete Clinical Trial	Perceived Value of Clinical Data	Commercialization Approval Process	Reimbursement Process
United States	Difficult	High	High	Difficult	Fair
European Union	Moderate	Moderate to high	High	Moderate	Moderate to difficult
South America	Low to moderate	Low to moderate	Low to moderate	Moderate	Fair to difficult
East Asia	Low to moderate	Low to moderate	Low to moderate	Moderate	Fair to difficult

3.1 *Asia*

While regulatory laws are constantly shifting to address both political pressures and innovation, the trend has generally been toward stiffer regulation and more broad regulatory authority. Perhaps the most significant change to this trend has been legislative changes in Japan and South Korea. Most notably, Japan introduced legislation to eliminate the need for Phase III trials, favoring instead a rigorous post-marketing surveillance period. This effectively allows companies to transition to commercial sales to finance what would otherwise be considered a Phase III trial. The requirement to advance to this conditional approval is clear demonstration of evidence and a scientifically valid rationale for efficacy. While Japan is not the cheapest place to do a Phase I/II clinical trial, there are few companies today that do not have formal efforts to establish clinical trials in Japan. Given the difficulties in navigating and penetrating the unique Japanese consumer market, most companies run clinical trials in conjunction with commercial distribution agreements. Indeed, partnering in Japan is perhaps the most significant trend in commercialization of cell-based therapies over the last decade, as evidenced by several high-profile mergers and acquisitions.

South Korea has similarly embraced the translation of cell-based therapies, though perhaps driven more by economic development boards than regulatory agencies. With facilities, financing, and partnership readily available, South Korea has become another key global destination for cell-based research and clinical trials.

3.2 *The United States*

The US market is a double-edged sword for regenerative medicine. On the plus side, the United States offers a single, centralized regulatory agency for both clinical trials and commercialization, with a now well-established regulatory framework for cell-based therapeutics. A very clear set of guidance documents, with "Guidance for Human Somatic Cell Therapy and Gene Therapy" and "Sterile Drug Product Produced by Aseptic Processing" at its core, define this framework.[5,6] In addition, the United States offers both a high rate of reimbursement and a huge potential market for sales, with a population of roughly 300 million people and a large fraction of elderly inhabitants.[7] However, due to a highly litigious society and huge political pressures to avoid scandal, the FDA is shifting more and more to a policy of complete risk avoidance. Thus, the FDA's approval processes are well established and predictable but they are inflexible and require large volumes of experimental data. While the FDA may attempt to evaluate products on a case-by-case basis, the truth is that the regulations are largely fixed and the agency typically errs on the side of extreme caution. As mentioned earlier, the US system seems to encourage the rise of globalized clinical trials for first-in-man studies. Even in the medical device markets, companies now tend to seek commercialization in Europe first, and then later, if at all, in the United States. As an example, there are approximately 15 repair devices for abdominal aortic aneurysms (AAAs) CE-marked for sale in the EU, but only about five in the United States.

One important change to this setting is the passage of the 21st Century Cures Act. While this new legislation covers broad topics, at its core is a new pathway intended for personalized cell therapies, typically utilized by physician innovators. Following a strong public outcry relating to reduced access to a patient's own cells, this legislation provides a reduced regulatory hurdle for products that fall between the traditional 361 and 351 pathways. It is important to note, however, that the FDA is not obligated to place products into this new category. While there is considerable optimism surrounding this legislation, we believe that little will change. Indeed, we believe that the most likely outcome will be a reduced effort by the FDA to challenge the use of fat transfer and SVF by point-of-care clinics, but that commercial development of cell-based therapies will be largely unchanged.

3.3 *European Union*

The European regulatory framework is advantageous due to its unique combination of regional and centralized controls. Clinical trials are approved and performed under regional authority, taking advantage of cost/approval advantages in countries such as Poland. Commercialization to all EU member states is then controlled by a single centralized entity, the European Medicines Agency (EMA), following the framework set up by the Advanced

Therapy Medicinal Products (ATMP) directive and governed by expert committees, such as the Committee for Human Medicinal Products.[8] This myriad of regulations can be difficult to navigate. For example, in Germany, a company must apply for a clinical trials approval through one state agency, approval of its GMP processes through another, and then a manufacturing permit through the EMA. In France, a company producing cell-based therapeutics must be registered as a national tissue bank, requiring the manufacturer to set up a facility in France. A flow chart summarizing the regulatory process in Europe is presented in Fig. 1. While the regional bureaucracies can create a complicated process, overall, the access to initial clinical trials and then a large commercial market (over 500 million) can be significantly more efficient and less costly than the US pathway.

3.4 *South America*

Unlike in Europe and the United States, there is no centralized regulatory body in South America. The advantage to running studies in South America is that the regulatory bodies here are significantly more accepting of risk and less litigious than the United States or even Western Europe. Thus, South America's various approval processes, especially for clinical trials, usually are completed much faster than in the United States and EU. In our experience, it took approximately six months to obtain approval to perform a Phase I trial in Argentina, versus the two plus years it has taken in Germany and three plus in the United States. However, this lack of a rigid system can also be a detriment, as the various regulatory systems do not necessarily correspond with one another. Thus, the pathway to approval for clinical trials or commercialization in Argentina will not be the same as Brazil or Ecuador. Finally, reimbursement rates in these countries are much lower than in the United States or Western Europe. As such, it has been our experience that it is far more attractive to perform clinical trials in these regions, but to hold off on commercialization.

3.5 *Other Regions*

It should be noted that there are other regions where federal regulation is almost non-existent. Clinical approval in Honduras, for example, is driven

 T. McAllister, C. Iyican and N. L'Heureux

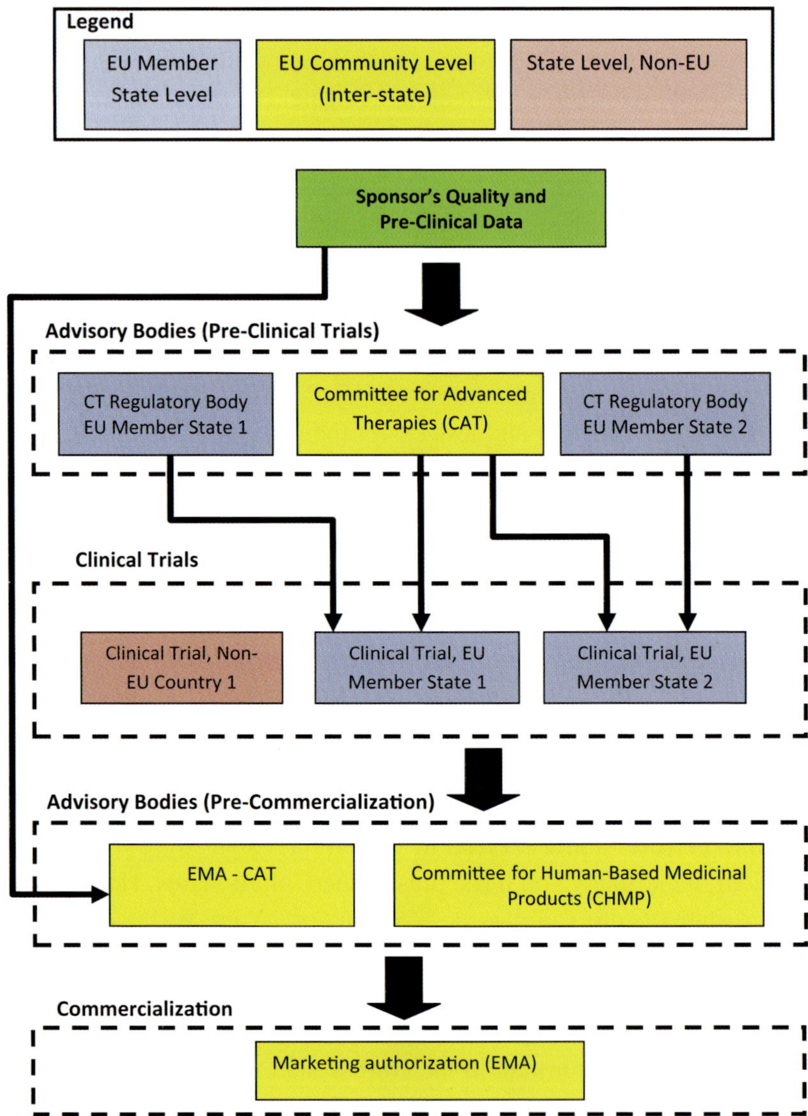

Fig. 1. Regulatory pathway for cell-based therapy clinical trials and marketing authorization in the European Union. Note that this pathway refers to advanced therapy medicinal products (ATMPs) placed on the market in Europe after 31 December 2008 (and *all* ATMPs on the market in Europe after 31 December 2012). This pathway is *not* applicable for therapies designated as medicinal devices. Also, this pathway does not include potential post-marketing Phase IV clinical trials that may be requested by the EMA.

almost entirely by local authorities that are generally linked to local/regional universities. This presents an incredible opportunity for rapid transition to First in human studies. However, clinical trials support infrastructure is almost non-existent in these locations, making patient follow-up extremely challenging. Very clearly the lack of infrastructure diminishes the value of this data when trying to transition to the United States or Western EU trials. With robust oversight, however, this early access to humans remains a viable commercialization pathway, as investors (and therefore follow-on financing) are not usually so critical of geography when it comes to the de-risking event associated with first clinical use.

4. Use of a Clinical Research Organization

The primary driver for conducting clinical studies in emerging countries like Brazil, Russia, or India is cost. While many sponsors are hopeful that these countries represent significantly accelerated access to human use, the actual review processes are typically more thorough and take longer than some may think. When one factors in the additional work of replicating regulatory efforts in Western Europe or the United States, the overall time benefit is marginal. So while there may be some near-term advantages in rapid initial access to human trials (Phase I/II), the long-term benefits relate to cost savings. Somewhat surprisingly, the costs of running a clinical study in South America or Eastern Europe can be dominated by the fees paid to a CRO. Careful selection of the CRO and the responsibilities they will be given is an excellent way to significantly reduce the overall costs associated with the regulatory approval process and the clinical study itself. There are two basic drivers that force the use of a CRO. The first is a general requirement in most countries for legal representation to submit regulatory documents, import medical products, or conduct clinical studies. The second is a general requirement to have a clinical monitor to check source data. While established CROs often perform or manage these roles, there are a variety of other options detailed below.

Given the uniqueness and complexity of running regenerative medicine studies, in our experience, most CROs are ill-equipped to handle trials outside the scope of the classic drug or device. Even those that promote themselves as experts in regenerative medicine often have little real expertise in the regulatory, clinical, or logistical requirements of running a study that is based upon the use of a living product. While this knowledge gap is closing as more and more clinical studies are performed in this field, it is safe to say that the sponsor should expect to play a direct and engaged role in the day-to-day management of the clinical trial. Indeed, we have hired our own local employees to help accelerate and manage our regulatory and clinical activities performed outside the United States. In most instances, these same local employees, under the tutelage and management of the sponsor's clinical trials manager, can perform all of the roles normally performed by a CRO at a fraction of the cost. It is important to note that this includes data monitoring. There is generally no requirement to use a CRO or even an independent monitor to perform these tasks. However, the risk of inadvertent mistakes that can be perpetuated when the sponsor is directly responsible for both study design and data monitoring should be evaluated and balanced against the cost savings associated with this strategy. Similarly, the potential for a regulator's negative perception of an "internal" biased monitor must be evaluated.

CROs can also provide localized legal representation. As an alternative to large global CROs, there are more and more smaller contractors that perform this role. We have used entities such as MARES Ltd. in Western Europe, and Access Medical Research in South America and Poland, with success. This provides a low-cost alternative to a full-time, local or regional CRO, but puts significantly more responsibility and workload on the sponsor's clinical and regulatory teams. On one hand, this may result in additional direct costs and delay due to lack of experience but, on the other hand, this will avoid paying for the indirect costs of the CRO, will allow better control over timeline and expenses, and develop in-house expertise that can be used in future trials. Given the relative inexperience with CROs in this field, we have found this to be an appropriate trade-off.

5. Universal Regulatory Considerations for Cell-Based Therapeutics

Tissue engineering and cell-based therapies have been hailed as the third pillar of future medicine alongside devices and drugs. While our field as a whole has tremendous support in academia and has reached important milestones over the last decade, translation to widespread clinical use will be crippled by the current regulatory framework. It is incumbent upon us all to collectively lobby for rational regulations that are more relevant to regenerative medicine and result in an appropriate risk/benefit ratio for the patient. The regulatory framework in almost every country with established guidance documents on cell-based therapeutics has been derived from aseptic drug manufacturing requirements. The fundamental problems with this approach are the dramatic disparities in cost per "dose" and lot sizes between pharmaceuticals and cell-based therapeutics. In the drug industry, lot sizes in the hundreds of thousands, and cost per dose, allow for an enormous profit margin; the drug industry can easily absorb the QA/QC costs associated with extensive manufacturing controls and lot release testing. In cell-based therapeutics, however, even most allogeneic products have lot sizes of a few thousand at most, and costs per dose that are orders of magnitude higher than pharmaceuticals. While companies like Organogenesis and Advanced Biohealing have brought manufacturing costs down to the point of profitability (if one ignores the hundreds of million of dollars in R&D expenses forgiven by bankruptcy and the later erosion of reimbursement), many companies are burdened with more complex products, smaller lot sizes, challenging cell sourcing, etc. While it is not in the FDA's mandate to establish a regulatory system that promotes profitability for the industry, it *is* their responsibility to enable patients to have access to new therapies. The question, of course, is what level of risk are we willing to tolerate in order to bring QA/QC costs down? There is a valid argument to be made that the current environment of complete risk aversion does a disservice to both patients and industry. For example, an immunocompromised, terminally ill cancer patient would likely accept the small risk of death from a contamination in order to have access to the potentially life-saving benefit of the treatment. With our own tissue-engineered vascular grafts, we are required to test for sterility, mycoplasma, and endotoxins at three separate time points in the manufacturing process. In nine years of production (most of which were done under strict but non-GLP conditions), we have had one test come back as "inconclusive." Thus it would seem that our process controls have more than adequately controlled the risk of contamination. Similarly, scores of hospital labs have been producing cultured epithelium for burn patients under non-GLP conditions without problems.

Ironically, the most sterile and perfectly characterized *in vitro*-produced tissue will be surgically implanted in an operating room environment that, by aseptic manufacturing standards, has a high risk of contamination and procedural error. We certainly could apply the same QC/QA system proposed for cellular therapies to surgical procedures to insure "zero risk" of contamination and error. The result would be that the already exorbitant cost of interventions would become astronomical and available to an even smaller portion of the population. In fact, this "zero risk" attitude is already a contributing factor to the steep rise of health care cost in the United States. A similar conclusion could be drawn when looking at organ transplant procedures where organs are implanted "as is" with some but limited QC evaluation. While historical reasons probably largely account for the privileged status of surgical interventions, there is also an important justification when considering the risk/benefit ratio and this assessment must be included in the reasoning behind the regulations. The term "*in vitro* surgery" has been coined to highlight the parallels between surgical procedures and autologous tissue production. In both cases, the living tissue of a patient is manipulated (often with surgical methods and instruments) and transformed into a construct that can be implanted to resolve a serious health problem. Are we proposing to produce tissues under "relaxed" standards? Certainly not. But if we want to see the day when complex tissue-engineered organs can be produced and implanted at an acceptable cost, regulatory agencies will have to accept some safety trade-offs to reach an acceptable cost-risk-benefit balance.

Another fundamental problem is that the regulatory agencies demand that the sponsor perform every test that has cumulatively encumbered other applicants. The responsibility then lies with the sponsor to whittle down this all-inclusive list that gives an appropriate level or risk-benefit for the particular patient population. While the regulators have the benefit of knowing all the data from prior applications, industry has not typically shared regulatory or manufacturing data so that we can turn the tables and argue for the *least* common denominator. In other words, knowing what lot release criteria, for example, for a similar product is required to demonstrate safety and efficacy, might have dramatic implications in a successful argument for a similarly streamlined lot release strategy. Despite the secrecy that has historically shrouded the development of medical devices and pharmaceuticals, we believe it is critical to the survival of our industry to begin an active data sharing program with respect to regulatory QA/QC requirements. Whenever possible, the FDA and other agencies should provide the data on which they base their regulatory and guidance policies and lay out their decision-making processes to the public.

References

1. E. Arias, Center for Disease Control and Prevention, United States Life Tables, 2004, *National Vital Statistics Report* **54**: 1–40 (2007).
2. S.W. Glickman, J. G. McHutchison, E. D. Peterson, C. B. Cairns, R. A. Harrington, R. M. Califf and K. A. Schulman, Ethical and scientific implications of the globalization of clinical research, *N Engl J Med* **360**: 816–23 (2009).
3. FDA. Guidance for Industry: cGMP for Phase 1 Investigational Drugs. US Department of Health and Human Services (2008).
4. Directive 2001/20/EC of the European Parliament and the Council of 4 April 2001 of the Approximation of the Laws, Regulations and Administrative Provisions of the Member States Relating to the Implementation of Good Clinical Practices in the Conduct of Clinical Trials on Medicinal Products for Human Use. *Official Journal of the European Communities*, L 121/34–44 (2001).
5. FDA/CBER. Guidance for Human Somatic Cell Therapy and Gene Therapy. US Department of Health and Human Services (March 1998).
6. FDA, Center for Biologics Evaluation and Research, Center for Drug Evaluation and Research. Guidance for Industry: Sterile Drug Products Produced by Aseptic Processing: Current Good Manufacturing Practice. US Department of Health and Human Services (September 2004).
7. US Census Bureau. Age: 2000. United States Census 2000. C2KBR/01-12.
8. Regulation (EC) No 1394/2007 of the European Parliament and the Council of 13 November 2007 on Advanced Therapy Medicinal Products and amending Directive 2001/83/EC and Regulation (EC) No 726/2004. *Official Journal of the European Union*, L 324/121–137 (2007).

INDEX